D1499932

CETACEAN BEHAVIOR

Cetacean Behavior:
MECHANISMS AND FUNCTIONS

Edited by Louis M. Herman

University of Hawaii
Honolulu, Hawaii

A WILEY-INTERSCIENCE PUBLICATION

JOHN WILEY & SONS
New York • Chichester • Brisbane • Toronto

Library of Congress Cataloging in Publication Data:

Main entry under title:

Cetacean behavior.

   "A Wiley-Interscience publication."
   Includes index.
   1. Cetacea—Behavior.   2. Mammals—Behavior.
I. Herman, Louis M.

QL737.C4C37   1980   599.5′0451   80-11772
ISBN 0-471-37315-X

Printed in the United States of America

10 9 8 7 6 5 4 3 2 1

To Kea and Puka

Contributors

William W. Dawson, Department of Ophthalmology, University of Florida College of Medicine, Gainesville, Florida.

R. H. Defran, Department of Psychology, San Diego State University, San Diego, California.

Thomas P. Dohl, Center for Coastal Marine Studies, University of California, Santa Cruz, California.

Louis M. Herman, Department of Psychology, University of Hawaii, Honolulu, Hawaii.

A. Blair Irvine, National Fish and Wildlife Laboratory, Gainesville, Florida.

Carolyn J. Madsen, Department of Psychology, University of Hawaii, Honolulu, Hawaii.

Kenneth S. Norris, Center for Coastal Marine Studies, University of California, Santa Cruz, California.

Arthur N. Popper, Department of Anatomy, Schools of Medicine and Dentistry, Georgetown University, Washington, D.C.

Karen Pryor, Department of Biology, New York University, New York, New York.

Michael D. Scott, Department of Zoology, University of Florida, Gainesville, Florida.

William N. Tavolga, Mote Marine Laboratory, Sarasota, Florida.

Randall S. Wells, Center for Coastal Marine Studies, University of California, Santa Cruz, California.

Preface

As an outgrowth of the heightened concern for the protection and conservation of marine mammals, the decade of the 1970s witnessed greatly increased efforts in the study of whales, dolphins, and porpoises. In organizing this volume, the goal was to provide a broad, tutorial review of current data and theory on selected aspects of cetacean behavior, combined, where possible, with new information generated by the authors. It was hoped that the collection of reviews would provide a source of background material and some new insights for workers in the field and for students interested in the study of the cetaceans.

Behavior is influenced and shaped by the perceived world, and hence a description of the major sensory mechanisms for extracting information from the world is given. Audition, a sense that appears to be primary in the life of all cetaceans is reviewed by Popper; and vision, a sense that may be more important than normally suspected, is reviewed by Dawson. Comments are given on both the structural and functional aspects of these sensory systems. Briefer discussions of the tactile and gustatory senses are contained within portions of the chapter on communication by Herman and Tavolga.

Behavior takes place within, and reflects the pressures of the ecological and social mileau, which may differ greatly from one cetacean species to the next. The modulation of behavior or of behavioral interactions among animals that may accompany changes in ecology and sociality are discussed in the chapters by Norris and Dohl; Wells, Irvine, and Scott; Madsen and Herman; as well as in portions of the chapter on communication. Norris and Dohl discuss the organization of cetacean schools and their significance to cetacean life. Wells, Irvine, and Scott consider the variations in structure and social groupings that occur among inshore odontocete cetaceans, in the context of differences in resource distribution and predation pressure. Madsen and Herman focus on the visual adaptations of the cetaceans to their underwater photic environment and the functions that vision and visual appearance may serve in cetacean life. Much of the thrust of the chapter on communication is toward an understanding of the communication systems of the cetacean species as a response to the matrix of ecological and social pressures confronting them.

Although field studies of cetacean populations, oriented toward the analysis of behavior and social organization, are appearing with increasing frequency, thanks to the increased availability of support for this difficult but important research, the cetacean maintained in the oceanarium or laboratory continues to provide the simplest point of access to the study of behavior. The chapter by Defran and Pryor reviews materials on behaviors of captive animals, and illustrates the variability in behavior and trainability that may be species-related. Commonalities in behaviors that may reflect convergent behavioral adaptations to similar ecological niches are briefly noted. The dolphin, usually meaning the ubiquitous bottlenosed dolphin *Tursiops* spp. or the common dolphin *Delphinus delphis*, has been a creature of myth and marvel throughout history. In very recent times popular interest has often centered on communication and cognition in the large-brained, vocal *Tursiops* spp. Given the limited number of scientific studies on these topics, free-wheeling speculation by press and public, and occasionally by the scientist himself, has often stepped into the gap of knowledge. The chapter on communication is in part an attempt to provide some firmer underpinnings for the understanding of cetacean communication and to relate it, where possible, to the attractive fabric of concepts created through the study of the communication systems of other, more accessible animals. The chapter on cognition describes the information-processing specializations, capabilities, and limitations of the dolphin, as revealed through laboratory studies of memory and conceptual abilities. The question of natural language is taken up here, as it is in the chapter on communication, and a distinction drawn between an extant natural language and the capability for acquiring some language fundamentals.

Most of the material in this volume is on the odontocete cetaceans, and much of that centered about the family of dolphins (Delphinidae). This is in keeping with the greater weight of behavioral data available for the delphinids, as compared with other cetacean groups. Some discussion of behavioral and social mechanisms and processes in the baleen whales (mysticete cetaceans) is to be found in several of the chapters, but, overall, the material is comparatively sparse. Field studies of the sociobiology of baleen whales are being actively pursued by many investigators today, so that knowledge of the behaviors of these animals and of their societies should expand greatly in the near future.

I wish to thank the contributors to this volume for their admirable patience with the editor, in what proved to be a longer-term undertaking than any of us had anticipated. The tragic loss in 1977, through theft and abandonment in the open ocean, of the dolphins Kea and Puka, with whom I had worked closely for many years, created an all-consuming burden that postponed work on this volume for well over a year. The postponement came at a time when much was happening in the cetacean field, so that, ironically, in the eventual resumption of work there was a greater scope to the material than might otherwise have been possible.

A number of reviewers commented on early versions of contributed chapters I sent to them. I am grateful for the time they gave to the task and appreciative of the wisdom of their comments. Reviewers included S. A. Altmann, Whitlow Au, George Barlow, Donald Dewsbury, A. D. G. Dral, William Evans,

Howard Gilliary, Sam Gruber, James McCormick, Keith Ronald, and William Stell. Additional reviewers solicited by the contributors themselves are acknowledged within each chapter.

There are many who offered support during the preparation of this volume, and by so doing made its completion easier, even possible. My wife, Hannah, has been many times a colleague in my studies and a counselor in my efforts. Her crisp intelligence helped me see more clearly through many a muddy issue, and her patience and understanding left me with energy to carry out the work. My good friends and research associates, Michael Yunker and Ronald Antinoja, provided invaluable assistance throughout much of the period of my work on this volume. Roger Thompson contributed insights and a great deal of hard work toward our research efforts during this same period. Among my current associates Paul Forestell was of continuous help and provided useful editorial comments on many of the chapters in this volume. Joe Mobley worked closely with me throughout the editorial process. His grasp of the material and of procedures made my work a lot easier. Douglas Richards and James Wolz, postdoctoral associates, commented helpfully on some of the material in this volume, and discussions with them have expanded my concepts and understanding of animal behavior. To the many other graduate and undergraduate students who helped in our research projects over the years, a fond Hawaiian *"Mahalo."*

Finally, the support of the National Science Foundation, Psychobiology Branch, for my research has been of fundamental importance in my efforts to learn more about cetacean behavior and has greatly facilitated the completion of this volume.

Louis M. Herman

April 1980
Honolulu, Hawaii

Contents

CETACEAN BEHAVIOR

CHAPTER 1

Sound Emission and Detection by Delphinids

Arthur N. Popper

INTRODUCTION

The delphinid sound-production and hearing systems include well-adapted and complex mechanisms that may be used for communication or for acquisition of information about the environment through passive listening or active biosonar (echolocation). The evolution of these systems involved the development of new, specialized mechanisms, as well as the modification or enhancement of systems

1

that were present in the terrestrial ancestors of the modern Cetacea. For example, new auditory pathways to the inner ear have been developed in the cetaceans because the mechanical ossicular system of terrestrial mammals, specialized as a matching network for overcoming the acoustic impedance differences between the external air medium and the fluid-filled inner air (van Bergeijk, 1967; Wever, 1974), is no longer suitable in the water medium. For humans, or other terrestrial mammals in water, the large impedance mismatch at the tympanic membrane junction of the fluid-filled external ear canal and the air-surrounded ossicular chain greatly reduces hearing sensitivity (Adolfson & Berghage, 1974). The development of echolocation in the toothed cetaceans has been accompanied by further changes in the hearing system, including enhanced frequency discrimination capabilities and high frequency hearing. These and related acoustic developments are treated further in discussions in this chapter. The reader is referred to other sources (e.g., Fleischer, 1976a; Norris, 1968; Reysenbach de Haan, 1957) for detailed discussions of the evolution of hearing mechanisms in the cetaceans.

Sound-production mechanisms have similarly undergone extensive adaptive changes in the cetaceans. The entire peripheral respiratory system has been greatly modified, not only to facilitate breathing, but as a sound-production mechanism. Special air sacs and valves have been developed that allow sound to be produced effectively without loss of air, and among the odontocetes, new structures for amplifying and beaming sound appear.

In the following sections the sound-producing, -detecting, and -analyzing systems of the cetaceans are discussed in greater detail. The material is principally restricted, however, to the family of dolphins, the delphinids, for which most information exists. Both structural and performance characteristics of the systems are described. Performance characteristics are expressed in terms of energy producing, detecting, or discriminating capabilities of the species, as estimated through physiological or psychophysical methods. In many cases, currently available data are insufficient to fully describe performance or to decide among hypothesized underlying mechanisms. Also, extrapolation of the findings on delphinids, primarily the bottlenosed dolphin *Tursiops truncatus*, to other cetaceans may not be valid. Crossing the line from odontocete cetacean to mysticete cetacean may be especially tenuous, and one may also expect that delphinid acoustic systems are very different from those described for pinnipeds or sirenids (cf. Bullock, Ridgway & Suga, 1971; Moore, 1975; Norris, 1968; Repenning, 1972; Reysenbach de Haan, 1957; Schusterman, Balliet & Nixon, 1972).

SOUNDS

Delphinid sounds are diverse and often complex (see Table 1.1). While a detailed review is beyond the scope of this chapter, a general overview of the sounds and their associated behaviors is given to provide a context for the later discussion of acoustic mechanisms.

Sounds produced by odontocetes generally fall into two functional and three acoustic categories. The functional categories may be broadly defined as echolocation, in which the animal emits and detects its own sounds reflected from objects in the environment (see Griffin, 1968; Sales & Pye, 1974), and communication. The three acoustic categories describe the spectral composition, structure, or quality of the sounds and include the "creaky" buzzes, consisting of trains of generally broad-spectrum clicks associated with echolocation; the relatively pure-tone, often frequency-modulated (FM), whistles; and the burst-pulse sounds, also comprised of trains of clicks, but whose envelope gives a peculiar and variable "squawking," "blatting," "squeaking," or even "groaning" quality to the sounds. Some of the burst-pulse sounds may be produced by expelling air through the blowhole, but most seem to be generated internally without loss of air. Not all delphinids whistle (see Herman & Tavolga, this volume), but all seem to produce some forms of burst-pulse sounds. Whistles and burst-pulse sounds are most often implicated in communication.

Before considering these various sounds in greater detail, it is important to remind ourselves of some chronic problems in analysis that may limit the description and interpretation of sounds. One problem derives from the acoustical properties of the recording environment. Interfaces between water and other materials can significantly affect sounds by altering their duration, amplitude, or spectral characteristics (Parvulescu, 1964). Although tank and field studies may agree on the general nature of a sound produced by a particular species, the signal length and some aspects of the spectrum have been found to differ in the two situations (e.g., Busnel & Dziedzic, 1966; Norris, Evans & Turner, 1967; Norris & Evans, 1967). Whether this is wholly artifactual or at times indicative of adaptive changes made by the animals to maximize acoustic information is often not clear.

A second problem in analysis was stressed by Diercks, Trochta, and Evans (1973) in their discussion of how the recording system may alter the apparent characteristics of sounds, when that system has a narrower bandwidth than the sound under study. Diercks et al. contended that the echolocation clicks from certain delphinids may have been artifactually reported as narrow-band because the recording system itself was of narrow bandwidth. The spectral composition of the clicks in "narrow-band" species such as the harbor porpoise (*Phocoena phocoena*) may in fact be broad-band, similar to the clicks emitted by the bottlenosed dolphin and allied species. However, Watkins (1974) countered that narrow-band recording systems accurately represent the portion of the spectrum within the bandwidth of the system and that the apparent wide frequency range of harbor porpoise clicks reported by Diercks et al. (1973) may have resulted from hydrophone overload and noise in the system. The issue remains poised. While additional data by Møhl and Andersen (1973) and Dubrovskii, Krasov, and Titov (1971) indicate that harbor porpoise clicks may indeed have energy up to 150 kHz, Ford and Fisher (1978), recording wild nar-whals *Monodon monocerus*, have recently confirmed earlier work of Watkins, Schevill, and Ray (1971) that the clicks of these animals are very narrow-band. As Ford and Fisher noted, however, it still remains to be shown empirically that these clicks are used for echolocation.

Whistles and Burst-Pulse Sounds

Nonecholocating sounds produced by odontocete cetaceans have been given a wide variety of verbal labels by different workers. The energy in these sounds is often below 20 kHz, and in some mysticete species the energy may be as low as 20 Hz (e.g., Payne & Webb, 1971; Schevill, Watkins & Backus, 1964; Winn & Perkins, 1976). The whistle sounds are very narrow-band tones of continuous frequency (CF) or else are frequency varying. Harmonic components are nor-mally present. The click structure of the burst-pulse sounds may be shifted in amplitude and/or rate, resulting in the variations of perceived sound quality described earlier as "squeaks," "squawks," and so on (cf. Lilly & Miller, 1961). The whole known acoustic repertoire of several delphinid species, including the killer whale (*Orcinus orca*), the Amazon River dolphin or "boutu" (*Inia geof-frensis*), and the harbor porpoise, consists of clicks produced at various rates (e.g., Busnel & Dziedzic, 1966; M. Caldwell, Caldwell & Evans, 1966; Evans, 1973; also see discussion in Herman & Tavolga, this volume).

Whistle sounds and burst-pulse sounds each vary greatly across or within animals of a given species (e.g., M. Caldwell & Caldwell, 1967; Dreher & Evans, 1964; Ford & Fisher, 1978). There may also be cases in which the sound production of individuals is highly stereotyped, especially for the whistles (Caldwell & Caldwell, 1968, 1971; but see discussion of whistle sounds in Her-man & Tavolga, this volume). Caldwell and Caldwell (1968) hypothesized that the stereotyped character of the whistle of individual bottlenosed dolphins or of Pacific white-sided dolphins (*Lagenorhynchus obliquidens*) functioned as a "signature" to identify individuals within a school. Stereotype indicative of a possible signature function may also be found in the click "codas" of the sperm whale (*Physeter catodon*) (Backus & Schevill, 1966; Watkins & Schevill, 1977) or in the pulsing pure-tone sequences produced by the narwhal (Ford & Fisher, 1978).

A number of workers have attempted to catalog the sounds produced by in-dividual delphinid species. A few examples can be given. Busnel and Dziedzic (1966) recorded a variety of sounds, including echolocation clicks, from a pilot whale (*Globicephala melaena*) at sea. The sounds included whistles up to 1 sec in duration with several harmonics, and a creak which appeared to be a rapid repetition of the echolocation click for 100 to 200 msec. In addition, there were a number of sounds combining various whistles and creaks, emitted simultaneously or sequentially. Recordings of the common dolphin (*Delphinus delphis*) by Busnel and Dziedzic (1966) revealed at least five types of sounds, each varying in frequency, FM, and harmonic components. Both whistles and clicks were heard. There was considerably more FM within the whistle of the common dolphin than in the pilot whale.

The most extensively studied delphinid, the bottlenosed dolphin, produces a wide range of whistles and burst-pulse sounds. The burst-pulse sounds can be highly variable (e.g., M. Caldwell & Caldwell, 1967; Lilly & Miller, 1961). The whistles have few harmonics (Dreher & Evans, 1964; Evans & Prescott, 1962; Lilly & Miller, 1961) but may have FM components ranging from 4 to 20 kHz (Burdin, Reznik, Skornyakov & Chupakov, 1975; Evans & Prescott, 1962). Evans and Prescott (1962) observed 18 different whistle "contours" with durations from 0.10 to 3.6 sec and also identified a series of "barks" having energy from 0.2 to 16 kHz with abundant harmonics. Burdin et al. (1975) analyzed sounds of two Black Sea bottlenosed dolphins maintained in two separate tanks connected by hydrophones. The animals primarily used whistles during exchanges of vocalizations, with each animal using a different whistle. Similar results were reported earlier by Lang and Smith (1965).

Though all may agree that sound is an important channel in delphinid (and cetacean) communication (e.g., M. Caldwell & Caldwell, 1966, 1967, 1968; D. Caldwell & Caldwell, 1972, 1977; Dreher & Evans, 1964; Evans, 1967; Kinne, 1975; Lilly & Miller, 1961; Tavolga, 1965; Tavolga & Essapian, 1957; Titov & Tomilin, 1970), the correlations between sound and behavior are tenuous. The limitations in knowledge of dolphin acoustic communication contrast sharply with what is known about sound and its use in communication in many other animals, including insects, birds, and primates (for example, see papers in Busnel, 1963; Sales & Pye, 1974; Sebeok, 1968, 1977). The main problem in assessing delphinid acoustic communication is that in most cases, except for some oceanarium observations, it is difficult to correlate sounds detected with hydrophones with the specific animal making the sounds, much less with the behavior associated with the sounds (see, for example, Cummings & Thompson, 1971a). At sea the lack of suitable directional hydrophones makes it difficult to ensure that the observer is actually listening to the animal being observed visually unless the animals are close to the observers and the hydrophones, as during captures (e.g., Busnel & Dziedzic, 1966). Arrays of hydrophones help solve the directional problem, though they are expensive and often difficult to analyze (Watkins, 1976). An additional problem in interpreting the social value of sounds is that environmental and diurnal changes may as easily determine or modulate sound production as do social variables (Busnel & Dziedzic, 1966; M. Caldwell & Caldwell, 1967; M. Caldwell et al., 1966; Titov & Tomilin, 1970).

A few observations at sea have uncovered some tentative correlations between sound and behavior, although much remains to be done in all aspects of field observations. Playback studies at sea have demonstrated that the sound of several species may have dispersal effects on conspecifics and, in some cases, on other species. The whistle sound of a common dolphin recorded during capture increased activity by conspecifics for up to 15 min, while the same sound did not affect the behavior of schools of bottlenosed dolphins (Busnel & Dziedzic,

1968). The "screams" of the killer whale have a more general effect and elicited flight reactions or produced alerted states lasting long periods of time in several species of cetaceans (e.g., Cummings & Thompson, 1971b; Fish & Vania, 1971). However, the screams were ineffective in dispersing Pacific spotted dolphin schools (*Stenella atenuata*) in Hawaii, which have little or no experience with killer whales (J. Fish, personal communication). More recent field work, in which the fabric of the delphinid society is studied in great detail, as in the material reported in this volume by Norris and Dohl and by Wells, Irvine, and Scott provides the necessary context in which acoustic communication may be understood. It seems now that we are at a stage in the collection of descriptive information about cetacean sounds where it is of diminishing value to record and catalog without quantification or without taking firmer note of environmental and social correlates. Some questions that might be easily addressed include the effects of diurnal or seasonal changes on sound production in wild animals (cf. Powell, 1966), more work on the responses to conspecific and heterospecific sounds, and the ecology of sound production. Herman and Tavolga (this volume) have raised some of these same questions and others about cetacean acoustic communication and discuss them at greater length.

Echolocation Sounds

The echolocation sounds of most of the delphinid species studied are short, broadband pulses produced at a variable rate (Evans, 1973). All of the broadband types have energy extending into the ultrasonic region, though the degree of penetration may vary greatly across species. The possibility that the echolocation sounds of some species may be narrow-band (and of relatively low frequency) has already been noted. In addition to the variation in pulse characteristics across species, changes in pulse characteristics may also occur within species, depending on the echolocation task (e.g., Turner & Norris, 1966) or the environment (Au, Floyd, Penner & Murchison, 1974). Delphinid echolocation sounds may be contrasted with those of bats, which may be FM, or CF, or a combination of these two types (Diercks, 1972; Möhres, 1967; Novick, 1973; Pye, 1967; Sales & Pye, 1974; Simmons, Howell & Suga, 1975).

Interspecific differences in echolocation sounds mainly involve click length and frequency spectrum. Pulse duration in the bottlenosed dolphin is about 100 μsec, according to Diercks, Trochta, Greenlaw, and Evans (1971), who recorded from transducers attached to the animal's head, while Au et al. (1974) reported shorter pulse durations of 35 to 45 μsec in a long-range echolocation task in a high-noise environment (Fig. 1.1). Echolocation clicks in other delphinids are generally less than 1-msec long (see Table 1.1). The duration of killer whale clicks reported by Schevill and Watkins (1966) was 10 to 25 msec; however, Diercks et al. (1971), using a wide-band recording system, observed click durations of 0.5 to 1.5 msec in this same species. Whether or not the differences in click length may be artifactually derived from the different recording systems, as discussed earlier, is unresolved.

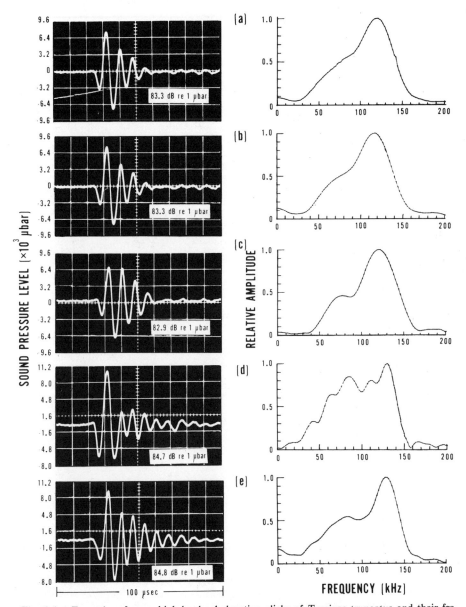

Fig. 1.1. Examples of some high-level echolocation clicks of *Tursiops truncatus* and their frequency spectra. The sounds were recorded in open waters during experiments on target discrimination, with the hydrophone placed 73.2 m from the test animal. From W. Au et al., *Journal of the Acoustical Society of America*, **56**(4) (1974), 1280–1290. Reproduced by permission.

Table 1.1. Sound Emission Characteristics of Selected Odontocete Species

Species	Type of Sound	Frequency Range (kHz)	Maximum Energy (kHz)	SPL-dB re: 1 $v$bar (distance)	Signal Duration	Comments	Reference
Delphinus delphis	Click		4-9	45-80 (1 m)		Paired, clicks &	Busnel & Dziedic 1976
	Whistles	4-16				whistles at same time 5 types of whistles	
	Click	0.2-150	30-60	40-80 (1 m)			Gurevich, reported in Evans, 1973
Inia geoffrensis	Click	25-200	100			Single or multiple clicks	Norris et al., 1972
	Click				65 msec		Fish et al., 1976
Lagenorhynchus obliquidens	Click	0.06-80		70 (1 m)	0.25-1 msec		Evans, 1973
	Whistle	1-12			200-1200 msec	Data from 3 animals	M. Caldwell & Caldwell, 1971
Orcinus orca	Click Scream	0.25-0.5 2		60 (1 m)	10-25 msec	Multiple clicks	Schevill & Watkins, 1966
	Click	25	0.1-30	78 (1 m)	0.5-1.5 msec	Transducer on head tank	Diercks et al., 1971
	Click	to 35	12	80 (1 m)	0.1-0.5 msec	tank	Diercks, 1972

Species	Signal						Reference
Phocoena phocoena	Pulses	41	2	0 (1 m)	0.5–5 msec	Length depends on train cycle, limited recording system	Schevill et al., 1969
	Pulses	< 100–160	110–150	40 (1 m)	0.1 msec		Møhl & Andersen, 1973
Steno bredanensis	Click	0.1–200			50–250 μsec		Norris, 1969
	Whistle		3–10 2–14		100–900 msec		Evans, 1967
Tursiops truncatus	Click	120–130		80–85 (69 m) 128.6 (1 m)	35–45 μsec	Calculated at 1 m	Au et al., 1974
	Click	>octave	35				Diercks et al., 1971
	Click	0.2–150	30–60	40–80 (1 m)			Evans, 1973
	Click Bark Whistles	0.1–>30 0.2–16 4–20			1–10 msec 100 msec 100–3600 msec	Narrow bands, 18 contours	Evans & Prescott 1962
	Whistles	2–20			800–900 msec	Signature whistles, mostly pure tones	M. Caldwell & Caldwell, 1967
Stenella attenuata	Pulse	to 150			0.075–0.2 msec		Diercks, 1972
	Whistle						Evans, 1967

For the Amazon bouto Table 1.1 shows that the spectral energy of the click lies entirely within the ultrasonic range (Evans, 1973; Dubrovskii et al., 1971; Møhl & Andersen, 1973). Other river dolphins, *Platanista gangetica* and *P. indi,* also produce very high-frequency components in the echolocation click, but there seem to be discrepancies in the literature as to whether sonic as well as ultrasonic components are also present (cf. Herald, Brownell, Frye, Morris, Evans, & Scott, 1969; Pilleri, Gihr, Purves, Zbinden & Kraus, 1976). Killer whale clicks, extending barely into the ultrasonic region, contrast strongly with the clicks of the river dolphins. Bottlenosed dolphin clicks generally have a rapid rise-time with spectral energy ranging from 0.2 to over 150 kHz (Evans, 1973). However, the major portion of the energy is concentrated in a relatively narrow band centered around 35 kHz in most circumstances (Diercks et al., 1971). Au et al. (1974) reported major energy between 120 to 130 kHz for two bottlenosed dolphins tested in an open-water, highly noisy environment. This dramatic increase in the location of the peak spectral energy of the clicks of the bottlenosed dolphins was interpreted as an adaptive response to masking background noise having significant energy in the 35-kHz region. Some of the variants in the clicks of different specimens of a given species may result from the different activities of the animals at the time when clicks were recorded (Ayrapet'yants & Konstantinov, 1974; Turner & Norris, 1966) or from changes in the position of the hydrophone relative to the head.

Source Levels

Delphinid echolocation sounds may contain a great deal of energy (Fish & Turl, 1976). Table 1.1 summarizes source level data, generally obtained at 1-m distance from the animal, for several species. Sound-pressure levels (SPL) for the killer whale (Diercks et al., 1971), harbor porpoise (Møhl & Andersen, 1973), and Pacific white-sided dolphin (*Lagenorhynchus obliquidens*) and common dolphin (*Delphinus delphis*) (Evans, 1973) range from 40 to 80 dB (re: 1 μbar).

The most impressive data on energy emission during echolocation was reported in the Au et al. (1974) study referred to earlier. SPL, measured at targets 65- to 75-m distance from two dolphins, were approximately 80 dB. Source levels derived from the theoretical transmission losses over these target distances were estimated to be 120 dB (re: 1 μbar) at 1 m from the two dolphins, with a peak level of 128.6 dB. Evans (1973) earlier reported source levels from bottlenosed dolphins of only 40 to 80 dB at 1 m, in an undescribed experimental situation, as did Fish and Turl (1976) in open-water conditions with relatively low background noise levels. The discrepancies in source levels of these different experiments are best viewed as reflecting the flexibility of response of the bottlenosed dolphin under various background noise conditions, even to the point, in the Au et al. study, of emphasizing portions of the audible spectrum—120 to 130 kHz—well outside the range of best auditory sensitivity (Johnson, 1966, 1967a), if more sensitive regions are preempted by noise.

That the source levels in the Au et al. study were so high may in fact have reflected the need to compensate for the relatively poor hearing sensitivity at these high frequencies.

Another factor that ought to correlate well with click energy is target distance. However, Yablokov, Bel'kovich, and Borisov (1972) found no changes in signal levels as a bottlenosed dolphin approached a target during an echolocation task. This result needs further study. More generally, data are needed to separate out effects of target distance, background noise level (and its masking effect on echolocation signals), and hearing sensitivity on the echolocation signal level. The target echo strength must also be considered.

SOUND-PRODUCTION MECHANISMS

There has been much speculation about sound-production mechanisms in delphinids. Evans (1973) argued that many of the mechanisms proposed are not really suitable for producing the complex sounds of delphinid species. He proposed that any hypothesized mechanism purporting to account for delphinid sound production, particularly that of *T. truncatus,* fulfill four functional "design criteria." These were (*a*) the ability to generate 10- to 100-μsec-long pulses at a rate of up to 600/sec for 3 to 5 sec, as is typical of a dolphin performing an echoranging task; (*b*) a capability for producing, for up to 3 sec, both CF and FM signals spanning the range from 2 to 30 kHz, with few harmonics; (*c*) an ability in at least some species for generating clicks and whistles simultaneously (e.g., Lilly & Miller, 1961; Evans & Prescott, 1962); and (*d*) a capability for patterned or beamed emissions, but with different patterns for clicks than for whistles. Though not noted by Evans, amplification and spectral composition might be additionally imposed design criteria. In the bottlenosed dolphin, and perhaps other delphinid species, the echolocation mechanism should be capable of source levels of approximately 120 dB (re: 1 μbar at 1 m), of varying the location of the peak of the power-density spectrum, and perhaps of varying the spectral content of the echolocation click.

Sites of Sound Production

Several anatomical sites have been proposed for sound production in delphinids (see reviews in Evans & Maderson, 1973; Kinne, 1975; Norris, 1969; Yablokov et al., 1972). The two most commonly mentioned are the tubed extension of the larynx (often erroneously called the arytenoepiglottic tube—see Dormer, 1974) and the nasal plugs associated with the nasal sacs. The early work of Purves (1967) favored the laryngeal area, while that of Evans and Prescott (1962) implicated the nasal sac region. Purves (1967) placed a whistle in the laryngeal region of a dead harbor porpoise, mapped the resulting sound field, and argued that both whistles and clicks were produced at the larynx, with the nasal sacs serving principally to recycle air to the sound-producing system. Evans and

Prescott (1962) produced several dolphinlike sounds from a spotted dolphin, *Stenella attenuata,* carcass when air was delivered to the nasal sac region, and suggested that whistles were produced at the larynx and clicks at the nasal plugs. In later work, Diercks et al. (1971) and Norris and Harvey (cited by Norris, Dormer, Pegg & Liese, 1971) placed transducer arrays on the heads of live bottlenosed dolphins and reported that sounds emanated from the dorsal tip of the rostrum and from the region of the melon. Both Diercks et al. and Norris and Harvey suggested that whistles were produced at the nasal plugs, although Norris et al. (1971) cautioned that the experiments by Norris and Harvey did not totally rule out laryngeal sound production, with sound then transmitted to the nasal plug region.

While each of the earlier experiments were somewhat indirect in their approach to determining the sites of sound production in the living animal, recent experimental evidence supports the suggestion of Norris (1964, 1969; Norris et al., 1971) that the primary site for production of both clicks and whistles may be the nasal plug region (see also Evans & Maderson, 1973).

The nasal sacs include at least three (and often four) pairs of large sacs attached to the unpaired bony nares leading from the region of the larynx to the blowhole (Fig. 1.2). The sacs vary in shape and in number of diverticula (Evans & Maderson, 1973). There is also significant asymmetry, at least in delphinids, with the three left sacs being more elaborate than the right. Just dorsal to the

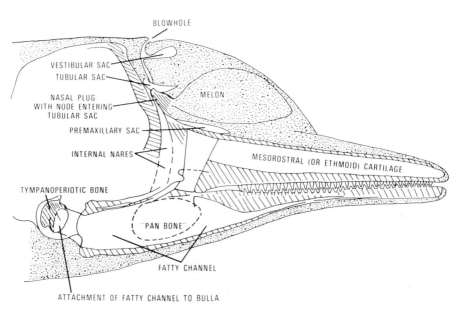

Fig. 1.2. Section of the head of a delphinid showing various structures associated with sound detection and production. From K. S. Norris, with slight modifications, *Evolution and Environment* (Ellen T. Drake (Ed.), 1968, New Haven: Yale University Press. With permission. Copyright © 1968 by Yale University.

most ventral sac, the premaxillary, are a pair of muscular nasal plugs which seal the nasal tube and fit tightly into the external bony nares.

Although the functional mechanism of the plugs is not clear, one suggestion is that they may act as relaxation oscillators to produce sounds (Evans, 1973; Evans & Prescott, 1962; Norris, 1968). The studies of Norris et al. (1971) and Dormer (1974, 1979), using X-ray cinematography to determine the functional correlates of sound production in several awake but restrained dolphins, indicated that the nasal plugs act as pneumatic mechanisms. Compressed air passes from the nasal sacs ventral to the plugs to the dorsal sacs in a "metered" fashion, with an associated production of whistles from the left side. Clicks may be produced from the right side (Dormer, 1974, 1979). Air is stored in the dorsal nasal sacs and recycled to the lower sacs for the next burst of sounds. It was pointed out that all events related to sound production appeared to take place above the level of the epiglottic spout, indicating that the diaphragm and intercostal muscles are not involved in sound production. Further, the premaxillary sacs were slightly inflated during sound production, suggesting that they may act as acoustic mirrors, focusing sound or blocking its transmission into deeper portions of the head (see Dormer, 1974, 1979, for more extensive discussions).

Sound Paths to the Water

The sound-transmission paths from animal to water have become better understood over the past few years. Evidence from hydrophone arrays placed on the head indicates that sound emanates from the lipid-filled melon and the rostrum (Diercks et al., 1971). Wood (1961, cited in Norris, 1969) and Norris (1964, 1968) have proposed that the melon in delphinids acts as an "acoustic lens" focusing sound into the appropriate beamed pattern. Norris and Harvey (1974) provided supporting physiological evidence for this hypothesis. The topographic distribution of the melon lipids (see Litchfield, Karol & Greenberg, 1973) forms a core of low-velocity fat surrounded by a shell of higher-velocity blubber and connective tissue. Norris and Harvey postulated that sound is maintained within the lower velocity core by refraction at the interface with the blubber, in much the same fashion as sound travels with little loss of energy through the less dense SOFAR channels in the sea (see Albers, 1965). Norris and Harvey further suggested that, as sound travels anteriorly, it passes into increasingly dense (higher-velocity) fat, changing the characteristic impedance of the sound and allowing it to enter the water with minimal loss at the interface.

How clicks and whistles, produced simultaneously and in close proximity to each other, may still have different beam patterns is an intriguing question. A possible answer is that the separate beam patterns are produced by different-sized reflective surfaces around the sound-production sites. Larger surfaces reflect both high- and low-frequency sounds, while smaller surfaces reflect only the low frequencies, like the whistles. One might hypothesize a two-layer system of reflectors. An inner layer of small reflectors, acoustically transparent to low frequencies, beams high-frequency energy (as in clicks) into a narrow pattern

that is further tightened by the channeling properties of the melon described by
Norris and Harvey (1974). An outer reflective surface of larger proportions
might then direct the lower frequencies, as well as reflect any "stray" high-
frequency energy. It is possible that one function of the asymmetries and extensive diverticula in the air-filled nasal sacs (air is an excellent sound reflector in
water) is to provide the type of reflective mechanism proposed here.

Observations by Dormer (1974) indicate that the anterior vestibular sac, and
possibly other sacs, can change shape (in tenths of a second) without addition of
air. This suggests that there could be rapid changes in the directional patterns
of emitted sounds without the animal changing its direction of movement. Such
observations of beam directivity were, in fact, made by Ayrapet'yants, Voronov,
Ivanenko, Ivanvo, Ordovskii, Sergeev, and Chilingiris (1974), although the
changes in directivity observed were apparently faster than the times noted by
Dormer. Ayrapet'yants et al. contended that a mechanical system could not be
responsible for the changes in beam patterns they detected. Further work is
necessary for deciding whether the air sacs can indeed change subtly and quickly enough to provide the noted directionally changes.

SOUND DETECTION

Auditory Sensitivity and Discrimination

Kellogg (1953) and Schevill and Lawrence (1953a,b) provided the first
behavioral evidence that *Tursiops truncatus* could detect high-frequency
sounds; however, a complete audiogram was not obtained until over a decade
later. Johnson (1966, 1967a), in a carefully controlled psychophysical study,
showed that the bottlenosed dolphin was sensitive to frequencies between 75 Hz
and 150 kHz. The maximum sensitivity, approximately −50 to −60 dB (re: 1
μbar) was in the region from 40 to 70 kHz (Fig. 1.3). The regions between 15
and 40 kHz and 70 and 100 kHz were only slightly less sensitive. Figure 1.3 additionally shows behavioral audiograms for the Amazon River dolphin or bouto
(Jacobs & Hall, 1972) and the killer whale (Hall & Johnson, 1972), which were
obtained using methods similar to those developed by Johnson (1966).
Behavioral data for the harbor porpoise are also graphed (Andersen, 1971a).
Andersen's data, obtained using operant conditioning techniques in conjunction with the psychophysical method of constant stimuli, differ from data of the
Russian workers Supin and Sukhoruchenko (cited in Ayrapet'yants & Konstantinov, 1974), who determined thresholds for the harbor porpoise using the
"galvanic skin response," though without indication of how this method was
applied. The Russian study credits the harbor porpoise with an ability to
detect sounds between 4 and 180 kHz, with maximum sensitivity of about −52
dB (re: 1 μbar) occurring at 128 kHz. This is very different from Andersen's
finding of maximum sensitivity between 8 and 30 kHz and a sensitivity of only
−35 dB at 128 kHz. The different results may be largely methodological artifacts, but this is difficult to judge until more is known about the Russian work.

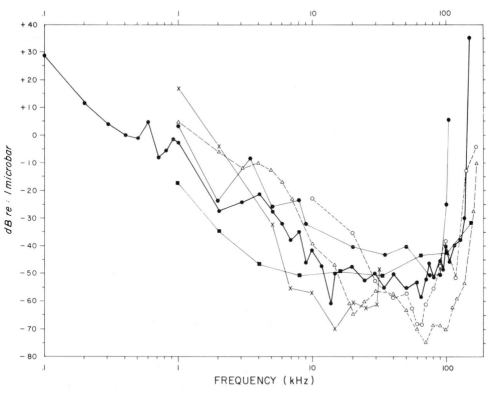

Fig. 1.3. Auditory sensitivity of several odontocete cetaceans. Unless noted otherwise, thresholds were behaviorally determined in water. △, *Tursiops gilli* (electrophysiological data and not absolute thresholds, Bullock et al., 1968); ◐, *Inia geoffrensis* (Jacobs & Hall, 1972); ×, *Orcinus orca* (Hall & Johnson, 1972); ■, *Phocoena phocoena* (Andersen, 1971a); ○, *Stenella coeruleoalba* (electrophysiological data, and not absolute thresholds (Bullock et al., 1968); ●, *Tursiops truncatus* (some data points at higher frequencies left out for clarity (Johnson, 1967a).

The Russians have also obtained an audiogram for the common dolphin (*Delphinus delphis*) (Bel'kovich & Solntseva, 1970). Reportedly, hearing sensitivity ranged over a remarkably broad region, from 10 Hz to 280 kHz, but no detailed data were given nor was information on methods provided. If the data indicating sensitivity to 280 kHz are accurate, this species would have an auditory ability exceeding that for any other delphinid (indeed any other animal) by almost one-half octave. The behavioral significance of such extensive sensitivity is unclear, considering that the bulk of the energy in the echolocation click of *Delphinus* is reported to be between 30 and 60 kHz (Evans, 1973).

Additional data in Figure 1.3 show auditory sensitivity for the Pacific bottlenosed dolphin (*Tursiops gilli*) and the striped dolphin (*Stenella coeruleoalba*, measured using evoked potentials from the inferior colliculus, a midbrain region involved in mammalian hearing in general and in high-frequency hearing in echolocating delphinids in particular (Bullock, Grinnel, Ikezono, Kameda, Katsuki, Nomoto, Sato, Suga & Yanagisawa, 1968; Bullock & Ridgway,

1972a). Care *must* be taken in interpreting the evoked-potential data in Figure 1.3, since the sensitivity levels on the ordinate are arbitrary, and represent values in decibel attentuation relative to the maximum intensity level obtained during stimulation (Bullock et al., 1968). Though absolute values cannot be compared, Figure 1.3 does reveal the similarity of form among audiograms obtained with psychophysical and physiological methods. Also the total range of sensitivity for the two species studied by Bullock et al. (1968) are reasonably similar to data for animals studied behaviorally. However, for any species, hearing thresholds obtained using cochlear potentials are generally higher than behaviorally determined thresholds (Raslear, 1974). The same is likely to be true for thresholds based on responses at higher neural centers, such as the inferior colliculus.

Figure 1.3 shows that *Tursiops, Stenella,* and *Phocoena* have a wider range of sensitivity than does *Inia* or *Orcinus.* The maximum sensitivity for *Orcinus* occurs in a region over one octave lower than the area of best sensitivity for *Inia,* suggesting that *Inia* is able to resolve smaller targets than *Orcinus* (cf. Penner & Murchison, 1970). In absolute sensitivity *Orcinus* appears to have the best detection capabilities of any of the species studied: -72 dB re: 1μbar at 16 kHz. The significance of the heightened sensitivity but reduced audible range for *Orcinus* in unclear; perhaps it reflects the diminshed use of echolocation by *Orcinus* as compared with some other delphinids, the compensatory development of hunting by passive listening, and the apparent wide-ranging use of sound in social signaling (Diercks, 1972; also see discussions in Herman & Tavolga, this volume). In drawing inferences from Figure 1.3, it must be kept in mind that all data were based on the study of a single specimen of the species. There are indications that the data have high validity, however, including internal consistency of results in replicated studies with a given animal (cf. Johnson, 1967a, 1968) and, more generally, a high correlation between the audible spectrum and the spectrum of emitted sounds within the species (e.g., Evans, 1973, and Table 1.1).

Delphinid auditory sensitivity can be contrasted with that of other aquatic vertebrates. Figure 1.4 shows that *Tursiops* has a much wider range of sensitivity than do teleost fishes or man. The range of sensitivity of *Tursiops truncatus* is approximately the same as for all of the pinnipeds, except *Zalophus*. Below 10 kHz, absolute sensitivity of *Tursiops* and the pinnipeds is comparable. Above 10 kHz and more distinctly, above 30 kHz *Tursiops* is considerably more sensitive than are the pinnipeds. Møhl (1967) has pointed out that the harbor seal cannot discriminate between frequencies above 60 kHz, though it can detect them. This is unlike the case for *Tursiops* (see next section) and suggests that the true upper frequency limit for the seal is more realistically set at 60 kHz.

Frequency Discrimination

Frequency discrimination can be of value in communication among delphinids, permitting recognition of whistle variants, for example. It can also contribute to the detection and processing of echolocation signals (e.g., Thompson & Herman, 1975).

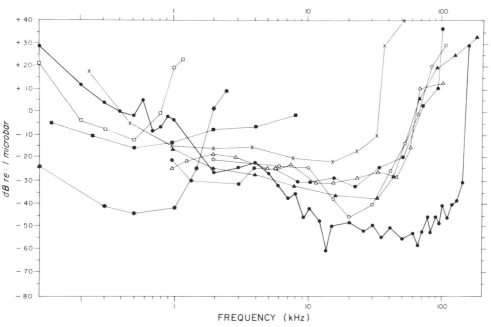

Fig. 1.4. Underwater audiograms for several aquatic vertebrates and man as compared to *Tursiops*. ✕, California sea lion (*Zalophus californianus*) (Schusterman *et al.,* 1972); ▲, harbor seal (*Phoca vitulina*) (Møhl, 1964); △, ringed seal (*Pusa hispida*) (Terhune & Ronald, 1975a); ◕, harp seal (*Pagophilus groenlandicus*) (Terhune & Ronald, 1972); ○, gray seal (*Halichoerus grypus*) (Ridgway & Joyce, 1974); ■, man (Brandt & Hollien, 1967); ★, goldfish (*Carassius auratus*) (Popper, 1971); □, yellowfin tuna (*Thunnus albacares*) (Iversen, 1967); ●, *Tursiops*.

Physiological and morphological studies of the bottlenosed dolphin (Bullock et al., 1968; Wever, McCormick, Palin & Ridgway, 1971b,c, 1972) reveal structures for fine frequency analysis, such as the unusually high degree of variation in width and stiffness of the basilar membrane from basal to apical end, or demonstrate responsivity at the cochlear and cortical levels to very small changes in frequency. Behavioral data underscore the precision of the frequency-discrimination capabilities (Fig. 1.5). Jacobs (1972) determined frequency difference limens (DLs) from 1 to 90 kHz for a bottlenosed dolphin. The relative DLs (DL/F $\times$ 100 where F is frequency in Hertz) ranged from 0.28 to 1.4%, with the smallest values of 0.3 to 0.4% occurring between 5 and 20 kHz. Herman and Arbeit (1972) and Thompson and Herman (1975) studied frequencies from 1 to 140 kHz in the same species. The best discrimination capabilities were between 2 and 53 kHz; relative DLs ranged from 0.2 to 0.4% within this frequency range. Discrimination remained at better than 0.9% outside this range except at 1 and 140 kHz, where relative DLs rose to 1.4% (Thompson & Herman, 1975).

It is noteworthy that in this, the first dolphin psychophysical investigation replicated independently on two animals of the same species, there was broad agreement in the data for the species. There are some differences in the

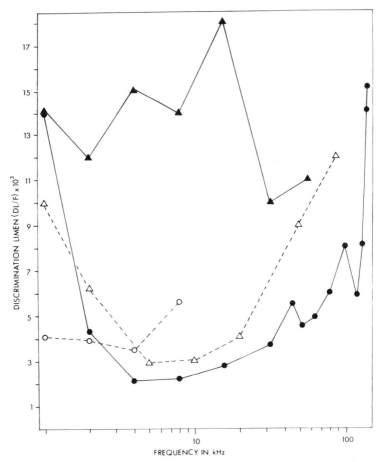

Fig. 1.5. Underwater frequency discrimination capabilities compared for bottlenosed dolphin, human, and harbor seal. ●, *Tursiops truncatus* (Thompson & Herman, 1975); △, *Tursiops truncatus* (Jacobs, 1972); ○, human (Thompson & Herman, 1975); ▲, harbor seal, *Phoca vitulina* (Møhl, 1967).

specification of the extent of the region of maximum sensitivity, Jacobs' data showing a more limited sensitivity at the higher frequencies. However, Jacobs' data are relatively incomplete, as his animal died before all planned replications were accomplished.

Sukhoruchenko (1973) reported relative DLs of 0.1 to 0.2% for the harbor porpoise for all frequencies between 3 and 225 kHz. These data appear highly suspect. The flat psychometric function described is strikingly unlike frequency-discrimination data for other vertebrate species, including *Tursiops*, where the function typically rises steeply at each end of the range of auditory sensitivity (see Fay, 1974, and Fig. 1.5). Moreover, hearing-sensitivity data already described for this species by Sukhoruchenko (Supin & Sukhoruchenko, cited in Ayrapet'yants & Konstantinov, 1974), as well as those of Andersen (1971a), credit the harbor porpoise with an upper frequency limit of at most 180 kHz.

Figure 1.5 also shows that frequency-discrimination capabilities of the bottlenosed dolphin are below those of the human for frequencies of 1 and 2 kHz, but are progressively better than the human at higher frequencies. Frequency-discrimination capability is principally centrally determined (at the cochlear level or beyond) and thus, for humans, is as good in water as in air (Thompson & Herman, 1975). Capabilities of the dolphin are approximately 2½ times better than capabilities reported for the seal at any frequency above 1 kHz. Fay's (1974) review shows that the best frequency-discrimination capability in *Tursiops* is superior to any other vertebrate studied except man. However, some species likely to have excellent frequency-discrimination capabilities, including echolocating bats, have yet to be studied directly. Indirect data indicate that one species of bat, *Rhinolophus ferrumequinum,* can detect frequency deviations of 50 Hz from a CF signal of 83 kHz (Simmons, 1973), suggesting a frequency-discrimination ability (relative DL of 0.1%) of roughly the same order as *Tursiops*.

Intensity Discrimination

Investigations of intensity discrimination with dolphins have been very limited. Burdin, Markov, Reznik, Skornyakov, and Chupakov (1973) determined behaviorally that the bottlenosed dolphin could discriminate 4 to 7% changes in amplitude of a broadband (50 to 100 kHz) white noise signal. Other data seem to suggest better capabilities. Evans (1973) reported sensitivity to 1-dB changes in target strength, based on the bottlenosed dolphin's ability to differentiate by echolocation between targets of the same material but different reflectivity (target strength). Although the dolphin may have been integrating energy information over time to increase the apparent differences between targets, the 1-dB value is in keeping with extrapolations from behavioral findings of Johnson (1967a) and with physiological evidence of Bullock et al. (1968) showing responsivity of some units of the inferior colliculus to signal amplitude changes as small as 1 dB. More generally, the dolphin data are similar to human psychophysical data (e.g., Riesz, 1928; Pollack, 1954) showing capabilities for discriminating amplitude changes of approximately 1 dB. If the 1-dB value for the dolphin is accepted, its intensity-discrimination capabilities seem somewhat better than those reported for the California sea lion (*Zalophus californianus*) (Moore & Schusterman, 1976).

Intensity discrimination, together with frequency discrimination, may play a role in the analysis of target signatures. The relative intensities in frequency bands within the frequency spectrum of the echo will shift differentially according to target composition or resonance and provide clues for target recognition. Intensity discrimination may also play a role in sound localization. As suggested by Norris (1969), an echolocating dolphin can determine target location through the variations in intensity information obtained while scanning. The location of the target should correspond to the direction of the head at approximately the time that echo strength is greatest.

Critical Bands and Masking

Measurements of signal-detection capabilities in a quiet environment provide the necessary upper boundary limits on absolute hearing sensitivity. Together with investigations of frequency and intensity discrimination, sensitivity studies help gauge the limits of the acoustic analysis system. However, a good analysis system must be able to operate efficiently not only in the quiet of the laboratory but under the more typical noisy background condition. Delphinids confront a myriad of background ocean noises (see Wenz, 1964) during acoustic communication and echolocation and must also be able to discriminate their own signals from those of nearby conspecifics.

A good working model for the discrimination of signals in noise is the critical-band concept of Fletcher (1940). The notion of the critical band derives from findings that a signal is only masked by noise when the frequency spectrum of the noise invades a relatively narrow region (the critical band) about the signal (see review by Scharf, 1970). Fletcher suggested that the auditory mechanism can be likened to a series of narrow band-pass filters (also see Green & Swets, 1966). Each filter passes signals within its range and rejects other frequencies. Measurements of the critical band attempt to determine how finely tuned the ear is—i.e., how narrow the band-pass filters—through studies of the ability to detect signals imbedded in noise of various bandwidths (Gourevitch, 1970).

Fletcher (1940) continuously measured thresholds for pure-tone signals as the width of a band of masking noise of constant noise-spectral level (noise level per Hertz) was decreased uniformly about the signal. Thresholds remained constant until at some "critical bandwidth" there was a relatively abrupt decrease in threshold. As the width of the noise band was narrowed further, so that its width was less than the critical band, thresholds continued to decrease.

Fletcher's procedure for estimating the critical band is today called the critical-ratio method, since it is based on the ratio of signal-power:noise-power density (Scharf, 1970). Direct measurements of the critical band derive from other types of experiments and may yield slightly different but highly correlated results. For example, the critical band may be given as the width at which a band of noise that is slowly being expanded is first perceived as noticeably louder (Scharf, 1970); or, the threshold for a narrow band of noise lying between two masking tones may be determined as a function of the separation of the tones. The threshold for the noise remains constant until the critical band is reached, and then decreases sharply (Zwicker, 1954).

Over the entire audible spectrum of an animal, a series of nonoverlapping critical bands can be demonstrated. For humans, 24 such bands have been described within the approximate frequency range from 50 Hz to 16 kHz (see Table 1 in Scharf, 1970). The critical bands are generally thought to represent distinct loci corresponding to constant distances along the basilar membrane (Greenwood, 1961). The critical band is intimately related to the DL for frequency. Critical-band and DL functions parallel each other, but at any given frequency the critical band may be 30 times greater (humans) or even 50 times greater (cats) than the frequency DL (Gourevitch, 1970).

Johnson (1968) measured critical ratios for the bottlenosed dolphin by determining differences in threshold levels for selected pure-tone frequencies as a function of the spectrum level of the masking noise at that same frequency. The critical ratios for the dolphin are shown in Figure 1.6 along with similar measures for the ringed seal (*Pusa hispida*) (Terhune & Ronald, 1975b), humans (Hawkins & Stevens, 1950), and the domestic cat (Watson, 1963). There is strong similarity in form among the critical-ratio functions for these four species, although the bottlenosed dolphin appears to have narrower critical ratios than the other species at those frequencies where data overlap. Data in Gourevitch (1970, Fig. 3) show the similarity of the critical-ratio function of the dolphin with that of chinchilla and rat, and again the dolphin appears to have the narrower critical ratios. The apparently more finely tuned characteristic of the dolphin auditory analyzer, as judged from these critical-ratio data, is in keeping with its enhanced frequency-discrimination capabilities, as determined

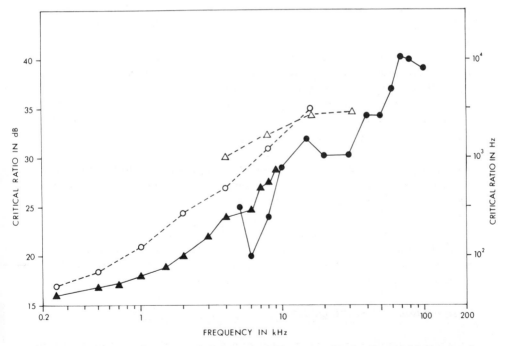

Fig. 1.6. Critical ratios, expressed in frequency and dB, compared for bottlenosed dolphin, ringed seal, human, and cat. ●, *Tursiops truncatus* (Johnson, 1968); △, ringed seal, *Pusa hispida* (Terhune & Ronald, 1975b); ▲, human (Hawkins & Stevens, 1950); ○, cat (Watson, 1963).

in the frequency DL studies discussed earlier. One area of uncertainty for the dolphin data is the finding by Bullock et al. (1968) that some masking noises as much as one octave removed from the stimulating tone will decrease the evoked response to that tone at the inferior colliculus. Clearly, interference at the level of the inferior colliculus may be complex and needs further study.

Johnson's data on criterial ratios indicate that the peripheral auditory analyzer in *Tursiops* is similar functionally to that of other mammals, although in consonance with its greatly expanded audible range, there are an estimated 40 critical bands in *Tursiops*, as compared with 24 in humans. Though Johnson assumed a simple relationship between the critical band and critical ratio—in humans the critical band is 2.5 times the empirically determined critical ratio (Scharf, 1970)—recent work on monkeys and chinchillas suggests a more complex relationship (Seaton & Trahiotis, 1975). Consequently, it would be informative to replicate and extend Johnson's work to direct measures of the critical band, especially in light of suggestions by Altes and colleagues (Altes, 1971, 1974; Altes, Evans & Johnson, 1975) that critical bands are an inherent component of the echolocation signal analyzer in delphinids.

Auditory Temporal Discrimination

One mechanism for determining absolute target distance during echolocation is through comparison of the elapsed time for the return of the leading edge of a pulse with some internalized temporal standard (e.g., Norris, 1969). Yunker and Herman (1974), in tests of absolute judgment, found that the bottlenosed dolphin could discriminate between pure tones of "standard" duration (300, 600, or 1200 msec) and comparison tones that were 8% or more longer in duration. Thus, a 25-kHz pure tone of 324-msec duration or longer was correctly classified as "long" in the dolphin, while one of the same frequency, but of 300-msec duration, was categorized correctly as "short." This 24-msec (8%) resolution capability implies an ability to discriminate between distances differing by 35 m, the distance that sound in water travels in 24 msec. This seems a moderately coarse capability for real-world manipulations, but the 8% resolution capability is nevertheless distinctly superior to human capabilities in comparable tests of temporal judgment (Yunker & Herman, 1974). Extending the Yunker-Herman procedures to test capabilities for resolving durations of very short broadband signals more in keeping with dolphin echolocation sounds seems imperative. If very small differences in duration of pulsed signals can be resolved, gauging distance by elapsed time would become more practical and more likely to be employed as a mechanism.

Discrimination of Sound Qualities

Of further significance to an echolocating animal is the ability to learn and recall the echoic signatures of ensonified targets. For example, it is of obvious advantage to be able to learn and later recognize the echoes returned from a preferred prey species. It is also important to generalize learning, in order to recognize that variants in signatures arising from changing aspects of the target are not in fact changes in the target itself. Herman and his colleagues, as reviewed in the chapter on cognitive characteristics in this volume, have demonstrated that the bottlenosed dolphin is exceptionally adept at learning and remembering sounds heard (Herman & Arbeit, 1973; Herman & Gordon, 1974; Thompson & Herman, 1977). Also variants in the duration of a given sound or even alterations in its frequency are easily recognized as derivatives of the previously learned sound. New sounds heard very briefly—e.g., 2 to 4 sec—were remembered reliably even after several minutes delay. Though these studies all dealt with the recognition of electronically generated sounds through passive listening, it seems likely that the impressive capability for learning, remembering, and generalizing sound characteristics extends to the recognition of echo signatures and their variants.

Sound Localization

A fundamental auditory task is the determination of the spatial location of sound source (Erulkar, 1972; Mills, 1972). For delphinids determination of the location of ensonified targets, as well as the position of active emitters, is important. During echolocation variations in echo intensity arising from variations in the angle of incidence of the echolocation beam to the target give clues to target location, as was noted earlier. There have been some studies of capabilities for resolving angular differences between targets via echolocation, but most of the work has been directed toward spatial accuracy in the passive listening mode. The available data show that passive sound localization by the bottlenosed dolphin is highly precise, in both the horizontal and vertical plans (McDonald-Renaud, 1974; Renaud & Popper, 1975). The pure-tone localization ability in the horizontal plane is about as good as the in-air capabilities of several mammals, including man (e.g., Mills, 1958) and the echolocating bats (Shimozawa, Suga, Hendler & Schuetze, 1974). The minimum auditory angle (MAA) between two spatially separated underwater speakers resolvable by the dolphin ranged from 2° to 3° for pure tones from 20 to 90 kHz (Fig. 1.7). At 6, 10, and 100 kHz the MAAs were approximately 4°, indicating a decline in resolution capability toward both extremes of the animal's audible spectrum. The MAAs for echolocation-like clicks having a center frequency of 64.4 kHz were remarkably precise—0.7° to 0.8°. Clearly, the onset or offset transients of these sounds and/or their broadband characteristics greatly aided localization.

When the two sound sources were moved 30° or 45° to the left or right of the animal, rather than being symmetrically located on each side of the animal (the animal's position was always fixed by requiring it to bite and hold a bar), the MAA for a 40-kHz pure-tone signal was elevated to 4° to 5°. Surprisingly, with 15° displacement left or right the MAA was decreased to an impressive 1.3°. Thus localization ability for pure tones was best for signals 15° off-center, next best for signals directly ahead, and poorest for signals at 30° or 45° displace-

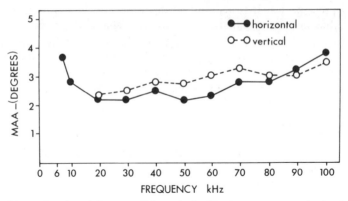

Fig. 1.7. Mean values for minimum audible angles for *Tursiops truncatus* on horizontal and vertical planes at 0° azimuth (data from Renaud & Popper, 1975).

ment. The heightened localization ability at 15° offset is interesting in that it corresponds with the degree of offset of a sound source from the midline of the dolphin that will produce maximum sound levels in the middle ear region (Norris & Harvey, 1974). The relation of this finding to hypothesized sound pathways is discussed in a later section.

To measure sensitivity in the vertical plane, the speakers were maintained in the horizontal plane, but the animal was trained to grasp the bite-bar while turned 90° on its side. This posturing ensured that the emitted sound characteristics were the same as during testing of horizontal localization ability. Vertical sound localization is a difficult task for humans (Roffler & Butler, 1968a,b), and the vertical localization mechanisms may differ from those for horizontal localization. However, vertical localization thresholds for the bottlenosed dolphin proved easy to determine and were comparable in precision with horizontal localization thresholds (Fig. 1.7) at all frequencies tested from 6 to 100 kHz (McDonald-Renaud, 1974; Renaud & Popper, 1975). Renaud and Popper pointed out that the vertical localization data may, to some degree, reflect intensity rather than directional discrimination. The region of maximum hearing sensitivity for *Tursiops* ranges from 5° above to 20° below the horizontal plane (Bullock et al., 1968). Consequently, sounds arriving from above the head, i.e., from the dorsal side of the 90° rotated dolphin, might have been perceived as less intense than sounds arriving from below (ventral side). With practice, the animal may have learned to classify low-intensity sounds as arriving from the dorsal side and high-intensity sounds as arriving from the ventral direction. Further experiments, perhaps varying the intensity of the sounds heard by the dolphin, are needed to tease out the roles of different cues in vertical localization.

In humans the pinnae provide some asymmetries to the arriving signals at each ear, thereby facilitating sound localization (e.g., Batteay, 1967; Searle, Braida, Cuddy & Davis, 1975). Though the pinna is absent in dolphins, cranial asymmetries might provide analogous differential localization cues (see Norris, 1968, 1969), as has been suggested for owls (Payne, 1971).

Diyachenko and colleagues (cited in Ayrapet'yants & Konstantinov, 1974) reported successful localization by *Tursiops* of long-duration signals separated by an auditory angle of 1.5°. Experimental details are lacking, but the data line up reasonably well with those of Renaud and Popper (1975). It is more difficult to compare the localization data for *Tursiops* with the limited data for other delphinids. Dudok van Heel's (1959, 1962) studies of the harbor porpoise yielded MAAs of 11° for a 3.5-kHz pure tone and 8° for a 6-kHz pure tone. Andersen (1971b) reported an MAA of 3° for a 2-kHz pure tone for the same species. Both experiments with the harbor porpoise used unrestrained animals, stimulus control was weak, and the frequencies used were far below the range of best auditory sensitivity of this species (see Fig. 1.3). In a different type of experimental approach, Bel'kovich, Borisov, and Gurevich (1970) reported an angular resolution capability, via echolocation, of 1.6° for the common dolphin for both the horizontal and vertical planes. The animal was free to move its head, although discrimination had to be made at a controlled distance. Further details of the techniques used or of the data obtained are lacking. Other Russian work summarized by Bullock and Gurevich (1979, Table 1) attributes surprising echolocation angular resolution capabilities of much less than 1° to the bottlenosed dolphin. Bullock and Gurevich note that the echolocating bat *Rhinolophus* has a much coarser capability—over 4°.

Taking all of the results for *Tursiops* at face value, it appears as if very small differences in target or source location can be discriminated in both the sonar mode and the passive listening mode. The bulk of the data suggests that sound-localization ability in *Tursiops* is somewhat better than that in pinnipeds. Møhl (1964) reported an in-water pure-tone MAA of 2° for the harbor seal *Phoca vitulina,* while the MAA for a click (Terhune, 1974) was about 9° in water and 3° in air. Gentry's (1967) data gave MAAs of 15° at 3.5 kHz and 10° at 6 kHz for the California sea lion (*Zalophus californianus*). Moore (1975) studied *Zalophus* in the same experimental site used in the *Tursiops* experiments by Renaud and Popper (1975) and reported an MAA of approximately 8.5° for a click centered at 1000 Hz. The best pure-tone-localization ability of this pinniped was 3.5° for a 1000-Hz signal (Moore & Au, 1975). The apparent superior localization capabilities of pinnipeds for pure tones over clicks is puzzling and unlike data for humans or the dolphin. The absolute differences in pinniped and delphinid underwater sound-localization ability, regardless of type of sound, may reflect the differences in the suspension systems of the ears in the two groups. In delphinids, there appears to be complete acoustical isolation of the two ears by surrounding air-filled cavities; in pinnipeds, which retain a good aerial hearing capability, the isolation is less complete.

Central Auditory System

The delphinid auditory system central to the inner ear is highly developed with considerable hypertrophy in regions associated with audition (Ayrapet'yants & Konstantinov, 1974; Dailly, 1972; Kruger, 1966; Morgane & Jacobs, 1972).

Bullock and his colleagues (Bullock et al., 1968; Bullock & Ridgway, 1972a,b) have studied evoked auditory reponses in the inferior colliculus, an important analysis and way station in the central auditory system, in *Tursiops gilli, Stenella coeruleoalba, Stenella attenuata*, and *Steno bredanensis*. Responses among these different species were very similar. A number of units in the inferior colliculus have response characteristics that closely parallel the behaviorally determined capabilities of the animals.

Bullock et al. (1968) also observed units that gave changes in response with changes in signal rise-times; these units showed an increase in response level with a decrease in rise-time from 50 to 20 μsec. Other units evidenced discrete off-responses and discrete on-responses to signals 1 to 3 msec in duration. These response patterns are reminiscent of some characteristics of the echolocation click of *Tursiops* (see next section), which include an abrupt on-time and signal duration of generally 1 msec or less.

The evoked response from the inferior colliculus varies greatly in recovery rate following stimulation (Bullock & Ridgway, 1972b). Recovery unfolds over 3 to 5 msec in some portions of the inferior colliculus, but is less than 0.5 msec in other parts. The rate of collicular response found by Bullock et al. (1968) is sufficient to follow click repetition rates faster than the highest recorded during echolocation tests—600/sec for *Tursiops* immediately after the final approach to a target (e.g., Norris, Prescott, Asa-Dorian & Perkins, 1961). Bullock and Ridgway (1972a), studying an awake *Tursiops,* showed that recovery is faster if the signal stimulating the animal comes soon after the animal has emitted an echolocation click. Consequently, the level of response to a signal following click emission is greater than that to a signal following no self-produced click emission. This seems to provide an interesting mechanism for enhancing the amplitude of response to echoes that immediately follow clicks and may also serve to partially filter out some cortical responses to the echolocating sounds of other animals.

Perhaps the most significant of the physiological findings to date indicate that *Tursiops* may have at least two functionally distinct groups of auditory responses. One group of units (Type II), observed at the level of the inferior colliculus, has a wide frequency range and short response latency (Bullock et al., 1968; Bullock & Ridgway, 1972a). A second group (Type I), observed in the cerebral cortex rather than in the inferior colliculus, has long response latencies and seems restricted to frequencies of less than 10 kHz, with the best response at about 5 kHz (Bullock & Ridgway, 1972a). Bullock and Ridgway (1972a,b) concluded that Type I units responded primarily to whistles or other "social" signals, and that these signals may only be processed at the level of the cerebral cortex. No evidence for Type I responses was found at the level of the inferior colliculus, which may explain why the sensitivity range reported for several delphinids by Bullock et al. (1968) did not often include frequencies below 6 kHz.

Type II responses seem to be involved in echolocation. Bullock and Ridgway

(1972b) found no evidence that *Tursiops* echolocation signals are processed through to the cerebral cortex. However, it should be noted that recording was restricted to only a few units outside of the inferior colliculus and was accomplished more by accident than by intention. In contrast, Gapich and Supin (1974) reported single units in the cerebral cortex of the harbor porpoise that were responsive to ultrasonic frequencies. They also reported cortical units that showed a number of different types of responses to clicks, as well as units that responded to long-duration signals.

Morphology of the Auditory System

The delphinid auditory system shows a number of morphological adaptations for the enhancement of sound detection in water and for the detection and initial processing of high-frequency signals.

Inner Ear

The inner ear structure has been studied in several delphinid species (Fleischer, 1976a,b; Reysenbach de Haan, 1957; Wever et al., 1971a,b,c, 1972). The delphinid cochlea closely resembles that of terrestrial mammals, although certain changes are present presumably reflecting adaptations for high-frequency sound detection and/or processing. The number of hair cells are approximately the same as in man (about 17,000 vs. about 15,000 for man), but the ganglion cell : hair cell ratio is 5 : 1 in *Tursiops* (Wever et al., 1971c) and 4 : 1 in *Lagenorhynchus* (Wever et al., 1972), compared with only 2 : 1 in man. Wever et al. (1971c) suggest that the large number of ganglion cells in the dolphin reflects an increased number of neutral paths for transmission of high-frequency information to the brain or, possibly, that more details about cochlear events are supplied to the auditory system in the dolphin than in man. The increased number of ganglion cells in *Tursiops* seems consonant with this species' excellent and wide-ranging frequency-discrimination capabilities (e.g., Thompson & Herman, 1975) and may provide a morphological basis for the capability.

High-frequency hearing is enhanced by an increased overall stiffness of the basilar membrane, as well as by an increased stiffness variation along its length (von Békésy, 1960). Increases in both aspects of stiffness, relative to that of most terrestrial mammals, has been observed in *Tursiops* and in *Lagenorhynchus* (Fleischer, 1976b). Both species have a bony support on the external side of the basal region of the basilar membrane, in addition to the more typical mammalian condition of anchoring of the inner edge of the basilar membrane through the internal osseus spiral lamina. Stiffness variation along the length of the basilar membrane is a function of changes in the width of the membrane; the width increases 14 times from basal to apical end in *Tursiops*, 11 times in *Lagenorhynchus*, but only 6.25 times in man (Wever et al., 1971b, 1972). While comparable data are not available for other delphinids, Reysenbach de Haan

(1957) found extensive stiffness and extra support in the basal region of the basilar membrane in both *Phocoena* and *Globicephala*, suggesting similar high-frequency adaptations in these species.

External Auditory Pathways

Several pathways for sound conduction to the middle ear have been proposed. One suggested path is via the lumen of the external auditory meatus (EAM), an S-shaped canal about 7 to 8 cm long, opening to a tiny external pore posterior to the eye. Fraser and Purves (1954; also see Purves, 1966; Purves & van Utrecht, 1963) claimed that sound transmission loss through the EAM was significantly less than through surrounding tissue, and that the lumen and the ligamentlike tympanum at its distal end that connects to the middle ear was therefore a functional sound pathway in odontocetes. Dudok van Heel (1962) and Reysenbach de Haan (1957) argued that the transmission differences reported by Fraser and Purves (1954) between lumen and surrounding tissue were not significant when experimental error was considered, and that large transmission differences did not occur at all frequencies tested. Furthermore, several experiments have demonstrated that the blubber layer surrounding the EAM is in fact a good sound transmission pathway in living animals, suggesting that sound is conducted to the middle ear through these fatty tissues (Norris & Harvey, 1974; Reysenbach de Haan, 1957).

Physiological experiments support the hypothesis of sound conduction via the blubber. Bullock et al. (1968) and McCormick, Wever, Palin, and Ridgway (1970) recorded from the midbrain or the cochlea of the auditory system of bottlenosed dolphins while stimulating various points on the body with vibratory signals (also see McCormick, 1968). Both groups of workers reported that transmission to the middle ear through blubber in the vicinity of the EAM was very good for sounds below 20 kHz. Additionally, McCormick et al. (1970) severed the lumen of the EAM, separating its proximal and distal ends, and found little decrement in the response measured in the inner ear, demonstrating that the EAM was no more efficient in transmitting sound to the inner ear than was the surrounding tissue in the side of the head.

The bulk of evidence thus supports the hypothesis that the whole region around the EAM is an effective sound pathway to the inner ear. However, it is still possible that in the living animal sounds are conducted along the lumen in a somewhat different fashion than through the surrounding tissue, providing supplementary or additional information about the acoustic environment. Whatever its function, it is certain that blockage of the lumen does not have nearly the effect on audition in the bottlenosed dolphin as it has in terrestrial mammals.

A very different pathway for sound transmission was proposed by Norris (1964, 1968, 1969) and has subsequently received experimental support from a number of workers (e.g., Ayrapet'yants & Konstantinov, 1974; Bullock et al., 1968; McCormick et al., 1970; Norris & Harvey, 1974). Norris hypothesized that sounds were conducted to the middle ear of the odontocete cetacean via the fat-filled mandibular canal in the lower jaw. This intramandibular fat body terminates at the tympanic bulla of the middle ear (Fig. 1.2). Norris suggested that sound enters this channel through the "pan bone," a thin (1.0-mm) region on the lateral wall of the mandible about midway between the tip of the rostrum and the eye. Corroborative evidence for this hypothesis was provided by Bullock et al. (1968) and McCormick et al. (1970), who found that evoked cortical or cochlear responses to sounds above 20 kHz were significantly greater when stimulation occurred along the mandible or in the region of the tongue than when stimulation occurred on the upper rostrum or in other regions of the head (except for the area around the EAM). Norris & Harvey (1974) emplaced small hydrophones internally in the region of the middle ear and found that maximum sound levels occurred when stimulation arrived from a source directed 15° to the midline of the animal in a location where sound would be expected to enter the mandibular channel through the pan bone. Much lower signal levels were observed when the sound was directly in front of the animal. These findings closely approximated those of Bullock et al. (1968) and are further confirmed by work of Ayrapet'yants and Konstantinov (1974). These latter workers conditioned the heart rate of five restrained delphinids (species unspecified) to sounds. Thresholds were determined with speakers positioned opposite the rostrum, opposite the opening of the ear canal, and behind the ear opening at 135° from the longitudinal axis of the animal. In all cases acoustic sensitivity was unaltered for sounds below 30 kHz, indicating that the area around the EAM, as well as the mandibular region, were functional to about 30 kHz. Above 30 kHz there was a decrease in sensitivity, as much as 7dB at 90 kHz when the sounds came from the side opposite the EAM. These findings confirm work cited earlier that the tissue around the EAM is involved in audition to approximately 20 to 30 kHz, while at higher frequencies the mandibular canal appears to be the primary region for sound detection. That the EAM and the mandibular regions overlap somewhat in sensitivity suggests they may be considered an integral hearing unit. More extensive investigations are now needed to corroborate and extend these findings to include a three-dimensional analysis of sound-detection paths to the ears for a variety of signals.

Norris & Harvey (1974) have also shown that there is a good impedance match between the fat in the mandibular canal and water, resulting in minimal transmission loss as sound enters the mandibular wave guide from the water (also see Norris, 1968). The angle at which sound enters the mandibular canal is of importance. With sound at nornal incidence to the canal—i.e., to the pan bone—there is little transmission loss, but at other angles the loss becomes significant. This attenuation effect with sound direction provides a usable localization mechanism, permitting the animal to detect intensity peaks while head scanning and impute source or target direction.

Middle Ear

The middle ear (Fig. 1.8) has been described for several delphinid species (e.g., Fleischer, 1973a; Fraser & Purves, 1960; McCormick et al., 1970; Purves & van Utrecht, 1963). The middle ear bones—malleus, incus, and stapes—are located in a cavity made up of periotic and tympanic bones connected to one another by cartilage (Fleischer, 1973a). The middle ear cavities of delphinids are independently suspended by ligaments and surrounded by air-filled spaces (Reysenbach de Haan, 1957; Purves, 1966). Consequently, each cavity is acoustically isolated from the rest of the skull and can act as an independent sound detector (Reysenbach de Haan, 1957). This suspension and isolation system is markedly different from that of pinnipeds, where there are firm bony attachments of ear cavities to the skull, producing poor isolation of the ears (Repenning, 1972). The odontocete suspension also differs from that of mysticetes in which some connections between the skull and ear cavity remain (Fraser & Purves, 1960).

According to Reysenbach de Haan (1957), the middle ear bones in delphinids

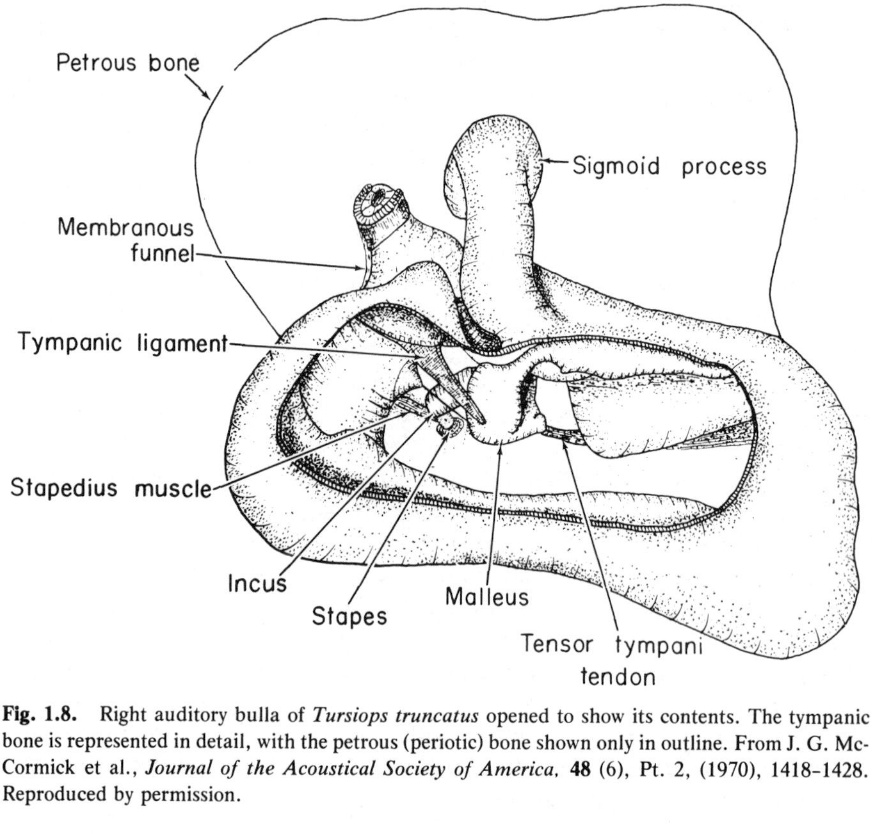

Fig. 1.8. Right auditory bulla of *Tursiops truncatus* opened to show its contents. The tympanic bone is represented in detail, with the petrous (periotic) bone shown only in outline. From J. G. McCormick et al., *Journal of the Acoustical Society of America,* **48** (6), Pt. 2, (1970), 1418–1428. Reproduced by permission.

have two primary functions. One is to provide a degree of stiffness to the transmission system, thereby enhancing the passage of high-frequency signals. The second is to provide for impedance matching between the external environment and the fluids of the inner ear. In terrestrial mammals the low mass of these bones is adequate to transform the low-level pressure signal to a sufficiently high level (a gain of approximately 60) to stimulate the inner ear. Sound in water, however, has an acoustic pressure 60 times greater than the same sound intensity in air (and 1/60th the amplitude). For good underwater hearing the ossicles must be massive enough to withstand the heightened pressure while still retaining sensitivity in the system (Reysenbach de Haan, 1957).

Reysenbach de Haan (1957) suggested that the independent suspension of the ear relative to the skull allowed the middle ear to respond independently from the rest of the skull as a resonant system in a pendulum-like manner. Fleischer (1973a) countered that the pendulum model was infeasible because of the relatively stiff suspension of the ear. He proposed, instead, that acoustic stimulation of the middle ear bones was through the tympanic bone, which, while acoustically uncoupled from the periotic bone (and thus from the rest of the ear cavity), is firmly attached to the processus gracilis of the melleus (also see Fleischer, 1973b). From experiments on preserved or frozen tissue Fleischer (1973a,b) concluded that sounds impinge on the outer part of the tympanic bone, which is in contact with soft tissues and then are transmitted through the middle ear to the cochlea in the usual manner of the terrestrial mammalian ear. Fleischer stressed that the position of the tympanic bone is such that it can receive sound stimulation whether the acoustic pathway is through the region around the EAM or through the lower jaw. However, J. McCormick (personal communication) contends that relative movement between tympanic and periotic bones is necessary to Fleischer's hypothesis but is not possible because these bones are firmly fused together.

McCormick et al. (1970) recorded cochlear potentials at the round window before and after the malleus was removed from the ossicular chain. Only a 4-dB decrement in response level was found after elimination of the malleus from the transmission pathway, although there was approximately an 18-dB loss when the whole ossicular chain was damped. The minimal loss of sensitivity after removal of the malleus suggested that sound detection was by translational bone conduction rather than by the more typical ossicular stimulation method (also see McCormick, 1968). In ossicular stimulation the cochlear capsule remains stationary and the ossicles move, resulting in fluid displacements of the inner ear by the action of the pistonlike stapes on the oval window. In translational bone conduction (Wever & Lawrence, 1954; Tonndorf, 1970) the cochlear capsule is set in motion by acoustic stimulation arriving through the animal's tissues. The fluids in the capsule also begin to move but lag the capsule in time and phase because of the inertial force of the stapes at the oval window. Relative movement is thus produced between the fluid and the cochlear capsule and its components, including the basilar membrane, resulting in stimulation of the hair cells. McCormick et al. (1970) argued additionally that the mass of the ossicular chain yields a high-resonance frequency, enhancing sensitivity to high frequencies.

Fleischer (1973a) contended that the severing of the malleus by McCormick et al. had no effect because the processus gracilis had already been severed during preliminary surgery; the middle ear bones were therefore already inoperative before baseline data had been obtained. McCormick (personal communication) maintains that Fleischer misinterpreted the surgical procedure, that the surgical approach for the initial recordings of cochlear potentials was not through the bulla, the bony capsule enclosing the middle ear (see McCormick et al., 1970; Ridgway, McCormick & Wever, 1974), and therefore that appropriate baseline data were indeed available to support the lack of a response change after isolation of the malleus.

The divergent hypotheses of Fleischer and of McCormick et al. are based on data gathered with very different materials. Fleischer used specimens frozen after death and preserved in alcohol. McCormick et al. used living specimens, but removal of the malleus was accomplished for only two animals. Clearly these experiments need replication in a larger number of animals and under conditions where no damage to middle ear structures can be assured—a most difficult procedure.

Overview

The hearing sensitivity of the bottlenosed dolphin and of other delphinids tested and their capabilities for frequency and intensity discrimination, sound localization, temporal resolution, detection of sounds in noise, and discrimination of sound quality are uniformly excellent. Where comparative data are available, delphinid auditory performance meets or exceeds the capabilities of the pinniped and terrestrial species so far studied. While capabilities for resolving frequency changes on the order of 0.2% may be found in both humans and delphinids, or ultrasonic hearing capabilities comparable to that of the delphinids may be found among echolocating bats (e.g., Brown & Pye, 1975), the total collection of highly refined delphinid auditory capabilities is impressive and underscores the immense auditory specializations of these animals. The excellent auditory analysis system reflects adaptations at both the peripheral and central levels. New sound pathways appear to have been developed and old structures have been modified substantially to meet the demands for high-frequency hearing, acoustical isolation of one ear from the other, impedance matching, etc. Though physiological data are limited to a few studies at the level of the inferior colliculus and the auditory cortex, they do reveal an excellent ability at these levels for detection of weak signals, as well as an exquisite sensitivity to changes in signal amplitude or frequency. Additionally, onset and offset of signals elicit responses, and very small temporal intervals between signals are resolvable by central mechanisms. Data relevant to the integrative activity of the central auditory system are lacking. Some suggestions as to the types of central processing to be expected might be developed by extrapolation from the fairly extensive data for bats (e.g., Simmons, 1973; Simmons et al., 1975; also see Altes, 1974), but at the same time the very different echolocation signals of bat and dolphin suggest that the two groups may have evolved different methods for extraction of data from echolocation signals.

A chronic problem in evaluating data on behavioral capabilities of marine mammals is the paucity of replicated work. As noted earlier, observations on more than one specimen are rare (cf. Jacobs, 1972; Herman & Arbeit, 1972). While the hearing sensitivity data for a given *Tursiops* specimen may show high internal validity (Johnson, 1967a; 1968), only preliminary studies exist for other specimens. The preliminary studies (e.g., Herman, personal communication; McDonald-Renaud, 1974) were intended only to determine whether the animals to be tested in other than auditory-sensitivity tasks had roughly "normal" hearing and do not provide the necessary fine-grain replicate data. The audiograms for the bouto (Jacobs & Hall, 1972) and the killer whale (Hall & Johnson, 1972) seem reliable and demonstrated that there can be large divergences in hearing characteristics across different odontocete species. The reliability of the sensitivity data for the common dolphin (Bel'kovich & Solntseva, 1970) or the harbor porpoise (Andersen, 1971a; Supin & Sukhorvchenko cited in Ayrapet'yants & Konstantinov, 1974) is questionable because of poorly reported data or what appear to be poorly controlled experiments. Additional areas for replicate or extended study include frequency and intensity discrimination for additional species, a more extensive study of the critical band, and further studies on temporal resolving capabilities, especially with respect to the use of this information in target-distance estimation.

It is still unclear how sound is detected and routed through to the central nervous system. Though there are excellent data showing sensitivity to acoustic signals in the region of the EAM and the mandibular fat channel, additional findings of heightened sensitivity to stimulation in the region of the insertion of the pectoral fin (McCormick et al., 1970) and the melon (Bullock et al., 1968) are difficult to interpret morphologically as sound-reception pathways (cf. Norris & Harvey, 1974). If they do function as effective pathways it may be, as suggested by Russian work cited in Bullock and Gurevich (1979), that the array of sound channels contributes to the complex acoustical imaging of the target and its surround.

The role of the middle ear in stimulation of the inner ear needs continued study. The resolution of the different views of Fleischer (1973a,b) and McCormick et al. (1970) must await further stimulation studies on live tissue, or possibly behavioral studies which involve tracing changes in sound-detection capabilities after carefully controlled impairment of the middle ear. The function of the independent suspension of the two ears and their mutual acoustic isolation is another open question. Reysenbach de Haan (1957) and others have suggested that the independent suspension and isolation provide for sound-localization capabilities in water. However, experiments with sea lions and seals (Moore, 1975; Moore & Au, 1975; Terhune, 1972) have demonstrated reasonably good underwater sound-localization abilities, though these species have ears firmly attached to the skull. Possibly, the air spaces around the ears in delphinids primarily function to provide an impedance mismatch, protecting the animal from its own high-level signals. These signals may be of sufficient energy, e.g., 126 dB re: 1 μbar according to Au et al. (1974), to damage the hair cells of the inner ear were they to arrive at the detection sites of the emitting animal without attenuation.

The role of the central auditory system in the processing of acoustic information has been studied, using both evoked potentials and single-unit responses at various levels of the auditory system. Bullock et al. (1968) studied responses at the inferior colliculus, a midbrain region implicated as a major center for sound processing in mammals, including bats. Cortical responses were studied in lesser detail. In general, evoked-response studies are difficult to undertake on delphinids because of the enormous surgical support required and the restriction on delphinid subjects in American laboratories. Bullock and Gurevich (1979) have recently summarized the extensive Russian literature on the nervous system of cetaceans. Those studies add many data on hearing mechanisms, though some of the findings may remain questionable until more is known about experimental methods. As these data are sifted more thoroughly and integrated with Western work, it is hoped that an enhanced image of auditory processing by the delphinids will emerge.

ECHOLOCATION

The initial suggestion that delphinids might use echolocation for orientation, navigation, or food finding was made by McBride in 1942 (see McBride, 1956). Kellogg and his co-workers (Kellogg, 1959; Kellogg, Kohler & Morris, 1953) provided the first demonstrations of echolocation capability in the bottlenosed dolphin, while the first measures of echolocation with vision eliminated as a potentially confounding detection mechanism was accomplished by Norris et al. (1961). Norris and his colleagues showed that a blindfolded bottlenosed dolphin was able to swim between an array of 2.54-cm diameter pipes, pick up food without touching a nearby pipe, and discriminate a gelatin pill capsule from a piece of fish of the same size. Subsequently, echolocation mechanisms and behaviors in delphinids have been investigated extensively, and refined discrimination capabilities have been demonstrated in a number of species (see reviews by Ayrapet'yants & Konstantinov, 1974; Bullock & Gurevich, 1979; Evans, 1973; Kinne, 1975; Norris, 1969; Yablokov et al., 1972).

Resolution Capabilities

Busnel and Dziedzic (1967) reported that a harbor porpoise could swim between 0.5-mm diameter wires without contact 90% of the time. Larger-sized objects were rarely touched. This is an impressive capability, but it may be too good.

Penner and Murchison (1970), working with the Amazon bouto, were aware that air bubbles adhering to thin wires in water can increase the effective diameter and target strength of the wires. Accordingly, they coated the thin copper wires used as targets with liquid detergent, to prevent the adhesion of air bubbles. The wires were then presented behind a visually opaque but acoustically transparent acrylic screen. Under these conditions, the limit for detection, at the 75% correct level, was estimated as a 1.32-mm diameter wire, a detectability threshold roughly 2½ times coarser than that given by Busnel and Dziedzic (1967) for the harbor porpoise. Given that the bouto is an almost constant echolocater and that its clicks are exceptionally rich in high-frequency energy (Penner & Murchison, 1970, and Table 1.1), its capabilities are likely to be as good as or better than other odontocete echolocaters. If so, this is strong presumptive evidence that the data of Busnel and Dziedzic overestimated the capabilities of the harbor porpoise.

The ability of dolphins to resolve size differences by echolocation has also been studied. Turner and Norris (1966) found that a bottlenosed dolphin, early in training, could discriminate between steel ball bearings 6.35-cm (2.5-in.) and 5.08-cm (2-in.) in diameter, a diameter ratio of 1.25. With further training, discrimination between diameter ratios of 1.19 became possible. Similar improvements with training were reported for the harbor porpoise by Ayrapet'yants et al. (1974).

Ayrapet'yants and Konstantinov (1974) evaluated data from a number of Russian workers on the discrimination of cylinders of different heights by the harbor porpoise and concluded that discrimination was dependent on the ratio of heights. A minimum ratio between paired cylinders of 1.1 to 1.3 was required for successful discrimination. It is noteworthy that these values bracketed the final ratio reported by Turner and Norris.

In other studies, Konstantinov, Mel'nikov, and Titov (1968) reported, not surprisingly, that a common dolphin was able to discriminate a fish with an attached metal plate from a fish without such an attachment. Evans (1973) reported that at a 3-m distance the killer whale could detect air-filled 10-mm diameter plastic rings, while at the same distance the Pacific white-sided dolphin detected 2-mm diameter targets. Evans (1973) also cited a Russian paper by Gurevich reporting that the common dolphin can discriminate between 50- and 100-cm² styrofoam pyramids.

Distance Determination

A basic question in evaluating echolocation systems is the distance at which targets can be detected. In a recent experiment by Murchison and Penner (E. Murchison, personal communication), a bottlenosed dolphin was able to detect a water-filled 7.62-cm diameter steel sphere, with reflective strength of -23.8 dB, at about 70 m. Ayrapet'yants, Golubkov, Yershova, Zhezherin, Zvorykin, and Korolev (1969, cited in Bullock & Gurevich, 1979) estimated the range of useful echolocation to be approximately 5 m for targets 5 to 15 cm in length, while Konstantinov et al. (1968) determined behaviorally that a 13.5-cm-long fish swimming perpendicular to the direction of the dolphin was detectable at 9.0-m distance. If the fish were swimming in front of the dolphin, but in the same direction, the detection distance decreased substantially to 3.4 to 3.8 m. Considering that many delphinids typically pursue schools of fish rather than single animals (see Norris & Dohl, this volume), Ayrapet'yants and Konstantinov (1974) calculated that a dolphin could detect a school of over 4000 fish, of the same size as used in the experiments by Konstantinov et al. (1968), at over 100 m, assuming that as the distance between dolphin and target increased the amplitude of the echolocation pulse was increased by the animal. This collection of results suggests that echolocation is basically a short to intermediate distance sense.

Acoustic Cues

The acoustic cues used by bottlenosed dolphins while performing echolocation discriminations are not well understood. Evans and Powell (1967) studied target strength (strength of echo) as a possible cue, while controlling for target signature (echo-frequency spectrum). Copper targets of identical size and composition but of different diameters and thickness were used in pairs, resulting in the same signature for each target but different target strengths. Under these conditions, a bottlenosed dolphin discriminated strength differences as small as 6%. Evans (1973) reported similar reflectivity thresholds for the Pacific white-sided dolphin.

Evans and Powell (1967) also studied the usefulness of target signature as a cue by requiring a discrimination between a copper and an aluminum target of the same target strength but of different thicknesses. The discrimination was easily accomplished. While the cues used for this discrimination are not known, MacKay (1965) earlier suggested the delphinids may make use of differences in phase information in the echoes from targets of different thicknesses. Johnson (1967b) subsequently calculated that phase information was indeed available from the targets used by Evans and Powell. Norris et al. (1967), addressing the same problem, pointed out that hollow and solid spheres each produced a two-part echo, the first resulting from surface reflectivity and the second, somewhat delayed in time, from resonances of the material. The time interval between the two parts and their spectral content gave information on the composition of the target. Kamminga and van der Ree (1976) have advanced a similar hypothesis in tests of capabilities of the bottlenosed dolphin for detecting differences between hollow and solid aluminum spheres of the same external dimension. Russian work cited by Bullock and Gurevich (1979) supports the idea that analysis of the intensity and spectrum of the secondary echo is important in discriminating between metal spheres of different composition or diameter.

A number of workers have pondered the role of pulse (click) repetition rate and interpulse interval in echolocation tasks. The interpulse interval is typically adjusted to be longer than the time required for a full round trip of the emitted signal from animal to target and back to animal (Au et al., 1974; Morozov, Atopian, Burdin, Zaitseva & Sokovykh, 1972; Norris et al., 1967). Au et al. (1974) observed that during discrimination of a distant target by the bottlenosed dolphin the interpulse interval averaged 30 to 50 msec longer than the two-way transit time for the signal, ensuring that signal emission and echo return did not overlap. The surplus 30 to 50 msec over transit time may reflect processing time for the echo (Morozov et al., 1972) and preparation time for initiation of the next click.

Pulse repetition rates remain relatively slow when the animal appears to be broadly scanning the environment (orientation clicks) (Norris et al., 1961; Schevill & Watkins, 1966; Turner & Norris, 1966), allowing for long echo return times from distant targets (Norris, 1969). Repetition rates may increase as the animal approaches a target. However, there is sizable intra- and interanimal variability in rate of increase as well as in the number of pulses at any absolute distance to the target (Konstantinov et al., 1968; Morozov et al., 1972; Norris et al., 1967). Furthermore, there does not appear to be any clear correlation between rate of pulse emission and the nature of the target (Johnson, 1967b), except for an increase in rate during discriminations that border on the limits of detectability (Evans & Powell, 1967; Norris et al., 1967). There is also a decrease in swimming speed toward the target during difficult discriminations and the distance at which the discriminations are made is decreased (Evans & Powell, 1967). Greatly increased scanning movements of the head are observed during these difficult discriminations, suggesting that the animal is attempting to maximize the amount and types of information returned from the target.

Adjustments of click spectra by the animal, if possible, can increase the information returned from echoes. For example, target resonances may be detectable through variations in click spectra. The data are, however, contradictory. Diercks et al. (1971), recording from several transducers placed directly on the head of *Tursiops*, found little variation in click composition at any one site during an echolocation task, indicating that the apparent click variation noted by others may have resulted from tank reverberations or from head movements relative to the hydrophones, as somewhat different spectra are obtained at different points around the head (e.g., Norris & Evans, 1967). Au et al. (1974) could not determine whether click variations heard during the initial 1.5 sec of a click train were animal-generated or due to propagation anomalies. If animal-generated, it might be that clicks of different spectra were being "tested" to select those that maximized information return. Ayrapet'yants et al. (1974) found that the click spectrum was shifted to higher frequencies for new targets or for more complex problems, suggesting that high-frequency energy was used for fine-grain information about unfamiliar targets.

Several Russian workers have observed changes in click spectra that are contingent on the type of background masking noise. Babkin and Dubrovskii (1971, cited in Ayrapet'yants & Konstantinov, 1974) reported shifts in pulse spectrum with changes in the spectral composition of the background noise. Paradoxically, the spectrum shift was into the region of maximum noise, but Au et al. (1974) could not confirm this. Instead, more appropriate shifts into the region of least noise were observed. Abramov et al. (1971, cited in Ayrapet'yants & Konstantinov, 1974) also found spectral shifts toward regions of lowest noise level.

Sound Emission Patterns

Sound is projected from the dolphin in patterns that are highly selective directionally for energy and spectral content. Norris et al. (1961), using a hydrophone with bandwidth limited to approximately 100 kHz, found that high-frequency energy (approximately 100 kHz) was directed in a narrow beam forward of the head of the echolocating bottlenosed dolphin. Fish lying below the level of the mandible were not detected, suggesting that sound emission or sound detection was greatly reduced in the animal's vertical plane. Later experiments, using both delphinid carcasses and living animals, confirmed and extended the original observations of highly structured sound beams. Using a carcass of *Stenella* sp., Evans, Sutherland, and Bell (1964) demonstrated significant horizontal signal directivity when an emitting transducer was placed in the vicinity of the nasal sacs. The bulk of the energy was contained within an area 15° to 30° of either side of the midline of the rostrum. When the transducer was placed in the vicinity of the larynx the major portion of the energy was emitted in the vertical plane. Similar results were found using a carcass of *Tursiops*, though the beamwidths were narrower than for *Stenella*.

Norris and Evans (1967) found a highly structured sound field in the rough-toothed dolphin (*Steno bredanensis*), measured in the living animal. The field was directed forward of the animal, with the highest frequency component, 208 kHz, located directly in front of the rostrum. The high-frequency components decreased progressively at points further from the rostrum: 150 kHz maximum at 5° from the midline, 100 kHz at 12°, and 50 kHz at 17°. Measurements of sound levels on the same coordinate plane demonstrated corresponding decreases in acoustic energy away from the midline of the animal: approximately 6 dB down at 10° from the midline, 10 dB at 25°, and 15 dB at 35°. Little or no energy was found directly to the sides or behind the animal. Similar results for *Tursiops* were reported by Reznikov (cited in Yablokov, et al., 1972), who found 6-dB reductions in sound levels at 20° on either side of the midline of the animal. Yablokov et al. also reported that sound levels in the vertical plane were considerably below those in the horizontal plane; at 30° to 40° from the midline, the levels in the vertical plane were only 20 to 30% of the levels in corresponding points on the horizontal plane. Evans (1973) traced emitted sound levels in three dimensions around the head of *Tursiops* and found large variations in click spectrum on, above, and below the horizontal plane.

These findings confirm observations at sea of changes in click-spectrum levels of odontocetes swimming past stationary hydrophones. Norris, Harvey, Burzell, and Krishna Kartha (1972) suggested that this effect results from the sound beam being directed predominantly forward of the animal and from changes in the animal's position relative to the hydrophones as it swims by.

One of the obvious features of actively echolocating bottlenosed dolphins are the jerky, lateral movements of the head through 25° to 30° of arc (Norris et al., 1961). Circular movements of the head, including pronounced bending of the neck, also occur (Kellogg, 1960). These movements may increase in frequency as the target is approached (Evans & Powell, 1967), apparently directing the high-amplitude, high-frequency components of the click over the boundaries of the target to increase the precision and rate of information gained. Lateral and circular scanning movements have also been noted in *Stenella* (Evans et al., 1964) and *Delphinus* (Konstantinov et al., 1968). The Amazon bouto performs a rectilinear scan, consisting of abrupt right-angle transitions between the horizontal and vertical planes (Penner & Murchison, 1970). Bullock and Gurevich (1979) note some Russian work imputing that the direction of the echolocation beam can be changed without any head movement.

Echolocation Mechanisms

The data so far reviewed indicate a highly refined echolocation discrimination capability in several delphinid species. The underlying mechanism and processes are largely unresolved. In asking about mechanisms and processes, it may be helpful to lay out the types of echoic information needed by an animal to cope with its world and then to consider ways that an echolocation system might extract such information. Some obvious information needs include target distance, velocity, direction, size, composition, and, possibly, number of targets.

For target distance a relatively simple mechanisms might measure the time between emission of a click and the returned echo, as was noted earlier. Behavioral temporal resolution capabilities of the dolphin (Yunker & Herman, 1974), and especially the resolution capabilities of the central nervous system (Bullock et al., 1968), provide the basic underpinnings for a reasonably precise distance estimation mechanism. The marked increase in the rate of echolocation clicks when the animal is close to the target allows for the rapid updating of information and may improve the animal's precision, including distance estimates, just prior to the final moments of closure with the target. The oceanic dolphin may also make use of its excellent visual system (Herman, Peacock, Yunker & Madsen, 1975; Dawson, this volume; Madsen & Herman, this volume) to obtain distance data and other information during final closure.

Temporal resolution mechanisms lose value to some degree if there is significant movement of the target during the interval between click and echo (presumably the dolphin can include its own velocity and movement direction in calculations of target distance). Several workers, including Thompson and Herman (1975), have suggested that the dolphin might use its highly precise frequency-discrimination capability to measure Doppler shift in the returned signal, as do some CF bats (see Simmons, 1973). From Dobbler shifts, estimates of target relative movement can be made, and target distance calculated or predicted. However, Altes and colleagues (Altes et al., 1975; Altes, 1974) point out that at the high frequencies of the echolocation click the Doppler shift produced by any reasonable target relative movement would be smaller than the frequency-discrimination capability of the dolphin, making Doppler-shift information of questionable value. For example, although at 50 kHz the bottlenosed dolphin can discriminate on the average a frequency shift as small as 200 Hz (Thompson & Herman, 1975), the Doppler shift at that same frequency for a target moving at 4.5 m/sec relative to the dolphin is 146 Hz. Possibly, however, the redundancy afforded by hearing many successive clicks and their echoes, as takes place in the real world of echolocation, might improve the frequency-discrimination capabilities of the dolphin further and allow for a useful analysis of Doppler shift.

Nordmark (1960) proposed another mechanism, time-separation pitch, for distance determination in echolocating animals. When hearing two pulses presented in rapid succession humans perceive a tone whose apparent frequency is a function of the time difference between the pulses. The effect implies the existence of a mechanism that measures the time interval between successive pulses and converts it into a pitch (Nordmark, 1970). A bat or dolphin might make use of the perceived time-separation pitch produced by the interval between the emitted sonar pulse and its echo to gauge the distance of the target, correlating apparent pitch with distance. Norris (1969) contended that such a system is suitable for a dolphin over a rather limited range of approximately 9 to 15 m; however, these values do fall within the boundaries of the effective echolocation range of delphinids, as discussed earlier.

Target-velocity information may be obtained through comparisons of the time between click and echo for two or more clicks (Norris, 1969) as long as the velocity of the dolphin itself relative to the target were accounted for. If the direction of movement of the target is at an angle to the dolphin's path, analysis can become even more complicated. The demonstrably good sound localization ability of the dolphin should enable it, however, to determine the relative position of the target with high accuracy. The integration of localization information with data on comparative return times of successive echoes may be sufficient for estimates of target velocity under most conditions.

The most likely source of information for the analysis of target size and composition is the spectral configuration of the echo. The spectral "signatures" of key targets are likely to be learned and remembered easily by the dolphin (see Herman, this volume). Spectra perceived while echolocating are perhaps compared with a stored "library" of spectra to identify the target, as in a template-matching procedure. Also specific acoustic "search images" may guide the echolocating animal is well-defined contexts, as when searching for preferred prey species. Norris (1969) noted that there may be changes in the relative-frequency spectra of the echo with different targets, that the loss of low-frequency energy will depend on target size, and that secondary echoes will vary with target composition. Johnson (1967b), Norris et al. (1967), Kamminga and van der Ree (1976), and some of the more recent Russian work described by Bullock & Gurevich (1979), have provided evidence that those types of cues are indeed important to the echolocating animal. There seem to be indications from the Russian work, as well as from some of the American work, that the spectral content of the emitted clicks may be adaptively altered to maximize reflected information, although it was noted that not all data are wholly supportive of this view. Herman (this volume) has suggested that the rate of improvement with practice of an animal in laboratory echolocation tasks may principally trace its success in modifying its emissions for maximizing the information return from the target.

Altes and his colleagues (Altes, 1971, 1974; Altes et al., 1975) have suggested that the dolphin performs a spectral analysis of the returned echo, using a bank of filters having a constant sharpness (constant Q). The spectral composition of the echo, after analysis by the filters, describes the target qualities. The confusing changes in an echo that may arise from target motion are guarded against in Altes' model by filters with bandwidths that increase with increases in frequency of the echo spectrum, making the system insensitive to Doppler shift. Altes et al. (1975) noted that a filter bank of the type specified in their system was inherent within the well-studied critical-band system of the mammalian ear and, for the dolphin, could provide the first level of analysis of the returned echo. This may be followed by more central analyses using auto-correlational or cross-correlational techniques to compare echo spectra with the emitted signal or with replicas of various target spectra previously experienced (cf. Simmons, 1973). Some of the Russian work summarized by Bullock and Gurevich (1979) has emphasized the cross-correlation of spectra of pulse and echo signals, although the Russians primarily discussed its use in resolving target direction rather than target quality.

The vast bulk of the available data on echolocation comes from studies of dolphins in oceanariums or open-ocean pens during ensonification of artificial targets. Relatively little is known about the ethology of echolocation behavior in the wild. What are some of the social and ecological correlates of the use of echolocation signals; what differences in regularity or conditions of use can be found among different species; and how is the form of the echolocation scanning pattern or the structure of content of the echolocation signal adaptive for the species? While some suggestive data relevant to these questions are available—e.g., an inverse correlation may exist between the upper frequency limits of the echolocation click and the size of prey taken, and platanistid species inhabiting murky river waters may echolocate much more frequently, sometimes constantly, than do oceanic delphinids—only a bare beginning has been made into the understanding of comparative echolocation. More generally, there is a pressing need, as in all animal behavior work, to bring together the results of laboratory investigations with observations from the field to appreciate more fully the functional significance of the behavior.

SUMMARY AND CONCLUSIONS

Delphinid sounds are complex and variable within and across species. They may have an underlying pulsed structure or be continuous in form. The short clicklike pulses implicated in echolocation are typically rich in ultrasonic energy, while the fundamental frequencies of the continuous tones fall most commonly within the sonic range. For most delphinid species studied, as well as the platanistid dolphins, the echolocation clicks are broadband, of short duration, and rich in ultrasonic energy. Narrow-band clicks of predominantly lower frequencies have been reported for some delphinids. Possibly some species may be able to modify the echolocation clicks adaptively in intensity or spectral composition to maximize the signal level of the echo or its information content. The pattern of the emitted sound beam may also be modifiable adaptively. There are many differences in echolocation signals and echolocation behavior across species, but their ecological, or possibly social, correlates are still poorly understood.

Burst-pulse sounds, comprised of clicks but with acoustic features perhaps best describable by the pulse envelope, occur commonly in social or emotional contexts and almost certainly have value as signals in intraspecific communication (see Herman & Tavolga, this volume). In some delphinid species narrowband, nearly pure-tone whistles changeable in frequency, duration, or amplitude appear. These are most often continuous (unbroken) but in a few species may be expressed as tone bursts. Whistles, like burst-pulse sounds, seem to have social/emotional value. The characteristics of the whistle may vary greatly within and between species. That different individuals of a species may have different "characteristic" whistles has suggested to some that whistles may function as individual identifiers or signatures. The degree of stereotypy is, however, still a matter at issue. More generally, there is a pressing need for better evidence on the degree to which specific signals are associated with particular social situations.

A variety of hypotheses have been proposed for the site of sound production in delphinids, but the bulk of recent evidence indicates that both whistles and clicks are produced in the region of the nasal plugs. Significantly, both types of sound can be produced simultaneously, although they may be projected in very different patterns about the animal's head. It was suggested that sound reflectors of different size, selective therefore for wavelength, could account for the different sound beams characterizing whistle and click emissions. Sound emission through the melon may provide a directionalization mechanism as well as an impedance-matching system for the more efficient transmission of internally generated sound into water.

Most delphinids, as well as the Amazon bouto, can detect a wide range of sound frequencies—up to 150 kHz or higher in some species—with impressive sensitivity. The killer whale stands apart with an upper frequency limit reaching only to approximately 32 kHz. Experiments, mostly with the bottlenosed dolphin, also demonstrate impressive abilities for frequency and intensity discrimination, temporal resolution, and sound localization. The auditory system of delphinids is in many respects morphologically similar to that in terrestrial mammals, although there are middle or inner ear adaptations for impedance matching and high-frequency detection as well as hypertrophy in auditory sections of the central nervous system. The entry paths for sound are highly modified; sound may arrive at the cochlea through general tissue conduction or through a special lipid-filled channel in the lower jaw. The array of sites of heightened sensitivity to sound—the region of the EAM, the melon, the area near the insertion of the pectoral fin, and the pan-bone region of the mandible—provide a tantalizing puzzle to those trying to understand the complexity of acoustic imagery in the dolphin.

The delphinid echolocation system is well constituted for gathering information about the environment, as is that of the Amazon bouto. Laboratory experiments indicate that through echolocation dolphins can resolve differences in size, distance, or texture of very similar targets. Additionally, targets of very small area are detectable at near distances, as are larger targets at distances of tens of meters. It appears that the click emission rate is always adjusted so that the echo from the ensonified target appears in the interspace between successive clicks. Since the return from near targets is sooner, click-emission rate can be increased in these cases to increase the rate of information return. Mechanisms used in extraction of target information from the echo remain to be clarified. A number of mechanisms have been proposed for distance and velocity information and for assessing target size and composition, including time-measuring and spectral-analysis devices. Suggestions have been made that the critical-band mechanism of the ear is important as a first step in spectral analysis of the signal.

The preponderance of data on sound emission and detection by delphinids has come from the bottlenosed dolphin, especially in the areas of echolocation capabilities, the psychophysical and physiological responsivity to sounds, and descriptions and theories of underlying structure and processes. The sound emissions of other species can be substantially different from *Tursiops,* suggesting that many special adaptations may be uncovered as these other species are investigated in detail. The study of comparative emission and detection capabilities and underlying mechanisms and their correlation with the social and ecological living conditions of the species seems an important next step in attempting to understand more fully the extent and significance of the acoustic adaptations of the cetaceans.

ACKNOWLEDGMENTS

Portions of the work reported and preparation of the manuscript were supported by Office of Naval Research contract N00123-74-C-0660. Preparation of the manuscript was also supported in part by Public Health Service grants NS-09374 and NS-15090 from the National Institute of Neurological and Com-

municative Disorders and Stroke. The author would like to express his
gratitude to Dr. Donna McDonald-Renaud for valued discussions relating to
cetacean audition and to Dr. Earl Murchison for permitting use of unpublished
data. Drs. L. Herman, J. McCormick, W. C. Stebbins, and W. N. Tavolga
kindly read and commented on the manuscript.

REFERENCES

Adolfson, J. & Berghage, T. (1974). *Perception and Performance Underwater.* New York: Wiley.

Albers, V. M. (1965). *Underwater Acoustics Handbook II.* University Park: Pennsylvania State University Press.

Altes, A. (1971). Computer derivation of some dolphin echolocation signals. *Science,* 173, 912–914.

Altes, R. A. (1974). Study of animal sensory systems with application to sonar. Sunnyvale, Calif.: ESL Inc. Report No. ESL-PR144.

Altes, R. A., Evans, W. E. & Johnson, C. S. (1975). Cetacean echolocation signals and a new model for the human glottal pulse. *J. Acoust. Soc. Am.,* 57, 1221–1223.

Andersen, S. (1971a). Auditory sensitivity of the harbour porpoise *Phocoena phocoena.* In: *Investigations on Cetacea,* Vol. III (G. Pilleri, Ed.), pp. 255–259. Berne, Switzerland: Institute of Brain Anatomy, University of Berne.

Andersen, S. (1971b). Directional hearing in the harbour porpoise *Phocoena phocoena.* In: *Investigations on Cetacea,* Vol. III (G. Pilleri, Ed.), pp. 260–264. Berne, Switzerland: Institute of Brain Anatomy, University of Berne.

Au, W. W. L., Floyd, R. W., Penner, R. H. & Murchison, A. E. (1974). Measurement of echolocation signals of the Atlantic bottlenose dolphin, *Tursiops truncatus* Montagu, in open waters. *J. Acoust. Soc. Am.,* 56, 1280–1290.

Ayrapet'yants, E. Sn., Golubkov, A. G., Yershova, I. Z., Zhezherin, A. R., Zvorykin, V. N. & Korolev, V. I. (1969). Echolocation differentiation and characteristics of radiated pulses in dolphins. *Doxlady Akademii Nauk SSSR,* 188, 1197–1199. (In Russian.)

Ayrapet'yants, E. Sn. & Konstantinov, A. I. (1974). Echolocation in nature. Arlington, Va.: Joint Publication Research Service (JPRS 63326-1-2).

Ayrapet'yants, E. Sn., Voronov, V. A., Ivanenko, Y. V., Ivanvo, M. P., Ordovskii, D. L., Sergeev, B. F. & Chilingiris, V. I. (1974). The physiology of the sonar system of black sea dolphins. *J. Evol. Biochem. Physiol.,* 9, 364–369.

Backus, R. H. & Schevill, W. E. (1966). *Physeter* clicks. In: *Whales, Dolphins, and Porpoises* (K. S. Norris, Ed.), pp. 510–528. Berkeley: University of California Press.

Batteau, D. W. (1967). The role of the pinna in human localization. *Proc. Roy. Soc. Lond.,* 168, 158–180.

Békésy, G. von (1960). *Experiments in Hearing.* New York: McGraw-Hill.

Bel'kovich, V. M., Borisov, V. I. & Gurevich, V. S. (1970). Angular resolution by echolocation in *Delphinus delphis. Proc. Sci.-Tech. Conf., Ministry of Higher and Secondary Specialized Education RSFSR,* Leningrad.

Bel'kovich, V. M. & Solntseva, G. N. (1970). Anatomy and function of the ear in the dolphin. Arlington, Va.: Joint Publication Research Service (JPRS 50253).

Bergeijk, W. A. van (1967). The evolution of vertebrate hearing. In: *Contributions to Sensory Physiology,* Vol. II (W. D. Neff, Ed.), pp. 1–41. New York: Academic.

Brandt, J. F. & Hollien, H. (1967). Underwater hearing thresholds in man. *J. Acoust. Soc. Am.,* 42, 966–971.

Brown, A. M. & Pye, J. D. (1975). Auditory sensitivity at high frequencies in mammals. In: *Advances in Comparative Physiology and Biochemistry*, Vol. VI (O. Lowenstein, Ed.), pp. 1–73. New York: Academic.

Bullock, T. H., Grinnel, A. D., Ikezono, E., Kameda, K., Katsuki, Y., Nomoto, M., Sato, O., Suga, N. & Yanagisawa, K. (1968). Electrophysiological studies of the central auditory mechanisms in cetaceans. *Z. Vergl. Physiol.*, **59**, 117–156.

Bullock, T. H. & Gurevich, V. S. (1979). Soviet literature on the nervous system and psychobiology of cetacea. *Int. Rev. Neurobiol.*, **21**, pp. 47–127.

Bullock, T. H., Ridgway, S. H. & Suga, N. (1971). Acoustically evoked potentials in midbrain auditory structures in sea lions (Pinnipedia). *Z. Vergl. Physiol.*, **74**, 732–787.

Bullock, T. H. & Ridgway, S. H. (1972a). Evoked potentials in the central auditory system of alert porpoises to their own and artificial sounds. *J. Neurobiol.*, **3**, 79–99.

Bullock, T. H. & Ridgway, S. H. (1972b). Neurophysiological findings relevant to echolocation in marine mammals. In: *Animal Orientation and Navigation* (S. R. Galler, K. Schmidt-Konig, G. J. Jacobs, & R. E. Belleville, Eds.), pp. 373–395. Washington, D.C.: NASA.

Burdin, V. I., Markov, V. I., Reznik, A. M., Skornyakov, V. M. & Chupakov, A. G. (1973). Determination of the just noticeable intensity difference for white noise in the Black Sea bottlenose dolphin (*Tursiops truncatus ponticus* Barabasch). In: *Morphology and Ecology of Marine Mammals* (K. K. Chapskii & V. E. Sokolov, Eds.), pp. 169–173. New York: Wiley.

Burdin, V. I., Reznik, A. M., Skornyakov, V. M. & Chupakov, A. G. (1975). Communication signals of the black sea bottlenose dolphin. *Sov. Phys. Acoust.*, **20**, 314–318.

Busnel, R.-G. (1963). *Acoustic Behavior of Animals*. Amsterdam: Elsevier.

Busnel, R.-G. & Dziedzic, A. (1966). Acoustic signals of the pilot whale *Globicephala malaena*, and of the porpoises *Delphinus delphis* and *Phocoena phocoena*. In: *Whales Dolphins & Porpoises* (K. S. Norris, Ed.), pp. 607–646. Berkeley: University of California Press.

Busnel, R.-G. & Dziedzic, A. (1967). Resultants metrologiques experimentaux de l'echolocation chez le *Phocoena phocoena*, et leur comparison avec ceux de certaines chauves-souris. In: *Animal Sonar Systems: Biology and Bionics*, Vol. I (R.-G. Busnel, Ed.), pp. 307–338. Jouy-en-Josas, France: Laboratoire de Physiologie Acoustique.

Busnel, R.-G. & Dziedzic, A. (1968). Etude des signaux acoustiques associés à des situations de détresse chez certaines cétacés odontocétes. *Ann. de l'Institut Oceanograph.*, **46**, 109–144.

Caldwell, D. K. & Caldwell, M. C. (1972). Senses and communication. In: *Mammals of the Sea* (S. H. Ridgway, Ed.), pp. 466–502. Springfield, Ill.: Thomas.

Caldwell, D. K. & Caldwell, M. C. (1977). Cetaceans. In: *How Animals Communicate* (T. A. Sebeok, Ed.), pp. 794–808. Bloomington: Indiana University Press.

Caldwell, M. C. & Caldwell, D. K. (1966). Epimeletic (care-giving) behavior in cetacea. In: *Whales, Dolphins, and Porpoises* (K. S. Norris, Ed.) pp. 755–788. Berkeley: University of California Press.

Caldwell, M. C. & Caldwell, D. K. (1967). Intraspecific transfer of information via pulsed sound in captive odotocete cetaceans. In: *Animal Sonar Systems: Biology and Bionics*, Vol. II (R.-G. Busnel, Ed.), pp. 879–936. Jouy-en-Josas, France: Laboratoire de Physiologie Acoustique.

Caldwell, M. C. & Caldwell, D. K. (1971). Statistical evidence for individual signature whistles in Pacific whitesided dolphins, *Lagenorhynchus obliquidens*. *Cetology*, **3**, 1–9.

Caldwell, M. C., Caldwell, D. K. & Evans, W. E. (1966). Sounds and behavior of captive Amazon freshwater dolphins, *Inia geoffrensis*. *Los. Ang. City. Mus. Nat. Hist. Contrib. Sci.*, **108**, 1–24.

Cummings, W. C. & Thompson, P. O. (1971a). Underwater sounds from the blue whale, *Balaenoptera musculus*. *J. Acoust. Soc. Am.*, **50**, 1193–1198.

Cummings, W. C. & Thompson, P. O. (1971b). Gray whales, *Eschrichtius robustus*, avoid the underwater sounds of killer whales, *Orcinus orca*. *Fish. Bull.*, **69**, 525–530.

Dailly, M. (1972). Contribution to the study of the cochlear apparatus in dolphins. In:

Investigations on Cetacea, Vol. II (G. Pilleri, Ed.), pp. 215-230. Berne, Switzerland: Institute of Brain Anatomy, University of Berne.

Diercks, K. J. (1972). Biological sonar systems: A bionics survey. Applied Research Laboratories, University of Texas. ARL-TR-72-34.

Diercks, K. J., Trochta, R. T. & Evans, W. E. (1973). Delphinid sonar: Measurement and analysis. *J. Acoust. Soc. Am.,* **54,** 200-204.

Diercks, K. J., Trochta, R. T., Greenlaw, R. L. & Evans, W. E. (1971). Recording and analysis of dolphin echolocation signals. *J. Acoust. Soc. Am.,* **49,** 1729-1732.

Dormer, K. J. (1974). The mechanism of sound production and measurement of sound processing in delphinid cetaceans. Ph.D. thesis, University of California, Los Angeles.

Dormer, K. J. (1979). Mechanisms of sound production and air recycling in delphinids: Cineradiographic evidence. *J. Acoust. Soc. Am.,* **65,** 229-239.

Dreher, J. J. & Evans, W. E. (1964). Cetacean communication. In: *Marine Bio-Acoustics* (W. N. Tavolga, Ed.), pp. 373-393. New York: Pergamon.

Dubrovskii, N. A., Krasov, P. S. & Titov, A. A. (1971). On the emission of echolocation signals by the Azov harbor porpoise. *Sov. Phys. Acoust.,* **16,** 444-447.

Dudok van Heel, W. H. (1959). Audio-direction findings in the porpoise *Phocoena phocoena. Nature,* **183,** 1063.

Dudok van Heel, W. H. (1962). Sound and cetacea. *Neth. J. Sea Res.,* **1,** 407-507.

Erulkar, S. D. (1972). Comparative aspects of spatial localization of sound. *Physiol. Rev.,* **52,** 237-360.

Evans, W. E. (1967). Vocalizations among marine mammals. In: *Marine Bio-Acoustics,* Vol. II (W. N. Tavolga, Ed.), pp. 159-186. New York: Pergamon.

Evans, W. E. (1973). Echolocation by marine delphinids and one species of freshwater dolphin. *J. Acoust. Soc. Am.,* **54,** 191-199.

Evans, W. E. & Maderson, P. F. A. (1973). Mechanisms of sound production in delphinid cetaceans: A review and some anatomical considerations. *Am. Zool.,* **13,** 1205-1213.

Evans, W. E. & Powell, B. A. (1967). Discrimination of different metallic plates by an echolocating delphinid. In: *Animal Sonar System: Biology and Bionics,* Vol. I (R.-G. Busnel, Ed.), pp. 363-383. Jouy-en-Josas, France: Laboratoire de Physiologie Acoustique.

Evans, W. E. & Prescott, J. H. (1962). Observations of the sound production capabilities of the bottlenose porpoise: A study of whistles and clicks. *Zoologica,* **47,** 121-128.

Evans, W. E., Sutherland, W. W. & Bell, R. G. (1964). The directional characteristics of delphinid sounds. In: *Marine Bio-Acoustics* (W. N. Tavolga, Ed.), pp. 353-370. New York: Pergamon.

Fay, R. R. (1974). Auditory frequency discrimination in vertebrates. *J. Acoustic Soc. Am.,* **56,** 206-209.

Fish, J. F., Johnson, C. S. & Ljungblad, D. K. (1976). Sonar target discrimination by instrumented human divers. *J. Acoust. Soc. Am.,* **59,** 602-606.

Fish, J. F. & Turl, C. W. (1976). Acoustic source levels of four species of small whales. Naval Undersea Center Report NUC TP 547.

Fish, J. F. & Vania, J. (1971). Killer whale, *Orcinus orca,* sounds repel white whales, *Delphinapterus leucus. Fish. Bull.,* **69,** 531-535.

Fleischer, G. (1973a). On structure and function of the middle ear in the bottlenosed dolphin (*Tursiops truncatus*). In: *Proc. 9th Annu. Conf. Biol. Sonar Diving Mammals,* pp. 137-179. Menlo Park, Calif.: Stanford Research Institute.

Fleischer, G. (1973b). Structural analysis of the tympanicum complex in the bottle-nosed dolphin (*Tursiops truncatus*). *J. Aud. Res.,* **13,** 178-190.

Fleischer, G. (1976a). Hearing in extinct cetaceans as determined by cochlear structure. *J. Paleont.,* **50,** 133-152.

Fleischer, G. (1976b). On bony microstructures in the dolphin cochlea relating to hearing. *N. Jb. Geol. Palant. Abh.,* **151,** 166-191.

Fletcher, H. (1940). Auditory patterns. *Rev. Mod. Phys.,* **12,** 47-65.

Ford, J. K. B. & Fisher, H. D. (1978). Underwater acoustic signals of the narwhal (*Monodon monoceros*). *Can. J. Zool.,* **56,** 552-560.

Fraser, F. C. & Purves, P. E. (1954). Hearing in cetaceans. *Bull. Br. Mus. Nat. Hist.,* **2,** 103-116.

Fraser, F. C. & Purves, P. E. (1960). Hearing in cetaceans. *Bull. Br. Mus. Nat. Hist.,* **7,** 1-140.

Gapich, L. I. & Supin, Y. A. (1974). The activity of single neurons of the acoustical region of the cerebral cortex of the porpoise *Phocoena phocoena. J. Evol. Biochem. Physiol.,* **10,** 182-183.

Gentry, R. L. (1967). Underwater auditory localization in the California sea lion (*Zalophus californianus*). *J. Aud. Res.,* **7,** 187-193.

Gourevich, G. (1970). Detectability of tones in quiet and in noise by rats and monkeys. In: *Animal Psychophysics* (W. C. Stebbins, Ed.), pp. 67-97. New York: Appleton-Century-Crofts.

Green, D. M. & Swets, J. A. (1966). *Signal Detection Theory and Psychophysics.* New York: Wiley.

Greenwood, D. N. (1961). Critical bandwidths and the frequency coordinates of the basilar membrane. *J. Acoust. Soc. Am.,* **33,** 1344-1356.

Griffin, D. R. (1968). Echolocation and its relevance to communication behavior. In: *Animal Communication* (T. A. Sebeok, Ed.), pp. 154-164. Bloomington: Indiana University Press.

Hall, J. D. & Johnson, C. S. (1972). Auditory thresholds of a killer whale *Orcinus orca* Linnaeus. *J. Acoust. Soc. Am.,* **51,** 515-517.

Hawkins, J. E. & Stevens, S. S. (1950). The masking of pure tones and of speech by white noise. *J. Acoust. Soc. Am.,* **22,** 6-13.

Herald, E. S., Brownwell, R. L., Jr., Frye, F. L., Morris, E. J., Evans, W. E. & Scott, A. B. (1969). Blind river dolphin: First side swimming cetacean. *Science,* **166,** 1408-1410.

Herman, L. M. & Arbeit, W. R. (1972). Frequency discrimination limens in the bottlenosed dolphin: 1-70 KC/S. *J. Aud. Res.,* **2,** 109-120.

Herman, L. M. & Arbeit, W. R. (1973). Stimulus control and auditory discrimination learning sets in the bottlenosed dolphin. *J. Exp. Anal. Behav.,* **19,** 379-394.

Herman, L. M. & Gordon, J. A. (1974). Auditory delayed matching in the bottlenosed dolphin. *J. Exp. Anal. Behav.,* **21,** 19-26.

Herman, L. M., Peacock, M. F., Yunker, M. P. & Madsen, C. J. (1975). Bottlenosed dolphin: Double slit pupil yields equivalent aerial and underwater diurnal acuity. *Science,* **189,** 650-652.

Iversen, R. T. B. (1967). Response of yellowfin tuna (*Thunnus albacares*) to underwater sound. In: *Marine Bio-Acoustics,* Vol. II (W. N. Tavolga, Ed.), pp. 105-122. New York: Pergamon.

Jacobs, D. W. (1972). Auditory frequency discrimination in the Atlantic bottlenose dolphin, *Tursiops truncatus* Montagu: A preliminary report. *J. Acoust. Soc. Am.,* **52,** 696-698.

Jacobs, D. W. & Hall, J. D. (1972). Auditory thresholds of a freshwater dolphin, *Inia geoffrensis* Blainville. *J. Acoust. Soc. Am.,* **51,** 530-533.

Johnson, C. S. (1966). Auditory thresholds of the bottlenose dolphin (*Tursiops truncatus* Montagu). U.S. Naval Ordinance Test Station NOTSTP 4178, 25 pp.

Johnson, C. S. (1967a). Sound detection thresholds in marine mammals. In: *Marine Bio-Acoustics,* Vol. II (W. N. Tavolga, Ed.), pp. 247-260. New York: Pergamon.

Johnson, C. S. (1967b). Discussion to paper by Evans and Powell. In: *Animal Sonar Systems: Biology and Bionics,* Vol. I (R.-G. Busnel, Ed.), pp. 384-398. Jouy-en-Josas, France: Laboratoire de Physiologie Acoustique.

Johnson, C. S. (1968). Masked tonal thresholds in the bottlenosed porpoise. *J. Acoust. Soc. Am.,* **44,** 965-967.

Kamminga, C. & van der Ree, A. F. (1976). Discrimination of solid and hollow spheres by *Tursiops truncatus* (Montagu). *Aquat. Mammals,* **4,** 1-9.

Kellogg, W. N. (1953). Ultrasonic hearing in the porpoise, *Tursiops truncatus. J. Comp. Physiol. Psychol.,* **46,** 446-450.

Kellogg, W. N. (1959). Auditory perception of submerged objects by porpoises, *J. Acoust. Soc. Am.,* **31,** 1-6.

Kellogg, W. N. (1960). Auditory scanning in the dolphin. *Psychol. Rec.,* **10,** 25-27.

Kellogg, W. N., Kohler, R. & Morris, H. M. (1953). Porpoise sounds as sonar signals. *Science,* **117,** 239-243.

Kinne, O. (1975). Orientation in space: Animals: Marine. In: *Marine Ecology,* Vol. II (O. Kinne, Ed.), pp. 709-852. London: Wiley.

Konstantinov, A. I., Mel'nikov, N. F. & Titov, A. A. (1968). On the abilities of dolphins to recognize objects. *Tez. Dokl. II Republ. Konf. po Bzon ike, Kiev,* pp. 57-59.

Kruger, L. (1966). Specialized features of the cetacean brain. In: *Whales, Dolphins, and Porpoises* (K. S. Norris, Ed.), pp. 232-254. Berkeley: University of California Press.

Lang, T. G. & Smith, H. A. P. (1965). Communication between dolphins in separate tanks by way of an electronic acoustic link. *Science,* **150,** 1839-1844.

Lilly, J. C. & Miller, A. (1961). Sounds emitted by the bottlenose dolphin. *Science,* **133,** 1689-1693.

Litchfield, C., Karol, C. & Greenberg, A. J. (1973). Compositional topography of melon lipids in the Atlantic bottlenose dolphin (*Tursiops truncatus*): Implications for echolocation. *Ma. Biol.,* **23,** 165-169.

MacKay, R. S. (1965). Telemetering from within the body of animals and man: Endoradiosondes. In: *Biomedical Telemetry* (C. Caceres, Ed.), pp. 148-233. New York: Academic.

McBride, A. F. (1956). Evidence for echolocation by cetaceans. *Deep-sea Res.,* **3,** 153-154.

McCormick, J. G. (1968). Theory of hearing for delphinids. Ph.D. thesis, Princeton University.

McCormick, J. G.., Wever, E. G., Palin, J. & Ridgway, S. H. (1970). Sound conduction in the dolphin ear. *J. Acoust. Soc. Am.,* **48,** 1418-1428.

McDonald-Renaud, D. L. (1974). Sound localization in the bottlenose porpoise, *Tursiops truncatus* (Montagu). Ph.D. thesis, University of Hawaii.

Mills, A. W. (1958). On the minimum audible angle. *J. Acoust. Soc. Am.,* **30,** 237-246.

Mills, a. W. (1972). Auditory localization. In: *Foundations of Modern Auditory Theory* (J. V. Tobias, Ed.), pp. 303-348. New York: Academic.

Møhl, B. (1964). Preliminary studies on hearing in seals. *Vidensk. Medd. Dansk Naturh. Foren.,* **127,** 283-294.

Møhl, B. (1967). Frequency discrimination in the common seal and a discussion of the concept of upper hearing limit. In: *Underwater Acoustics* Vol. 2 (V. M. Albers, Ed.), pp. 43-54. New York: Plenum.

Møhl, B. & Andersen, S. (1973). Echolocation: High-frequency component in the click of the harbor porpoise (*Phocoena ph.* L.). *J. Acoust. Soc. Am.,* **57,** 1368-1372.

Möhres, F. P. (1967). General characters of acoustic orientation signals and performances of sonar in the order Chiroptera. In: *Animal Sonar Systems: Biology and Bionics* (R. -G. Busnel, Ed.), pp. 401-408. Jouy-en-Josas, France: Laboratoire de Physiologie Acoustique.

Moore, P. (1975). Underwater localization of click and pulsed pure tone signals by the California sea lion (*Zalophus californianus*). *J. Acoust. Soc. Am.,* **57,** 406-410.

Moore, P. & Au, W. (1975). Underwater localization of pulsed pure tone by the California sea lion (*Zalophus californianus*). *J. Acoust. Soc. Am.,* **58,** 721-727.

Moore, P. & Schusterman, R. J. (1976). Discrimination of pure tone intensities by the California sea lion. *J. Acoust. Soc. Am.,* **60,** 1405-1407.

Morgane, P. J. & Jacobs, M. S. (1972). Comparative anatomy of the cetacean nervous system. In: *Functional Anatomy of Marine Mammals*, Vol. I (R. J. Harrison, Ed.), pp. 117–244. New York: Academic.

Morozov, V. P., Atopian, A. I., Burdin, V. I., Zaitseva, K. A. & Sokovykh, Y. A. (1972). Tracking frequency of the location signals of dolphins as a function of distance to the target. *Biofizika*, **17**, 139–145.

Nordmark, J. O. (1960). Perception of distance in animal echolocation. *Nature*, **183**, 1009–1010.

Nordmark, J. O. (1970). Time and frequency analysis. In: *Foundations of Modern Auditory Theory*, Vol. I (J. V. Tobias, Ed.), pp. 57–83. New York: Academic.

Norris, K. S. (1964). Some problems of echolocation in cetaceans. In: *Marine Bio-Acoustics* (W. N. Tavolga, Ed.), pp. 317–336. New York: Pergamon.

Norris, K. S. (1964). Some problems of echolocation in cetaceans. In: *Marine Bio-Acoustics* (W. N. Tavolga, Ed.), pp. 317–336. New York: Pergamon.

Norris, K. S. (1969). The echolocation of marine mammals. In: *The Biology of Marine Mammals* (H. T. Anderson, Ed.), pp. 391–423. New York: Academic.

Norris, K. S., Dormer, K. J., Pegg, J. & Liese, G. J. (1971). The mechanisms of sound production and air recycling in porpoises: A preliminary report. In: *Proc. 8th Annu. Conf. Biol. Sonar Diving Mammals*. Menlo Park, Calif.: Stanford Research Institute, pp. 113–129.

Norris, K. S. & Evans, W. E. (1967). Directionality of echolocation clicks of the rough-toothed porpoise, *Steno bredanensis* (Lesson). In: *Marine Bio-Acoustics*, Vol. II (W. N. Tavolga, Ed.), pp. 305–316. New York: Pergamon.

Norris, K. S., Evans, W. E. & Turner, R. N. (1967). Echolocation in an Atlantic bottlenose porpoise during discrimination. In: *Animal Sonar Systems: Biology and Bionics*, Vol. I (R.-G. Busnel, Ed.), pp. 409–437. Jouy-en-Josas, France: Laboratoire de Physiologie Acoustique.

Norris, K. S. & Harvey, G. W. (1974). Sound transmission in the porpoise head. *J. Acoust. Soc. Am.*, **56**, 659–664.

Norris, K. S., Harvey, G. W., Burzell, L. A. & Krishna Kartha, D. K. (1972). Sound production in the freshwater porpoise *Sotalia* cf. *fluviatilis* Gervais and Deville and *Inia geoffrensis* Blainville in the Rio Negro Brazil. In: *Investigations on Cetacea*, Vol. IV (G. Pilleri, Ed.), pp. 251–262. Berne, Switzerland: Institute of Brain Anatomy, University of Berne.

Norris, K. S., Prescott, J. H., Asa-Dorian, P. V. & Perkins, P. (1961). An expermental demonstration of echolocation behavior in the porpoise, *Tursiops truncatus* (Montagu). *Biol. Bull.*, **120**, 163–176.

Novick, A. (1973). Echolocation in bats: A zoologist's vies. *J. Acoust. Soc. Am.* **54**, 139–146.

Parvulescu, A. (1964). Problems of propagation and processing. In: *Marine Bio-Acoustics* (W. N. Tavolga, Ed.), pp. 87–100. New York: Pergamon.

Payne, R. (1971). Acoustic location of prey by barn owls (*Tyco alba*). *J. Exp. Biol.*, **54**, 535–573.

Payne, R. & Webb, D. (1971). Orientation by means of long-range acoustic signaling in baleen whales. *Ann. N.Y. Acad. Sci.*, **188**, 110–141.

Penner, R. H. & Murchison, A. E. (1970). Experimentally demonstrated echolocation in the Amazon River porpoise *Inia geoffrensis* (Blaineville). Naval Undersea Center Technical Publication, **187**, 1–25.

Pilleri, G., Gihr, M., Purves, P. E., Zbinden, K. & Kraus, G. (1976). On the behaviour, bioacoustics, and functional morphology of the Indus River dolphin (*Platanista indi* Blyth, 1859). In: *Investigations on Cetacea*, Vol. VI (G. Pilleri, Ed.), pp. 13–151. Berne, Switzerland: Brain Anatomy Institutue, University of Berne.

Pollack, I. (1954). Intensity discrimination thresholds under several psychophysical procedures. *J. Acoust. Soc. Am.*, **26**, 1056–1059.

Popper, A. N. (1971). Effects of size on the auditory capacities of the goldfish. *J. Aud. Res.*, **11**, 239–247.

Powell, B. A. (1966). Periodicity of vocal activity of captive Atlantic bottlenose dolphins, *Tursiops truncatus. Bull. South. Calif. Acad. Sci.,* **65,** 237-244.

Purves, P. E. (1966). Anatomy and physiology of the outer and middle ear in cetaceans. In: *Whales, Dolphins, and Porpoises* (K. S. Norris, Ed.), pp. 321-376. Los Angeles: University of California Press.

Purves, P. E. (1967). Anatomical and experimental observations on the cetacean sonar system. In: *Animal Sonar Systems: Biology and Bionics* (R.-G. Busnel, Ed.), pp. 197-270. Jouy-en-Josas, France: Laboratoire de Physiologie Acoustique.

Purves, P. E. & van Utrecht, W. L. (1963). The anatomy and function of the ear of the bottlenosed dolphin, *Tursiops truncatus. Beaufortia,* **9,** 241-256.

Pye, J. D. (1967). Synthesizing the waveforms of bat's pulses. In: *Animal Sonar Systems: Biology and Bionics,* Vol. I (R.-G. Busnel, Ed.), pp. 43-65. Jouy-en-Josas, France: Laboratoire de Physiologie Acoustique.

Raslear, T. G. (1974). The use of the cochlear microphonic responses as an indicant of auditory sensitivity: Review and evaluation. *Psychol. Bull.,* **81,** 791-803.

Renaud, D. L. & Popper, A. N. (1975). Sound localization by the bottlenose porpoise, *Tursiops truncatus. J. Exp. Biol.,* **63,** 569-585.

Repenning, C. A. (1972). Underwater hearing in seals: Functional morphology. In: *Functional Anatomy of Marine Mammals,* Vol. I (R. J. Harrison, Ed.), pp. 307-331. New York: Academic.

Reysenbach de Haan, F. W. (1957). Hearing in whales. *Acta Oto-Laryngol. Suppl.* **134,** 1-114.

Ridgway, S. H., McCormick, J. G. & Wever, E. G. (1974). Surgical approach to the dolphin's ear. *J. Exp. Zool.,* **188,** 265-276.

Riesz, R. R. (1928). Differential sensitivity of the ear for pure tones. *Phys. Rev.,* **31,** 868-875.

Roffler, S. K. & Butler, R. A. (1968a). Factors that influence the localization of sound in the vertical plane. *J. Acoust. Soc. Am.,* **43,** 1255-1259.

Roffler, S. K. & Butler, R. A. (1968b). Localization of tonal stimuli on the vertical plane. *J. Acoust. Soc. Am.,* **43,** 1260-1266.

Sales, G. & Pye, D. (1974). *Ultrasonic Communication by Animals.* London: Chapman and Hall.

Scharf, B. (1970). Critical bands. In: *Foundation of Modern Auditory Theory,* Vol. I (J. V. Tobias, Ed.), pp. 157-202. New York: Academic

Schevill, W. E. & Lawrence, B. (1953a). Auditory response of a bottlenosed porpoise, *Tursiops truncatus,* to frequencies above 100 kc. *J. Exp. Zool.,* **124,** 147-165.

Schevill, W. E. & Lawrence, B. (1953b). High-frequency auditory responses of a bottlenosed porpoise, *Tursiops truncatus* (Montagu). *J. Acoust. Soc. Am.,* **25,** 1016-1017.

Schevill, W. E. & Watkins, W. A. (1966). Sound structure and directionality in *Orcinus* (killer whale). *Zoologica,* **51,** 71-76.

Schevill, W. E. & Watkins, W. A. (1971). Pulsed sounds of the porpoise, *Lagenorhynchus australis. Brevoria,* **366,** 1-10.

Schevill, W. E., Watkins, W. A. & Backus, R. H. (1964). The 20-cycle signals and *Balaenoptera* (fin whales). In: *Marine Bio-Acoustics* (W. N. Tavolga, Ed.), pp. 147-152. New York: Pergamon.

Schevill, W. E., Watkins, W. A. & Ray, C. (1969). Click structure in the porpoise, *Phocoena phocoena. J. Mammal.,* **50,** 721-728.

Schusterman, R. J., Balliet, R. F. & Nixon, J. (1972). Underwater audiogram of the California sea lion by the conditional vocalization technique. *J. Exp. Anal. Behav.,* **17,** 339-350.

Searle, C. L., Braida, L. D., Cuddy, D. R. & Davis, M. F. (1975). Binaural pinna disparity: Another auditory localization cue. *J. Acoust. Soc. Am.,* **57,** 448-457.

Seaton, W. H. & Trahiotis, C. (1975). Comparison of critical ratios and critical bands in the monaural chinchilla. *J. Acoust. Soc. Am.,* **57,** 193-199.

Sebeok, T. (1968). (Ed.). *Animal Communication.* Bloomington, Indiana University Press.

Sebeok, T. (1977). (Ed.). *How Animals Communicate.* Bloomington: Indiana University Press.

Shimozawa, T., Suga, N., Hendler, P. & Schuetze, S. (1974). Directional sensitivity of echolocation system in bats producing frequency-modulated signals. *J. Exp. Biol.,* **60,** 53-69.

Simmons, J. A. (1973). The resolution of target range by echolocating bats. *J. Acoust. Soc. Am.,* **54,** 157-173.

Simmons, J. A., Howell, D. J. & Suga, N. (1975). Information content of bat sonar echoes. *Am. Sci.,* **63,** 204-215.

Sukhoruchenko, M. N. (1973). Frequency discrimination in dolphins (*Phocoena phocoena*). *Fiziol. AH. SSR. IM. M. Sechenova,* **59,** 1205-1210.

Tavolga, M. C. & Essapian, F. S. (1957). The behavior of the bottlenosed dolphin (*Tursiops truncatus*): Mating, pregnancy, parturition, and mother-infant behavior. *Zoologica,* **42,** 11-31.

Tavolga, W. N. (1965). Review of marine bio-acoustics, state of the art, 1964. Tech. Rep. 1212-1, U.S. Naval Training Device Center, Orlando, Florida.

Terhune, J. M. (1974). Directional hearing of a harbor seal in air and water. *J. Acoust. Soc. Am.,* **56,** 1862-1865.

Terhune, J. M. & Ronald, K. (1972). The harp seal, *Pagophilus groenlandicus* (Erxleben, 1777), X. The underwater audiogram. *Can. J. Zool.,* **50,** 565-569.

Terhune, J. M. & Ronald, K. (1975a). Underwater hearing sensitivity of two ringed seals (*Pusa hispida*). *Can. J. Zool.,* **53,** 227-231.

Terhune, J. M. & Ronald, K. (1975b). Masked hearing thresholds of ringed seals. *J. Acoust. Soc. Am.,* **58,** 515-516.

Thompson, R. K. R. & Herman, L. M. (1975). Underwater frequency discrimination in the bottlenosed dolphin (1-140 kHz) and human (1-8 kHz). *J. Acoust. Soc. Am.,* **57,** 943-948.

Thompson, R. K. R. & Herman, L. M. (1977). Memory for lists of sounds by the bottlenosed dolphin: Convergence of memory processes with humans? *Science,* **195,** 501-503.

Titov, A. A. & Tomilin, A. G. (1970). Acoustic activity of common dolphins and harbor porpoises in different situations. *Bionika,* **4,** 88-94.

Tonndorf, J. (1970). Bone Conduction. In: *Foundations of Modern Auditory Theory,* Vol. II (J. V. Tobias, Ed.), pp. 195-227. New York: Academic.

Turner, R. N. & Norris, K. S. (1966). Discriminative echolocation in a porpoise. *J. Exp. Anal. Behav.,* **9,** 535-544.

Watkins, W. A. (1974). Bandwidth limitations and analysis of cetacean sounds with comments on "Delphinid sonar: Measurement and analysis" [K. J. Diercks, R. T. Trochta & W. E. Evans, *J. Acoust. Soc. Am.,* **54,** 200-204 (1073)]. *J. Acoust. Soc. Am.,* **55,** 849-853.

Watkins, W. A. (1976). Biological sound-source locations by computer analysis of underwater data assay. *Deep-Sea Res.,* **23,** 175-180.

Watkins, W. A. & Schevill, W. E. (1977). Sperm whale codas. *J. Acoust. Soc. Am.,* **62,** 1485-1490.

Watkins, W. A., Schevill, W. E. & Ray, C. (1971). Underwater sounds of *Monodon* (narwhal). *J. Acoust. Soc. Am.,* **49,** 595-599.

Watson, C. (1963). Masking of tones by noise for the cat. *J. Acoust. Soc. Am.,* **35,** 167-172.

Wenz, G. M. (1964). Curious noises and the sonic environment in the ocean. In: *Marine Bio-Acoustics* (W. N. Tavolga, Ed.), pp. 101-120. New York: Pergamon.

Wever, E. G. (1974). The evolution of vertebrate hearing. In: *Contributions to Sensory Physiology,* Vol. V (W. D. Kiedel & W. D. Neff, Eds.), pp. 423-454. Berlin: Springer-Verlag.

Wever, E. G. & Lawrence, M. (1954). *Physiological Acoustics.* Princeton, N.J.: Princeton University Press.

Wever, E. G., McCormick, J. G., Palin, J. & Ridgway, S. H. (1971a). The cochlea of the dolphin *Tursiops truncatus:* General morphology. *Proc. Natl. Acad. Sci., ***68,** 2381-2385.

Wever, E. G., McCormick, J. G., Palin, J. & Ridgway, S. H. (1971b). The cochlea of the dolphin *Tursiops truncatus:* The basilar membrane. *Proc. Natl. Acad. Sci., ***68,** 2708-2711.

Wever, E. G., McCormick, J. G., Palin, J. & Ridgway, S. H. (1971c). The cochlea of the dolphin *Tursiops truncatus*: Hair cells and ganglion cells. *Proc. Natl. Acad. Sci.,* **68,** 2908-2912.

Wever, E. G., McCormick, J. G., Palin, J. & Ridgway, S. H. (1972). Cochlear structure in the dolphin, *Lagenorhynchus obliquidens. Proc. Natl. Acad. Sci., ***69,** 657-661.

Winn, H. E. & Perkins, P. F. (1976). Distribution and sounds of the minke whale, with a review of mysticete sounds. *Cetology,* **19,** 1-12.

Yablokov, A. V., Bel'kovich, V. M. & Borisov, V. I. (1972). *Whales and Dolphins.* Jerusalem: Israel Program for Scientific Translations.

Yunker, M. P. & Herman, L. M. (1974). Discrimination of auditory temporal differences in the bottlenose dolphin and by the human. *J. Acoust. Soc. Am.,* **56,** 1870-1875.

Zwicker, E. (1954). Die Verdeckung von Schmalbandgerauschen durch Sinustone. *Acustica,* **4,** 415-420.

CHAPTER 2

The Cetacean Eye

William W. Dawson

INTRODUCTION

Visual Nature

There seems no doubt that the delphinidae are highly auditory animals and use echolocation extensively in their natural habitat. The auditory capabilities and underlying mechanisms of many of the delphinidae have been studied in detail (see review by Popper, this volume). In contrast, the visual nature of the dolphin is poorly understood and often underestimated. Observations of dolphin performances at marine exhibits will convince even the most skeptical observer, however, of the extent of visual development and function. On occasion dolphins appear to use vision from choice rather than necessity. Figure 2.1 shows dolphin 232 (Marineland of Florida), an adolescent male *Tursiops* who has lived most of his life in the holding tanks at the Marineland facility. The figure shows 232 in his "visiting hour" position. Having learned to chin himself on the edge of his concrete tank, 232 may stay in this position for extended periods apparently watching those who come by his living area and fre-

Fig. 2.1. *Tursiops* #232, in the aerial observation position.

quently participating in a game of pitch and catch with his striped ball. Such observational information on cetaceans and inferred visual function is the most common class of material found in earlier reviews (Slijper, 1962; Walker, 1964) with the recent addition of gross anatomical measures (Kinne, 1975) and cell quantification (Dral, 1977). The first behavioral quantification of visual function appears to be very recent (Kellogg & Rice, 1966). Compared to the magnitude of reviews of eye function in terrestrial mammals (e.g., rabbit, Prince, 1964), which require hundreds of text pages, the typical review of marine mammal eye research is small, rarely exceeding 20 pages. Since even eye mobility has been a topic of dispute (Walls, 1942), there is doubt that some reviewers ever observed a living cetacean eye. It appears that there is almost a logical paradox between the ancient and intense human interest in cetaceans, traced by Fraser (1977) to the Minoan culture approximately 2000 BC, and the weak parallel development of scientific understanding. According to Fraser, cetaceans were not distinguished scientifically from fishes until AD 1693.

The slow accumulation of data on the cetacean eye over the years is in part due to the limitation in availability of specimens, well-preserved tissue, and a practical laboratory model. These problems continue through to the present day. The early work on the eye was carried out from samples obtained from whaling expeditions or occasionally from beached animals and includes nineteenth century work by Hulke (1868) on the retina of the harbor porpoise and Matthiessen (1893) on the eye of the fin whale and the narwhal. Recent reviews treating the cetacean eye (Peers, 1971; Dral, 1977) list 13 and 18 publications, respectively, since 1860. Roughly half the cited papers have been published in the last 30 years. This suggests that the availability of modern techniques has not greatly enriched the total literature.

Organization of Chapter

This chapter provides a review of new and older findings on the structure and physiology of cetacean eyes, their specializations, their limitations, and how they vary. Because there is more quality material available, *Tursiops* will be emphasized as a representative member of the Odontoceti. It is recognized, however, that there are likely to be differences in visual capability or function among many of the families of odontocetes, reflecting in part adaptations to ecological niches with different visual demands and opportunities. Some of these differences in visual apparatus and presumed function will be discussed. New comparative information will be provided on the eye of the dwarf sperm whale (*Kogia simus*) and the Amazon dolphin (*Inia geoffrensis*). The suborder Mysticeti will be represented, for comparative purposes, by the sei whale (*Balaenoptera borealis*) eye. Most of the tissue which will be pictured and discussed was obtained either from animals under anesthesia, prior to sacrificing for some humane reason, or was obtained within a few minutes post mortem. Only the tissue condition of the *Balaenoptera* may be questioned, since the animal had been dead some 8 to 12 hours before the eyes were removed and fixed. All of the animals, except *Inia*, were found in the Gulf of Mexico. Three *Kogia* have been examined, two *Inia*, and, to date, eight *Tursiops*. Fixation was by gluteraldehyde, continued for 10 or more hours prior to washing and dehydration. The most routine procedures included embedding in celloiden with staining either by the Golgi method, our modification of the Bodian (1936) silver technique, cresyl violet, or, occasionally, hematoxylin and eosin.

Condition of Animals

All the animals that have contributed tissue or physiological information have been older adolescents or adults. All of those that we have had an opportunity to examine physiologically were in good health, except two *Tursiops*, which had symptoms of Lobo's disease (a terminal skin fungus communicable to humans).

The chapter includes reviews of the scientific literature supplemented by material from personal communications, unpublished theses, and some unpublished findings. This supplementation is justified by the particular problems which face workers in the anatomy and physiology of marine mammals. Because of scarcity, federal controls, and the need for fresh, well-preserved tissue numerous replications of findings are difficult to generate. Some species and good tissue are available only through chance beachings, good communications, the cooperation of many, and a great amount of luck. The probability of obtaining several repeated measures (within one lifetime) on, for instance, the dwarf sperm whale (*Kogia simus*) is quite low.

Visual Adaptation

After 33 years the most generally referenced discussion of marine mammal eyes is still in the book by Gordon Walls (1942). Walls discriminated between the visual nature of the "herbiverous" Mysticeti and the predatory Odontoceti and related the requirements of these behavior patterns to eye development. He blamed the apparent optical inferiority of Mysticeti eye on regressions, which occurred through a loss of emphasis on vision secondary to the trawling method of feeding. Walls states that the reversion of the cetacean eye to water has not been as toward "an ancient friend" but to "a new enemy." The cetacean eye has not shed its terrestrial modification but has faced several environmental problems. Optical requirements are different in air and water. It is subjected to frequent, rapid, extreme changes in temperature and pressure. The light level between the surface and the depths ranges from full sunlight, through color filtering of light near the surface, to near total darkness. The eye is buffered by water at relatively high swimming speeds. It must maintain osmolarity in a saline environment and must contend with turbidity, suspended particles and microorganisms. In the conclusion of this chapter adaptations to these environmental problems will be discussed.

GROSS ANATOMY

Corneas

Sketches of eye shapes, internal detail, and vascular patterns for *Tursiops*, *Balaenoptera*, *Kogia*, and *Inia* are shown in Figure 2.2. When viewed in the living animal the eye size is deceiving. Usually only a small portion is seen through the partially closed lids. This is particularly true for *Inia*, the Amazon River dolphin, whose observable eye is often so small that some describe it as regressed. The amount of eye surface devoted to cornea, on a relative basis, is probably somewhat larger in *Tursiops* than any of the other species shown. In life the *Tursiops* cornea is approximately 18 mm across at the horizontal meridian. Thus the axial length of the eye, from the corneal surface to the retina, is only slightly larger than the size of the cornea. This provides a very favorable condition for light gathering. A similar situation exists in the *Kogia*, but in the

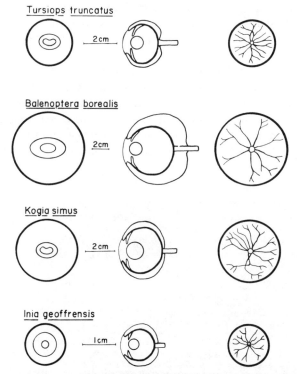

Fig. 2.2.   Sketches for gross anatomical comparison of eyes from some whales. Dorsal aspect is up. *Left column*: View from directly in front of eye. *Center column*: Transverse section of eye through geometric vertical meridian. *Right column*: Fundus pattern, drawn from vessels after enucleation. Sketches of other whale eyes may be found in Walls (1942) and Pilleri and Wandeler (1970).

Balaenoptera the corneal area is much smaller than the length of the eye, and less light (per unit area) should be available at the retinal surface. All of these corneas have the typical mammalian layering, which includes an epithelium, a thin Bowman's membrane, a relatively thick stroma layer, Descemet's membrane, and, finally, a thin endothelial layer. By terrestrial standards the sclera is massive. It becomes especially thick on either side of the optic nerve. This thickening is most pronounced in the *Balaenoptera* but can also be found in *Tursiops* and in *Kogia*.

Pupils

We have observed the living pupil in *Tursiops*, The pupil is reactive to light, but unlike other mammals it is partially consensual (Barris, Dawson, Adams & Litzkow, 1978). Simultaneous measures of both dolphin pupils show that in the unstimulated eye pupil response is about 60% of the pupil reflex in the highly stimulated eye. For low-level stimulation (approximately $\bar{1}$ mL), the reflex response is much less. In terrestrial binocular mammals both pupils are always the same size (consensual). In terrestrial monocular mammals the pupils operate independently (nonconsensual); that is, one may constrict while its fellow pupil dilates. In both *Tursiops* and *Kogia* the pupillary opening has a small operculum or "excess" tissue in the dorsal-central aspect. Figure 2.3 is a photograph of an isolated anterior segment, including sclera, cornea, and pupil from *Tursiops*. In its relaxed state this pupil is almost a centimeter in its largest dimension, while the human pupil in active dilation rarely exceeds 8 mm. When the *Tursiops* pupil constricts the central area may become completely covered by the operculum or be reduced to at most a narrow slit. In very bright light there remain only two openings on the nasal and temporal sides (Herman,

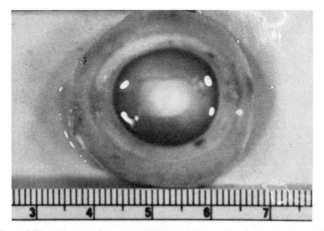

Fig. 2.3. View of *Tursiops* anterior segment. Dorsal is up, showing operculum of iris. Scale is in mm.

Peacock, Yunker & Madsen, 1975). This general configuration is also seen in the *Kogia*. In our fixed specimens we have not seen an operculum in the *Balaenoptera* or in the *Inia*. Both of these seem to have essentially oval pupil openings in the relaxed state. In the *Inia* a most unusual situation is found, since even the central optical region, through which the pupil admits light, is partially obscured by small blood vessels in the cornea. We have found consistent vascularizations of the corneas of the four eyes which we have examined. The vessels are radial in orientation and send small arterioles and venules into the central cornea. The remainder of the cornea is transparent. We believe that this is the natural condition in this eye, since vascularization of corneas, due to pathology, usually arises only after corneal damage has produced an opacity. We have not found evidence of any corneal damage or opacity in four eyes, although the vascularization was consistently present. Corneal vessels similar to these have been found in *Platanista* by Purves and Pilleri (1973) and by Dral and Beumer (1974). No other marine mammal corneas have been shown to be normally vascular (Purves & Pilleri, 1973).

Lens

The lenses of these eyes and the muscles of the iris are arranged so that the anterior chamber, between the iris and cornea, is much shallower than found in most terrestrial forms (Jamieson & Fisher, 1972; Walls, 1942). The drawings (Figure 2.2) do not fully communicate the relative shallowness of the chamber. The lenses themselves are almost spherical, except for the *Tursiops* lens, which appears ellipsoidal. The lenses are all transparent and essentially colorless, except that in the *Inia* the lens has the appearance of a small yellow sphere. The deep yellow of this lens has been found consistently in the four eyes which we have examined.

In distinction, *Platanista indi* and *P. gangetica* have no lens at all. A flat membrane, which may be the remnant of the capsule, is all that remains (Purves & Pilleri, 1973; Dral & Beumer, 1974). To the rear of the lens is the vitreous body, which in each of these animals is unorganized, transparent, highly gelatinous, and firmly attached in many places to the retina. On the surface of the retina lies the retinal vascular tree. In these cetaceans the tree is typically mammalian, some vessels issuing from the periphery and some from the central portion of the optic papilla. The pattern of vessels is essentially radial and uniform, which is the holangeotic pattern described by Prince (1956). The same pattern is found in the domestic cat. The density of vascularization is approximately as shown in Figure 2.2. The most notable departure from a uniform distribution is found in the *Kogia* fundus pattern (general appearance of the vascular tree and internal eye features as seen through the pupil) but is shared to some degree by the *Tursiops*. This exception is seen in the ventral retina, where the vessels are particularly large in diameter. In those *Kogia* eyes which we have examined microscopically we have found that lateral to the ventral vessels are arcuate confluences of small veins and arteries, which are highly suggestive of the vascular pattern seen in the macular area of humans, nonhuman primates, and in the area centralis of the cat. At this time we have been unable to complete the microanatomical studies which would be necessary to establish that these two retinal areas are areas of specialization such as are found in other mammals. These areas are inferior in position to the areas of high ganglion cell density described in *Tursiops* by Dral (1975a).

Fundus and Optic Nerve; Vascular Sheath

The same general increase in vascular caliber is seen in the ventral portion of the *Tursiops* fundus. However, the arcuate areas have not been found. The fundus pattern of the *Tursiops* is shown in some detail in a retouched photograph of the dolphin eye cup in Figure 2.4. The paired vein and artery, which is similar to that in the *Kogia*, may be seen more clearly. This photograph was retouched to enhance the contrast of the vascular tree in a fixed specimen colored by residual blood. This approach has the disadvantage that many of the smaller vessels are not visible.

The vessels of the fundus enter the cetacean eye through the optic nerve head or papilla. In the papilla they join to form two major vessels, the central retinal vein and artery. These travel through the optic nerve for several millimeters,

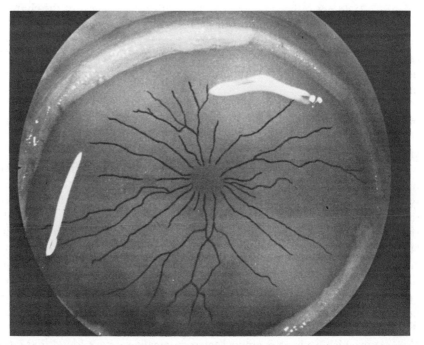

Fig. 2.4. Posterior segment of *Tursiops* eye. Fundus vascular pattern has been retouched to provide as much detail as possible. Very fine vessels do not reproduce. (×4.)

then exit into the orbit. They do not pass through the optic foramen with the optic nerve. Their exact destination has not been seen by us, for they tend to become lost in a large vascular structure at the rear of the eye, which fills a good portion of the orbit in the vicinity of the optic nerve. This appears to be an ophthalmic division of the rete mirabilia (Ridgway, 1972, p. 267). According to Ridgway the function of the retia are not fully understood, but the close association between the ophthalmic retia and optic nerve is interesting. When fixed the tissues of the retia become very white and hard (Burne, 1952; Yablokov, Bel'kovich & Borisov, 1972). This is probably the "thick, stiff sheath of the optic nerve" described by Walls (1942). Walls blamed the mechanical properties of this sheath for eye "immobility,"which he incorrectly reported in whales. In fresh tissue the optic nerve of whale is very pliable and typically mammalian. In the deep-diving mammals the high volume and surface area of the ophthalmic rete should serve to stabilize the temperature of the retina and optic nerve and to fill relatively void areas of the orbit. Nagel, Morgane, McFarland, and Galliano (1968) have shown a marked pressure-damping effect on dolphin cerebral circulation as a major physiological function of the thoracic rete. The ophthalmic retia should provide a similar nonpulsatile circulation for the retina. The plexiform nature of the ophthalmic rete is shown in an excellent drawing, Figure 25 by Pilleri and Wandeler (1964). They show that the ophthalmic rete acts as the arterial stem supply for choroidial circulation and the long and short ciliary artery systems. It also provides the pial circulation which envelops the outer layers of the optic nerve as it is found between the eye and the optic foramen.

We found no ophthalmic rete in *Inia*, nor was one reported by Dral and Beumer (1974) or by Purves and Pilleri (1973) in descriptions of the eye of *Platanista*. In these freshwater species, the rete location was occupied by a large orbital fat pad similar to but of greater magnitude than that frequently seen in larger terrestrial mammals. Both papers on *Platanista* also decribe a hyaloid artery. This is a major embryonic vessel of mammals which usually regresses with development. It serves the nutritional needs of the core of the eye during the period of development of the choriocapillaris and retinal circulation. In contrast, we did not find a hyaloid artery in our *Inia* material. Body lengths of our specimens were 2 m or over. Nor have we seen a hyaloid artery in any other adult cetaceans we have examined. One immature *Kogia* (140-cm length) had a hyaloid system. The most thorough development study that is available on visual anatomy was done by Pilleri and Wandeler (1964). They mention the presence of a hyaloid artery in the fin whale at the 350-mm embryonic stage. However, their description of the adult eye does not include it. We must therefore assume that, at least in *Tursiops*, *Kogia*, and *B. physalus*, the hyaloid artery eventually regresses much as we have learned to expect in terrestrial mammals. The *Platanista* eyes described by Purves and Pilleri (1973) and by Dral and Beumer (1974) were from animals with body lengths ranging between 105 and 120 cm. Perhaps these animals were immature to the extent that the adult anatomy had not yet been expressed. Herald and Dempster (1965) mentioned that as adults *Inia* can obtain weights of 136 kg and lengths up to 230 cm. Yet, none of the three *Platanista* described by Herald, Brownell, Frye, Morris, Evans, and Scott (1969) exceeded 120 cm in body length. If these are representative adult examples of *Platanista* eyes, there appears to be a rough, general relationship among dolphins, between eye development (in the adult condition), eye size, and body dimensions. This is probably true for cetaceans generally, but more data are needed. A rough comparison is presented in Table 2.1.

Orbit and Mobility

The dimensions of the whale eyes studied, except for the *Inia*, are large compared to terrestrial mammals. A cross-species comparison may be made with the aid of Table 2.1. Our measures showed that the optic nerve diameters for the *Balaenoptera*, *Kogia*, *Tursiops*, and *Inia* averaged 8.5, 5.75, 4.88, and 3.62 mm, respectively. Since the optic nerve appears to be highly compressed (Figure 2.5) as it enters the sclera, these measures have been taken from cross sections further toward the rear of the orbit. In the *Tursiops* the size of the orbit was also larger than expected. The heavy external musculature of the eye is capable of retracting the eye into the orbit for at least a distance of 2 cm so that it is covered with soft tissue, and the thick lid may be closed over it. There is also the possibility for even further retraction on requirement, since the orbit is at least 8 cm in depth. A protractor–retractor muscle was described in several cetaceans by

Table 2.1   Gross Measures of Structures in Some Cetacean Eyes[a]

Genus, species	BL	AL	CT	LT	ST	ON	Source
Balaneoptera physalus	22.9	85.0	1.3	15.0	40.0	—	Pilleri & Wandeler (1964)
Megaptera novaeangliae	13.7	20.0	0.5	11.5	—	6.0	Rochon-Duvigneaud (1940)
Physeter catodon	13.1	48.0	1.5	11.0	—	5.0	Rochon-Duvigneaud (1940)
Kogia simus	3.0	28.0	0.3	10.0	5.0	5.7	Dawson (1976)[b]
Tursiops truncatus	2.4	25.0	0.8	6.0	3.2	5.0	Dawson et al., (1972)
Delphinus delphis	2.1	24.0	0.5	7.3	5.0	5.0	Rochon-Duvigneaud (1940)
Inia geoffrensis	2.1	13.0	0.4	3.0	2.0	3.6	Dawson (1976)[b]
Platanista gangetica	2.1	5.7	0.5	1.5	0.3	0.6	Purves & Pilleri (1972) Dral and Beumer (1974)

[a]AL, axial length; CT, corneal thickness; LT, axial lens thickness; ON, optic nerve diameter; ST, scleral thickness, maximum; BL, adult body length (m). Other measures in mm.
[b]Unpublished measures.

Yablokov et al. (1972). Through its use the eyes may be extended (as in Figure 2.1) so that when the animal is viewing an object in front or below it both eyes may receive some of the image, offering the possibility of binocularity (Dral, 1972). A capability for binocular vision (with its numerous advantages) is consistent with animal's high accuracy of localization of objects in air. D. Caldwell and Caldwell (1972) have suggested binocularity in the downward gaze direction in *Stenella plagiodon*, *Lagenorhynchus obliquidens*, *Grampus griseus*, and *Globicephala scammoni*. The use of ocular motility by the *Tursiops* is not limited to movements of the eye in and out of the orbit. Both eyes are movable in the vertical and anterior-posterior direction (D. Caldwell & Caldwell, 1972), although we have never seen cyclotorsional movements. In the laboratory we have observed horizontal movements of one eye with the other eye stable. A vertical nystagmus has been seen on occasion. Recent computer analysis of eye movements from two *Tursiops* showed binocular correlations in less than 10% of 2000 sample time epochs and strongly suggest much less than a chance relationship (Dawson, Carder, Ridgway and Schmeisser, 1980). In an interesting degeneration study Jacobs, Morgane, and McFarland (1975) removed the left eye from one *Tursiops* and observed the degenerating nerve fibers by the Nauta method. They found degeneration in the left optic nerve and extending across the optic chiasm only into the contralateral optic tract. The degeneration was followed through the right optic tract and into the lateral geniculate nucleus. There was no degeneration found in the optic tract of the same side as the enucleated eye. This is strong evidence to support complete crossing of optic fibers in the *Tursiops*. This fact, coupled with the apparently separate motor control of the extraocular muscles of the eye, correlates nicely with the separation of motor and afferent pathways found in some teleosts and in rabbits. Separation, in the absence of any binocular option, is not unusual.

OPTICS AND IMAGES

Light Gathering—Cornea and Tapetum

Dawson, Birndorf, and Perez (1972) examined, ophthalmoscopically and with the streak retinoscope, the eight eyes of four normal *Tursiops*. Upon examination of the dolphin eye one is impressed with the size of the pupillary opening, particularly when the pupil is dilated with a standard clinical compound such as phenylephrine. An attempt to view the fundus is almost always met with failure. First, in the laboratory the eye is covered with a highly viscous transparent film, which is secreted from glands around the conjunctival sac. The flow of protein mucus (Yablokov et al., 1972) may be quite copious and form an irregular corneal layer up to 1 mm or more in thickness. In order to view the interior of the eye this must be washed off because of its irregular nature. After washing the cornea negative spherical lenses of 14 to 19 diopters were required before the proper image was obtained. Even then, retinoscopic endpoint was not distinct, and polaroid filters were used to reduce the high degree of internal light scatter. We later found that the scatter was due to the complete tapetalization of the retinal rear surface. The tapetum is a reflective specialization of the epithelium which lies behind the retina. The tapetum of *Tursiops* has been described by Peers (1971) as partially tapetalized (as in cat). However, we cannot confirm this. In eight dolphins (four *Tursiops*, two *Stenella*, two *Inia)* and four *Kogia* we examined the tapetum covered the whole fundus but varied in the *coloration* across the surface. Frequently there was a near 50/50 division, silver-green inferior retina and silver-blue superior retina. There is distinct variation between species and individuals. The tapetum provides additional useful light to the receptors and a distinct filtering for color enhancement (in most cases, blue enhancement for images from below and green for above). However, some degree of image quality must be lost from the additional scattering of the light. No tapetum was found in the *Inia* eye, although Slijper (1962) considered it a general feature of the cetacean eye. Direct measurement of the reflection of light from the dolphin tapetum would be of little comparative value because of edema and other optical effects caused by death and fixation. Nevertheless, the amount of light reflected by the tapetum appeared much greater than we see when looking at the cat fundus. Some of this may be an artifact, since the cat's fundus is only 50% tapetalized. However, the impression remains that the amount of light reflected per unit area of tapetum was greater in the *Tursiops*. Spectral analysis and quantification are needed in the living eye.

The Cornea Problem, Secretion(s), Astigmatism

The retinoscope refracts with an accuracy better than ±0.5 diopters. With corrective lenses we were able to take several fair-quality photographs of the fundus pattern, notwithstanding much difficulty due to the respiratory movements of the animal. These fundus pictures showed an area approximately two-disc diameters in extent. The great magnification was due to the additional lenses which were required to correct the animal's extreme in-air myopia. This magnitude of myopia would put the animal's far point approximately 6 to 10 cm in front of the cornea. This was surprising because the animals did not appear to have any visual difficulty in routine daily activities. It is very unlikely that any simple muscular accommodative mechanism could correct for such extreme myopia. Although histological results frequently contain artifacts, and cetacean eyes are frequently necrotic, some reviewers have concluded that no accommodative muscles exist (Dral, 1975b; Yablokov et al., 1972). In the live eye we never observed an accommodative movement in a *Tursiops* under study. Consequently we discontinued using any cycloplegic drugs to relax accommodation and did not treat the eye at all. In the absence of medication we were still unable to detect accommodative movements in *Tursiops*. If no medication was provided to human children or adults, in the absence of extensive training, it would be impossible to refract or even ophthalmoscopically examine a fundus. The focus would drift in and out as the patient's gaze and attention shifted.

Yablokov et al. (1972) and Dral (1972, 1974, 1975a) have pointed to regional variations in the curvature of the *Tursiops'* cornea as a possible means for overcoming this in-air myopia in the rostroventral direction of gaze. Our measures by retinoscope and ophthalmoscope do not preclude the presence of peripheral corneal departures from the axial curvature. These are common in humans but have no significance because the peripheral cornea is not used. Indeed, the ophthalmoscopic instruments which we use require only a small optical pathway through this large cornea, and, unless diligently pursued, peripheral variations probably would not be discovered. Dral (personal communication) informed me of his observations prior to his 1975 publication, and we have looked for it in *Tursiops*. Our method was to dust the wetted cornea in the living animal with a very fine white powder. Upon this we projected a checkerboard pattern and looked for distortions in dimensions of the pattern elements in excess of those which would be expected on a spherical surface. Our observations tend to verify Dral's findings. However, we have not quantified the optical results. The reasons lie in the confused optical picture which we see produced by the heavy secretions on the cornea and what they may do to image formation via corneal curvature. Even if the secretions are of the same density as seawater (Yablokov et al., 1972), in air they produce an irregular optical surface. Also there is the possibility of a ramp retina as illustrated in Figure 2.5. By ramp retina I mean only that the retinal surface does not conform to a simple sphere with a uniform radius. If the distances from the cross-point of light rays A, B, and C to the retinal surface are measured it will be found that they are all very different.

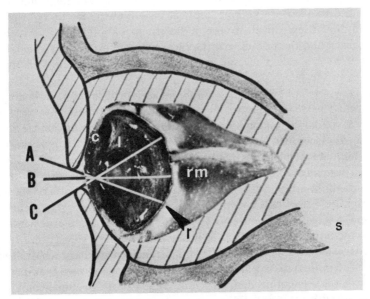

Fig. 2.5. Montage of frozen *Tursiops* eye and orbital position. Dorsal is up, nasal half of eye is shown. The ophthalmic rete (rm), corneal margin (c), lens (l), and retina (r) are marked. (×1.)

Even if the nodal point for the three rays does not fall as drawn, the distance from any nodal point to the retina along these three rays will be different in length. Consequently it seems unlikely that parts of an object at some distance from the eye can be in focus both at point A of the retina and point C. The tissue shown in the montage in Figure 2.5 was removed from the *Tursiops*, frozen in carbon dioxide, and kept at the temperature of solid carbon dioxide until it was cut on a high-speed electric saw. The plane passed through the center of the cornea, the center of the lens, and bisected the optic nerve. In the approximate geometric center of the terrestrial mammal retina lies the area for maximal image processing. It is in this region that the dolphin optic nerve head lies and constitutes a blind spot. Behaviorally meaningful measures of the eye optics can be made then only if the preferred direction of gaze is established and if the cornea is determined to have a particular layer of secretion on it. The corneal astigmatism is an optical artifact of departure from a spherical surface. But this departure may have hydrodynamic benefits, as pointed out by Walls (1942, p. 281) and Jamieson (1971). Many fast-swimming fishes have "streamlined" corneas which reduce drag and uneven pressures and distortion of the eye itself during rapid motion. Walls further proposed that scleral ossicles served to reduce eye distortion during rapid motion in fishes. The massive sclera of whales may be the mammalian counterpart of the teleost scleral ossicle.

The cornea and even the secretion problems may "disappear" under water. But visual performance in air must still be explained. Our solution to the corneal astigmatism–myopia–secretion problem of dolphins (Dawson, et al., 1972) essentially does away with the cornea as an optical device and relegates its role primarily to keeping out the water. In this paper we explained that the pupil was reduced to a slit or "pinhole" optical system in the relatively bright aerial state and should serve nicely to correct the myriad of optical problems produced by the cornea and the refractive differences between air and seawater. Herman et al. (1975) has made several extensions and improvements on this basic mechanism.

Our (Dawson et al., 1972) solution to the aerial vision problem in dolphins is not unique. The stenopaic pupil hypothesis has been proposed by several to correct for a similar aerial myopia and astigmatism in pinnipeds (reviewed by Jamieson & Fisher, 1972). The major problem with such an hypothesis is the loss of luminance incurred by the reduced pupil opening. However, some enhanced luminance advantage may be obtained from the *double* "pinhole" pupil in the light-adapted *Tursiops*, reported by Herman et al. (1975). Because of the highly efficient optics of the dolphin eye and the tapetal reflector behind the retina it seems that some light could be lost with no great difficulty in the brighter medium, air. However, in water the pupillary correction of the corneally generated myopia would be less important, since the cornea does not contribute as an optical surface. Our previous calculations showed that in water the cornea's power was not needed, since one could approach emmetropia within a few diopters with only the residual power of the lens. Clearly, all of the eyes (except *Inia*) of this group are specialized for efficient dim-light function by virtue of their totally tapetalized fundus.

Area Centralis

Future serious work on the refractive situation in *Tursiops* and other marine mammals must consider the functional site of the area(s) centralis by virtue of the position in which the animal places the retinal image. There may be one or more specialized areas. Until these locations are better known there can be only limited progress. Dral (1977) has made a beginning by charting two regions of maximum ganglion cell density in the rostroventral and caudodorsal areas of the intermediate retina.

THE ANTERIOR SEGMENT

Cornea

Anterior segments cut from celloiden blocks are shown in Figure 2.6. From left to right these are *Balaenoptera, Kogia, Tursiops*, and the rhesus macaque monkey. In our experience the cetacean eye undergoes large pressure changes immediately after death. Even with the most careful fixation the cornea is often drawn down against the lens and iris. This tends to make the depth of the anterior chamber appear even more shallow than it is normally. Relative to the primate anterior segment the anterior chamber is quite shallow even when the eye is under full intraocular pressure.

   The general structural characteristics of the corneas in Figure 2.6 are similar. Each contain the major mammalian divisions, the epithelium, stroma, and inner limiting membrane. Bowman's membrane, lying between the epithelium and stroma, was clearly present in the three cetacean eyes, but the presence of the Descemet's membrane was not so clearly observed. The cornea of *Balaenoptera* was 400 μ thick with a 50-μ epithelium. In *Kogia* the cornea was 275 μ thick with a 25-μ epithelium. The monkey cornea (Fig 2.6) has almost exactly the same dimensions as that in *Kogia*. Most distinct from the other corneas was *Tursiops*. Here the central cornea was 600 μ in thickness with a massive, 200-μ epithelium.

Chamber Angle and Zonular Fibers

The structure of the anterior chamber of the cetacean eyes in Figure 2.6 is not unusual. A typical mammalian trabecular meshwork (the filter system through which aqueous humor flows) was found in all of the chamber angles, and the surface of the irides were quite usual. Dral (1972) mentioned the interesting possibility that the iris might play a role in the accommodative mechanism

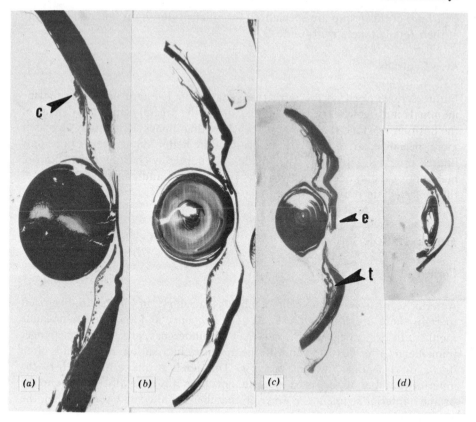

Fig. 2.6. Anterior segments of (*a*) *Balaenoptera*; (*b*) *Kogia*; (*c*) *Tursiops*; and (*d*) Rhesus monkey. Arrows indicate ciliary body (c), corneal epithelium (e), trabecular meshwork (t). (×2.3.)

through control of lens shape, orientation, or position. If this were the case it would seem that some relatively strong mechanical attachment between the iris and the lens capsule should be found, since these lenses are indeed formidable. No such attachment could be seen. In fact, in Figure 2.6*a,b* clear separation between the iris and lens capsule may be found due to disturbance in tissue processing. The separation did not appear to be artifactual, since in all cases the relatively delicate zonular ligament attachments to the lens capsule could be seen clearly and were undisturbed. The zonule attachments run, in typical mammalian fashion, from the pars plana just posterior to the ciliary body and attach to the lens capsule. At this point the arrangement differs markedly from that of the primate lens zonules where the zonular attachments are confined to an area approximately 1.5 to 2 mm on either side of the lens equator. In *Balaenoptera, Tursiops,* and *Kogia,* the zonular attachment covered almost the entire posterior portion of the lens capsule, attaching uniformly from the equator all the way to the posterior aspect of the lens axis.

Ciliary Muscles

In all three cetaceans shown in Figure 2.6 the attachments between the sclera and the ciliary body seem quite tenuous. Particularly noticeable is the frailness of the ciliary body–scleral relationship in *Kogia* (Fig. 6*b*). This is quite different from the situation reported in pinnipeds by Jamieson and Fisher (1972), in which the large ciliary body is firmly fused to the sclera and is useful in preventing pulling on the choroid during strong accommodative movements. In fact, in the cetaceans we have studied there has been little evidence for accommodative musculature. No clear muscle fibers could be found in either *Balaenoptera* or *Kogia*. However, some relatively sparse structures, which would not be identified as other than ciliary muscles, were found in the more dense portion at the base of the ciliary body below the trabecular meshwork in *Tursiops*. If this tissue was indeed muscular it was exceedingly sparse and weak in comparison to the dense ciliary muscular structures seen in the monkey eye, with a comparatively tiny lens (Fig 6*d*). It seems unlikely that such frail muscular structures could have any particular influence over such relatively large lens masses as are in the three cetacean eyes. Dral and Beumer (1974) do not report a ciliary muscle in *Platanista*, but Pilleri and Wandeler (1970) describe the ciliary muscle of the fin whale as "reduced."

Intraocular Pressure

On one occasion we measured intraocular pressure directly in an anesthetized. *Tursiops*. The animal was maintained on a mixture of oxygen–halothane, which was delivered by tube into the larynx–trachea. Each eye was cannulated directly by a 26-gauge needle introduced at the limbus. The right eye read 68 mm Hg, and the left eye gave a reading of 75 mm Hg. Later the cannuli were removed, and the corneas sealed themselves immediately. The eyes remained full and rigid. The animal was subsequently sacrificed (because of a serious skin disease) by an intravenous dose of barbiturate. Within 5 minutes the pressure in the eyes was reduced to such an extent that the corneas had begun folding.

The living eye pressures in this *Tursiops* were at least 50 mm Hg above the normal human levels (Duke-Elder, 1971). When the eyes from this animal went to histology careful examination of the retinal vascular tree and of the optic nerve head gave no indication whatsoever that there was pathology from glaucoma. If such high intraocular pressures are common among dolphins or cetaceans in general it would be most interesting to find what physiological adaptations are gained by such high pressures. Perhaps equally interesting is the rapid decline in pressure following death. This would suggest that the intraocular pressure is maintained by a more powerful active process than that which is seen in terrestrial mammals, where pressure is reduced more slowly after death. If aqueous humor is produced rapidly in the *Tursiops* eye it may lead to some suspicion of the accuracy of previous calculations of optical constants based on the refractive index of "normal" intraocular media, such as those for fin whale by Matthiessen (1893). We have suspected these refractive indices on other grounds. In one living *Tursiops* we measured intraocular distances by ultrasonic "B" scan technique. An SKF Ultrasonograph was used to transmit very-high-frequency clicks through the cornea. The reflected images of the lens surfaces, retina, and choroid scleral junctions could be seen in the ultrasonic echos. The ultrasonic device had been calibrated by standard procedure, using known distances in human blood serum. The echograms showed clearly that the corneal surface to retinal surface distance was approximately 33.7 mm. This was in disagreement with the actual distance (22 mm) measured in the excised eye by over 1 cm. No surgical error or technical artifact could account for such a large disparity. We suspect that the density of the intraocular media of *Tursiops* was markedly different from the densities in terrestrial mammalian eyes which are duplicated almost exactly by the density of human serum. Since Matthiessen's refractive indices from the postmortem materials are almost the same as those in primates, one may question their accuracy in cetaceans.

THE RETINA AND OPTIC NERVE

General Histological Methods

Retinal anatomy shows remarkable consistency across large phylogenetic barriers. Although differences exist, in many ways the primate retina appears to have the same gross cell types and general organization as that found in teleost fishes. The literature on marine mammal retinal anatomy includes mainly those citations in Table 2.2, with a few additions (Walls, 1942; Dral & Beumer, 1974; Dral, 1975a, 1977).

Though even the most casual examination of the *Tursiops* retina (Fig. 2.7a,b) discloses the exceedingly large cells in the ganglion cell layer, these cells have seldom been referred to in detail in the literature (Hulke, 1868; Rochon-Duvigneaud, 1940; Mann-Fischer, 1946; Peers, 1971, Dral, 1975a). We have combined interference contrast microscopy (Fig. 2.7) (Allen, David & Nomarski, 1969) with silver techniques to obtain, simultaneously, an overall picture of the microanatomy and the relative organization of single cell types within the larger framework of cellular structure and layering (Perez, Dawson & Landau,1972). The celloiden sections may be examined directly with interference contrast optics or may be stained with Golgi or a modified (Bodian, 1936) technique. The results of Figure 2.7a,b are typical of the interference contrast microscope, while Figures 2.7c and 2.13 use the Golgi and our modified silver technique, respectively. The remaining discussion of retinal cells and layers will describe primarily those retinal layers where there are basic questions or where structure departs from that found in the terrestrial vertebrates (Polyak, 1941): the receptor layer (Figure 2.7c and the ganglion cell and innerplexiform layer (Figure 2.7a). The other retinal layers found in *Tursiops*, *Kogia*, *Balaenoptera*, and

Table 2.2   Reports of Cones and/or Rods for Selected Marine Mammals

Species	"Cones" and "Rods"	"Rods" only	Source
Cetacea		X	Walls (1942)
Balaenoptera physalus	X		Pilleri & Wandeler (1964)
Delphinapterus leucas	X		Pilleri & Wandeler (1964)
Tursiops truncatus	X		Peers (1971)
	X		Perez et al. (1972)
	X		Dawson & Perez (1973)
	X		Dral & Beumer (1974)
Platanista gangetica	X		Dral & Beumer (1974)
Platanista indi		X	Purves & Pilleri (1973)
Pinnipeds		X	Walls (1942)
Pagophilus groenlandicus		X	Nagy & Ronald (1970)
	X		Nagy & Ronald (1975)
Phoca vitulina		X	Landau & Dawson (1970)
	X		Jamieson & Fisher (1971)
Zalophus californianus		X	Landau & Dawson (1970)
Callorhinus ursinus		X	Landau & Dawson (1970)
Mirounga angustirostris		X	Landau & Dawson (1970)

Inia are (at the light-microscope level of resolution) much the same as in many terrestrial mammals. The ultra-structural connectivity remains obscure in all marine mammals.

In Figure 2.7b, retinal divisions are identified as the receptor layer, the outer nuclear layer, the outer plexiform layer, the inner nuclear layer, the plexiform layer, the ganglion cell layer, and, finally, the nerve fiber layer which converges to produce the optic nerve. Conceptually it is convenient to think of the receptor, outer nuclear layer, the inner nuclear layer, the ganglion cell, and fiber layers as being way stations through which information passes from the periphery toward the central nervous system. Functionally, the outer plexiform layer and the inner plexiform layers are conveniently considered as processing layers in which information spreads laterally and where data manipulation takes place on a large scale.

Receptor Layer

Mann-Fischer (1946) suggested that cetaceans might have "duplex" retinas. To some mammalian physiologists this would indicate the presence of rod and cone cells with different anatomical characteristics. To others "duplexity" would require more than a single functional receptor system and implies both nocturnal (high-sensitivity) and diurnal (high-resolution) function. Some shift of spectral

Fig. 2.7. Transverse sections of the *Tursiops* retina made within 2 mm. (*a*) Marked are portions of the giant cell system, dendrite (D), nucleolus (NC), nucleus (N), and axon (A). (*b*) Arrows in receptor layer (R) mark oval bodies presumed to be receptor ellipsoid. Layers are outer nuclear layer (OL), outer plexiform layer (OPL), inner nuclear layer (INL), inner plexiform layer (IPL), ganglion cells (G), and fiber layer (F). (*c*) Golgi-stained outer layers showing outer segment (OS), ellipsoid (E), and cell body (C). (*a, b*: X 240); (*c*: ×800.)

sensitivity between the two systems is usual, while color analysis is a common duplex retinal attribute in many mammals. The meaning of anatomical duplexity in the absence of functional duplexity is difficult to interpret.

A simplistic approach to this complex question is to look at the receptor layer morphology and to attempt to determine if there is more than a single distinguishable cell type. Figure 2.7*a, b* were from the same section of retina but are separated by 2 mm. Figure 2.7*a* shows a receptor layer of great uniformity with the outer segments appearing almost as bristles on a brush. The outer limiting membrane between the outer segments and the receptor cell bodies is present, and the junction between the outer nuclear layer and outer plexiform layer is not remarkable. In contrast, Figure 2.7*a, b* shows numerous expansions in the receptor layer, a prominent outer limiting membrane and numerous expansions at the junction between the outer nuclear layer and outer plexiform layer. One characteristic of cones, as found in peripheral mammalian retinas (Walls, 1942), is that their ellipsoid is larger than their outer segment. The rod ellipsoid is essentially the same diameter as the outer segment and would pro-

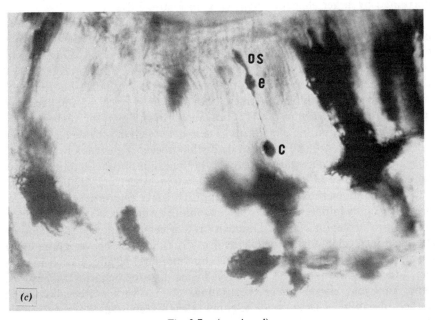

Fig. 2.7. (*continued*)

duce a palisaded appearance, as in Figure 2.7a. Larger ellipsoids would produce an irregular appearance of the receptor layer, as in Figure 2.7b. Staining
of cells by the Golgi technique is unspecific. The probability of finding a cell
type with the Golgi approach is relatively small, unless the cells are common.
Figure 2.7c shows a receptor structure, Golgi stain, which has the enlarged
ellipsoid and conical outer segment normally associated with a conelike receptor. Enlargement in the synaptic region of cone cells is also common. These are
called "expanded pedicles" and could account for the lumpy appearance of the
outer nuclear layer—outer plexiform layer junction in Figure 2.7b. We have
previously reported this larger pedicle type, using the Golgi approach (Perez et
al., 1972). The expanded pedicle was reported in the harbor seal by Jamieson
and Fisher (1972). Cell nuclear characteristics have been used to differentiate
receptor cell types (Purves & Pilleri, 1973). On this basis, Dral and Beumer
(1974) wrote that cone and rod types were found in *Tursiops*.

The receptor mosaic pattern may be seen by flat mount, using the interference microscope (Figure 2.8b). Here the receptors in the parafovea of the
rhesus macaque monkey may be seen. Close examination shows that there are
two distinct "grain" sizes. Here the larger elements are "cone" outer segments,
and the smaller elements are the more numerous rod outer segments packed
between the "cones." Tangential views of the surface of the receptor mosaic in
the inferior nasal and temporal regions of the *Tursiops* retina are shown in
Figure 2.8a,c. When flatness is lost, round receptor cell bodies (marked S)
come into focus. Above this plane is a much finer mosaic, which is occasionally
interrupted by larger elements, marked C in Figure 2.8a.

Figure 2.9 is an electron photomicrograph of a slightly tangential section,
just through the region of the outer limiting membrane. Junctions may
be seen between (a) adjacent Müller cells (marked by small arrowhead), (b) portions of the same Müller cell (large arrowheads), and (c) photoreceptors and
Müller cells. A "conelike" receptor process (marked cr) may be seen surrounded, first by a Müller cell (mc), and then by a circle of smaller, rod-
receptor processes. Receptor myoids (rm) are above and receptor nuclei (rn) are
below the external limiting membrane. These data seem to provide fairly strong
evidence that there are two structurally distinct receptor elements in *Tursiops*.
Behavioral data are necessary to answer the question of functional duplexity.
The usual sequence of discovery includes anatomy, electrophysiology, and finally behavioral methods, usually in approximately this order. Anatomical
duplexity usually initiates the physiological inquiry which validates the observation of "two" receptor classes. Duplexity is a necessary but not sufficient requirement for "color" vision, which involves discrimination between pure lights
of equal brightness but different color. This final level of information is most
often added to the anatomical and physiological data by behavioral measures.
Many years may elapse before the findings of one worker are validated in
another discipline. Dodt and Elenius (1960) reported "duplex" elec-
troretinograms in the rabbit. It was many years before Bunt (1978) showed with
electron microscopy that there are two receptor types in the rabbit retina. The

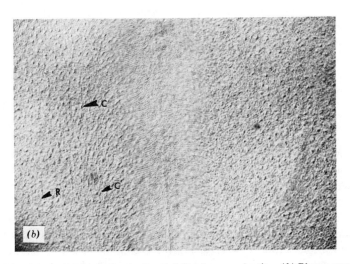

Fig. 2.8. Mosaic pattern of retinal receptors. (*a*) *Tursiops* nasal retina. (*b*) Rhesus monkey para-fovea. (*c*) *Tursiops* temporal retina. Rod and cone outer segments are marked (R) and (C). Receptor cell bodies are marked (S). Presence of "rods" and "cones" in *Tursiops* is established on the basis of cell size. (Unstained ×340.)

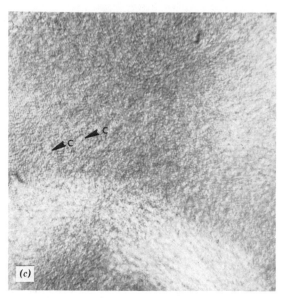

Fig. 2.8. (*continued*)

extent of "color" vision exhibited by this "nocturnal" mammal will not be known until extensive behavioral and cellular electrophysiological research is done. "Nocturnality" or "diurnality" of behavior pattern is a weak basis for the making of assumptions about receptor cell classes. A related controversy revolved around retinal receptors in the cat eye until behavioral studies were done. For many years there were electrophysiological and anatomical suggestions of duplex function. These questions were put to rest in 1973 by Brown et al. by an elegant (but very difficult) behavioral demonstration of color discrimination. Carefully designed experiments on *Tursiops* will be necessary to accomplish the same goal.

Examination of the absorption spectra of receptor pigments is a forceful means of settling questions about duplex retinal function. Microspectrophotometry has established the presence of multiple pigments in single receptor cells of several mammals, including humans and monkeys. This highly specialized technique has not been applied to marine mammals, though McFarland (1971) has published the difference spectra of gross extracts of pigment from whole retinas of several marine mammal species. This is a very difficult undertaking in the marine environment. To record a difference spectrum, the retinal pigment extract is divided into two parts. One is strongly bleached with white light and is used as a reference solution. The difference spectrum is then measured between the unbleached sample and the reference. Such procedure tends to reduce error due to sample impurities. McFarland (1971) provides data on Mysticeti (*Eschrichtius, gibbosus* and *Megaptera, novaengliae*), Ziphiidae (*Berardius bairdii*), and Delphinidae (*Pseudorca crassidens, Cepha-*

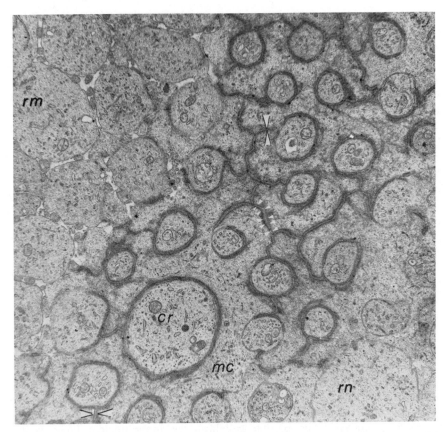

Fig. 2.9. Electron micrograph of a tangential section through the outer limiting membrane of the *Tursiops* retina. Arrows mark junctions between Müller cell processes. Müller cell (mc), conelike receptor (cr), rod myoid (rm), rod nucleus (rn). (×10,000.)

lorhynchus commersonii, Lagenorynchus australis, L. obliquidens, Stenella attenuata, and *Tursiops gilli*). The maximum absorptions (λ_{max}) occurred between 497 and 481 nm. The λ_{max} values among Delphinidae ranged from 489 to 485 nm. McFarland concluded that the pigment was a rhodopsin based on retinal -1, as in terrestrial mammals. However, terrestrial rhodopsin λ_{max} values cluster around 498 nm across many species. The shift toward greater blue sensitivity was explained as an environmental adaptation and is similar to the λ_{max} range of many deepsea fishes (McFarland, 1971). These are very important results on the major visual pigment system of dolphins. However, it would be improper to assume that they suggest an absence of other pigment (receptor?) types. In many animals the preponderance of rods tends to swamp out suggestions of other λ_{max} values. Frog has a great many cones but extracts such as these routinely have λ_{max} values around 500 nm (Dartnall, 1957, p. 185).

We have not seen convincing evidence for receptor duplexity in *Inia, Kogia,* or *Balaenoptera.* Apparently due to tissue autolysis of the highly susceptible receptors, nuclear characteristics have been used in the treatments of receptor types in *Platanista* by Purves and Pilleri (1973) and by Dral and Beumer (1974). The former authors report two receptor types. Some high-quality sections that contain cells with the large ellipsoids of *Tursiops* cones are shown by Peers (1971). Peers cautiously called these "atypical" receptors. The clearest finding of cone receptors has been reported in the developmental study of the fin whale (*Balaenoptera physalus*) by Pilleri and Wandeler (1964). In this species the receptors have differentiated by the 9-cm embryonic stage! Later, other retinal layers appear. In terrestrial mammals the typical development pattern consists of earlier development of the inner retinal layers, while the receptors develop last. Usually the receptor development is after birth, around the time when the eyes are opened.

Table 2.2 summarizes the receptor class positions of several authors. Many different criteria have been used to determine rod and cone presence. Some of these are discussed by Jamieson and Fisher (1972). Nevertheless, there appears to be a reasonable homogeneity, indicating receptor cell-type duplexity in whales and monoplexity in pinnipeds. Functional duplexity may be demonstrated in the absence of clear anatomical duplexity.

Intermediate Layers

Information on the fine structure of the outer plexiform layer, the inner nuclear layer, and the inner plexiform layer is confined primarily to the Golgi stained sections from *Tursiops.* The horizontal cell layer is not remarkable by light-microscope analysis. The bipolar cell layer, or inner nuclear layer as it is sometimes called, may be described in a similar fashion (Perez, et al., 1972). The most notable deviations to common terrestrial mammalian retinas begin to be seen in the inner plexiform layer and in the "ganglion cell and nerve fiber layer." A relatively common neuron type found in the inner plexiform layer of mammals is the amacrine cell (Perez et al., 1972). Amacrine, meaning "axon-less," is an apt description because the cell appears as a circular cell body at the margin of the inner nuclear layer and spreads interconnecting fibers in all directions. Polyak (1941) indicates that the amacrine cells serve to control information transfer to the ganglion cell level and, by virtue of the extent of its arborizations, must do so over a large retinal area. Dawson and Perez (1973) reported the presence of several extremely rare amacrine cells in this layer of the *Tursiops* retina. These cells appear to fit the description of the "interplexiform" cell (Gallego, 1971). The interplexiform cell appears only rarely in the visual literature, although a similar cell was described as rare by Cajal. The distinction between the interplexiform cell and other amacrine cells is an ascending process which leaves some portion of the cell body region and courses across the inner nuclear layer toward the outer plexiform layer (Figure 2.10). Of particular relevance is the fact that in a few *Tursiops* eyes we saw several variations on this

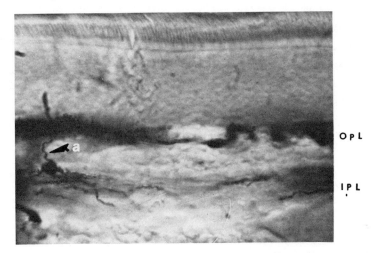

OPL

IPL

Fig. 2.10. Transverse section of *Tursiops* retina, Golgi stain. Layers are outer plexiform layer (OPL), inner plexiform layer (IPL), transplexiform cells, ascending process (a). (×200.)

configuration of amacrine cells, which have been seen before only in cat and monkey (Gallego, 1971; Laties, 1972). The interplexiform cell structure may possibly function as the major component of a feedback loop and provide for superior stability and control of numerous functions.

Comparison of Retinas

Sections of peripheral and central retina for *Inia, Balaenoptera,* and *Kogia* may be found in Figure 2.11. The appearance of these sections will be different from those seen above from *Tursiops* (Figure 2.7). This is due to the use of hematoxylin (Figure 2.11*a,c*) and cresyl violet (Figure 2.11*b*) stains, which have the advantage of densely staining all cell bodies, a facility necessary for quantification. Receptor cell bodies and ganglion cells, or cell bodies in the layers most commonly identified with ganglion cells, were counted at 10 randomly picked sites, 3 mm from the limits of the optic papilla. Sites counted on this circumference are called "central" counts. A second circumference, 6 mm from the edge of the optic papilla, was identified, and 10 more sites were counted along this circumference. The results of these counts, corrected for shrinkage and overestimation (Landau & Dawson, 1970), were used to calculate overall receptor–ganglion cell ratios and are presented in Table 2.3.

The literature does not give a great deal of aid in resolving the presence of a "specialized" area for high resolution on the *Tursiops* retina. Based on ganglion cell counts of large and small cells Dral (1975a, 1977) identified two areas with concentrations exceeding 500 cells/mm². One area was found dorsal to the geometrical median and close to the temporal equator, while the other was located ventrally of the median and at the rostral side of the fundus. Also for

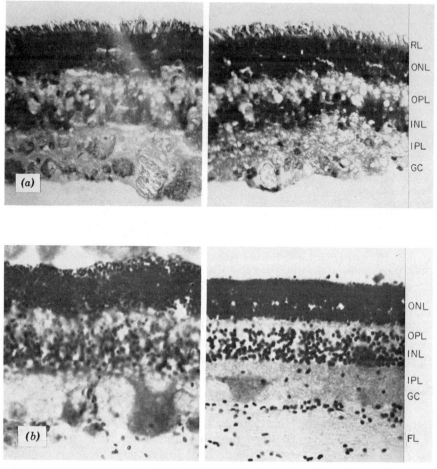

Fig. 2.11. Sections from central (left column) and more peripheral retina (right column) from (a) *Inia*, (b) *Balaenoptera*, (c) *Kogia*. Layer identifications are receptor layer (RL), outer nuclear layer (ONL), outer plexiform layer (OPL), inner nuclear layer (INL), inner plexiform layer (IPL), and ganglion cell layer (GC). Stains: a, b, cresyl violet; c, hematoxylin and eosin. (×200.)

Tursiops Peers (1971) shows a region of maximum ganglion cell density extending from nasal to temporal across the geometric center of the retina. Peers reported densities of 271–193/mm². The full effect of correction for tissue shrinkage is difficult to predict. Neither Dral nor Peers applied such a correction nor can fixation techniques be compared. It seems probable that Peers' (1971) receptor counts could benefit from some correction. He reported 2,225,000/mm² in central retina and 302,500/mm² peripherally. The maximum density of human cones in the fovea is about 150,000/mm², while rod-density maximum is only about 160,000/mm² (Pirenne, 1948). Cetaceans can

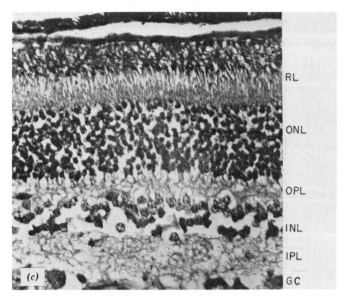

RL

ONL

OPL

INL

IPL

GC

(c)

Fig. 2.11. (*continued*)

produce relatively low receptor/ganglion cell (R/G) ratios (Table 2.3) of the order found in the perimacular regions of primates. Further, it appears that the eye of *Inia* is not so impoverished as one might suspect from its relative size and poor optics. But the amount of information that can be removed from the *Inia* eye must be relatively modest. The number of information channels provided by the ganglion cell layer is comparatively small. It is also interesting that the *Inia* has the least dense population of giant ganglion cells. Large cells were seen in all of the cetacean retinas. Those seen in Figure 2.11*b, Balaenoptera,* range up to 110 μ in the longest dimension.

The "Ganglion" Cells

"Giant" ganglion cells are not unique to the marine mammal retina. Receptive fields up to 700 μ in diameter are described in the peripheral retina of the cat (Leicester & Stone, 1967). The cell bodies were in the 15–35-μ diameter range. Sharks also have ganglion cells with large dendritic fields, which were fully described for the smooth dogfish (*Mustelus canis*) by Stell and Witkovsky (1973) and for the hammerhead (*Sphyrna lewini*) by Anctil and Ali (1974). The hammerhead giant ganglion cell is multipolar and very similar to those in dolphins and *Kogia*. However, the authors above apply the term "giant" primarily as a descriptor of the large size of the dendritic fields. Indeed, the cell body dimensions of the cetacean giant cells, up to 110 μ are as large as the whole cells identified as "giant" in other species. This cell type is a major departure from common terrestrial mammal retinal structure. Even pinniped ganglion

Table 2.3. Average Density of Receptor Nuclei and Ganglion Cells in Peripheral (p) and Central (c) Retinal Regions and Receptor—Ganglion Cell (R/G) Ratios

| Species | Eyes Counted | Receptor Nuclei Density (cells/mm^2) | | Cell Bodies in "Ganglion" Cell Layer Density (cells/mm^2) | | | | Ratio Overall R/G | |
| | | | | <20-μ diam. | | >20-μ diam. | | | |
		p	c	c	p	c	p	c	p
Tursiops truncatus	2	287,103	388,416	4377	4547	533	684	80	55
Kogia simus	2	83,016	87,910	4624	533	286	38	18	145
Balaenoptera borealis	2	381,246	260,429	312	3740	470	5337	332	42
Inia geoffrensis	2	38,932	123,552	1201	533	261	6	84	72

cells (28–50 μ) do not reach these dimensions (Hulke, 1868; Nagy & Ronald, 1970). For fin whale Pilleri and Wandeler (1964) drew a retinal section which shows both large (>100-μ cell bodies) and small ganglion cells much as pictured (for *Kogia*) in this chapter. The term "giant," based on cell body size greater than 60 μ, must be reserved tentatively for whales. Peers (1971) classified *Tursiops* ganglion cells as large and small. Further elaboration has been provided by Dral (1975a), who described large cells as oval to round with pale nuclei and a fine tigroid. Dral emphasized that there was a full range of intermediate appearances.

The cresyl violet stain was much more effective at uncovering the dimensions and stems of extensive dendritic trees of the giant cells and even the smaller cells adjacent to the giant cells. The presence of the smaller cell bodies and their function has undoubtedly added to the uncertainty of the interpretation of the numerical values provided in Table 2.3. The numbers of large (>20 μ) cells agree well with Dral (1975a). With some stains the small cells are not seen. The small cells and interconnecting dendritic processes were not seen by us in earlier studies, using interference contrast, which reveals only the cell bodies and larger portions of the dendrites (Fig. 2.7).

The layers of ganglion cells and the retinal organization internal to the ganglion cells of *Tursiops, Kogia,* and *B. physalus* are more elaborate than found by many specialists in terrestrial mammals. In primates, these retinal layers have been called the "least complex layer" by Polyak (1941). The situation in whales is a sharp departure from Polyak's generality. Exclusive of the receptors and their cell bodies the ganglion cells and the more inner layers are 35% of the neural retina's thickness in the fin whale (Pilleri and Wandeler, 1964, Fig. 28). This layering would correspond to primate retinal layers 8 and 9 as drawn by Polyak. Layers 8 and 9 comprise about 25% of the retinal thickness in the primate perimacula when layers 1 through 4 (receptors and nuclei) are excluded. In *Tursiops* the equivalent comparison shows that about 40% of the thickness is devoted to layers 8 and 9. The extensiveness and complexity of these layers has been the subject of papers from the Russian literature on selachians (Shibkova, 1971) and dolphins (Shibkova, 1969). The organization of ganglion cells as interpreted in terrestrial forms is challenged by Shibkova, who points out a range of cell sizes where there is no demonstration of axons given off to the optic nerve, and there is frequent indication of interconnection between cells. Classically, ganglion cells have been defined as located in the innermost retinal cell layer and as contributing an axon to the optic nerve. In *Delphinus delphis* Shibkova (1969) reported groups of small neural cells in "layers 8 and 9" often clustered around a larger cell with its axon in the fiber layer. She characterizes these as ensembles with integrative function, a function reserved by most authors for the inner plexiform layer of terrestrial species. Cells with "recurrent" dendrites were seen in these layers. These formed a loop and returned to the cell body, ending in a terminal spray. Similar organization and interchange were reported in the shark retina, but numbers of cells were smaller than in the dolphin (Shibkova, 1971). In both papers the author was careful to discuss the presence of supporting (nonneural) glial cells and to discriminate between them and members of the "neural cell colonies." She reported both oligodendrocytes and Müller cells.

Structures similar to those described by Shibkova may be found in *Kogia* and *Tursiops* but not *Inia*. In cetacean retinas some cells do not stain as terrestrial mammalian tissue. In our hands the Golgi technique readily stains all retinal cells except the larger ganglion cells. The chemistry of the Golgi stain is not understood, but Bodian's silver (1936) technique produces a fair result in our thick sections. Results (similar to those of Shibkova) are shown in Figure 2.12. Section A shows the inner nuclear layer, inner plexiform layer, ganglion cell layer, and finally the nerve fiber layer of the *Kogia* retinal intermediate zone, which is a circumferential zone in the vicinity of 3 to 6 mm from the optic nerve head. Section B of this figure shows a similar group of laterally spreading, interwoven small and large fibers. The large lateral fasciculi, which stain heavily with silver, are seen to the left. Section C is an ink drawing of relationships seen in different focal planes of the material in Section A. The primary source of difficulty is to identify which cells contribute the axons that form the optic nerve and therefore should be called ganglion cells. This is the same problem identified in *Delphinus* by Shibkova (1969).

Ganglion cells with other than unipolar output process are not frequently described in the literature on terrestrial species (see review in Rodieck, 1973). The closest approximation is the displaced ganglion cell described by Boycott and Dowling (1969). However, its cell body should rest in the inner portion of the inner nuclear layer, as distinct from the cell locations described above. The other ganglion cells classified by Boycott and Dowling (1969) appear to be most usually characterized by a single axon and one or two dendrites, which may later branch in the inner plexiform layer. Those cells pictured in Figure 2.12 and described by Shibkova (1969, 1971) do not fit this classification. The ma-

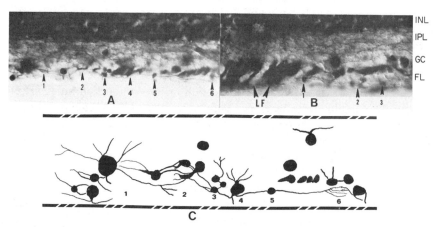

Fig. 2.12. Inner layers of the whale (*Kogia*) retina. Sections A and B, lateral processes of cells in ganglion cell–fiber layer interface (Bodian silver stain). LF, lateral fasciculi (1–6, see text); other notations as in Figure 2.11. Section C, ink drawing of organization in A. (×240.)

jority of new research on mammalian ganglion cell layer contents has been pursued by electron microscopy in the last few decades. However, Hogan, Alvarado, and Weddell (1971, p. 482) conclude that it is not possible, by electron microscopy, to differentiate the various types of ganglion cells that can be cataloged with the light microscope in silver-stained sections. The most likely source for error in classification of cells in this region (layers 8 and 9) is the possible confusion with glial cells. But these cells do not fit glial systems described by Anderson (1967), Wolter (1955, 1957), Vrabec (1970), or Uga (1974). Identification of the smaller cells in layers 8 and 9 of the whale cetacean retina as glia would be more consistent with current ideas of retinal structure. Without much more study this could be premature.

The "mosaic" of giant cells and smaller cells in the inner layer are shown in Figure 2.13 from a flat-mounted *Tursiops* retina. In addition to ganglion cells the previously described lateral fascicular bundles of fibers may be seen together with some vessels of the retinal fundus. Typical red blood cells (r) are

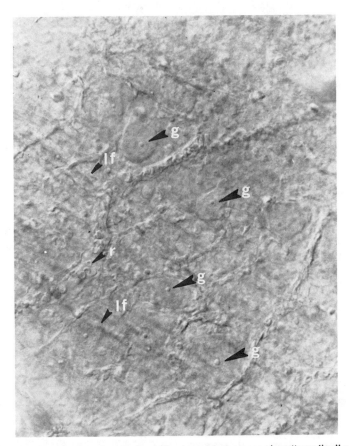

Fig. 2.13. Mosaic of inner retinal surface, *Tursiops.* Markers are giant "ganglion" cell nucleoli (g), lateral fasciculi (f), red cell in fundus vessel (r). Unstained, interference contrast. (×300.)

present in the vessel, running from the lower left to the upper right portion of the figure. The arrows marked "g" point directly toward the nucleoli of the larger ganglion cells. The nuclei are apparent in good light but are somewhat less distinct. At least three processes may be seen emerging from the ganglion cell in the middle right of the figure. Considering the focal depth of these optics, these processes must be almost on the inner retinal surface, although it is not possible to determine if they contribute to the lateral fasciculi.

Optic Nerve

Table 2.4 summarizes the optic nerve axon counts for *Balaenoptera, Kogia, Tursiops,* and *Inia.* These counts are the means (corrected for 20% shrinkage) taken from randomly selected areas from both the central and peripheral portions of the optic nerve. The optic nerves were stained with silver or the Weil stain. Figure 2.15 illustrates the relatively large number of large-diameter axons in the *Tursiops* nerve. Our count of total fibers in the optic nerve of *Tursiops* is larger than that published by Morgane and Jacobs (1972). However, they made no mention of the large number of axons in the 6 to 15 μ class. Also we tried to extend our count down to the optical limit.

Sections of optic nerves taken from the posterior portions of the orbit are shown in Figures 2.14 and 2.15. Only the section from *Inia* (Figure 2.14c) was taken close enough to the eye so that the central retinal artery and vein were seen. The sections from *Balaenoptera, Kogia,* and *Tursiops* all show an extensive fasciculation of optic nerve fibers. Although the *Inia* nerve is fasciculated, it is much less prominent than in the other cetaceans we have examined. Because of magnification problems all of the sections in Figures 2.14 and 2.15a are direct prints of the histologically prepared tissue. Consequently they are negative. The darkly stained cytoplasm in the center of a thinly stained axis cylinder (as in Fig. 2.15b would appear in the other negative sections as black dots. A few of the largest fibers can be found distinctly in the periphery of the *Balaenoptera* eye section (Fig. 2.15a). There appeared to be no segregation of large and small fibers in any of the optic nerves except *Kogia*. In *Kogia* the majority of the large axons appeared toward the peripheral portion of the optic

Table 2.4. Optic Nerve Axon Counts

Species	Number of Eyes	Corrected Across Sectional Nerve Area (mm²)	Total Axons	
			$<8\ \mu$	$>8\ \mu$
Balaenoptera borealis	2	56.7	367,789	54,781
Kogia simus	2	25.9	1,195,755	154,290
Tursiops truncatus	3	18.7	112,996	72,243
Inia geoffrensis	2	10.3	61,548	5,146

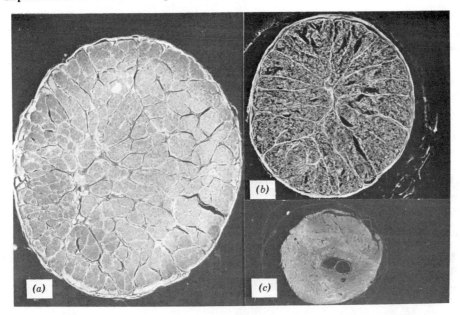

Fig. 2.14. Optic nerve sections. (*a*) *Balaenoptera*; (*b*) *Kogia*; (*c*) *Inia*. These are enlargements of the histological sections and are negative. (×10.5.)

nerve. In humans there is a fixed relationship between retinal field and optic nerve region.

The central terminations of the optic tract in *Tursiops* would be largely unknown except for the paper by Jacobs et al. (1975), which has been mentioned earlier. These authors cut one optic nerve, and degeneration was found in the contralateral optic tract, lateral geniculate nucleus, and superior colliculus. No degenerating retinofugal fibers were found in the posterior commissure or passing to the pretectum. There were no accessory optic system fibers found. Jacobs et al. (1975) conclude that the distribution of retinofugal fibers in dolphin more closely resembles that found in terrestrial primates than carnivores. An earlier paper by Langworthy (1932) described the gross brain anatomy well but inferred function only from anatomical correlation.

IMPLICATIONS FOR ENVIRONMENTAL ADAPTATION AND SEEING

Walls (1942) painted a picture of the whale eye as a relatively inadequate compromise of adaptations of a basically terrestrial eye to the marine environment. The eyes of all mammals are confronted by many of the same problems. Those in the Pinnipedia must serve an even more dual function because of their amphibious nature. The adaptations of marine mammal eyes begin with those at the junction between the physiological system and the physical medium.

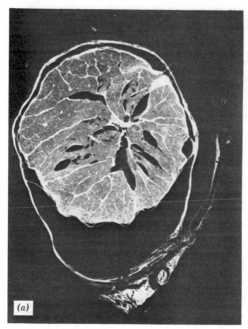 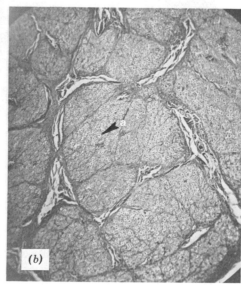

Fig. 2.15. Optic nerve sections from *Tursiops*. (*a*) Photonegative as in Figure 2.14 (×10.2.) (*b*) Section from peripheral area, marker *a* indicates a giant axon with darkly stained axis sheath and shrunken cytoplasmic center, also stained. (Photopositive, Bodian silver, ×111.)

Cetacea have done an excellent job of isolating the transparent cornea from the cold, salty environment. Though the corneal epithelium is relatively impermeable in terrestrial animals, the corneas of cetacea are further isolated from their environment by a relatively thick, viscous secretion, which originates in the glands of the conjunctival sac. This transparent secretion may also act as a lubricant and provide some degree of thermal isolation. Its presence in the air is well documented, although its rate of secretion and removal in seawater is unknown (Yablokov et al., 1972). Isolation is probably also afforded from the various impurities which are found in seawater at all depths. The most extreme adaptation would be the vascularization of the corneas seen in river dolphins, where microorganisms and particulate matter are a particular threat to corneal well-being. Vascularization is critical for rapid cellular repair and defense against pathogens. The adaptations in some river species have been so extensive that image formation may not be possible. Freshwater dolphins may be found in several large rivers, the Amazon, La Plata, and Yangtze, and the Ganges-Brahmaputra-Indus systems (M. Caldwell & Caldwell, 1969). The visual problems confronted by the freshwater dolphin are variable but seem to revolve around underwater impediments, turbidity, and debris. From the limited information available it would be possible to subdivide the river dolphins into Old World dolphins, typified by *Platanista gangetica* and New

World dolphins, typified by *Inia geoffrensis*. The two species superficially are quite similar but are visually distinct. A smaller species (*Sotalia fluviatilis*) has also been located in the Amazon system and looks more like a small *Tursiops* than her freshwater neighbor the *Inia*. *Sotalia* is mentioned briefly and pictured in the article by Herald (1967), but no detailed information was provided. According to Herald most Delphinidae have 44 chromosomes. But the Amazon River dolphins and the sperm whales have only 42. This is disputed by Kulu (1972), who found 44 chromosomes in *Inia*. It is not clear how many chromosomes are found in the cells of the Old World freshwater dolphins.

The marine dolphins such as *Tursiops* have superior visual apparatus to any of the freshwater dolphin species. The intermediate quality visual system is found in *Inia*, while the most regressed system is in the *Platanista*, which often swim on their side. An interesting theory was advanced by Purves and Pilleri (1973) and Pilleri (1974) to account for side-swimming behavior of *Platanista* (Herold et al., 1969). The eye is proposed to be a light detector and, in cooperation with one pectoral fin and the "sonar" system, provides for navigation in a world which has a light top and a dark bottom and very turbid extension in all other directions. The *Platanista* have been reported to emit clicks at 40 to 80 pulses/sec regardless of time of day or condition of the aquarium (Anderson & Pilleri, 1970). That is, it appears that they use echolocation almost continually. In contrast, *Inia* emits "echolocating" pulses primarily at night or in the presence of food or strange objects (Caldwell et al., 1966; Herald et al., 1969). Although there has been no demonstration that the *Platanista* see images, Purves and Pilleri (1973) propose a "pinhole" camera hypothesis to account for possible image formation in the lensless eye. In other forms (birds and lizards) lensless (pineal or parietal) eyes provide for entrainment of behavior to diurnal cycles (Gundy, Ralph & Wurst, 1975). Consequently, another possible function can be listed for the rudimentary eyes of *Platanista*.

Walls' analysis of the function of the "teardrop"-shaped corneas of both Pinnipedia and small whales was discussed again by Jamieson (1971). The shaping of the cornea provides for a more uniform distribution of pressures which the eye encounters during high-speed aquatic maneuvers. Together with the extremely thickened sclera and high intraocular pressure the shaping would tend to minimize distortion of the eye due to differential water pressure at the rostral edge of the cornea. Indeed, the marine mammal eye is confronted both by the pressure of the water due to depth of the dive as well as a differential pressure produced by movement of the animal through the water at various speeds. The pressure due to living depth (which can be quite extreme) may not be a problem for the eye. It was argued earlier by Walls that the eye is physically the same density as the seawater which surrounds it. Further, the orbit is filled with equal density material, and the eye is cushioned by the ophthalmic rete. Consequently there may be no voids and no pressure differentials. Compared to the problems produced by blood gases and the vacant spaces in the viscera and thoracic cavities or marine mammals, the eye is essentially "trouble-free" in any water depth. The situation is similar to the filling of a thin plastic bag with seawater and then carrying it several meters down into the sea. Under these circumstances, the bag would suffer no distortions or damage. There is simply an interface between two media of equal density. On this topic Walls pointed out that whales and elephants have scleral tissue of approximately equal thickness. But the elephant rarely gets his head below water.

This explanation assumes that the eye tissues are all incompressible and of nearly equal density to seawater. The possibility that this is an incorrect assumption cannot be ignored, since each tissue and fluid in the eye has not been evaluated. Publications spanning an extended period have related scleral thickness and adaptation to hydrostatic pressure at depths (Burne, 1952; Yablokov et al., 1972). Kellog (1938) cited damage to an undersea communications cable from an entangled sperm whale at a depth of 3240 (988 m) off the coast of Columbia. Pressure at that depth is approximately 1400 psi. If it may be assumed that 3240 feet is a "safe" depth for large whales one may estimate the scleral shell thickness that would be required if the eye depended on structural strength alone. Table 2.1 shows that whales of this size may be expected to have an internal eye radius near 3.3 cm. If a spherical eye shell were to be constructed of cast iron, calculations under American Society of Mechanical Engineers (1971) standard codes require a wall thickness of 0.77 cm to withstand 1400 psi, external. One of the largest eyes reported (*Balaenoptera physalus*; Pilleri and Wandeler, 1964) had a maximum scleral thickness of 0.40 cm. The cornea was much thinner. Although the sclera is a relatively tough tissue, it seems unlikely that its structural strength or that of the cornea can have more than a minor role in the adaptations of the eye to such intense hydrostatic pressures.

The solution to eye temperature as a problem to the marine mammalian eye is not so clear. Even though the cornea is relatively thick (about 0.6 mm), it may not provide sufficient insulation. The nervous system "shuts down" or may be irreversibly damaged by ion leakage when its temperature begins dropping below 15 to 20°C. In deep dives the temperature outside must drop below these limits. The neural tissue of the eye is probably the most exposed of tissues with an information-processing function. The only means available to aid the eye in the deep-diving mammal that is not available to its terrestrial counterparts is the ophthalmic rete and, to some degree, the intraocular fat. The rete covers the posterior 40% of the eye and may act as a radiator to prevent the loss of heat from the retina and optic nerve. Such a function would require a controlled, vigorous perfusion of the rete during dives. This has not been demonstrated. Pilleri and Wandeler (1970) propose a pressure-regulator function to the ophthalmic rete system. It is equally probable at this time that the massive eyelids are closed during dives and that these serve an insulating function and together with the rete maintain the eye temperature. Some mechanism of thermostatic control for the eye seems necessary in its exposed condition. The remaining signal-processing components of the nervous system are very well encased in insulating blubber and probably need no extensive additional thermal control.

Adaptation to solve one problem appears to bring on a second, so far as the cornea is concerned. The streamlining of the cornea produces a marked deviation from the spherical optical surface necessary for uniform refraction of light. One of the first to call attention to this problem was Johnson (1893). Johnson's publication identified an extreme corneal astigmatism in the harbor seal. Walls (1942) wrote as though the astigmatism was a general problem in all fast-swimming mammals. Piggins (1970) clarified the issue by demonstrating that in most seals the astigmatism disappears when the eye is underwater. This localizes the astigmatism to the cornea. Astigmatism could originate at any irregular refractive surface. The refractive condition in the pinniped eye has been carefully reviewed by Jamieson and Fisher (1972).

For dolphins Walls's comments about astigmatism have been elaborated by Dral (1972, 1974, 1975b). As reported also by Dawson et al. (1972), Dral found a high myopia in air, but Dral noted that the myopia disappeared in some quadrants. Earlier measures by Dawson et al. (1972) had been made as in humans and were confined to the central regions of the cornea. If it were clearly demonstrated that the dolphin's direction of vision in air was confined largely to "special" regions of the cornea a very functional adaptation could be demonstrated.

There remains still another major optical problem for those animals with optics like the dolphin. There appears to be no mechanism for adjustment of the refractive power of the eye. As noted by Walls, the lenses are relatively large and, although not as rigid as supposed by many, have no identifiable attachments to any elaborate intraocular musculature required for lens accommodations typical of terrestrial mammals. It is not surprising that Walls took such a negative view of the aerial acuity vision in marine mammals. Yet, the functional acuity in air appears to be good.

Earlier papers have described the results of preliminary behavioral studies of underwater visual acuity in the Pacific white-sided dolphin (*Lagenorhynchus obliquidens*) (Spong & White, 1971) and the killer whale (*Orcinus orca*) (White, Cameron, Spong & Bradford, 1971). Underwater acuities of approximately 6 min of arc were obtained for both species in a task requiring a discrimination between a single black line on a white background and two parallel black lines of variable spacing on a white background. These acuities are roughly of the same order of magnitude as reported for the California sea lion by Schusterman (1972). There is also general agreement of these values with the predicted acuity, based on anatomical grounds, of the sea lion (Landau & Dawson, 1970) and of *Tursiops* (Perez et al., 1972). However, Herman et al. (1975) noted that the target types used in the dolphin acuity tests contained brightness cues, which could have artifactually yielded resolution thresholds several minutes of arc smaller than those that might be obtained with grating targets, which control for brightness cues.

Pepper and Simmons (1973), using grating targets and testing for in-air acuity, recorded a minimum visual angle of about 18 min of arc for *Tursiops*. A more comprehensive study of *Tursiops* acuity was published by Herman et al.

(1975). Using grating targets, both aerial and underwater acuity was measured at various distances with results in the 8 to 12 min of arc range for best resolution threshold. At this point the reader is confronted with an apparent paradox of equal aerial and underwater visual resolution in an animal that is naturally astigmatic, has a high aerial myopia, and has no clear physiological mechanism for lens accommodation. The first embryonic indications of a solution to this complex problem was provided by Johnson (1893), who proposed that the pinniped iris constricts to a small (pinhole) opening in air offering a correction for the optical problems produced by the cornea. Schusterman (1972) further elaborated on this mechanism to solve the in-air acuity problems of pinnipeds. The basic idea was used once again by Dawson et al. (1972) to account for the apparently good (but unquantified at that time) aerial acuity in *Tursiops*. Variations on the "pinhole" corrective mechanism (sometimes referred to as the stenopaic pupil) have been discussed by many writers (Cobb, 1915; Rubin & Walls, 1969). The most elaborate development of this concept is provided by Herman et al. (1975). Herman described the *Tursiops* pupil, which under high light levels proceeds from an elongated slit to a "dumbbell" shape and eventually to two separate openings separated by approximately 5.3 mm (see Figure 3.2 in Madsen & Herman, this volume). Herman suggested that the aerial acuity of *Tursiops* is dependent on the large depth of field gained by these small apertures ("double slits"), which serve to overcome the gross myopia in air and also serve to account for improved acuity in air with distance as was demonstrated.

The physiological utility of a "pinhole" correction mechanism might be demonstrated by the dependence of aerial-visual behavior on relatively high illumination. Schusterman and Balliet (1971) tested such a hypothesis for the stenopaic pupil of the sea lion. Though no direct correlation of acuity with pupil size was attempted, it was found that the visual-resolution capabilities decreased much faster in air than in water with decreasing background luminance levels, especially at very low luminance levels. Dawson, Adams, Barris, and Litzkow (1978) determined that for *Tursiops* the "pinhole" size is not achieved at levels of less than 2 log mL but that visual behavior in air does not suffer until light levels fall below $\bar{2}.5$ log mL. Hence the pinhole or double-slit mechanism may not be the major mechanism for aerial acuity in *Tursiops*, though it may contribute importantly to acuity and depth of field at very high light levels. In absence of proven mechanisms for accommodation it should be pointed out that the highly developed extraocular muscles of the cetacean eyes may be used to vary the diameter of the eye and thereby adjust its focal length. There are no direct data which bear on this hypothetical mechanism.

The aerial-acuity results of Herman et al. (1975) have been explained recently by Rivamonte (1976) who proposed a "bifocal" lens model. The model describes a lens with a zone of low optical power in the periphery, selectively used during aerial viewing. The center of this lens model has more power, producing axial myopia in the air and emmetropia in the water. Data on lens characteristics are not available to support or refute the bifocal hypothesis. Though Rivamonte suggests that there is no known animal lens with these particular properties, the eye of the "four-eyed" fish *Anableps* sp. is bifocal (Munk, 1969; Schwassman & Kruger, 1965), and adult human lens heterogeneity was recognized before Helmholtz (Southall, 1962). The mammalian lens is laid down concentrically around its nucleus. New cortical material is added each year, and growth continues throughout life. Severe illness or metabolic changes are indicated as zone changes in the human lens much as dry periods are reflected in tree growth rings. This sort of dynamic heterogeneity should be part of any comprehensive model involving the lens periphery but will make accurate modeling very difficult. In humans the lens outer zones have no special optical significance, since the light passes in the axial direction, and the extreme periphery is not used. Clearly more research is required to identify all factors which provide for aerial-visual function in *Tursiops*.

The tapetum lucidum of cetaceans has received relatively less attention than other eye structures and is considered in only a few sentences by some authors (Pilleri & Wandeler, 1970; Yablokov et al., 1972; but see Slijper, 1962 for exception). Prince (1956, p. 253) states that the tapetum never lies far ventrally to the equator in animals active in the day and at night. However, our observations in fresh *Tursiops* and *Kogia* material seem to establish a ventral reflective tapetum. In some teleosts the reflectance from the tapetum covers a broad spectral region and ranges from 50 to 60%. In selachians the reflectance may reach more than 85% in a narrow spectral region during dark adaptation (Nichol, 1975). With light adaptation the same tapetum may reflect 5% or less in the spectral region of maximum efficiency (Nichol, 1975). The reflectance loss originates in the screening effects of migrating melanophores, which cover the tapetal cells in bright light (Nichol, 1975). Pigment migration probably does not occur in mammals (Rodieck, 1973, p. 183). As pointed out earlier in the chapter, the dolphin retina with its tapetalized reflector and great receptor density probably compensates for the light loss from water depth or from the pinhole or double-slit pupil. Important questions arise from this material, such as: Why has the dolphin "rejected" the more usual mammalian accommodative mechanisms? Perhaps even more important: Why does any cetacean need to see well in air? This question and other questions on visual adaptation are considered more fully by Madsen and Herman (this volume).

Other deviations from structure of the eye of common mammals have been described in the organization of the retinal inner layers of seals (Nagy & Ronald, 1970), whales (Pilleri & Wandeler, 1964), and marine dolphins (Perez et al., 1972). A major deviation is seen as the development of an elaborate giant-cell system in layers 8 and 9 (the ganglion cell and fiber layer) of the retina. The giant-cell system and smaller cells in this traditionally "simple" layer may reach its greatest development in the marine dolphins. The large visual fields served by this elaborate network suggest that it transmits information incident on extended regions of the receptor surface. But a most striking aspect of this system is the large size of each element. Each of the great cells generates an axon of large dimensions, which continues in the optic nerve "cable." It is difficult to propose a function for the extremely large cell bodies. However, it is well established that conduction velocity along neurons is directly proportional to the cross-section diameter and to the extent of axonal myelination. Consequently the giant-fiber system seemingly provides the marine mammal with an uncommonly rapid communication pathway with the central nervous system. It has been recently established in terrestrial mammals that ganglion cells are divided roughly into "sustained" and "transient" systems (Rodieck, 1973, p. 562). The transient system appears to provide information about movement of images on the peripheral retina. That is, it appears to be a detector system which provides information about target acquisition. The transient system in terrestrial mammals has a higher conduction velocity and therefore larger fibers than the sustained cell system. Thus it is analogous, on an anatomical basis, to the giant-cell system which we have described in marine mammals and is highly elaborated in dolphins. Functionally such a system would be of great value for the detection of movement in the peripheral visual fields and subsequent orientation toward the source of movement.

Color vision may be defined as the ability to detect and act upon differences in spectral hue in the absence of brightness cues. Many animals are adept in detecting subtle differences in brightness and can lead experimenters to believe that the detection has been based on hue rather than brightness. This behavioral propensity makes research in animal color vision difficult and is the primary reason that the domestic cat was thought to be color blind for many years (Brown, LaMott, Shively & Sechzer, 1973). Earlier in the chapter anatomical evidence has been presented, at both the light and electron-microscope level, showing the probable existence of at least two receptor configurations in the retinas of dolphins. Anatomical evidence is often strongly suggestive of the capability of functional color vision but is certainly not conclusive. The anatomical evidence for two receptor types (retinal duplexity) is even less strong in Pinnipedia. Several authors have been unable to find differences in receptor outer segment anatomy (Nagy & Ronald, 1970; Landau & Dawson, 1970; Jamieson & Fisher, 1972). A later examination of the harp seal retina (Nagy & Ronald, 1975) supported the older work which had found only rod-like receptor outer segments but reported two types of synaptic terminals associated with the receptor cells. There were both the larger pedicles and smaller spherules associated with cones and rods, respectively, in terrestrial mammals with demonstrable color vision. There are numerous anatomical criteria that may be applied toward the academic determination of the presence of retinal duplexity. These are never definitive, and behavioral data are an absolute necessity.

Some behavioral data from Pinnipedia are beginning to appear. A manuscript by Lavigne and Ronald (1975) describes measures of the behaviorally determined dark-adaptation curve for the harp seal. A monoplex retina typically produces a monotonic function showing gradually increasing sensitivity with time in dark. This harp seal dark-adaptation curve yielded a duplex function and suggests the presence of two receptor systems. This finding further amplifies an earlier report (Lavigne & Ronald, 1972), which reported the presence of a shift in spectral sensitivity as the seal eye proceeded from light to dark adaptation. This Purkinje shift is taken to indicate duplex retinal function in terrestrial mammals. A second bit of information was provided by Lavigne and Ronald (1973) as a reanalysis of Schusterman's (1972) data on visual acuity for the California sea lion as a function of background luminance. Older data on humans were quoted showing a duplex curve when acuity is plotted against luminance (Lavigne & Ronald, 1973). Since human has an acknowledged duplex system, a duplex acuity–luminance finding in the California sea lion would strengthen the evidence of retinal duplexity. Unfortunately, Schusterman's (1972) data do not seem to show a duplex curve. However, it is important to realize that even if duplexity were clearly demonstrated it is not identical with color vision, particularly trichromatic color vision of the type enjoyed by primates. Very difficult and carefully controlled behavioral experiments utilizing chromatic adaptation would be required to establish the presence of true color vision. A beginning is indicated by a recent work on color sensitivity in *Tursiops* (Madsen, 1976). The data indicated no evidence of discrimination between monochromatic lights at 485, 547, and 632 nm or between any of these and white light when the target stimuli were brightly illuminated (greater than 215 lux for unfiltered monochromatic lights) under nighttime viewing conditions. A scotopic spectral sensitivity curve was also obtained and peaked at 496 nm. This is a shift of 10 nm more toward the longer wavelengths than would be expected, based on the maximum photopigment absorption point (486 nm) reported for retinal extracts of *Tursiops* by McFarland (1971). The optical media of the eye tends to filter out short wavelengths, and this may explain some of the observed shift. A photopic spectral sensitivity curve obtained by Madsen peaked at 500 nm and was regarded as evidence of a duplex retina in *Tursiops*. The closeness of the peak sensitivities under photopic and scotopic conditions was taken by Madsen to indicate that the duplex system might be used to extend the range of sensitivity to ambient illumination rather than to provide a system for color analysis. But this closeness is very unusual among mammals and requires much more work before any broad interpretation can be widely accepted.

After more than three decades it is clear that some of the observations made by Walls relative to the visual function, anatomy, and physiology of marine mammals showed great insight and continue to hold. Other observations must be abandoned, particularly his statements about the inadequacy of the cetacean eye in air. In view of new data Walls' generalization is no longer correct that the marine mammal eye, particularly those of cetaceans, is an adequate adaptation of a basically terrestrial structure. Recent work has shown that there are numerous attributes of the cetacean eye which have no direct precedent in their terrestrial mammalian counterparts. The highly specialized systems or methods for image control, intraocular pressure maintenance, and retinal data processing offer a rich point of departure for future research which cannot be duplicated in terrestrial species.

ACKNOWLEDGEMENTS

The data we have shown would not be possible without generous aid from many people. Most notable has been the cooperation from the personnel and management of the Aquatarium at St. Petersburg, Florida, its manager, Mr. Michael Haslett, and its veterinarian, Dr. Richard Goldston. We have also had considerable aid from the Marineland of Florida, its management and associated scientists, Dr. and Mrs. David Caldwell. I wish to thank Dr. Calvin K. Adams, my close friend and associate, for providing some of the material for tissue flat-mount studies by interference contrast microscope and Dr. David Landau of the University of New Mexico for the electron microphotographs. Mrs. Minnie Hawthorne's skill and patience is responsible for the quality of silver and Golgi stain techniques. Special thanks are due to the University of Florida's Sea Grant Program for making funds available for the production of this chapter and to the National Science Foundation, Neurobiology Program, for financial support of the research. The constructive criticisms of A. Dral, K. Ronald, S. Gruber, and C. Madsen have been valuable.

REFERENCES

Allen, R., David, G. & Nomarski, G. (1969). The Zeiss–Nomarski differential interference equipment for transmitted light microscopy. *Z. Wissensch. Microsk. Microsk. Tech.*, **69**, 193.

American Society of Mechanical Engineers (1971). *Rules for Construction of Pressure Vessels. Division I.* New York: American Society of Mechanical Engineers.

Anctil, M. & Ali, M. (1974). Giant ganglion cells in the retina of the hammerhead shark. *Vision Res.*, **14**, 903-904.

Anderson, D. (1967). The fine structure of the astroglia in the human optic nerve head. *Trans. Am. Ophthal. Soc.*, **65**, 275-305.

Anderson, S. & Pilleri, G. (1970). Audible sound production in captive *Platanista gangetica.* In: *Investigations on Cetacea* (G. Pilleri, Ed.), pp. 83-86. Berne, Switzerland: Institute of Brain Anatomy, University of Berne.

Barris, M., Dawson, W., Adams, C. & Litzkow, C. (1978). The partially consensual pupillary light reflex of the dolphin (unpublished manuscript).

Bodian, D. (1936). A silverstain for nerve fibers and terminals. *Anat. Rec.*, **65**, 89-97.

Boycott, D. & Dowling, J. (1969). Organization of the primate retina: Light microscopy. *Proc. Roy. Soc. Lond. (Biol.)*, **255**, 109.

Brown, J., LaMott, J., Shively, F. & Sechzer, J. (1973). Color discrimination in the cat. *J. Comp. Physiol Psychol.*, **84**, 534-544.

Bunt, A. (1978). Fine structure and autoradiography of photo receptors. *Invest. Ophthal. Vision Sci.*, **17**, 90-104.

Burne, R. H. (1952). *Handbook of R. H. Burne's Cetacean Dissections.* London: British Museum.

Caldwell, D. K. & Caldwell, M. C. (1972). Senses and communication. In: *Mammals of the Sea* (S. H. Ridgway, Ed.), pp. 466–502. Springfield, Ill.: Thomas.

Caldwell, M. C. & Caldwell, D. K. (1969). The ugly dolphin. *Sea Frontiers,* **15,** 5–6.

Caldwell, M. C., Caldwell, D. K. & Evans, W. E. (1966). Sounds and behavior of captive Amazon fresh water dolphins, *Inia geoffrensis. Los Ang. Mus. Nat. Hist. Contrib. Sci,* **108,** 1–24.

Cobb, P. (1915). The influence of pupillary diameter on visual acuity. *Amer. J. Physiol,* **36,** 335–346.

Dartnall, H. J. A. (1957). *The Visual Pigments.* London: Wiley.

Dawson, W., Adams, C., Barris, M. & Litzkow, C. (1979). Static and kinetic properties of the dolphin pupil. *Amer. J. Physiol.,* **237,** 301–305.

Dawson, W., Birndorf, L. & Perez, J. (1972). Gross anatomy and optics of the dolphin eye. *Cetology,* **10,** 1–12.

Dawson, W., Carder, D., Ridgway, S. & Schmeisser, E. (1980). Synchrony of dolphin eye movements and their power density spectra. *Amer. J. Physiol.* (in press).

Dawson, W. & Perez, J. (1973). Unusual retinal cells in the dolphin eye. *Science,* **181,** 747–749.

Dodt, E. & Elenius, V. (1960). Change of threshold during dark adaptation measured with orange and blue light in cats and rabbits. *Experientia,* **16,** 313–314.

Dral, A. D. G. (1972). Aquatic and aerial vision in the bottle-nosed dolphin. *Neth. J. Sea Res.,* **5,** 510–513.

Dral, A. D. G. (1974). Problems of image-focusing and astigmatism in cetacea—A state of affairs. *J. Aquat. Mammals,* **2,** 22–28.

Dral, A. D. G. (1975a). Some quantitative aspects of the retina of *Tursiops truncatus. J. Aquat. Mammals,* **2,** 28–31.

Dral, A. D. G. (1975b). Vision in cetacea. *J. Zoo Anim. Med.,* **6,** 17–21.

Dral, A. D. G. (1977). On the retinal anatomy of cetacea. In: *Functional Anatomy of Marine Mammals,* Vol. 3 (R. Harrison, Ed.), pp. 81–134. London: Academic.

Dral, A. D. G. & Beumer, L. (1974). The anatomy of the eye of the Ganges River dolphin, *Platanista gangetica* (Roxburgh, 1801). *Z. Saugetierkd.,* **39,** 143–167.

Duke-Elder, S. (1971). *Systems of Ophthalmology,* Vol. XI. St. Louis: Mosby.

Fraser, F. (1977). Royal fishes: The importance of the dolphin. In: *Functional Anatomy of Marine Mammals,* Vol. 3 (R. J. Harrison, Ed.), pp. 1–44. London: Academic.

Gallego, A. (1971). Horizontal and amacrine cells in the mammal's retina. *Vision Res. (Suppl. 3),* pp. 33–50.

Gundy, G., Ralph, C. & Wurst, G. (1975). Parietal eyes in lizards: zoogeographical correlates. *Science,* **190,** 671–672.

Herald, E. S. (1967). Boutu and Tookashee—Amazon dolphins. *Pac. Discovery,* **20,** 2–9.

Herald, E. S., Brownell, R. L., Frye, F. L., Morris, E. J., Evans, W. E. & Scott, A. B. (1969). Blind river dolphin: First side-swimming cetacean. *Science,* **166,** 1408–1410.

Herald, E. S. & Dempster, R. (1965). Meet "whiskers." *Aquar. J.,* **36,** 213–215.

Herman, L. M., Peacock, M. F., Yunker, M. P. & Madsen, C. J. (1975). Bottle-nosed dolphin: Double-slit pupil yields equivalent aerial and underwater diurnal acuity. *Science,* **189,** 650–652.

Hogan, M., Alvarado, J. & Weddell, J. (1971). *Histology of the Human Eye.* Philadelphia, Pa.: Saunders.

Hulke, J. (1868). Notes on the anatomy of the retina of the common porpoise (*Phocoena communis*). *J. Anat. Physiol,* **2,** 19–25.

Jacobs, M., Morgane, P. & McFarland, W. (1975). Degeneration of visual pathways in the bottlenose dolphin. *Brain Res.,* **88,** 346–352.

Jamieson, G. (1971). The functional significance of corneal distortion in marine mammals. *Can. J. Zool*, **49**, 421-423.

Jamieson, G. & Fisher, H. (1971). The retina of the harbour seal, *Phoca vitulina. Can. J. Zool*, **49**, 19-23.

Jamieson, G. & Fisher, H. (1972). The pinniped eye: A review. In: *Functional Anatomy of Marine Mammals*, Vol. 1 (R. J. Harrison, Ed.), pp. 245-261. London: Academic.

Johnson, G. (1893). Observations on the refraction and vision of the seal's eye. *Proc. Zool Soc. Lond.*, p. 719.

Kellogg, R. (1938). Adaptation of structure to function in whales. *Coop. Res.*, **501**, 649-682.

Kellogg, W. N. & Rice, C. F. (1966). Visual discrimination and problem solving in a bottlenosed dolphin. In: *Whales, Dolphins, and Porpoises* (K. Norris, Ed.), pp. 731-754. Berkeley: University of California Press.

Kinne, O. (1975). Orientation in space: Animals: Mammals. In: *Marine Ecology*, Vol. II (O. Kinne, Ed.), pp. 702-852. London: Wiley.

Kulu, D. D. (1972). Evolution and cytogenetics. In: *Mammals of the Sea* (S. Ridgway, Ed.), pp. 502-527. Springfield, Ill.: Thomas.

Landau, D. & Dawson, W. (1970). The histology of retinas from the Pinnipedia. *Vision Res.*, **10**, 691-702.

Langworthy, O. (1932). A description of the central nervous system of the porpoise (*Tursiops truncatus*). *J. Comp. Neurol*, **54**, 437-499.

Laties, A. (1972). Specific neurohistology comes of age: A look back and a look forward. *Invest. Ophthal.*, **11**, 555-584.

Lavigne, D. K. & Ronald, K. (1972). The harp seal, *Pagophilus groenlandicus* (Erxleben, 1777). XXIII. Spectral sensitivity. *Can. J. Zool.*, **50**, 1197-1206.

Lavigne, D. K. & Ronald, K. (1973). Evidence of duplicity in the retina of the California sea lion (*Zalophus californianus*). *Comp. Biochem. Physiol.*, **50**, 65-70.

Lavigne, D. K. & Ronald, K. (1975). Visual sensitivity in the harp seal *Pagophilus groenlandicus. Rapp. P. Reun. Cons. Int. Explor. Mer*, **169**, 254-256.

Leicester, J. & Stone, J. (1967). Ganglion, amacrine and horizontal cells of the cat's retina. *Vision Res.*, **7**, 695-705.

Madsen, C. (1976). Tests for color discrimination and spectral sensitivity in the bottlenosed dolphin, *Tursiops truncatus*. Ph.D. thesis, University of Hawaii.

Mann-Fischer, G. (1946). Ojo y vision de las ballenas. *Biologica*, **4**, 23-28.

Matthiessen, L. (1893). Über den physikaish-optischen bau der angenoom knowal und finwal. *Z. Verl. Augenheilk*, **7**, 94.

McFarland, W. N. (1971). Cetacean visual pigments. *Vision Res.*, **11**, 1065-1076.

Morgane, P. & Jacobs, M. (1972). Comparative anatomy of the cetacean nervous system. In: *Functional Anatomy of Marine Mammals*, Vol. I (R. J. Harrison, Ed.), pp. 117-244. New York: Academic.

Munk, O. (1969). The eye of the "four-eyed" fish *Dialommus fuscus* (pices, Blennioidei, Clinidae). *Vidensk. Meddr. dansk naturh. Foren.*, **132**, 7-24.

Nagel, E., Morgane, P., McFarland, W. & Galliano, R. (1968). Rete mirabile of dolphin: Its pressure-damping effect on cerebral circulation. *Science*, **161**, 898-900.

Nagy, A. & Ronald, G. (1970). The harp seal, *Pagophilus groenlandicus* (Erxleben, 1777). VI. Structure of retina. *Can. J. Zool.*, **48**, 336-370.

Nagy, A. & Ronald, G. (1975). A light and electronmicroscopic study of the structure of the retina of the harp seal, *Pagophilus groenlandicus* (Erxleben, 1777). *Rapp. P. Cons. Int. Explor. Mer*, **169**, 92-96.

Nichol, J. (1975). Studies on the eyes of fishes: Structure and ultrastructure. In: *Vision in Fishes* (M. Ali, Ed.), pp. 579-607. New York: Plenum.

Peers, B. (1971). The retinal histology of the Atlantic bottlenose dolphin, *Tursiops truncatus* (Montagu, 1821). M.S. thesis, Guelph University.

Pepper, R. & Simmons, J. (1973). In-air visual acuity of the bottlenose dolphin. *Exp. Neurol*, **41**, 271-276.

Perez, J., Dawson, W. & Landau, D. (1972). Retinal anatomy of the bottlenosed dolphin (*Tursiops truncatus*). *Cetology*, **11**, 1-11.

Piggins, D. (1970). Refraction of the harp seal, *Pagophilus groenlandicus* (Erxleben, 1777). *Nature*, **227**, 78-79.

Pilleri, G. (1974). Side-swimming, vision and sense of touch in *Platanista indi* (Cetacea, Platanistidae). *Experientia*, **30**, 100-104.

Pilleri, G. & Wandeler, A. (1964). Developmental and functional anatomy of the eyes of the fin whale *Balaenoptera physalus*. *Acta Anatomica*, **57**, (Suppl. 57), 1-74.

Pilleri, G. & Wandeler, A. (1970). Ontogeny and functional morphology of the eye of the fin whale *Balaenoptera physalus*. In: *Investigations on Cetacea*, Vol. II (G. Pilleri, Ed.), pp. 179-229. Berne, Switzerland: Institute of Brain Anatomy, University of Berne.

Pirenne, M. (1948). *Vision and the Eye*. London: Pilot.

Polyak, S. (1941). *The Retina*. Chicago, Ill.: University of Chicago Press.

Prince, J. (1956). *Comparative Anatomy of the Eye*. Springfield, Ill.: Thomas.

Prince, J. (1964). *The Rabbit in Eye Research*. Springfield, Ill.: Thomas.

Purves, P. E. & Pilleri, G. (1973). Observations on the ear, nose, throat and eye of *Platanista indi*. In: *Investigations on Cetacea*, Vol. V (G. Pilleri, Ed.), pp. 13-57. Berne, Switzerland: Institute of Brain Anatomy, University of Berne.

Ridgway, S. H. (1972). *Mammals of the Sea*. Springfield, Ill.: Thomas.

Rivamonte, L. (1976). Eye model to account for comparable aerial and underwater acuities of the bottlenose dolphin. *Neth. J. Sea Res.*, **10**, 491-498.

Rochon-Duvigneaud, A. (1940). L'oeil des cetaces. *Arch. Mus. Nat. Hist. Paris*, **16,** 57-90.

Rodieck, R. (1973). *The Vertebrate Retina*. San Francisco, Calif.: Freeman.

Rubin, M. & Walls, G. (1969). *Fundamentals of Visual Science*. Springfield, Ill.: Thomas.

Schusterman, R. (1972). Visual acuity in pinnipeds. In: *Behavior of Marine Animals*, Vol. 2 (H. Winn & B. Olla, Eds.), pp. 469-492. New York: Plenum.

Schusterman, R. & Balliet, R. (1971). Aerial and underwater visual acuity in the California sea lion (*Zalophus californianus*) as a function of luminance. *Ann. N.Y. Acad. Sci.*, **188**, 37-46.

Schwassman, H. O. & Kruger, L. (1965). Experimental analysis of the visual system of the four-eyed fish *Anableps microlepis*. *Vision Res.*, **5**, 269-281.

Shibkova, S. (1969). Structure of internal layers of the retina in dolphins. *Arch. Anat. Histol. Embryol. (USSR)*, **57**, 68-74.

Shibkova, S. (1971). Retinal ganglionic cells in selachia. *Arch. Anat. Histol. Embryol. (USSR)*, **60,** 21-28.

Slijper, E. J. (1962). *Whales*. London: Hutchinson.

Southall, J. P. C. (1962). *Helmholtz's Treatise on Physiological Optics*. New York: Dover Press.

Spong, P. & White, E. (1971). Visual acuity and discrimination learning in the dolphin (*Lagenorhynchus obliquidens*). *Exp. Neurol.*, **31**, 431-436.

Stell, W. & Witkovsky, P. (1973). Retinal structure in the smooth dogfish *Mustelus canis:* General description and light microscopy of giant ganglion cells. *J. Comp. Neurol*, **148**, 1-32.

Uga, S. (1974). Some structural features of the retinal mullerian cells in the juxta-optic nerve region. *Exp. Eye Res.*, **19**, 105-115.

Vrabec, F. (1970). Microglia in the monkey and rabbit retina. *J. Neuropath. Exp. Neurol.*, **29**, 217–224.

Walker, E. P. (1964). *Mammals of the World.* Baltimore, Md.: Johns Hopkins University Press.

Walls, G. (1942). *The Vertebrate Eye and its Adaptive Radiation.* New York: McGraw-Hill.

White, D., Cameron, N., Spong, P. & Bradford, J. (1971). Visual acuity of the killer whale (*Orcinus orca*). *Exp. Neurol.*, **32**, 230–236.

Wolter, J. (1955). The astroglia of the human retina. *Am. J. Ophthal.*, **40**, 88–100.

Wolter, J. (1957). Glia of the human retina. *Am. J. Ophthal.*, **48**, 370–393.

Yablokov, A. V., Bel'kovich, V. M. & Borisov, V. I. (1972). *Whales and Dolphins.* Jerusalem: Israel Program for Scientific Translations.

CHAPTER 3

Social and Ecological Correlates of Cetacean Vision and Visual Appearance

Carolyn J. Madsen and Louis M. Herman

INTRODUCTION

The development of the visual system of the cetaceans varies with the characteristics of their underwater photic environment. In species of dolphins living in turbid river waters the peripheral and central visual mechanisms may be greatly reduced in comparison with species inhabiting the clearer oceanic waters (Dawson, this volume; Gruenberger, 1970; Pilleri, 1974, 1977). This correlation of visual development with photic environment is sufficient to indicate that under favorable photic conditions vision contributes to fitness, supplementing other senses in their support of a variety of biological and social functions.

Two main types of cetacean visual development can be identified: protective adaptations and optical adaptations. Protective adaptations are responses to the physicochemical stresses of the undersea environment, buffering the eye against the chemical properties of seawater, biological organisms, and extremes of temperature and pressure. Optical adaptations are responses to the optical properties of seawater. They enhance visual resolution or other aspects of visual performance. Discussion in this chapter is limited to the optical adaptations, which bear directly on the use of vision in the ecological and social world of the cetaceans.

A nonvisual development, but one related to the vision of others—prey, predators, and conspecifics or associates—is the evolution by cetaceans of characteristics promoting visual conspicuousness or visual deception, as appropriate for the animal's needs. Size, shape, special morphological features, and pigmentation patterns, supplemented by postures, movements, and other behaviors, determine the type and degree of conspicuousness or deception. Generally, being seen and recognized is important in conspecific interactions, while avoiding visual detection or recognition is important in predator-prey relationships. An exception is the development of frightening or warning coloration or of certain visually conspicuous behaviors, which may immobilize prey species or give pause to predatory or competing species. However, such developments are rare in the cetaceans.

In the material that follows the major characteristics of the underwater photic environment are described together with related optical adaptations of the cetaceans. Next, variations in coloration and other features of cetaceans which affect visual conspicuousness or deception are reviewed. Finally, some likely functions of vision in the underwater world of the cetaceans and in their above-surface aerial environment are considered. Although cetaceans are thought of as "auditory" animals, making their way about much of the time through sound production and hearing (e.g., see Popper, this volume; Herman & Tavolga, this volume), other senses, including vision, assist in coping with their ecological and social world (cf. Kinne, 1975). The contributions of the visual appearance or visually perceived behaviors of cetaceans to the communication of identity or of behavioral state are also taken up in this final section.

PHOTIC PROPERTIES OF THE UNDERSEA ENVIRONMENT AND ADAPTATIONS FOR SEEING

The photic properties of the undersea environment have been reviewed by a number of investigators (e.g., Adolfson & Berghage, 1974; Duntley, 1963;

Luria & Kinney, 1970; Lythgoe, 1975; Smith & Tyler, 1967), and only a brief summary is intended here. The emphasis is on those photic properties that greatly stressed cetacean vision, evoking adaptive visual developments. These properties include the increased optical density of water relative to air; the scattering and diffraction of light by suspended particles; the marked attenuation of light in water (loss of radiant energy); the differential brightness at different depths or at different viewing angles; and the alteration of spectral composition with depth through the selective absorption of wavelengths. Visual adaptations to these problems can be identified among some of the oceanic and coastal-dwelling cetaceans, strengthening the notion that they have evolved good underwater visual capabilities. Concurrently, some cetaceans seem to have retained, or reevolved, the capability for seeing well in air.

Optical Density and Visual Resolution

The principal refractive element of the terrestrial mammalian eye is the cornea. This optical effect is induced by the large difference in optical density and the corresponding difference in refractive index of air (1.00) and cornea (approximately 1.34). However, seawater, more than 800 times as dense as air, has a refractive index closely matching that of the cornea. The result is that in water the cornea is optically absent and fails to contribute significantly to the focusing of images on the retina. The large spherical or ellipsoidal fishtype lenses found in most cetacean eyes have greatly increased power over the typically thinner lenses of terrestrial mammals, compensating for the loss of corneal power in water (Dawson, Birndorf & Perez, 1972). The success of these steep-curvature lenses in focusing images on the retina is revealed by ophthalmoscopic measurements of in-water emmetropia or near-emmetropia for the bottlenosed dolphin (*Tursiops truncatus*) eye (Dral, 1972, 1975).

When the dolphin eye is raised above water for aerial viewing, the cornea is once again a strong refractive element and, together with the steep-curvature lens, produces a grossly overpowered focusing system, at least in the axial viewing direction (Dawson et al., 1972; Dral, 1972, 1974). However, Dral (1972, 1974) noted that a small emmetropic field of view remained in the rostroventral viewing direction in the bottlenosed dolphin eye, as well as in the eyes of the spinner dolphin (*Stenella longirostris*), rough-toothed dolphin (*Steno bredanensis*), and belukha (*Delphinapterus leucas*). Shifting the direction of gaze toward this emmetropic field may be the major accommodative mechanism for good aerial vision in these species. Rivamonte (1976) proposed that a lens of radially decreasing refractive index, similar to that found among teleost fish, could account for the nearly equivalent best acuities of the bottlenosed dolphin eye in water and in air, as measured behaviorally by Herman, Peacock, Yunker, and Madsen (1975), and illustrated in Figure 3.1.

The pair of tiny pupillary slits appearing in each eye in bright light (Fig. 3.2) may further facilitate aerial vision by acting as stenopaic apertures and, like

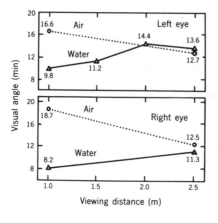

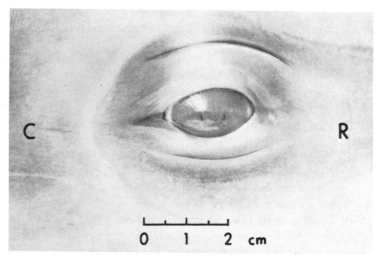

Fig. 3.1. The underwater and in-air monocular resolution acuity of the dolphin Puka as determined by her ability to resolve black and white lines on grating targets. The data show the mean width of a line, expressed in minutes of arc, resolvable as a function of viewing medium, viewing distance, and left and right eye. Note that viewing distance has opposite effects in the two media. (From L. M. Herman et al., *Science,* **189** 1975, 650–652. With permission. Copyright © 1975, by the American Association for the Advancement of Science.)

Fig. 3.2. Right eye of bottlenosed dolphin Puka photographed out of water in bright sunlight. The caudal (C) and rostral (R) orientation is as shown. The twin pupillary slits have been lightly retouched to improve contrast against dark iris. (Photograph by L. Herman.)

a pinhole camera, increase the depth of the field of view (Dawson et al., 1972; Dral, 1972, 1974; Herman et al., 1975). The pair together increase the total illumination on the retina over that obtainable with a single aperture, their lateral spread increases the horizontal field of view, and their locations may possibly correspond to the emmetropic fields of view delimited by Dral. Suggestions of dynamic accommodation in cetaceans by lens displacement or deformation (Dral, 1972; Yablokov, Bel'kovich & Borisov, 1972) were discounted by Dawson (this volume) because of poorly expressed ciliary musculature and the failure to observe any dynamic accommodative effects during ophthalmoscopic examination. Lens displacement is, however, a common accommodative mechanism in teleost fish (Schwassman, 1975; Sivak, 1973).

Behavioral measurements of the resolution acuity of the bottlenosed dolphin in clear seawater (Herman et al., 1975) and preliminary studies of underwater acuity in the Pacific white-sided dolphin (*Lagenorhynchus obliquidens*) (White, Cameron, Spong & Bradford, 1971) and the killer whale (*Orcinus orca*) (Spong & White, 1971) indicate that these species have acuity levels comparable with that of pinnipeds (reviewed by Jamieson & Fisher, 1972), the better teleost fish (Schwassman, 1975), and the better terrestrial nonprimate mammals (Walls, 1942). Clearly, the visual resolution problem introduced by the optically dense aquatic medium has been successfully overcome by these cetaceans.

Scattering and Diffraction, Viewing Distance, Movement Perception

The scattering and diffraction of light by small suspended particles limits underwater visual range and resolution. Diffraction diverts a portion of the light rays from their original path and produces complex interference patterns, blurring the outlines of objects. Scattering, or the spreading of light rays, adds to the loss of radiant energy in the path between eye and object and also blurs the contours of objects.

Scattering is virtually independent of wavelength in most natural waters because the planktonic organisms and detritus producing the major scattering effect are large compared with the wavelengths of light. Refraction and reflection at the surfaces of these particles result in a scattering function (volume/scattering/scattering angle) that is basically the same in all natural waters (Duntley, 1963). Diffraction and scattering together produce large losses in contour information and in contrast as viewing distance increases. The rate of loss depends on turbidity level (Luria & Kinney, 1970). Contour information and contrast are further reduced through small-scale variations in the local refractive index of water arising from phenomena such as thermoclines and the presence of transparent biological organisms ranging in size from microns to centimeters.

All of these factors dictate that underwater resolution will be better at nearer-viewing distances. Herman et al. (1975) verified that the underwater resolution acuity of the bottlenosed dolphin was, in fact, better at a 1-m viewing distance than at 2.5 m, as was seen in Figure 3.1. Importantly, the acuity values were invariant with changes in the light transmissibility of the water, which ranged from 49 to 80%, depending on algal growth between weekly tank cleanings. Hence, in this case, the improved acuity at nearer-viewing distances was an adaptive optical property of the dolphin eye rather than an environmentally imposed artifact. Humans wearing face masks in clear shallow water show no loss in visual acuity at viewing distances as great as 5 m (Kent, 1966). However, the human eye has fine lenticular accommodation, unlike the eye of the bottlenosed dolphin. The contrast in the effects of viewing distance within the 5-m limit for human and dolphin suggests therefore that the dolphin eye has two fixed foci, one adaptively specialized for under water near viewing, with a far point of 1 m or less, and the second specialized for distant viewing, with a near point of 2.5 m or greater.

Development of Movement Acuity

The limitations on contour (form) and contrast information with increased in-water viewing distance heightens the value of other types of visual cues. Mizusawa (1966) speculated that aquatic animals may rely heavily on color and motion stimuli, while terrestrial mammals place relatively greater emphasis on distance, form, and size cues. The speculation about aquatic animals may be nearer the truth if limited to longer-viewing distances or to turbid waters. Thresher (1976) showed that form discrimination was important for inter-specific recognition in territorial defense by the reef-dwelling three-spot damsel-fish. Goldfish easily learn to discriminate between abstract two-dimensional geometric forms at near-viewing distances (Sutherland, 1968), as do California sea lions (Schusterman & Thomas, 1966). Paradoxically, the bottlenosed dolphin, with its demonstrated high level of resolution acuity for visual gratings (Herman et al., 1975), nevertheless has difficulty in learning to make abstract-form discriminations, even at near distances (Herman, this volume). We have no ready explanation for this difficulty. Possibly the availability of echolocation for feature detection and recognition has led to some regressions in capabilities for processing visual-form information. In contrast, the goldfish and sea lion have no comparable alternative system to vision for form discrimination.

A discussion of color vision in cetaceans is deferred to a later section, but here we can consider evidence for a heightened responsivity to motion stimuli. While no psychophysical measurements of movement acuity have been made for any cetacean, trainers working with dolphins may place many behaviors under control of rather subtle moving-hand signals. A flick of the open hand out of water, for example, can be a sufficient cue for eliciting a leap or other behavior. Some physiological data also support a hypothesis of good movement acuity in cetaceans. Dawson (this volume) noted that the proportion and size of the giant cells in the cetacean retina are much greater than in other mammalian retinas studied. He postulated that these giant cells provided the cetacean with a high-speed neural pathway, allowing for rapid orientation toward movement detected in the peripheral visual field.

Another factor supporting the salience of motion stimuli in the cetacean eye is the shape of that eye. Walls (1942) remarked that horizontally elliptical eyes, like those of cetaceans, enhance movement-detection capabilities. Stimuli moving in the wide peripheral field of the horizontally elliptical eye can be detected readily, along with stimuli moving in the more central fields. Also the "barrel distortion" effect, unique to the peripheral field, lengthens and increases the speed of the sweep of an image moving through the peripheral field, enhancing both its perceived size and speed.

Real movement can be hard to detect in the open ocean where there are few stable visual referents as background for moving objects. However, movement relative to the eye of the observing animal does provide important cues. The flashing changes of illumination on the moving surfaces of swimming prey animals are excellent motion stimuli. The light-colored patches evident on many cetaceans may similarly enhance movement detection, aiding in the maintenance of school orientation or serving for intraspecific communication (see later sections). Detection of bright patches or of flashing changes of illumination is of course dependent on brightness sensitivity as well as movement sensitivity. The eye of the bottlenosed dolphin has excellent brightness sensitivity at the longer wavelengths appropriate to its aquatic habitat, as discussed in the following sections.

Walls (1942) contended that movement perception, together with brightness perception, was the most ancient aspect of vision and may show a progression in capability in more advanced visual systems. The various factors suggestive of the value of movement-detection capabilities in the aquatic environment and of developments in movement-detection capabilities in cetaceans should alert us to the further study of this interesting area.

Light Attenuation and Brightness Sensitivity

Only a fraction of the light energy striking the water surface penetrates the interface; the remainder is lost through reflection. The proportion of energy reflected increases as the angle of the rays incident to the water surface decreases, a diurnally and seasonally changing variable. Moreover, light penetrating the interface is rapidly attenuated by the scattering effects previously discussed, and through the absorption of energy (the conversion of radiant energy to heat). Scattering and absorption together result in an attenuation coefficient for light in water 1000 times greater than for light in air (Luria & Kinney, 1970).

Attenuation on a water column increases with depth, as increasing amounts of light energy are scattered and absorbed. Figure 3.3 compares light transmittance of the clearest oceanic water (JI) with the most turbid coastal water (J9) at several depths. The curves are derived from Dartnall (1975, Fig. 4). Dartnall used the spectral irradiance curves characterizing the eight Jerlov water types (three oceanic types, I, II, and III, and five coastal types, 1, 3, 5, 7, and 9) at 1-m depth to extrapolate values for 10-m depths and greater. Figure 3.3 shows that turbid coastal water loses irradiance much more rapidly than does clear oceanic water. Human photopic and scotopic depth limits for J9 water are approximately 20 m and 30 m, respectively, as contrasted with theoretical limits of approximately 550 m and 800 m for JI water.

Clearly, light intensity may change radically between the near-surface layers and deeper layers, especially in turbid waters. Consequently the eye of a cetacean may encounter large brightness shifts during dives and ascents, or when raised above the water surface, or when the direction of gaze is

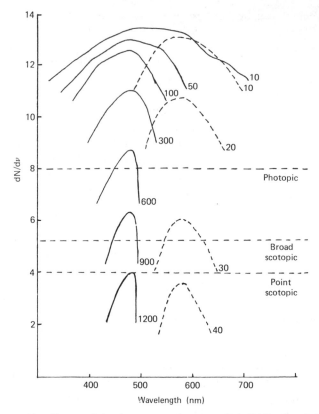

Fig. 3.3.   Spectral irradiances of the clearest oceanic waters, JI (solid lines) and the most turbid coastal waters, J9 (dashed lines). Wavelengths are in nanometers. The ordinate values are irradiances, measured as quanta/unit time/unit area/unit frequency. Parameter values are depths in meters. Human threshold values are shown for photopic, broad-field scotopic, and small-source (point) scotopic conditions, as derived from consideration of the percentage of surface irradiance penetrating to different depths in different waters and the function relating irradiance to visual threshold in the light-adapted and dark-adapted eye. (Modified from H. Dartnall, in: *Vision in Fishes* (M. Ali, Ed.) pp. 543–563, 1975, Fig. 4.) With permission. Copyright © 1975, by Plenum Press.

shifted in the vertical plane. Furthermore, all cetaceans seem active to some degree under both diurnal and nocturnal conditions. These highly variable brightness conditions must select for pupillary and retinal mechanisms capable of broad adjustments to brightness levels.

Pupillary Adaptations

The pupil of the bottlenosed dolphin eye and of some other delphinids is adapted for radical responses to variable brightness conditions (Dral, 1974). In dim light, the dilated pupil takes the shape of a horizontally oriented crescent approximately 5-mm wide from temporal to nasal extreme (Herman et al., 1975). As light intensity increases the centrally located operculum begins to close from above, progressively constricting the aperture. Its shape proceeds to a slit crescent with a major horizontal component and finally to two very small diagonal slits at the temporal and nasal extremes of the pupil (Fig. 3.2). A slittype pupil is an efficient light regulator capable of producing smaller apertures than can an elliptical or circular pupil (Prince, 1956). Additionally, a horizontal-slit pupil provides a natural sunshade for light from above. The partial pupillary independence of each eye of the bottlenosed dolphin, reported by Dawson (this volume), may permit adaptive pupillary responses to simultaneously different brightness conditions, as when an animal is swimming on its side with one eye oriented toward, and the other away from, the surface.

Retinal Adaptations

The conetype cells found in the retinas of cetaceans (Dawson, this volume) may be an adaptation for photopic viewing conditions that saturate rodtype receptors (Madsen, 1976). Conversely, the numerous rodtype cells enhance visual function under scotopic viewing conditions. The tapetum lucidem—the layer of reflecting material lying posterior to the retina—also contributes to scotopic viewing capability. Fishes with tapeta possess less visual pigment than those without, resulting in about the same visual sensitivity for either type of eye (Zyznar, 1975). However, cetaceans have large amounts of visual pigment and therefore must gain considerably in sensitivity under very dim light conditions. Behavioral tests show that the bottlenosed dolphin is particularly sensitive to light in the blue-green portion of the spectrum (Madsen, 1976). It is worth noting that in dim light a tapetum not only increases light absorption by the retina but may also enhance the differential brightness, or contrast, between figure and ground (Walls, 1942; Zyznar, 1975).

Easter (1975) postulated that the ventral retinas of pelagic fishes are dominated by dimming detectors (off-center cells), for increasing the detectability of dark overhead objects viewed against the background of downwelling light (cf. Lythgoe, 1972). Similarly, Easter hypothesized that the dorsal retinas are dominated by brightness detectors (on-center cells) to better detect light reflecting back from objects below the eye against the darkness of the surrounding water. No electrophysiological studies seem to have been done yet to confirm Easter's hypothesis nor have dimming or brightness detector cells been identified among cetaceans. However, Dawson (this volume) noted that the tapetum coloration is different in the superior and inferior retinal fields. Possibly this variation relates to the differential figure–ground contrast conditions in the dorsal and ventral fields of view, producing effects similar to those of dimming and brightness detectors. It may also be, as suggested by Dawson, that the different colorations function to increase sensitivity to the wavelengths predominating at different depths.

Spectral Composition, Retinal Pigments, Spectral Sensitivity

Figure 3.3 showed that long wavelengths are favored in turbid J9 water. With increasing depth there is a trend toward monochromacy at 580 to 590 nm. In contrast, the monochromatic trend in JI water is toward 465 to 470 nm. These different spectral shifts are brought about by the selective wavelength absorption effects of different natural waters. In pure fresh water increasing depth results in selective absorption of both long wavelengths (particularly above 575 nm) and short wavelengths, yielding a narrow-band spectral distribution with principal energy at approximately 425 nm (Smith & Tyler, 1967). In natural waters the effects are different. Chlorophylls in the living phytoplankton absorb light in a diffuse band between 400 and 500 nm and also between 670 and 680 nm (Lythgoe, 1975). Additionally, the Gelbstoff, which includes all of the stable products of decay of natural vegetation, absorbs light over a wide spectral range but with least absorption at 475 nm in the open ocean and at 530 to 550 nm in inshore waters (Lythgoe, 1975). The net result is that "average" oceanic waters have a peak transmission at about 480 nm, while in more turbid coastal waters the peak shifts by varying amounts toward longer wavelengths, since the Gelbstoff from river runoffs absorbs more of the shorter-wavelength light (Lythgoe, 1972).

Offset Retinal Pigments

In shallow water, where most of the illumination of objects is from down-welling light, it may be possible to improve the visual contrast of nearby small bright objects, e.g., small silver-colored fish, if the peak absorbance of the retinal pigments is offset from the peak transmission of the surrounding water. The bright object will reflect a broader range of wavelength than will the surrounding water (background light), because most of the light reaching the bright object has traveled a relatively short distance from the surface, while most of the background light has traveled farther (Muntz, 1975). Longer pathlengths result in an increasingly narrowed light spectrum because of selective absorption. Hence an offset pigment absorbs more irradiance from the object than from the surrounding water, especially at wavelengths differing from the peak wavelength transmitted by water.

Given an offset rather than a matching pigment the figure/ground contrast will be improved at near-viewing distances, the optimal viewing condition in water. However, an offset pigment would not improve contrast at long distances or in deep water nor would it improve the contrast for small dark objects. It would also fail to affect significantly the resolution of large objects, which typically have high intrinsic contrast (Muntz, 1975). Under all of these latter conditions maximum visibility is in fact given by a visual pigment with peak absorbance coincident with the maximum transmission of water.

The visual pigments in the rods of aquatic vertebrates seem to represent a compromise between an offset pigment for detecting small bright objects and a matching pigment for optimum detection under all other conditions (Lythgoe, 1972). Deepsea fishes, for which even downwelling light is relatively monochromatic, are the only teleost group with visual pigments more or less coincident (470- to 485-nm range) with the maximum irradiance transmitted by the water of their habitat (Dartnall, 1975). Surface-dwelling fishes, whether oceanic or coastal, tend to have pigments peaking at wavelengths (500- to 510-nm range) shorter than the maximum spectral irradiance transmitted by the waters they inhabit, though perhaps not sufficiently offset for maximum advantage in detecting small, bright objects nearby (Dartnall, 1975; Lythgoe, 1972; Muntz, 1975).

The absorption characteristics of rhodopsin have been examined for nine cetacean species (Lythgoe & Dartnall, 1970; McFarland, 1971). Absorption maxima ranged from 481 to 497 nm, with shorter-wavelength absorbances occurring for deeper-diving species. This may indicate a general trend toward a pigment coincident with the spectral irradiance transmitted by the surrounding water for cetacean species feeding in the deeper layers, as is the case for deepsea fish. The overall span of 481 to 497 nm demonstrates the compromise effect important for animals, like cetaceans, that range through a wide variety of photic environments.

Color Vision and Spectral Sensitivity

In many fish, multiple visual pigments occur and are related primarily, or secondarily, to color vision. Easter (1972), in discussing the cyprinid fish (e.g., carps), proposed that multiple-pigment vision developed to improve visual contrast, and was later secondarily adapted for color vision. He suggested that the red-sensitive mechanism of cyprinid fish improved contrast by making use of the long-wavelength spectral window, where the veiling luminance produced by the intensive scatter of shorter-wavelength light was weakest.

Behavioral studies have confirmed good color-discrimination capabilities in some fishes, e.g., in the goldfish (Yager & Thorpe, 1970). The brilliant coloring of many shallow-water fishes suggests that color vision plays an important role in intraspecific communication. In coastal and fresh water, which are green or yellow-green from the presence of Gelbstoff washed from the land, many fishes have brilliant red patches that figure prominently in their visual displays. Reef-dwelling fishes, living in water that is clear and blue, often have brilliant yellow markings (Lythgoe, 1975). Human studies have shown that in turbid water oranges are visible at greater range than other colors, while in clear water yellow-green has the longest visible range (Behan, Behan & Wendhausen, 1972; Kinney, Luria & Weitzman, 1967).

There is no evidence, however, that cetaceans or other marine mammals have color vision as effective as that of fishes. In fact such evidence as exists suggests that color vision is absent or, at best, weak in marine mammals.

Marine mammal pigmentation patterns are mainly limited to black, shades of gray or brown, and white, alone or in combination. In a few species tints of pale pink or yellow may be found in some body areas. In a preliminary study of color vision in the harbor seal (*Phoca vitulina*) an ability to discriminate between orange and blue hues was indicated (Wartzok & McCormick, 1978). However, the brightness match for these two hues by the seal yielded an orange judged as much brighter than blue for a human observer. This indicates that color vision in the seal was weaker than that found with human trichromatic vision and that sensitivity was greatly reduced at the longer wavelengths. The photopic spectral sensitivity function of the harp seal (*Pagophilus groenlandicus*) leads to similar conclusions for this species (Lavigne & Ronald, 1972). Sensitivity peaked at 525 nm. At longer wavelengths, sensitivity was very poor relative to that of a human observer.

A study of color vision in a bottlenosed dolphin (Madsen, 1976) suggests that this capability is even less developed than in the seal. As shown in Figure 3.4, under photopic conditions the dolphin's spectral-sensitivity function peaked at 500 nm, a very small shift away from the peak of 496 nm obtained under scotopic conditions. Like the seal, photopic sensitivity to red was very weak relative to that of a human observer. Intensive prolonged training by several different techniques yielded no success in teaching the dolphin to discriminate among blue (485 nm), green (547 nm), or red (632 nm) monochromatic light, or between any of these and achromatic (white) light. If the bottlenosed dolphin does possess color sensitivity it must indeed be weak.

VISUAL APPEARANCE OF CETACEANS

Coloration and Feeding Ecology

Yablokov (1963), as summarized in Yablokov et al. (1972), divided the cetaceans into three main color groups: (*a*) those with uniform, or nearly uniform, body coloration; (*b*) those with distinct patches or markings that contrast with the overall body coloration; and (*c*) those with dark dorsal fields, lighter sides, and very light ventral fields, a pattern which conforms to the principle of countershading. The distinctions between these color groups are not always precise, as some animals have coloration features that cut across two or even three categories.

Yablokov et al. (1972) postulated that color groups *a* and *c* correlated best with food-gathering habits. Species that feed mainly on organisms that do not flee, such as the passively drifting zooplankton, or that feed in near-darkness, at night, or at great depths, do not require camouflage and tend to be uniformly colored. Examples are the bowhead whale (*Balaena mysticetus*) which grazes on small zooplankton, and the sperm whale (*Physeter catodon*), the false killer whale (*Pseudorca crassidens*), and Cuvier's beaked whale (*Ziphius cavirostris*), all of which feed on cephalopods or deepsea fish

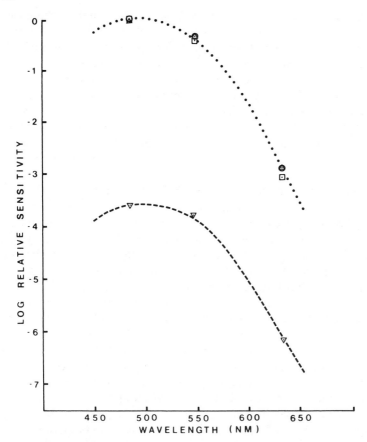

Fig. 3.4. Spectral sensitivity of the bottlenosed dolphin Puka. "Scotopic" data were obtained at night in illumination of less than 0.17 lux, and fitted with standard Dartnall 495 nm function. "Photopic" data were obtained in artificial illumination of approximately 250 lux and fitted with Dartnall 500 nm function. The scotopic data shown are based on absolute thresholds (erect triangles) for monochromatic light of different wavelengths, tested in a "Go" (light present) vs. "No-go" (light absent) paradigm with eye and light in water, and on difference thresholds, expressed as the minimum difference in intensity between two monochromatic lights of different wavelength that could just be detected, with eye and light in air (circles) and with eye and light in water (squares). Photopic data are based on absolute thresholds only, with eye and light in water. (From Madsen, 1976.)

at great depths. Other examples are the gray whale (*Eschrichtius robustus*), the narwhal (*Monodon monoceros*), the belukha, and some of the river dolphins, all of which may feed on the bottom or in murky water.

Cetaceans which feed on fearful or elusive prey under conditions of good visibility, such as during diurnal epipelagic fishing, tend to be cryptically colored. Examples are the Atlantic spotted dolphin (*Stenella plagiodon*) and the common dolphin (*Delphinus delphis*). In most cases cryptic coloration is expressed through countershading, sometimes in combination with other optically deceptive signals as discussed below. Color group *b*, those with areas of vivid coloration, tend to be schooling animals that travel in large groups. The Pacific white-sided dolphin (*Lagenorhynchus obliquidens*) and Dall's porpoise (*Phocoenoides dalli*) are species falling within this category.

Further details on the relationship of feeding ecology and coloration are given in Table 3.1. The 10 mysticete species and a selected 25 of the 65 or so odontocete species are classified either as uniformly colored or as primarily countershaded, and their primary and secondary food items are listed. Odontocete species not listed are those which are primarily patched and some intermediate forms. Some of these species are discussed later in the section on visual deception.

The coloration characteristics in Table 3.1 are taken from a number of sources (Gaskin, 1972; Leatherwood, Caldwell & Winn, 1976; Mitchell, 1975; Nishiwaki, 1972; Perrin, 1972; Tomilin, 1967; Yablokov et al., 1972), as well as our own personal observations. There is some inconsistency across the published sources in the descriptions of coloration of given species, so that a number of the classifications in Table 3.1 should be regarded as tentative. In particular, distinguishing countershaded animals from animals that seem primarily uniformly colored but which have white patching on areas of the ventrum, e.g., *Globicephala melaena*, can sometimes be difficult. Our criterion for countershading, however, has been a gradual lightening of coloration from dorsal to ventral surface, so that *G. melaena* appears in Table 3.1 as a primarily uniformly colored species. Additionally, for some species, e.g., *Ziphius cavirostris*, there are individual variations among animals such that some may have countershaded characteristics while others appear uniformly colored. In such cases we have classified the species as primarily uniform. Obviously, there is a greater subtlety to coloration among cetaceans than allowed by the two categories of Table 3.1.

The food items consumed by the listed species in Table 3.1 are culled from some of the sources listed for coloration features, plus additional sources (Berzin, 1972; Fitch & Brownell, 1968; Mitchell, 1978; Nemoto, 1970; Rice & Wolman, 1971). In many cases, the feeding information is scanty, and it may be difficult to judge whether the oceanic prey are taken principally in the euphotic (epipelagic) zone or in the dysphotic (mesopelagic) zone, or whether feeding occurs mainly at night or during daylight hours. The euphotic zone, extending to perhaps 100-m depth in oceanic water, receives sufficient light to support photosynthesis. Here, predator camouflage can contribute to the capture of fast-swimming elusive prey species. Camouflage has little value for species feeding exclusively in the deeper dysphotic zone, except possibly as an antipredator mechanism when passing through or swimming in the upper water layers.

It is assumed that the planktonic species (e.g., crustacea such as mysids, amphipods, copepods, or euphasids and the pteropod molluscs) are taken by the mysticete whales mainly in the euphotic zone. The benthic invertebrates (the cephalopod molluscs are listed separately), include a variety of organisms such as polychaete worms, shrimp and other decapods, and echinoderms such as starfish and sea cucumbers. The benthic invertebrates, together with demersal fish and the cephalopod molluscs, may occupy both the euphotic and dysphotic zones. Many of the river fish of the tropical and subtropical regions live in turbid waters that provide little opportunity for predator vision.

What regularities can be seen in Table 3.1? Among the mysticete whales, possibly all six of the countershaded group consume pelagic fish in the euphotic zone but none of the uniformly colored group do. The diet of the odontocete species tends to be more varied than that of the mysticetes, making food-related coloration regularities harder to detect. However, a trend in coloration similar to that noted for the mysticete whales occurs. Of the 14 odontocete species classified as primarily uniformly colored, only 6 (42.9%) consume pelagic and/or demersal fish in the euphotic zone. This contrasts with 9 of the 11 species classified as primarily countershaded (81.8%).

A few highly unusual feeding-related coloration patterns crop up here and there. The fin whale (*Balaenoptera physalus*) has distinct countershading, but rotated 90° on the body axis, so that its right side—including the lips, baleen, and tongue—is lighter in color than other body areas. This remarkable coloration pattern is almost certainly related to the fin whale's habit of skimming the surface layer for food while swimming on its side (Mitchell, 1972). The white coloration along the mouth lining of the sperm whale and the white coloration on the heads of many of the beaked whales, all deep divers and feeders on bioluminescent squid in the aphotic zone, may be a visual lure, reflecting bioluminescence and attracting squid. Possibly the white lining along the lips of the melon-headed whale (*Peponocephala electra*), a little-known pelagic species that is otherwise uniformly dark, signifies that it too feeds on bioluminescent prey in the aphotic zone.

Visually Deceptive Characteristics

Hailman's (1977) review of optical signaling in animals summarized those visual features or behaviors promoting deception or conspicuousness. Visually deceptive characteristics included countershading, disruptive coloration, imitation (mimicry), visual ambiguity (movement deception). Hailman also listed matching coloration, shadow minimization, and transparency as deceptive features, but for the most part these do not apply to cetaceans. One or more of the remaining characteristics can be found as predator camouflage among some cetaceans and, ocassionally, as antipredator camouflage.

Countershading has already been discussed, in general terms, as predator camouflage, but there are many variations on this pigmentation theme that may heighten deception. Overlays of stripes, spots, or other pigmentation on the countershaded pattern increase visual ambiguity or disrupt body contours and form.

The most extensive discussion of visual deception in cetaceans is that of Mitchell (1970), although his major analysis was limited to the delphinids.

Table 3.1. **Principal Types of Coloration and Categories of Food Items Taken by Cetaceans**

Coloration/Species	Other Coloration	Zooplankton	Benthic Invertebrates E	D	Cephalopods E	D	Pelagic Fish E	D	Demersal Fish E	D	Freshwater Fish
Mysticetes											
Primarily uniform											
Balaena mysticetus		x									
Balaena glacialis		x									
Eschrichtius robustus	S,M	x	x						s		
Balaenoptera musculus	S,M	x									
Primarily countershaded											
Carperea marginata		x					?				
Balaenoptera physalus		x			s		x				
B. borealis	S,M	x			s		x				
B. edeni		x			s		x				
B. acutorostrata		x					x				
Megaptera novaeangliae	S,M,P	x	s		s		x				
Odontocetes											
Primarily uniform											
Physeter catodon	P			s		x		s			
Kogia breviceps			s			x		s		s	
Ziphius cavirostris	S,M,P					x		x			

116

Monodon monoceros	S,M		s		x		s		s	
Delphinapterus leucas		x		x		x		x		x
Globicephala melaena	P			x	x	x		x		
Orcaella brevirostris						x		x		x
Pseudorca crassidens					x	s	x		x	
Lissodelphis borealis	P				x		x			
Grampus griseus	P			x	x					
Sousa teuszii						x		x		
Berardius bairdii	S,M,P		s		x		x	s	s	
Mesoplodon bidens	P				x		s			
Neophocaena phocaenoides	P	x		x				x		
Primarily countershaded										
Hyperoodon ampullatus	M,P		s		x					
Cephalorhynchus hectori	P,T	x		x		x				
Tursiops truncatus	T	x				x		x		
Stenella longirostris	S,M			s	s		x			
S. attenuata	S,M			s	s	x				
S. plagiodon	S,M			s	s	x				
Delphinus delphis	P,T	s		x	x	x	x	x	x	
Phocoena phocoena	T	s				s		x		x
Steno bredanensis	S,M			x		x				
Lagenorhynchus obliquidens	P,T			x		x		x		
L. albirostris	P,T	x				x		x		

[a]Food items marked by "x" are considered a major part of the diet and/or are frequently reported. Food items marked by "s" are considered supplementary, or of secondary importance in the diet, and/or are infrequently reported.
E = euphotic zone; D = dysphotic zone; S = spotted; M = mottled; P = patched; T = striped.

Mitchell divided the primarily countershaded delphinids into four subclasses of pigmentation pattern (Fig. 3.5). The basic coutershaded pattern, with sides considerably darker than the ventral field and closer in shade to the dorsal field, was termed a "saddled" pattern, typified by the pan-tropical spotted dolphin (*Stenella attenuata*). The saddled pattern was judged the most generalized and primitive of the delphinid pigmentation patterns. Other patterns add to or subtract from the saddled form. The "striped" pattern, typified by the striped dolphin (*Stenella coeruleoalba*), superimposes various darker and/or lighter stripes and blazes on the saddled pattern. The high-contrast markings on the torso of the striped dolphin may draw attention away from the dangerous head and teeth, as well as provide disruptive coloration for masking the eyes, mouth, pectoral fins, and dorsal fin. According to Hailman (1977), stripes promote visual ambiguity or movement deception, camouflaging the motion of the animal by detracting from the coherence of movement cues.

The third subclass, the "spotted" pattern, is typified by the Atlantic spotted dolphin. Spots may overlay the entire saddled pattern, possibly increasing the camouflage effect through mimicry of the dappling of sunlight on water (environmental imitation).

Finally, the "crisscross" pattern, seen only in the common dolphin, consists of four differently shaded and clearly demarcated areas, including a black dorsal field, gray flank patch, and brownish-yellow thoracic patch. The latter two patches form an X design between the dorsal and ventral fields. The crisscross pattern seems to have the same function as the striped pattern: the redirection of attention away from the head, masking of mouth and dorsal fin through disruptive coloration, and creation of movement ambiguity. The thoracic patch may function for conspecific recognition at close range. However, Mitchell also wondered whether the location, color, and fusiform shape of the thoracic patch of the adult might camouflage an accompanying calf, swimming in close-echelon formation. Mitchell also suggested that the shape of the saddle might have an antipredator function by approximating the shape of the shadow cast on the mother by the calf.

Mitchell's four countershaded patterns have limited applicability outside of the delphinid family. Most other cetaceans are either uniformly colored or primitively countershaded (saddled), with a few partially mottled or spotted species (Table 3.1). Also some have conspicuous rather than deceptive coloration. Mottling occurs on several nondelphinid species believed to feed under conditions of poor visibility, e.g., the gray whale, the narwhal, and the beaked whales *Berardius bairdii* and *Ziphius cavirostris*. Perhaps mottling or countershading serves in these cases as antipredator camouflage against deep-water sharks or killer whales, rather than as predator camouflage. The cryptic coloration of blue and sei whales might also provide concealment from these predators (Mitchell, 1972). However, Mitchell (1970, 1978) postulated that mottling on the blue and sei whales might also be predator camouflage, allowing these large rorquals to blend visually with the shoals of krill on which they

Fig. 3.5. Pigmentation patterns in the *Stenella–Delphinus* group. *Top to bottom*: Striped pattern from *S. coeruleoalba*; crisscross pattern from *D. delphis*; saddled pattern from *Stenella* sp.; spotted pattern from *S. plagiodon*. (From E. Mitchell, *Can. J. Zool.* **48** 1970, 717–740. With permission of the National Research Council of Canada.)

feed. The krill species *Euphasia superba,* which comprises all of the blue whale diet and part of the sei whale diet in southern-hemisphere waters, can be highly evasive. Tomilin (1967, p. 95) reports that krill will leap from the water when pursued by fish. The schooling fish that are additional items in the diet of the sei whale (Nishiwaki, 1972) are also evasive.

Some phases of the coloration of the baleen whales may be adaptations for concentrating prey, according to Brodie (1977). Brodie suggested that the long white pectoral fins of a North Atlantic humpback whale when extended forward may be used to herd a shoal of capelin into the mouth of the whale. The fish are concentrated by the flashing illumination from the white extended pectorals and "escape" into the blackness of the dark head of the whale. The pectorals are mobile and flexible enough for this purpose according to our own underwater observations (Herman & Antinoja, 1977) and as described more fully by Edel and Winn (1978). Brodie also hypothesized that the white patch on the pectoral fins of the minke whale might be used for fish herding. Northern forms of the minke whale feed heavily on pelagic fish and evidence the white coloration; Antarctic forms feed primarily on crustacea and relatively few of these whales have the white patch.

Mitchell (1970) did not discount the possibility that highly-contrasting patches, though conspicuous, might serve as a form of disruptive coloration, masking body contours. He cited the example of the piebald dolphin (*Cephalorhynchus commersonii*) in which a large area of white extends upward from the ventral field through to the dorsal field posterior to the blowhole, visually dividing the body in two. The large lateral white areas on Dall's porpoise may similarly function to disrupt body outlines. For most species, however, the lateral patches are too small to disrupt body contours effectively.

In at least one case, the conspicuous patchy coloration may be protective mimicry. The diminutive Heaviside's dolphin (*Cephalorhynchus heavisidii*) has a coloration pattern so similar to the killer whale, a probable predator on *Cephalorhynchus,* that it prompted Mitchell to suggest mimicry by *Cephalorhynchus* of killer whale calves as an antipredator mechanism. If the coloration is considered a form of Batesian mimicry it might reduce shark predation and/or competition from sympatric cetaceans. Batesian mimicry is, however, rare in mammals (Wickler, 1968), necessitating caution in inferring that the coloration similarity is indeed animal imitation.

Visually Conspicuous Characteristics

Hailman (1977) listed movement, reversed countershading, enhancement of shape, surface contrast, image size, repetition in space, and signal rarity as general features or behaviors enhancing visual conspicuousness. For the aquatic medium, Hailman noted that high-contrast visual patterns, such as patches of iridescence, bright yellows and oranges, white, and some shades of green promote conspicuousness. In general, complementary colors, such as yellow and blue, provide the best contrast in any medium. Yellow patches are found on many reef-dwelling fish living in shallow clear water of predominantly blue coloration. However, for cetaceans, a group that may have little or no capability for discriminating colors, the principal thrust for conspicuous coloration has been toward white patches on dark body areas. Table 3.1 shows that white patches can be found among a number of cetaceans that are otherwise primarily uniform in coloration or primarily countershaded. Brownish-yellow patches are seen in a few species—e.g., in the baleen of the gray whale and, as was noted, in the thoracic region of the common dolphin. The white patches are typically highlighted by large, highly contrasting dark areas. In some species, white patches may be confined to the abdominal region or to parts of it, such as the anal and genital vents. The northern right whale dolphin (*Lissodelphis borealis*) is mainly uniformly black, but an hourglass-shaped brilliant white ventral patch emphasizes the upper abdomen and the genital and anal vents. Other species, such as killer whales or the Atlantic white-sided dolphin (*Lagenorhynchus acutus*), have patches of white along the flanks or in other body areas.

Some species have evolved black patches, bordered by white, as conspicuous marks. The Atlantic white-sided dolphin has a small black spot near the anal vent. Commerson's dolphin (*Cephalorhynchus commersonii*) has a teardrop-shaped band of dark color surrounding the genital slit, as may Hector's dolphin (*C. hectori*) (Van Bree, 1972). Dark streaks on a light background, or the reverse, are also common and are usually found in patterns and places that may indicate some communication function, such as sexual signaling.

The exposure of patches to others may promote species recognition, indicate sexual state, or allow individual identification. Individual identification is facilitated by idiosyncratic differences in patches across animals, as in the killer whale (Mitchell, 1970; Yablokov et al., 1972) and the common dolphin (Mitchell, 1970). Exposure of patches may also enhance the visual detection of movements of schoolmates. These various communicative functions of conspicuous coloration are discussed in greater detail in a later section.

Other types of visual information besides high-contrast coloration patterns can contribute to conspicuousness. Paradoxically, uniform body coloration can promote conspicuousness by offering a clear, uninterrupted outline of the animal's shape. Also, when near the surface, uniformly dark animals are silhouetted against the downwelling light (reversed countershading effect), increasing their conspicuousness to animals below. The absence of selective pressure for cryptic coloration may lead to the evolution of uniform coloration. Mitchell (1970) hypothesized that uniform body coloration developed through a darkening of pigmentation in all body areas, obliterating the more primitive saddled form of pigmentation.

Enlarged image size increases conspicuousness. The large body size of most cetaceans results in large retinal images, contributing directly to visual conspicuousness. Many of the white areas of the patched cetacean species are large in size and, consequently, yield large retinal images promoting conspicuousness.

Signal rarity, the principle that infrequent signals are more likely to be perceived or attended to than familiar signals, may be expressed through the hidden patches of white coloration found under the pectoral fins or tail flukes of some species. Exposure of these bright surfaces attracts attention and may convey information about movement or movement intention. Movement, perhaps the most important type of optical signal in cetaceans, can promote conspicuousness in many ways. The speed of movement, its direction, repetitiveness, suddenness of onset, and rhythmicity, can all be varied to create novelty or to draw attention to specific body parts or behaviors.

FUNCTIONS OF VISION IN CETACEAN LIFE

The optical adaptations of the oceanic cetaceans and certain aspects of their visual appearance or visually perceived behaviors can help promote key functions in their life, including (*a*) orientation and navigation in space and time and the coordination of group movements; (*b*) prey detection and capture; (*c*) predator defense; (*d*) identification of conspecifics, of age and sex classes, and of individuals; and (*e*) communication of behavioral states. The actual or potential contributions of vision or visual appearance to these functions are considered in the following sections.

Orientation, Navigation, and Group Movements

The sensory cues controlling the orientation and migration of cetaceans and other marine mammals were reviewed by Kinne (1975). As is true for other vertebrate species, multiple sensory cues, together with endogenous factors, initiate and control the daily and seasonal movements of the cetaceans. Several types of visually perceived environmental cues may affect cetacean orientation and migration.

Photoperiods and Visual Entrainment

Photoperiods apparently trigger the migratory cycles of the mysticete whales that move between high-latitude summer feeding grounds and low-latitude winter breeding grounds (Dawbin, 1966; Pike, 1962). The motivation for migration may come from a complex of alimental, gametic, and climatic factors (Orr, 1970; Yablokov et al., 1972), but the only reliable starting cue for an animal is daylength. This implies that whales must possess a well-regulated circadian clock for timing the photoperiods and for ensuring a favorable arrival time at the breeding grounds. The optimal arrival time seems to vary with age and reproductive condition of the animal, at least for humpback whales (Chittleborough, 1965; Dawbin, 1966). The dependence of whales on photoperiods seems very much like that observed among migratory birds (e.g., see Farner & Lewis, 1971).

Norris (1967) suggested that daylength constrains the subpolar-tropical migrations of whales to one hemisphere, since an equatorial crossing would result in a reversal of daylength cues. This factor, together with the inversion of feeding and breeding seasons across hemispheres and the presence of an equatorial warm-water barrier, has apparently reproductively isolated northern- from southern-hemisphere stocks of whales. According to Davies (1963) hemispherically isolated stocks may have originated during the mid-Tertiary when cold-water forms appeared in response to the expansion of cold-water masses on both sides of the equator. Thereafter, a reenlarged equatorial warm-water belt acted as a thermal barrier to interhemispheric travel.

Herds of common dolphins may navigate by the sun, swimming easterly in the morning and returning westerly in the afternoon (Busnel & Dziedzic, 1966; Pilleri & Knuckey, 1968). Movements toward feeding areas of a number of species seem to be visually entrained by the day–night cycle. Schools of Hawaiian spinner dolphins leave their near-shore resting places in later afternoon or near dusk to begin a night-long search for food in deeper waters (Norris & Dohl, this volume). The Indian Ocean bottlenosed dolphin (*Tursiops aduncus*) feeds in the early morning and then again in the late afternoon (Saayman, Tayler & Bower, 1973). Subgroups of common dolphins assemble into large herds just before sunrise and sunset, presumably to feed (Evans, 1971). In all of these cases variables other than photic may also influence movements, but in many cases the photic condition predicts the other variable. For example, the nightly rise in the deep-scattering layer is associated with an increased noise level that could control the movements of cetaceans (Evans, 1971). However, dolphins may learn to predict the change in acoustic level by the change in light level, or the sun's position, and begin preparatory feeding movements in advance of the acoustic change.

Orienting to Surface Features

During migration, as well as in the terminal breeding and feeding grounds, some of the mysticete whales may "spy-hop" by raising themselves vertically out of the water head first. While in this posture, the whale may sometimes rotate on its longitudinal axis, as if visually scanning a 360° sector. Most often spy-hopping among cetaceans seems to orient the cetacean to nearby or distant surface features, although it may have social functions as well. Scammon's (1874) account of 19th-century whaling included a sketch of gray whales spy-hopping among a field of ice floes. Pike (1962) reported that gray whales may spy-hop when near land and suggested they were scanning the environment to reorient themselves. Spy-hopping has also been observed in the southern right whale (Cummings, Fish & Thompson, 1971) and in the humpback whale (Fig. 3.6) in its winter breeding ground in Hawaii (Herman &

Fig. 3.6. "Spy-hopping" or "head-rise" behavior of humpback whales in Hawaiian waters. *Top*: Left side of animal visible, top of jaw (maxilla) on right. Left eye, obscured by shadow, is located at approximately the same distance caudal of the jaw as is the blowhole, visible on right. *Bottom*: Ventral surface of calf showing ventral folds. Left eye visible at lower right side of figure. (*Photographs by University of Hawaii research team, courtesy L. Herman.* Top: D. McSweeney; bottom: P. Forestell.)

Forestell, 1977). Observations of the University of Hawaii humpback whale re-search team additionally suggest that when a whale breaches (leaps out of the water) in the presence of a nearby boat or low-flying aircraft, its body is often turned so that the eyes are oriented toward the observation vehicle. Among odontocete species, in addition to the Atlantic bottlenosed dolphin, spy-hopping has been reported for the sperm whale, the pilot whale, the Pacific bottlenosed dolphin (*Tursiops gilli*) (Norris, 1974), and the killer whale (MacAskie, 1966). Supposedly, killer whales may inspect ice floes for basking seals on which they prey (Norman & Fraser, 1949). Layne (1958) reported spy-hopping in the Amazon River dolphin or bouto, *Inia geoffrensis,* although this species lacks the highly developed visual system of the oceanic dolphins (Dawson, this volume). In the portion of the habitat studied by Layne, three small lakes were joined by a narrow channel to the main river. The bouto were seen to raise their heads out of water several times before swimming furiously through the channel. Layne suggested that the animals were visually scanning the channel for logs or other debris.

Orienting to Underwater Features

Where viewing conditions are favorable, bottom topography may alert ceta-ceans to prime feeding locations, or otherwise orient them in space. Evans (1971, 1974) noted that the daily surface movements of common dolphins in coastal waters of southern California and Mexico tended to trace the cliffs (escarpments) rising from the sea floor. The steep, nutrient-rich cliffs support an abundance of marine life. Evans postulated that the increase in biologi-cally produced noise near these cliffs provided an acoustic cue that could be tracked through passive listening. The presence of the cliffs and the abundance of associated marine organisms may also provide visual cues, either directly through observation during exploratory dives or indirectly through changes in water clarity or appearance brought about by the abundance of suspended particles in the area. Norris (1974), based on observations from a semisubmersible towed vessel, suggested that humpback whales in Hawaiian waters might gauge bottom depth by the cloud of particulate matter that surrounds every island like a halo. The shallower the water the greater the incidence of detritus. Humpback whales in Hawaiian and other tropical breeding grounds assemble on shallow banks or coastal waters of less than 100-fathoms (183-m) depth (Dawbin, 1966; Herman & Antinoja, 1977). Aerial surveys and boat observations in Hawaii (Herman & Antinoja, 1977; Herman, Forestell & Antinoja, 1977) reveal that tongues of deeper water (more than 183-m deep) extending into a shallow-water area are sparsely inhabited and that whales swimming in a shallow-water area will turn back when reaching a deep-water area. Why the shallow-water areas are preferred is not understood, but it may have to do with avoidance of deep-water sharks, or offer some acoustic advantage to singing whales, or possibly limit the space over which whales must search for other whales (i.e., the deep water can be excluded from the search area). Whatever the motivation, it is apparent that vision can contribute to orientation toward shallow-water areas.

In the captive circumstance the bottlenosed dolphin and many other delphinid species seem to increase their reliance on vision for orientation in space. Typically, the incidence of echolocation behavior and other acoustic activity decreases with the length of time an animal is maintained in the same environment (M. Caldwell & Caldwell, 1967), yet the animal has no problem orienting. Presumably it makes its way about by visual cues and through memory of the topography of its tank. This parallels the case for captive echolocating bats that learn to fly through openings in permanently emplaced nets by spatial memory alone (Neuweiler & Mohres, 1966). If the openings are shifted, the bat flies to where the openings were, rather than to the new locations. Schevill and Lawrence (1956) demonstrated a reliance on vision in the extreme case of a captive bottlenosed dolphin blind in one eye. The animal always approached objects with its good eye. Later, when its vision failed entirely, it proved clumsy in its approach to objects and had difficulty in locating food fish. Though its echolocation system was probably functional, it did not seem to use it. Observations in captivity also show that aerial vision may come into extensive use. After a period of habituation to the tank environment, many dolphins spend a great deal of time with one or both eyes out of water, watching human activities outside their tank (also see Dawson, this volume).

Maintaining Spacing and Controlling Movements

Many of the smaller cetacean species leap while traveling. The display of flashing bodies and splashing water produced by herds of leaping common dolphins provides a visual display that can be seen at great distances (Brown & Norris, 1956). While leaping, or through a series of leaps, an animal may be able to determine the location of other leaping or surfacing school members and estimate the direction and speed of school movement. At the same time the leap of the individual contributes this type of visual information for others. Of course there may be other functions of leaping. The reentry splash provides an acoustic as well as a visual signal, which can be especially useful for contact at night (Norris & Dohl, 1980). Also, prey species might be "acoustically herded" before a group of leaping dolphins (Würsig & Würsig, 1980).

Underwater vision can help maintain affiliations among individuals and coordinate the spacing and movement of the school, as was suggested earlier. Norris and Dohl (this volume) noted that the striking changes in the displayed pattern of coloration on the lateral and ventral portions of the dolphin's body as it shifts position were important cues to movement. A banking movement of a countershaded dolphin reveals the light-colored undersurface to a nearby schoolmate. The white coloration on the ventral surface of the tail flukes of the killer whale (Evans & Yablokov, 1978; Nishiwaki, 1972) flashes conspicuously as the tail beats, providing a vivid visual display to trailing schoolmates. The long mobile pectoral fins of the humpback whale are brilliant white on most of the ventral surface and often are white on major portions of the dorsal surface as well (Herman & Antinoja, 1977). The pectorals may be held in various positions, raised and lowered during banking movements and even used for sculling. To the human observer underwater the white-colored pectoral can be a cue to the orientation and movement of the whale at distances at which the rest of the dark body can barely be seen (Fig. 3.7). Such visual cues should be equally useful to the whales.

Another type of visual signal for humpback whales may be the bubbles of air that are often released in profusion in social settings, as well as in apparent defense or during feeding. The social functions are still largely obscure, but they may signal affiliation or location. Streams of fine bubbles arising from the blowhole and trailing behind the swimming animal are commonly seen among groups of whales. In many cases the bubbles are densely packed, and their trail extends well behind the animal. From observation aircraft, the stream of bubbles may sometimes be followed, like continuous markers on a path, to the bubbling animal ahead (Fig. 3.8). Occasionally, very large quantities of air may be released from the sides of the mouth of the whale as it swims by, producing a spectacular display of bubbles. Apparently the air was "gulped" into the mouth at the surface. All of these bubbling behaviors have strong visual components to the human observer. If they have acoustic components as well, the joint effect would make for a particularly potent short-range communication signal.

Although the visual apparatus of the humpback whale has not been extensively studied, it is probably well developed, like that of its fin whale cousin (see Dawson, this volume). Observations in Hawaiian waters suggest good visual capabilities. The eye of the humpback whale can sometimes be seen moving in its socket, fixating on the underwater diver as the whale glides by, and on one occasion a swimming whale gently raised its outstretched pectoral fin over a diver's head and then lowered it again after passing by (University of Hawaii unpublished field observations).

Prey Detection and Capture

Both underwater and aerial vision can help cetaceans detect and capture prey. The usefulness of underwater vision is limited by the clarity of the waters in which feeding occurs, the depth and time of day at which prey are taken, prey conspicuousness, and prey elusiveness. Underwater vision may be of greatest value for oceanic dolphins feeding in the clear upper layers of the sea. The tracking of fast-swimming elusive prey in clear water is an ideal task for a visual system with good resolution capabilities, a wide field of view, and movement enhancement mechanisms, all known or postulated

Fig. 3.7. Underwater views of humpback whales showing the bright white coloration on almost all of the top surface of the pectoral fin occurring in approximately one third of the population (Herman & Antinoja, 1977). *Top*: Mother with nearly all-white pectorals, calf with dark pectorals immediately above. *Bottom*: Bright pectorals reveal presence of dark whale. (Photographs by University of Hawaii research team, courtesy L. Herman. Top photograph: D. McSweeney; bottom: L. Herman.)

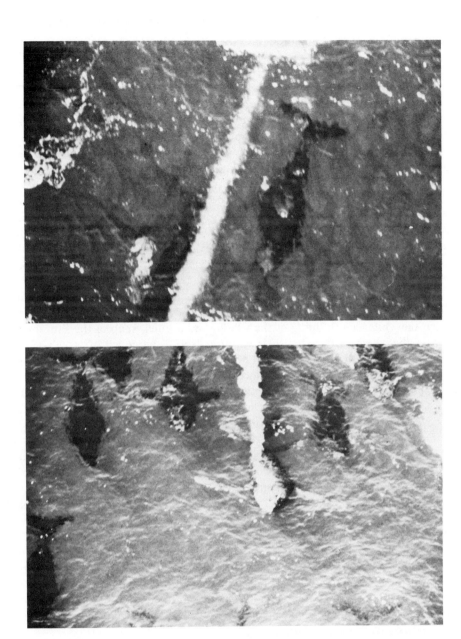

Fig. 3.8. Bubble trails from a single animal located centrally within a group of 15 humpback whales. Top and bottom photographs show the same group as seen from the air at different portions of the aircraft's orbit about the group. The bubble trail in the top photograph was produced by the central animal, still bubbling, in the lower photograph. Underwater observations verify that these bubble trails are highly visible from below the surface, as well as from above. (Photographs by University of Hawaii research team, courtesy L. Herman. Both by P. Forestell.)

features of the eyes of some delphinids. Other adaptations for enhancement of contrast and brightness, discussed earlier, may allow for visual search and capture of prey in less ideal circumstances.

The contrast-enhancing mechanisms improve visual detection capabilities in turbid waters and in the deeper, darker water layers. They also improve the chances for detecting prey above or below the dolphin, regions not well-covered by the sonar scan, which is primarily forward-directed (Norris, Prescott, Asa-Dorian & Perkins, 1961). The brightness-enhancing mechanisms almost certainly come into play during night hunting in epipelagic zones, in daylight hunting in deeper layers, or possibly when hunting in the aphotic layers for bioluminescent prey. An interesting hypothesis about the sperm whale's hunt for bioluminescent prey was given by Clarke (1970, 1979). Clarke postulated that a primary function of the spermaceti organ of the sperm whale was buoyancy regulation at the great depths to which these animals dive. According to Clarke, the whale takes in water through the blowhole, cooling the spermaceti oil and increasing its density, which promotes neutral buoyancy. The whale then remains motionless in the water, waiting for luminescent squid to swim past. Possibly the squid are attracted by reflections from the white mouth lining of the whale, and a quick snap of the thin lower jaw of the whale suffices for their capture. Not all agree with Clarke's views on the function of the spermaceti organ, nor with the presumptive feeding strategy of the sperm whale (cf. Norris & Harvey, 1972). Inferential evidence for the use of vision in prey detection or capture by the sperm whale might be obtained by determining whether the rhodopsin absorption spectrum of its eye seems adapted to the spectral bioluminescence of the squid (cf. Lythgoe & Dartnall, 1970).

Bottlenosed dolphins, and probably other delphinids, can see well in air, as noted earlier. This capability makes possible the capture of leaping fish and may contribute to prey detection as well. Seabirds circling in the distance, a perennial cue for fishermen searching for shoals of fish, may also be a cue for dolphins. Fish leaping in the air at a distance or creating white water may be additional useful visual detection cues. The Pacific bottlenosed dolphin is a known predator on flying fish (Ljunblad, Leatherwood, Johnson, Mitchell & Awbrey, 1977), and may be able to detect or track these leaping fish during their in-air flights. The dolphins themselves, in their pursuit of fleeing fish, or through the leaps accompanying active feeding, may provide visual assembly signals for associates at some distance from the feeding site (cf. Würsig & Würsig, 1980).

Documented cases of wild dolphins capturing leaping fish in midair include the description of Kritzler (1952) of wild bottlenosed dolphins herding mullet toward a steep bank. Once trapped against the bank the fish begin leaping to escape capture but in some cases are caught in midflight by the dolphins. Harris (1938) provided photographs (reproduced by D. Caldwell & Caldwell, 1972, p. 71) of bottlenosed dolphins catching fish in midflight. Yablokov et al. (1972) described similar behaviors for Black Sea bottlenosed dolphins.

Hoese (1971) reported an interesting variation on the theme. Pairs of bottlenosed dolphins in Georgia marshes were seen driving small fish onto exposed mudbanks. The dolphins followed the fish onto the mudbanks and, while sliding back into the water, grabbed fish in midair. Busnel (1973) described the unusual case of a cooperative fishing enterprise between natives of the west coast of Africa and schools of Kukenthal's dolphin (*Sousa teuszii*). Mullet migrating offshore past the coast were herded shoreward by the dolphins in apparent response to hearing the fishermen beating sticks against the water. When the dolphins were sighted moving shoreward the fishermen entered the shallow water with their gill nets, trapping the fish between them and the dolphins. Motion pictures show the frenzied leaping of the fish and some caught in midflight by the dolphins. This mutualism of fishermen and dolphin is apparently culturally transmitted in both species, allowing each to enjoy a more bountiful harvest.

Predator Defense

The lateral and downward visual field of the delphinids seems well suited for detecting deep-water sharks or other predators rising from below or approaching from the side, filling in sensory gaps in their sonar scan, as is the case for prey detection. The passive nature of vision and the wide visual field of the delphinids (and other cetaceans) provide an environmental monitoring system with little energy cost and many benefits in promoting survival. Furthermore, an echolocation system can be disadvantageous in interactions with those predators able to localize acoustic information accurately. It is not surprising, therefore, that belukhas and some other cetaceans cease vocalizing in the presence of killer whales (e.g., see descriptions in Kleinenberg, Yablokov, Bel'kovich & Tarasevich, 1964). When threatened by predators, the line of sensory defense may shift to vision and passive listening.

Defensive responses to a predator vary, depending on species and circumstances, but may include diving, fleeing, evading, group defense, counterthreat or fighting. A complex of sensory cues, including vision, may play a part in the closing of ranks or the placing of vulnerable calves in the center of the delphinid group (e.g., see Leatherwood, 1977). Charges at the predator by peripheral animals may be mainly visually guided (see descriptions in Wood, Caldwell & Caldwell, 1970). The mysticete, as well as the odontocete, may be attacked by killer whales (Tomilin, 1967). In response, the mysticete may turn on its back and slash out with its powerful tail. Chittleborough (1953) observed a humpback whale successfully warding off an attacking group of killer whales in this manner. Cummings et al. (1971) observed two southern right whales defending themselves against five killer whales by endless slashing with flukes and pectoral fins. It is hard to imagine effective defensive fighting of this type without nearly continuous visual monitoring of the predator's position and movements.

Among the inshore delphinids, such as the bottlenosed dolphin, shark predation is a major threat, as witnessed by the extensive scarring of live dolphins, the remains of dolphins in the stomach contents of sharks, and direct observations of attacks at sea (Wood et al., 1970). Extensive scarring by sharks has also been reported for Hawaiian spinner dolphins (Norris & Dohl, 1980). Clearly, there must be intensive selection pressures for the evolution of antipredator mechanisms in these vulnerable species. Social defense may be the most important adaptation, but its usefulness would seem to rely heavily on the visual detection and recognition of predators and on the visual monitoring of their movements or other behaviors.

The Identification Function

Conspicuous visual characteristics help animals recognize their own species and identify age and sex classes and even individuals. In cetaceans conspicuous visual characteristics include the general coloration patterns reviewed earlier (Table 3.1), as well as special coloration patterns or patches found on various parts of the body or head in some species. These special characteristics of 25 selected species are reviewed in Table 3.2, together with some sex- and age-dimorphic features. The table is not exhaustive in its description of known features, and there are undoubtedly additional special features or coloration patterns in these and other species which have not yet been cataloged. Prolonged field work with species frequently uncovers previously undetected variations among animals permitting individual identification or their sexual or age classification.

Species Identification

Table 3.2 does not include the multitude of features that permit cetaceans to recognize conspecifics (species identification). These features are pervasive, and whatever variations in gross morphology or coloration allow biologists to recognize a species may be a visual cue for the animals themselves. Typically there are many generalized visual features among related species as well as special characteristics within species. For example, the different species of *Lagenorhynchus* have various head shapes ranging from the poorly defined, almost absent, beak of *L. cruciger* to the well-defined, moderately long beak of *L. albirostris* (Mitchell, 1970, Plates X & XI). Their coloration patterns also differ markedly, though there are elements common to all species within the genus. Another striking example is the coloration of the genus *Sousa,* several species of which have been described by Pilleri and Gihr (1972). The plumbeous dolphin (*Sousa plumbea*) and Kukenthal's dolphin (*S. teuszii*) are a nearly uniform grayish color, while the speckled dolphin (*S. lentiginosa*) is uniformly spotted with dark and light elliptical spots, and the Indo-Pacific humpback dolphin (*S. chinensis*) is almost pure white.

    There are even examples of significant variations in external features among populations of one species. Perrin (1972, 1975) described three populations

Table 3.2. Species Having Visually recognizable Individual (I), Sexual (S), or Age-related (A) Features

Species	I	S	A	Comments
Mysticetes				
Balaena glacialis	x			Callosity configuration on head unique to individual
Eschrichtius robustus			x	Young uniform gray, adults mottled
Balenoptera musculus	x			Coloration on throat and flank area variable
Megaptera novaeangliae	x		x	Coloration on dorsal pectorals, ventral flukes, and abdomen variable; very young animals may be light colored
Odontocetes				
Physeter catodon	x	x		Variable light areas on lower jaws, ventrum; males 33% larger
Ziphius cavirostris	x	x	x	Variable white on head, variable blaze on dorsal field; tusks (1 pair) in adult males; white on head increases with age, entire head white on older males
Hyperoodon ampullatus		x	x	Males 25% larger, have protuberant melons that are white in older males, 1–2 pairs of tusks (a few females have 1 pair); yellowish/white mottling ventrum/flanks on older animals
Berardius bairdii	x		x	Variable white areas on adults; white on head increases with age
Mesoplodon bidens		x	x	Adult males have 1–2 pairs of tusks; the young are lighter colored on ventrum
Monodon monocerus		x	x	Adult males 1 (rarely 2) spiral tusks; the young are bluish-gray, adults spotted yellowish-gray
Delphinapterus leucas		x	x	Males 12% longer with longer beak, falcate pectorals; the young are slate gray, adults uniformly white, yellowish-white
Orcinus orca	x	x		Variable white underside, flank patch, postocular patch, blazes behind dorsal fin; males to 33% longer, very tall dorsal fin, broader flukes
Globicephala melaena	x	x		Variable light blazes on chin, belly, behind dorsal; males have larger melon, hooked dorsal
G. macrorhynchus		x		Males differentiated as in *G. melaena*
Pseudorca crassidens	x			Variable white streak ventrally

Table 3.2.   (*Continued*)

Species	I	S	A	Comments
Cephalorhynchus hectori	x		x	Variable thoracic patch, flank streaks; dorsal triangular in the young, rounded in adults
Grampus griseus			x	Young are slate gray, adults lighter colored, white head in older adults
Tursiops truncatus	x		x	Beak/head stripes vary, amount of white on ventrum varies; white chin, genital patches enlarge with age
Stenella longirostris		x	x	Females have longer beak; adult males have canted dorsal, prominent keel (except Hawaiian spinner); the young are countershaded, adults spotted
S. attenuata			x	Adults spotted over countershaded pattern of young
S. plagiodon			x	Same as *S. attenuata*
Delphinus delphis	x	x		Pattern of brownish-yellowish thoracic patch varies; males have broader flukes and hooked dorsal fin
Steno bredanensis	x	x	x	Variable scarring, spotting on body; adult males have prominent keel; the young have gray flank blaze, adults may be mottled with yellowish spots
Neophocaena phocaenoides	x		x	Variable diffuse gray or mottled patches on pectorals and genital areas and flukes; light areas decrease with age

of spinner dolphins in the North Pacific Ocean. Adult males of the eastern spinner races have a large ventral keel on the tail stock and a triangularly erect or even forward-canted dorsal fin (Fig. 3.9). The whitebelly spinner male has a smaller keel and a posteriorly recurved (falcate) to triangularly shaped dorsal fin. Finally, the Hawaiian spinner male lacks a ventral keel and has a falcate dorsal fin. The Hawaiian spinner dolphin also lacks the spotting of the other two populations. Perrin suggested that the Hawaiian spinner represented a paedogenic form of the species, retaining in adulthood some of the characteristics of the immature animal. Since the ventral keel and canted dorsal fin are secondary male characteristics, they may function for species recognition in sexual encounters between eastern spinners, which frequently school with the pantropical spotted dolphin (*Stenella attenuata*). The Hawaiian spinner does not school with spotted dolphins, so perhaps has

Fig. 3.9. Variations of ventral keel and dorsal fin shape in populations and sexes of spinner dolphins in the North Pacific Ocean. *Top to bottom*: Male adult eastern spinner, unknown length; male adult eastern spinner, 175 cm; subadult male whitebelly spinner, 164 cm; subadult female whitebelly spinner, 129 cm; and subadult Hawaiian spinner male, 179 cm. (From W. Perrin, *Fish. Bull.* **70** 1972, 983-1003. With permission.)

less need of strikingly different male visual characteristics. In any case its lack of spotting would be sufficient to distinguish between adults of the two species. In frontal view, the two species also have different beak colorations (see Norris & Dohl, this volume, Fig. 5.10).

Age Identification

Several types of visual characteristics may communicate age information, as noted in Table 3.2. Bodily proportions change as young animals mature. Secondary sexual characteristics, such as the ventral keel and canted dorsal fin of the eastern male spinners, are not present in subadults. Spotted species generally have unspotted young. The spotting of *S. attenuata*, for example, develops through several stages, beginning with a countershaded pattern in infancy (Perrin, 1970). The young of the gray whale and narwhal are un-mottled. The belukha presents an even more striking example of color change with age. The young are born slate gray and progress through several stages of lighter coloration to white or yellowish-white at maturity. The narwhal's color also lightens with age, as does that of *Sousa* spp. and of the Amazon bouto. In contrast, some humpback whale calves are born light colored and apparently darken with age (Chittleborough, 1953; unpublished field observations, University of Hawaii). Several species develop areas of white on the head with age, including the beaked whales *B. bairdii* and *Z. cavirostris* and Risso's dolphin (Tomilin, 1967). In the bottlenosed dolphin, the white chin and genital patches enlarge with age (D. Caldwell & Caldwell, 1972).

Many changes in body proportion with age are noted in the literature (e.g., see Tomilin, 1967). The skull of a young common dolphin comprises only one fourth of its total body length, while it comprises one third in the adult. Young killer whales have small pectoral fins, one ninth of their body length as compared to one fifth in adults. In adult harbor porpoises (*Phocoena phocoena*) the distance from the tip of the snout to the insertion of the dorsal fin is about 42% of the total length of the animal, while in juveniles it is much longer.

Sex Identification

Although there is little development of secondary sexual characteristics in many cetacean species, dimorphic characteristics are present or even exaggerated in some species, providing visual cues to gender. The most striking example of size dimorphism is in the sperm whale. In most other species size differences are relatively subtle, and in some cases females may be slightly larger than males (e.g., mysticete whales and beaked whales). Other types of secondary sexual characteristics may occur in the appearance or size of head, appendages, and teeth. Female spinner dolphins have longer beaks than males (Perrin, 1975). The canted dorsal fin and the pronounced ventral keel of male eastern spinner dolphins have already been noted. Adult male rough-toothed dolphins also have ventral keels (Norris, 1967).

Adult male killer whales have an extremely tall dorsal fin and wider flukes than the female, in addition to being as much as 33% longer in body length. The dorsal fin of the male common dolphin is more curved than that of the female, and the flukes are proportionately broader. Male pilot whales are reportedly identifiable by a long-based strongly hooked dorsal fin and a very protrusive melon (Sergeant, 1962). Male dense-beaked whales (*Mesoplodon densirostris*) have a large wedge-shaped prominence extending upward from the lower jaw and covering a portion of the upper jaw (Nishiwaki, 1972). The tip of a large tooth protrudes from this prominence. The prominence in females is much reduced. Similarly, the unique spiraled tusk(s) of the narwhal develops only in the adult male.

Sex-related color differences are not common in cetaceans—suggesting that coloration is more important for species and/or individual recognition—but some seasonal variations have been noted. Norris and Prescott (1961) stated that male common dolphins were more boldly marked in the breeding season than at other times. The bottlenosed dolphin females we have observed and the males observed by D. Caldwell & Caldwell (1977) tend to develop a pink coloration in the abdominal area during certain times of the year. In our experience in Hawaii this has been the spring or early summer, a period when females may come into estrus. On the female the pink coloration may be most highly saturated around the genital–anal vent. Given the insensitivity of the dolphin eye to the red portion of the spectrum (Fig. 3.4), the deep pink areas should appear dark, heightening their contrast with surrounding lighter-pink or white areas and, possibly, signaling sexual readiness.

Individual Identification

Idiosyncratic differences in coloration, dorsal fin shape, natural scars, and patterns of ectoparasites can all contribute to visual recognition of individuals. The detection of such natural "tags" has become important in the socio-biological study of wild groups of dolphins (see Wells, Irvine & Scott, this volume) and killer whales. There also are efforts underway to identify some of the mysticete whale species through individual markings. The ventral surfaces of the flukes in humpback whales (Fig. 3.10) are variably patterned in black, gray, and white, and uniquely identify individuals (Kraus & Katona, 1977). The extent and orientation of the diagonal band of white on the pectoral fins of the minke whale (*Balaenoptera acutorostrata*) vary across individuals (Leatherwood et al., 1976), and the callosity pattern on the head of a southern right whale is one of a kind (Payne, 1976). The unique variations may be useful to the cetaceans themselves for individual identification, supplementing identification information gained through acoustic or other sensory channels (see Herman & Tavolga, this volume).

Fig. 3.10.   Variations in coloration patterns on undersurface of tail flukes of humpback whales in Hawaii. Variations between animals may be subtle or distinct. *Top row*: The same whale photographed in March 1976 (left) and in March 1979 (right). *Middle and bottom rows*: Different whales photographed in March 1977 (center left), March 1978 (center right), February 1978 (bottom left), March 1979 (bottom right). Center left shows additionally a whale with its pectoral fin extended out of water, a commonly occurring behavior. (Photographs by University of Hawaii research team, courtesy L. Herman. Top row photographs: L. Herman left, S. Baker right; Center row: D. McSweeney left, R. Wells right; Bottom row: P. Forestell left, R. Antinoja right.)

Identification by Behavior

The visual components of stereotyped behaviors may be additional cues for species, age, or gender identification. A few illustrative examples can be given. The twisting leap of the spinner dolphin quickly identifies the species to human observers (Norris & Dohl, 1980) and may serve that function for the animals themselves. Pryor (1973) additionally reported what appear to be species-characteristic leaps among Pacific bottlenosed dolphins and Pacific spotted dolphins. Young animals typically perform some behaviors differently than adults. Young bottlenosed dolphins lift their snouts out of water when breathing, while adults expose only the blowhole. Humpback whale calves execute a much less vigorous and elegant breach than do adults (Herman & Forestell, 1977). A sigmoid-shaped courtship posture is adopted by male bottlenosed dolphins (Puente & Dewsbury, 1976; Tavolga & Essapian, 1957). To perform the sigmoidal posture, the male positions himself to one side of the female, within her lateral field of vision, and elevates his head and lowers his flukes.

Communication of Behavioral State

The emotional or motivational state of the cetacean may be communicated through the visual components of some behaviors already described in other contexts. An animal leaping nearby not only reveals its position visually or acoustically, but the form, vigor, and number of leaps may convey an excited state of the animal, express dominance or antagonism, display fitness to a potential mate, or even reveal hunger or a readiness to begin the hunt in the case of cooperative foragers (see Norris & Dohl, 1980). The display of ventral areas may diagnose sexual receptivity and excitement, and the exposure of hidden coloration may be a contact or affiliative signal. A tail slap may be a threat signal not because of the loud retort produced, but because it visually demonstrates or reminds one of the power of the tail as a weapon.

Several additional types of visual displays, mainly gross bodily postures or movements associated with agonism or sexuality, have been described in the literature. D. Caldwell and Caldwell (1972), referring to the bottlenosed dolphin, stated that threat may be expressed by facing another animal and by opening the mouth and exposing the teeth, sometimes with the back arched and the head down. Head-nodding and head-shaking behaviors as a sign of disturbance or aggression have also been observed in the bottlenosed dolphin (Herman & Gordon, 1974). McBride (1940) asserted that a dominant bottlenosed dolphin need only assume a posture like that of an animal preparing to get underway quickly to intimidate a subordinate animal. Bateson (1965) described threat behavior in Hawaiian spinner dolphins as a sudden movement toward another animal, or as a sudden turn of the head with the mouth open. Submissive displays often include visual cues opposite to those used in aggressive threat, such as turning away with the mouth closed and head up (D. Caldwell & Caldwell, 1972).

Aggressive behavior is rare among mysticetes (Norris, 1967). However, as with the odontocetes, tail-slapping may sometimes be a threat signal in humpback whales (see Herman & Tavolga, this volume), as may tail "swishing," a sidewise movement of the tail in water (Herman et al., 1977). A humpback whale may breach close to a vessel that is following it. The behavior is not uncommon, is intimidating, and may be so intended.

The courtship pattern in the bottlenosed dolphin has been described by a number of observers (e.g., McBride & Hebb, 1948; Puente & Dewsbury, 1976; Saayman et al., 1973; Tavolga, 1966; Tavolga & Essapian, 1957). Courtship is characterized by a complex of contactual, visual, acoustic, and possibly, chemical signals. The visual components include display of the white ventral field by both male and female, leaping out of water, rapid swimming, inverted swimming, and the sigmoid posture of the male. It is interesting that the described aggressive display (back arched and head down) and the sigmoid sexual display (tail down and head up) seem to be nearly opposite postures. The display of the ventral field reveals the unique configuration and positioning of the anal and genital slits in male and female.

Some of the precopulatory activities of mysticete whales were described or summarized by Herman and Tavolga (this volume). The display of the ventral surface, the inverted posture, and display or waving of pectorals or flukes are commonly observed in what appears to be precopulatory contexts. All of these behaviors have strong visual components.

Investigators have remarked that the immobility of the facial features of the cetaceans forces reliance on other means for visually signaling emotional or motivational state (e.g., D. Caldwell & Caldwell, 1972) or even on other communication channels, as in Bateson's (1966) theory about digital forms of vocally signaling behavioral state in dolphins. Coloration, bodily postures, movements, and other motor behaviors provide a complex matrix of possibilities for visual signaling of behavioral state, as demonstrated by the examples given. Though there would seem to be little selection pressure in the limited photic environment of the sea for the development of subtle facial display as communication signals, the open mouth displays of the formidable teeth, or head-nodding and head-shaking, may be considered gross facial displays. Also Norris (1965) remarked that exposure of the whites of the eyes accompanies fear in dolphins. Whether this subtle feature is an intraspecific signal is not known.

Much remains to be learned about the communication of behavioral states in cetaceans through any means, let alone vision. However, there are likely to be visual signaling components to many of the social behaviors described in the literature. Care-giving (M. Caldwell & Caldwell, 1967) may be triggered by observing the peculiar posture or immobility of the distressed animal, even in the absence of sound signals. Some of the games of dolphins include imitating another animal, heavily involving the visual system. The differentiation of play chase from aggressive chase, may partly depend on visual metacommunication signals about intention (cf. Bekoff, 1972), such as the vigor or persistence of the chase or on visually recognizing who is doing the chasing. More generally, recognizing by visual or other means who is performing a behavior, or its age or sex class, modulates the interpretation of the behavior and of the underlying emotional or motivational state.

SUMMARY AND CONCLUSIONS

The oceanic cetaceans appear to have adjusted well to the optical constraints of the underwater photic environment and to have exploited most of the visual opportunities. The extent of the optical adaptations is impressive. Sensitivity of the eye has been selectively shifted toward the wavelength absorption characteristics of the aquatic habitat; light-gathering and -enhancement devices have been developed; adjustments to the wide range of intensities experienced in the different water layers and throughout the day/night cycle have been made possible through special retinal and pupillary mechanisms; the field of vision has been greatly enlarged; resolution acuity has been maintained or reevolved in air with emphasis on distant vision, while becoming specialized for near-viewing distances in water; and movement enhancement and contrast enhancement mechanisms seem to have been evolved. Other adaptations at the retinal level and in the visual projection and association areas have been described (e.g., see Dawson, this volume; Dral, 1977), but we are only at the beginning of our understanding of all of these adaptive visual mechanisms, some of which may be unique to the cetaceans.

Equally impressive are the many adaptations promoting visual deception or visual conspicuousness. Conspicuous visual characteristics include coloration and the special anatomical features that identify species, age, and gender, or that enhance the communication of movement information or of other behavior. Conspicuous idiosyncratic variations in coloration, markings, or other features across individuals contribute to individual identification, thus aiding complex social interactions among animals. That these conspicuous characteristics exist implies that visual mechanisms sufficient for their detection and processing have evolved.

The correlations of coloration with feeding behavior and ecology are no less remarkable. Where important, predator camouflage has developed extensively. In some species visual deception seems also to have evolved as an antipredator strategy. There may even be a case of Batesian mimicry in the similarity of coloration of Heaviside's dolphin to that of the killer whale.

Vision can contribute significantly to the biological fitness of the individual cetacean. The perception of environmental features, orientation in space and in time, the performance of coordinated activities within the school, the recognition of individuals and of age and gender, food detection and gathering, predator defense, and the communication of behavioral state may all be assisted by vision or by visual displays. Aerial, as well as underwater, vision contributes to many of these functions.

We are not implying that the functions listed as visually assisted are restricted to that sense, nor that they are necessarily primarily supported by that sense. Neither do we claim to have been exhaustive in the analysis of the functions of vision. We do maintain that vision and the visual world should not be discounted as an important facet of life for many cetacean species. The development of the visual apparatus among the oceanic cetaceans underscores the significance of vision to their life. If the case were otherwise, vision would be expected to regress, as it has among the platanistid dolphins inhabiting turbid river waters.

ACKNOWLEDGMENTS

Preparation of this chapter and some of the research reported was supported by National Science Foundation Grants BNS 77-16882 and 77-24943 to Louis M. Herman. The comments of Jack Hailman on an earlier version of this chapter are greatly appreciated, as are additional comments of Edward Mitchell.

REFERENCES

Adolfson, J. & Berghage, T. (1974). *Perception and Performance Underwater.* New York: Wiley.

Bateson, G. (1965). Porpoise community research: Final report. Naval Undersea Research & Development Center Contract no. N 60530-C-1098.

Bateson, G. (1966). Problems in cetacean and other mammalian communication. In: *Whales, Dolphins, and Porpoises* (K. S. Norris, Ed.), pp. 569–596. Berkeley: University of California Press.

Behan, F. L., Behan, R. A. & Wendhausen, H. W. (1972). Color perception underwater. *Hum. Factors,* **14,** 41–44.

Bekoff, M. (1972). The development of social interactions, play and metacommunication in mammals: An ethological perspective. *Q. Rev. Biol.,* **47,** 412–434.

Berzin, A. A. (1972). *The Sperm Whale.* Jerusalem: Israel Program for Scientific Translations.

Brodie, P. F. (1977). Form, function and energetics of Cetacea: A discussion. In: *Functional Anatomy of Marine Mammals* (R. J. Harrison, Ed.), pp. 45–58. New York: Academic.

Brown, D. H. & Norris, K. S. (1956). Observations of captive and wild cetaceans. *J. Mammal.,* **37,** 311–326.

Busnel, R.-G. (1973). Symbiotic relationship between man and dolphins. *Trans. N.Y. Acad. Sci,* **35,** 112–131.

Busnel R.-G. & Dziedzic, A. (1966). Acoustic signals of the pilot whale *Globicephala melaena* and of the porpoises *Delphinus delphis* and *Phocoena phocoena.* In: *Whales, Dolphins, and Porpoises* (K. S. Norris, Ed.), pp. 607–646. Berkeley: University of California Press.

Caldwell, D. K. & Caldwell, M. C. (1972). *The World of the Bottlenosed Dolphin.* Philadelphia: Lippincott.

Caldwell, D. K. & Caldwell, M. C. (1977). Social interactions and reproduction in the Atlantic bottlenosed dolphin. In: *Breeding Dolphins: Present Status, Suggestions for the Future* (S. H. Ridgway & K. Benirshcke, Eds.), pp. 133–142. Washington, D.C.: U.S. Marine Mammal Commission Report MMC-76/07.

Caldwell, M. C. & Caldwell, D. K. (1967). Intra-specific transfer of information via the pulsed sound in captive odontocete cetaceans. In: *Animal Sonar Systems,* Vol. II (R.-G. Busnel, Ed.), pp. 879–936. Jouy-en-Josas, France: Laboratoire de Physiologie Acoustique.

Chittleborough, R. G. (1953). Aerial observations on the humpback whale *Megaptera nodosa* (Bonnaterre). *Aust. J. Mar. Freshw. Res.,* **4,** 219–226.

Madsen, C. J. (1976). Tests for color discrimination and spectral sensitivity in the bottlenosed dolphin, *Tursiops truncatus*. Unpublished Ph.D. thesis, University of Hawaii.

McBride, A. F. (1940). Meet Mr. Porpoise. *Nat. Hist.*, **45**, 16-29.

McBride, A. F. & Hebb, D. O. (1948). Behavior of the captive bottlenose dolphin, *Tursiops truncatus*. *J. Comp. Physiol Psychol.*, **41**, 111-123.

McFarland, W. N. (1971). Cetacean visual pigments. *Vision Res.*, **11**, 1065-1076.

Mitchell, E. (1970). Pigmentation pattern evolution in delphinid cetaceans: An essay in adaptive coloration. *Can. J. Zool.*, **48**, 717-740.

Mitchell, E. (1972). Whale pigmentation and feeding behavior. *Am. Zool.*, **12**, 655 (Abstract).

Mitchell, E. (1975). (Ed.). International Whaling Commission: Report of the meeting on smaller cetaceans: Montreal, April 1-11, 1974. *J. Fish. Board Can.*, **32**, 889-983.

Mitchell, E. (1978). Finner whales. In: *Marine Mammals of Eastern North Pacific and Arctic Waters* (D. Haley, Ed.), pp. 36-45. Seattle, Wash.: Pacific Search.

Mizusawa, K. (1966). Psychological aspects of underwater vision. In: *Seminar Proc. Underwater Photo-optics*. Santa Barbara, Calif., C-XVI, pp. 1-9.

Muntz, W. R. A. (1975). Visual pigments in the environment. In: *Vision in Fishes. Approaches in Research* (M. A. Ali, Ed.), pp. 565-578. New York: Plenum.

Nemoto, N. (1970). Feeding patterns of baleen whales in the ocean. In: *Marine Food Chains* (J. Steele, Ed.), pp. 241-252. Edinburgh: Oliver & Boyd.

Neuweiler, E. & Mohres, F. P. (1966). The role of spacial memory in the orientation. In: *Animal Sonar Systems*. Vol. I (R.-G. Busnel, Ed.), pp. 129-140. Jouy-en-Josas, France: Laboratoire de Physiologie Acoustique.

Nishiwaki, M. (1972). General biology. In: *Mammals of the Sea: Biology and Medicine* (S. H. Ridgway, Ed.), p. 3-204. Springfield, Ill.: C Thomas.

Norman, J. R. & Fraser, F. C. (1949). *Field Book of Giant Fishes*. New York: Plenum.

Norris K. S. (1965). Trained porpoise released into the open sea. *Science*, **147**, 1048-1050.

Norris, K. S. (1967). Aggressive behavior in Cetacea. In: *Aggression and Defense: Neural Mechanisms and Social Patterns* (C. D. Clemente & D. B. Lindsley, Eds.), pp. 225-241. Berkeley: University of California Press.

Norris, K. S. (1974). *The Porpoise Watcher*. New York: Norton.

Norris, K. S. & Dohl, T. (1980). The behavior of the spinner porpoise *Stenella longirostris* (Schlegel, 1841). *Fish. Bull.*, **77**, 821-849.

Norris, K. S. & Harvey, G. W. (1972). A theory for the function of the spermaceti organ of the sperm whale (*Physeter catodon* L.). NASA Special Publication SP-262, pp. 397-417.

Norris, K. S. & Prescott, J. H. (1961). Observations on Pacific cetaceans of Californian and Mexican waters. *Univ. Calif. Publ. Zool.*, **63**, 291-402.

Norris, K. S., Prescott, J. H., Asa-Dorian, P. V. & Perkins, P. (1961). An experimental demonstration of echolocation in the porpoise, *Tursiops truncatus* (Montagu). *Biol. Bull.*, **120**, 163-176.

Orr, R. T. (1970). *Animals in Migration*. New York: Macmillan.

Payne, R. S. (1976). At home with right whales. *Nat. Geogr.*, **149**, 322-339.

Perrin, W. F. (1970). Color patterns of the eastern Pacific spotted porpoise, *Stenella graffmani* Lonberg (Cetacea, Delphinidae). *Zoologica*, **54**, 135-142.

Perrin, W. F. (1972). Color patterns of spinner porpoises (*Stenella* cf. *S. longirostris*) of the Eastern Pacific and Hawaii, with comments on delphinid pigmentation. *Fish. Bull.*, **70**, 983-1003.

Perrin, W. F. (1975). Distribution and differentiation of populations of dolphins of the genus *Stenella* in the eastern tropical Pacific. *J. Fish. Res. Board Can.*, **32**, 1059-1067.

Pike, G. (1962). Migration and feeding of the gray whale (*Eschrichtius gibbosus*). *J. Fish. Res. Board Can.,* **19,** 815-838.

Pilleri, G. (1974). Side-swimming, vision, and sense of touch, in *Platanista indi* (Cetacea, Platanistidae). *Experientia,* **30,** 100-104.

Pilleri, G. (1977). The eye of *Pontoporia blainvillei* and *Inia boliviensis* and some remarks on the problem of regressive evolution of the eye in Platanistoidea. In: *Investigations on Cetacea,* Vol. VIII (G. Pilleri, Ed.), pp. 149-159. Berne, Switzerland: Brain Anatomy Institute, University of Berne.

Pilleri, G. & Gihr, M. (1972). Contributions to the knowledge of the Cetaceans of Pakistan with particular reference to the genera *Neomeris, Sousa, Delphinus* and *Tursiops,* and description of a new Chinese porpoise (*Neomeris asiaorientalis*). In: *Investigations on Cetacea,* Vol. IV (G. Pilleri, Ed.), pp. 107-162. Berne, Switzerland: Brain Anatomy Institute, University of Berne.

Pilleri, G. & Knuckey, J. (1968). The distribution, navigation, and orientation by the sun of *Delphinus delphis* L. in the western Mediterranean. *Experientia,* **24,** 394-396.

Prince, J. (1956). *Comparative Anatomy of the Eye.* Springfield, Ill.: Thomas.

Pryor, K. (1973). Behavior and learning in porpoises and whales. *Naturwissenschaften,* **60,** 412-420.

Puente, A. E. & Dewsbury, D. A. (1976). Courtship and copulatory behavior of bottlenosed dolphins (*Tursiops truncatus*). *Cetology,* **21,** 1-9.

Rice, D. W. & Wolman, A. A. (1971). The life history and ecology of the gray whale, *Eschrichtius robustus.* Stillwater, Oklahoma: *Am. Soc. Mammal., Spec. Publ. 3.*

Rivamonte, L. A. (1976). Eye model to account for comparable aerial and underwater acuities of the bottlenose dolphin. *Neth. J. Sea Res.,* **10,** 491-498.

Saayman, G. S., Tayler, C. K. & Bower, D. (1973). Diurnal activity cycles in captive and free-ranging Indian Ocean bottlenose dolphins (*Tursiops aduncus* Ehrenberg). *Behavior,* **44,** 212-233.

Scammon, C. M. (1874). *The Marine Mammals of the Northwestern Coast of North America.* New York: Putnam. (Republished in 1968.)

Schevill, W. E. & Lawrence, B. (1956). Food-finding by a captive porpoise (*T. truncatus*). *Breviora,* **53,** 15.

Schusterman, R. J. & Thomas, T. (1966). Shape discrimination and transfer in the California sea lion. *Psychon. Sci.,* **5,** 21-22.

Schwassman, H. O. (1975). Refractive state, accommodation and resolving power of the fish eye. In: *Vision in Fishes. New Approaches in Research* (M. A. Ali, Ed.), pp. 279-288. New York: Plenum.

Sergeant, D. (1962). The biology of the pilot or pothead whale *Globicephala malaena* in Newfoundland waters. *Fish. Res. Board Can. Bull.,* **132,** 1-84.

Sivak, J. G. (1973). Interrelation of feeding behavior and accommodative lens movements in some species of North American freshwater fishes. *J. Fish. Res. Board Can.,* **30,** 1141-1146.

Smith, R. C. & Tyler, J. E. (1967). Optical properties of clear natural water. *J. Opt. Soc. Am.,* **57,** 589-595.

Spong, P. & White, D. (1971). Visual acuity and discrimination learning in the dolphin (*Lagenorhynchus obliquidens*). *Exp. Neurol.,* **32,** 431-436.

Sutherland, N. S. (1968). Shape discrimination in the goldfish. In: *The Central Nervous System and Fish Behavior* (D. Ingle, Ed.), pp. 35-50. Chicago, Ill.: University of Chicago Press.

Tavolga, M. C. (1966). Behavior of the bottlenose dolphin, *Tursiops truncatus*: Social interactions in a captive colony. In: *Whales, Dolphins, and Porpoises* (K. S. Norris, Ed.), pp. 718-730. Berkeley: University of California Press.

Tavolga, M. C. & Essapian, F. S. (1957). The behavior of the bottlenosed dolphin, *Tursiops truncatus:* Mating, pregnancy, parturition, and mother–infant behavior. *Zoologica,* **42,** 11-31.

Thresher, R. E. (1976). Field experiments on species recognition by the threespot damselfish *Eupomacentrus planifrons,* (pisces: Pomacentridae). *Anim. Behav.,* **24,** 562-569.

Tomilin, A. G. (1967). *Mammals of the USSR and Adjacent Countries,* Vol. IX, *Cetacea.* Jerusalem: Israel Program for Scientific Translations.

Van Bree, P. J. H. (1972). On the validity of the subspecies *Cephalorhynchus hectori bicolor* Oliver, 1946. In: *Investigations on Cetacea,* Vol. IV (G. Pilleri, Ed.), pp. 182-186. Berne, Switzerland: Brain Anatomy Institute, University of Berne.

Walls, G. (1942). *The Vertebrate Eye and its Adaptive Radiation.* New York: McGraw-Hill.

Wartzok, D. & McCormick, M. G. (1975). Color discrimination by a Bering Sea spotted seal *Phoca largha. Vision Res.,* **18,** 781-784.

White, D., Cameron, N., Spong, P. & Bradford, J. (1971). Visual acuity of the killer whale (*Orcinus orca*). *Exp. Neurol.,* **32,** 230-236.

Wickler, W. (1968). *Mimicry in Plants and Animals.* New York: McGraw-Hill.

Wood, F. G. Jr., Caldwell, D. K. & Caldwell, M. C. (1970). Behavioral interactions between porpoises and sharks. In: *Investigations on Cetacea,* Vol. II (G. Pilleri, Ed.), pp. 264-277. Berne, Switzerland: Institute of Brain Anatomy, University of Berne.

Würsig, B. & Würsig, M. (1980). Behavior and ecology of dusky porpoises *Lagenorhynchus obscuris,* in the South Atlantic. *Fish. Bull.,* **77,** 871-890.

Yablokov, A. V. (1963). Types of cetacean coloration. *Byull. Mosk. Ob-va Ispyt. Prir. Otd. Biol.,* **68,** 27-41. (Nat. Biol. Fish. Res. Bd. Transl. Ser. No. 1239.)

Yablokov, A. V., Bel'kovich, V. M. & Borisov, V. I. (1972). *Whales and Dolphins.* Jerusalem: Israel Program for Scientific Translations.

Yager, D. & Thorpe, S. (1970). Investigations of goldfish color vision. In: *Animal Psychophysics: The Design and Conduct of Sensory Experiments* (W. C. Stebbins, Ed.), pp. 259-275. New York: Appleton-Century-Crofts.

Zyznar, E. (1975). Theoretical considerations about tapeta lucida. In: *Vision in Fishes. New Approaches in Research* (M. A. Ali, Ed.), pp. 305-312. New York: Plenum.

CHAPTER 4

The Communication Systems of Cetaceans

Louis M. Herman and William N. Tavolga

INTRODUCTION

The cetacean group radiated widely during its evolution, and species currently occupy most of the available aquatic subzones. While many general adaptations for communication appear across the different species, there are also divergent trends reflecting specializations for a particular ecological substrate or for a particular form of social living. This chapter reviews these adaptations and discusses the forms of cetacean communication and their social and ecological correlates, so far as current information allows. The relation of cetacean communication to data and concepts on communication in other taxa is also considered when practicable.

The communication systems of the cetaceans are analyzed as a set of component mechanisms and processes, stretching from the production of information to its reception and analysis by other animals. The sensory and signal-production capabilities and characteristics of cetaceans are considered within the context of the available pathways or channels for information transmission in the sea, and the use of these channels by different species or groups of species is reviewed and compared. Where inferences are possible, the functions of communication are discussed and the complexity of communication evaluated.

Also included is some discussion and evaluation of the likely influences on communication of sociality and dispersion, feeding ecology and feeding strategies, predatory pressure, and mating and parental systems. Ontogenetic factors in communication are also discussed, but the material here is especially sparse.

Both current knowledge and deficiencies in knowledge are noted. The data on cetacean communication are often constrained in scope or reliability because of only limited opportunities for observation and study of wild or captive groups and, not uncommonly, because of limitations in concept. Suggestions for research that might resolve discrepancies among data, clarify concepts, or add new information about communication among cetaceans are given.

The emphasis here on communication as a system contrasts with emphases of others that are mainly concerned with selected issues or with limited features of the communication process: for example, its influence on other animals (Altmann, 1967; Cullen, 1972; Frings & Frings, 1964; Scott, 1968); its value as information (Fuller, 1976; Mackay, 1972; but cf. Dawkins & Krebs, 1978); its adaptiveness (Burghardt, 1970; Marler, 1968; Wilson, 1975); or whether or not it may be intentional or purposeful (cf. Bastian & Bermant, 1973; W. Evans & Bastian, 1969; Mackay, 1972; Menzel, 1971; Smith, 1977, pp. 262-267). Hinde (1972) took note of these variations in emphasis and as an example illustrated (Hinde, 1974) how they may lead to different conclusions about the communication repertoire of a species. He noted that Hinde and Rowell (1962), restricting their definition of communication to adaptive behavior, tallied only 22 different visual displays among rhesus monkeys, while Altmann (1962), using a broader criterion of any social interaction affecting behavior, enumerated 50 visual displays.

The study of cetacean communication has all too often focused on only limited aspects of the communication process or been pointed toward narrow issues. Rarely has the intermesh of species, society, ecology, and communication been considered (but see W. Evans & Bastian, 1969). It is hoped that the material in this chapter and its discussion will contribute toward a broader image of cetacean communication, its functions, and its determining variables.

CHARACTERISTICS OF COMMUNICATION CHANNELS IN THE SEA

W. Tavolga (1968a), referring to fish, summarized the attributes of the channels available for communication in the sea (also see Fine, Winn & Olla, 1977). The useful channels were given as chemical (gustatory and olfactory), mechanical (tactual and acoustic), photic, and electrical. Marine mammals appear to rely on these same channels, with the exception of the electrical, and for cetaceans, the olfactory (D. Caldwell & Caldwell, 1977; Winn & Schneider, 1977).

The use of a channel for communication is likely if the species produces energy in the channel and receiver capabilities for detecting that energy exist. Assessing signal and receiver capabilities is therefore an important initial step in describing the communication system of a species. For cetaceans these capabilities are most uncertain for the chemical modality. However, it would be surprising if the cetaceans had abandoned this oldest of communication channels (Wilson, 1968) during their evolution. Water is an excellent chemical solvent and carrier. Substances released or deposited by an animal may remain detectable for long periods of time, and in favorable currents may disperse over great distances. More commonly, however, the slow rate of diffusion and the weakening of the diffusion gradient with increasing distances from the source restrict the detectability or usefulness of chemical signaling to relatively short distances. Other limitations are that chemical substances are not easily varied in structure or modulated in degree, and as a result chemical communication is typically simple and highly stereotyped.

The remaining communication channels offer their own unique advantages and disadvantages. Photic energy may be highly directional in water and limited to relatively short ranges, depending on turbidity levels or depth. However, visual information can be varied greatly, from simple, largely passive displays of coloration, features, or orientation, to complex forms or sequences of motor behaviors. Hence its information content can be large.

The tactual channel is limited to very short ranges—the length of a pectoral fin or of the caudal peduncle—though water displacement or turbulance produced by a swimming cetacean might be sensed by a school mate more than a flipper or fluke away. Fishes are specialists in such detection through their lateral-line system (Cahn, 1967; Dijkgraaf, 1947), while cetaceans may rely on more conventional types of dermal receptors. Touch signals can be varied in many ways to increase their information content, including how produced, where touched, and the intensity of contact.

The acoustic channel provides the most efficient means for long-range underwater communication, and is useful as well at near distances. Acoustic signals often lack the precise directional component of visual signals, although clicks and some high-frequency sounds can be highly directional. An omnidirectional signal can be of advantage in calling to dispersed schoolmates or in advertising one's presence. Acoustic signals may be of various types, are easily modulated, and given their short life may be patterned sequentially in many ways to greatly increase their information content.

How these channels are actually used by different cetaceans remains to be shown. Both ecological and social forces can be expected to exert strong effects on the exploitation of a channel. The visual channel is largely restricted to coastal or oceanic species inhabiting clearer waters. Also, visual communication may be more richly developed among cetaceans living in large schools than among the more solitary species (cf. Yablokov, Bel'kovich & Borisov, 1972). During acoustic communication highly social species may emphasize shorter-range signals that convey context-dependent information, while more solitary species may emphasize redundant, long-distance signaling with little information content other than direction, distance, and perhaps individual identification. These and other characteristics of cetacean signaling within the different channels are taken up in greater detail later.

CHEMICAL, TACTUAL, AND VISUAL COMMUNICATION

Data Implicating Gustatory Sensitivity and Chemical Communication

Receiver Capabilities

The cetaceans are probably anosmatic, or at best microsmatic, but some may have reasonably good taste sensitivity. Olfactory nerves, bulbs, and peduncles are absent in adult odontocete cetaceans and though present in adult mysticetes, are greatly reduced in comparison with the structures of terrestrial mammals (Jansen & Jansen, 1969; Morgane & Jacobs, 1972). Also there is little olfactory epithelium in the nasal passages of any cetacean (Yablokov et al., 1972). In water the nares are almost always closed, precluding olfaction, and in air the plosive, infrequent cetacean respirations seem ill-suited for olfaction. Though Purves (1967) suggested that mysticetes might "sniff the winds" in search of plankton-rich waters, this idea is not well supported by anatomical or behavioral data.

Waterborne chemicals, detectable by taste, can provide a rich source of social and environmental information for cetaceans, but little is known about central or peripheral taste structures in cetaceans. Yablokov et al. (1972), based in part on work of Kruger (1959), noted that gustatory sensibility can reside in the thalamus and nearby areas of the brain, which are well represented in the cetaceans, and that the distinctive fossae, or small pits, on the root of the tongue of some odontocete cetaceans might contain chemoreceptors. Berzin (1972) observed specialized epithelial cells near the root of the tongue of sperm whales, which he thought were chemoreceptors. Suchowskaja (1972) examined the pits in the tongue of the bottlenosed dolphin (*Tursiops truncatus*) and the common dolphin (*Delphinus delphis*) and reported small "buds" and gustatory papillae of typical mammalian form. Arvey and Pilleri (1970, 1972) also observed gustatory papillae in the common dolphin and, additionally, in the harbor porpoise (*Phocoena phocoena*), belukha (*Delphinapterus leucas*), and Chilean dolphin (*Cephalorhynchus eutropia*). These structures were not seen in the white-beak dolphin (*Lagenorhynchus albirostris,*) the Ganges susu (*Platanista gangetica*) nor, unlike data of Suchowskaja (1972) or Donaldson (1977), in *T. truncatus.* D. Caldwell and Caldwell (1972) cited a histological report suggesting that pits in the bottlenosed dolphin tongue were probably lingual salivary glands and not sensory organs. Hence existing anatomical reports are contradictory for the bottlenosed dolphin but suggest a gustatory capability for some species.

Food Preferences

Taste sensitivity might be more easily assessed behaviorally than physiologically. Some suggestive evidence of taste sensitivity comes from captive animals that often develop preferences for one type of fish, rejecting others or displaying decreased task motivation if offered nonpreferred fish as reward. At times preferences may develop for fish different from that comprising the bulk of the diet in the wild. Kellogg (1961) in fact used the preference of a captive bottlenosed dolphin for spot (*Leistomes* sp.) over its principal food in the wild, mullet (*Mugil* sp.), to make one of the first demonstrations of echolocation capability. He offered the dolphin a simultaneous choice between a mullet, a relatively large fish, and a spot, at night or in turbid water. The dolphin almost always chose the spot. The presence of clicking sounds, together with controls for extraneous cues, strongly implicated echolocation as the means for discrimination. Though the basis for the preference—in contrast to the basis for the discrimination—could have been texture or size, these variables seem less compelling as a primary explanation for food preference than does taste.

Taste Sensitivity

Sokolov and Volkova (1973) reported briefly on an experiment by V. B. Kuznetsov testing gustatory sensitivity in the common dolphin and the harbor porpoise. Different, but unnamed, chemical substances were introduced into a stream of water flowing through the mouth of the test animal. The animal was apparently restrained out of the water, and the galvanic skin response—a change in the electrical resistance of the skin—was used to gauge responsivity to changes in chemical stimulation. Kuznetsov claimed to find changes in skin resistance to the different chemical substances, with the amount of change different for different substances. Additional work on the bottlenosed dolphin by Kuznetsov (1974), described in Bullock and Gurevich (1979), reported discrimination between salt water and 0.04 to 2.5% solutions of valeric acid, oxalic acid, or bentaphenylethyl alcohol in salt water. The solutions were placed on the dolphin's tongue while it held its head out of water with mouth open. Responses were made on a lever, or levers, and correct discriminations, probably indicating the presence or absence of the substance, were rewarded with fish.

Some counterevidence for taste sensitivity comes from uncorroborated reports of Hawaiian fishermen unsuccessfully discouraging bait stealing by dolphins—including Pacific bottlenosed dolphins (*Tursiops gilli*), roughtoothed dolphins (*Steno bredanensis*), false killer whales (*Pseudorca crassidens*), and possibly pygmy killer whales (*Feresa attenuata*)—by lacing the bait with quinine. Also captive Pacific bottlenosed dolphins readily ate fish containing quinine (Shallenberger, 1979). Possibly, these observations merely indicate an absence of specific taste sensitivity for quinine or that the taste is not aversive to these cetaceans.

Little is known of taste sensitivity among mysticete whales. Yablokov et al. (1972) contended that these whales might be able to locate zooplankton concentrations through taste by searching for optimum salinity values for plankton growth, given as 33.5 to 33.9‰ salinity. They stated that this small range of salinity attracts baleen whales and that ocean areas having different salinity values contain only transient whales. The cited optimum salinity range is indeed narrow, and its influence on plankton productivity is debatable (George Losey, personal communication). A more reasonable hypothesis may be that excretions or other chemical substances produced by the zooplankton concentrations themselves, and not the salinity values, provide the important gustatory cues for the whales. Also, surface concentrations of zooplankton can sometimes be seen from aircraft as patches of dark material (Watkins & Schevill, 1976) and may be detectable by the whales visually as well as through gustation.

Chemical Signals

Kuznetsov (1978), also cited in Bullock and Gurevich (1979), studied taste responses of a bottlenosed dolphin to sexual and excretory compounds. Responses (unspecified) were observed to weak solutions of the prostate gland secretion and the excretory gland (perianal gland?) secretion. Kuznetsov believed this supported the hypothesis of chemical communication in dolphins. Berzin (1972) and Yablokov et al. (1972) had earlier suggested a possible signal function for the perianal glands found among male odontocete cetaceans. Ducts from these glands open directly into the water through paired anal pores just anterior to the anal vent and might emit chemical signals sensed through taste. D. Caldwell and Caldwell (1977) suspected as much, but cautioned that the perianal structures might be nothing more than underdeveloped mammary glands.

M. Caldwell and Caldwell (1967) hypothesized that waste products of cetaceans could provide taste cues for other members of their species. Berzin (1972) noted that the volume of the bladder in cetaceans was small relative to body weight, necessitating frequent urination. Accordingly, the presence of a herd of animals in a given area must leave a long-standing chemical trace in the water after their departure. Norris and Dohl (this volume) suggested that trails of both feces and urine deposited by schools of dolphins contained sexual pheromones. They reported that spinner dolphins (*Stenella longirostris*) at times appeared to swim deliberately through the dispersing excrements deposited by schoolmates. A similar observation for wild Pacific bottlenosed dolphins was made from the bow-viewing chamber of an oceanographic vessel by Paul Forestell of the University of Hawaii whale research team.

Whalers have reported that belukhas may suddenly panic and flee on entering areas where conspecifics had earlier been killed or frightened by gunshots (Yablokov et al., 1972). The killed or frightened animals were from different pods than those arriving later and, reportedly, there were no traces of blood in the water at the time of arrival of the new pods. In some of the cases of groups frightened by gunshots no animal had been struck. These descriptions hint at the possibility that the belukha releases a pheromone when alarmed.

Overview

The sum of evidence suggests that gustatory sensitivity, and probably gustatory communication, has been maintained or developed among some cetaceans. D. Caldwell & Caldwell (1977) noted that, unlike olfaction, which apparently gave way to more pressing demands for respiratory control in water, gustation had no comparable pressures forcing its exclusion.

A few data point to alarm or sexual pheromones in some species detected by taste, but these are only indications needing additional study. Other possible social functions of gustatory communication, such as species identification, individual identification, or trail following, should also be considered in further research. For the smaller odontocete cetaceans that can be maintained in captivity some issues might be settled easily by additional studies of taste sensitivity for urine and fecal extracts, glandular secretions, or extracts from prey fish. Much of the Russian work needs replication, and D. Caldwell and Caldwell (1977) have already briefly alluded to one replication failure. Based on the evidence reviewed here, the belukha might be an excellent candidate for gustatory studies. Although Arvy and Pilleri (1970, 1972) failed to find gustatory papillae in the susu, the river dolphins may be additional interesting subjects for study, given their visually limited environment and the dispersion of individuals over wide areas. Has chemical sensitivity possibly been enhanced in riverine species as a compensatory sensory development for visual limitations?

Data Implicating Cutaneous Sensitivity and Tactual Communication

Receiver Capabilities

Anatomical studies of cetacean integument and behavioral observations of responses to tactile stimuli support the idea of a well-developed, broad-ranging cutaneous sensitivity in cetaceans. Several receptorlike structures have been

identified in the integument of those cetacean species studied—mainly the bottlenosed dolphin, harbor porpoise, blue whale (*Balaenoptera musculus*), and fin
whale (*B. physalus*) (Agarkov, Khomenko & Khadzhinskiy, 1974; Ling, 1974;
Palmer & Weddell, 1964; Yablokov et al., 1972). Ling (1974) suggested that the
findings likely applied to most other cetacean species. Some studies (Agarkov et
al., 1974; Simpson & Gardner, 1972) indicate a sensitivity to water flow in the
turbulent boundary layer surrounding the swimming animal. Such sensitivity
may help detect schoolmates swimming nearby and preadapt the species for tactual communication through direct bodily contact.

Free-nerve endings, which in other mammals may function as mechanorecep-
tors, pain receptors, or thermoreceptors (Schmidt, 1977), are densely packed in
the skin of the bottlenosed dolphin, especially around the snout, nipple, genital
regions, and parts of the pectoral fin (Palmer & Weddell, 1964; also see Bullock
& Gurevich, 1979). These body parts are all areas of frequent contact among
dolphins. Some free-nerve endings are encapsulated with structures resembling
Pacinian corpuscles, a type of acceleration detector sensitive to vibration
(Schmidt, 1977). Recent analyses demonstrate that structures such as Pacinian
corpuscles may only indirectly produce sensations through the way they
modulate the responses of the free-nerve endings (Mountcastle, 1974). Hence it
may not necessarily follow that Pacinian corpuscles or other somatosensory
structures yield sensations for the dolphin corresponding to those found in some
terrestrial mammals.

The mysticete whales have short, immobile, but highly innervated vibrissae or
bristles along the jaws and portions of the head (Ling, 1977; Tomilin, 1967;
Yablokov et al. 1972). The number may range from 50 or so in some rorqual
species to 250 in the bowhead whale (*Balaena mysticetus*) (Slijper, 1962;
Yablokov et al., 1972). In the humpback whale (*Megaptera novaeangliae*)
bristles sprout from nodules spaced along the head and jaw. Bristles, if functional as somatosensory structures, may be sensitive to low-frequency vibrations
produced by water movement (Palmer & Weddell, 1964), or help in detecting
prey organisms through touch (Lillie, 1915; Ling, 1977). They may also play a
role in tactual communication, as discussed later.

Among the odontocetes vibrissae are present throughout life only in the river
dolphins (Yablokov et al., 1972). Their functions, if any, may overlap those
considered for the mysticete whales. Layne and Caldwell (1964) suggested that
the visbrissae of the Amazon bouto (*Inia geoffrensis*) might aid in the detection
of freshwater molluscs or other food items stirred up from the silty bottom with
the snout. In the bottlenosed dolphin and in the spotted and spinner dolphins
(*Stenella* spp.) vibrissae may remain until the end of the nursing period and are
then lost. The hair pits along the rim of the upper jaw remain well expressed
throughout life, however. The sperm whale and the pilot whales (*Globicephala*
spp.) retain vibrassae only while embryos, and the narwhal (*Monodon
monoceros*) and belukha lack vibrissae at any life stage. Roughly, there seems
to be a positive correlation between the degree to which vibrissae persist through

various life stages and the slimness and length of the upper and lower jaws, which in turn reflects feeding ecology.

Responsiveness to Touch

Any who have handled a living cetacean can attest to its sensitivity to touch. A dolphin out of water on a stretcher may flinch visibly when a few drops of water or even a light spray falls on its back. Many captive species appear to solicit stroking from their handlers, though some species, such as the Amazon bouto, are more responsive than others (see Defran & Pryor, this volume). Most dolphins seek self-stimulation. Bottlenosed dolphins may expose various body parts, including the opened mouth, to a stream of water from a suspended hose or rub against a brush attached to the tank floor for long periods of time. Both male and female dolphins will masturbate frequently by rubbing their genitals against objects in the tank, even in the presence of available mates (M. Caldwell & Caldwell, 1972; McBride & Kritzler, 1951). Tactile stimulation can also be an effective reinforcer in training situations with bottlenosed dolphins (Defran & Milberg, 1973) and with the bouto (Braden, Conklin & Friedman, 1973).

Affiliative and Sexual Contact

Among mysticete whales, the bristles on the jaw or the head itself may be used to rub the head or body of another whale. Head-rubbing of this type may contribute to simple affiliation, be used as a greeting, or be a component of the courtship process. Head-rubbing has been reported among bowhead whales (Marquette, 1978), southern right whales (*Balaena glacialis*) (Cummings, Fish & Thompson, 1972; Payne, 1976), gray whales (Norris, Goodman, Villa-Ramiriz & Hobbs, 1977), and humpback whales (Herman & Forestell, 1977).

Head-rubbing is less common among odontocetes, at least for those species with pronounced beaks. Rubbing of the flanks or head of another with the snout has been reported in killer whales (*Orcinus orca*), in what was surmised to be play or precopulatory behavior (Voisin, 1972). Head- or melon-butting has been observed in bottlenosed dolphins (M. Caldwell & Caldwell, 1967; Puente & Dewsbury, 1976) and pilot whales (Norris & Prescott, 1961) in the context of precopulatory behavior.

Extensive contact with pectorals, flukes, dorsal fin, or trunk during affiliative, play, or precopulatory behavior is common among captive dolphins (Bateson, 1965; Brown & Norris, 1956; Brown, Caldwell & Caldwell, 1966; D. Caldwell & Caldwell, 1972; Essapian, 1962; Evans & Bastian, 1969; McBride & Hebb, 1948; M. Tavolga, 1966; M. Tavolga & Essapian, 1957; Puente & Dewsbury, 1976). It has also been seen among wild specimens (Norris & Dohl, 1980; Saayman, Tayler & Bower, 1973; Saayman & Tayler, 1979). Intersexual contact may be gentle, mainly nudges and strokes of the fins and genitals and mild tooth-raking (M. Caldwell & Caldwell, 1977). Male-male sexual contacts may also be gentle and male-male sexual play has been observed frequently among dolphins in captivity (Bateson, 1965; M. Tavolga, 1966) and in the wild (Norris & Dohl, 1980).

Scammon (1874) observed humpback whales during the mating season rubbing and patting each other with their large pectoral fins while lying side by side. More recently, Herman and Forestell (1977) reported instances of humpback whales in Hawaiian breeding waters lying on their side with pectorals extended, gently contacting another animal. In two cases a whale swam between the outstretched pectorals of another, which appeared to stroke it from head to tail. During an interaction between a humpback whale and a rare North Pacific right whale, observed for some 14 minutes in Hawaiian waters, the humpback patted and stroked the back and sides of the right whale with its long pectorals and nudged the flanks and genital region with its head (Herman, Baker, Forestell & Antinoja, 1980). These were judged to be precopulatory behaviors of the humpback but were rewarded with only a single brief ventral presentation by the otherwise passive right whale.

Contact in Dominance and Aggression

Aggression by mature male dolphins toward subordinate animals is common (M. Caldwell & Caldwell, 1977; Norris, 1967), and may include not only tooth-raking and biting but ramming another with the jaw and striking with the powerful tail. Slijper (1962) and Norris (1967) observed scarring from tooth marks among many delphinids, with young males often appearing to be the major recipients of attack. Tomilin (1967) noted that male common dolphins were extensively scarred on trunk and fins during the mating season and that males have more teeth missing than females. The same seems true for sperm whales, in which male-male competition in apparent dominance contests includes extensive tooth-raking (Best, 1979) and reportedly may escalate to violent head-ramming and locking of the jaws by two animals while twisting and biting (Berzin, 1972; Caldwell, Caldwell & Rice, 1966). Kritzler (1952) described attacks by three male bottlenosed dolphins on a young short-finned pilot whale (*Globicephala macrohynchus*) that were severe enough to break several ribs. Milder forms of aggressive contact are more often the rule, however. In competition for access to food or in response to observing a handler stroking another animal one dolphin may displace another from a favorable position by settling on top of it. In some extreme cases one or several dolphins may pin another to the tank floor (M. Caldwell and Caldwell, 1967; M. Tavolga, 1966). In early May 1978 newspapers carried the story of a male killer whale, at Marineland of the Pacific in California, pinning its trainer to the tank bottom for some four minutes and nearly causing her death.

In the wild, vertically organized swimming arrangements have been seen among groups of pantropical spotted dolphins (*Stenella attenuata*) in Hawaiian waters, and Pacific white-sided dolphins (*Lagenorhynchus obliquidens*) and

Pacific bottlenosed dolphins in California. W. Evans and Bastian (1969) hypothesized that these were dominance arrangements, expressed as priority of access to the surface. It is also common to see dolphins jockeying with one another for favorable positions in the bow wave of a vessel (Norris & Dohl, 1980). Through the underwater viewing bubble of a large oceanographic vessel underway in the Northwest Hawaiian Islands, one of us (LMH) observed rough-toothed dolphins bodily displacing one another from the near-central bow position. In one instance, a large animal placed its mouth about the tail stock of another, causing it to leave the apparently favored position. Leatherwood and Reeves (1978) described apparent intergeneric hierarchies in bow-wave riding among common dolphins, bottlenosed dolphins, and Pacific white-sided dolphins ("lags"). The lags displaced the common dolphins and were in turn displaced by the bottlenosed dolphins. Displacement occurred through one species bodily forcing another species into less favorable or deeper positions on the bow and then driving them away entirely.

Little in the way of agonism has been reported among mysticete whales (Norris, 1967). However, in feeding areas groups of humpback whales may sometimes compete physically for access to food, pushing and shoving one another to get closer to concentrations of surface prey (Watkins & Schevill, 1979). "Escort" whales accompanying mother–calf humpback whale pairs may occasionally display intense aggression toward other humpback whales approaching the escorted pair, as described more fully in the later section on development.

Mother–Young Contact

Among both wild and captive dolphins mother and calf commonly swim in close echelon formation. The calf may at times touch the mother with its pectorals or the underside of its body (M. Caldwell & Caldwell, 1972; Norris & Dohl, 1980). The young calf typically remains above and to one side of the mother, perhaps riding the compression wave generated by the adult's swimming movements (Norris & Prescott, 1961), or drafting along in areas of decreased water resistance in the boundary layer surrounding the mother's body. Assisted swimming of these types can conserve energy in the calf, and contribute importantly to its survival (Prescott, 1977).

In both mysticete and odontocete species the calf may be physically retrieved by the mother when it strays or sheltered with the mother's body, as well as by other adults, when danger threatens (M. Caldwell & Caldwell, 1966; Herman & Antinoja, 1977; Leatherwood, 1977; Norris et al., 1977; M. Tavolga, 1966; M. Tavolga & Essapian, 1957; Payne & Payne, 1971, Pl. 1). During nursing the jaw, tongue, and palate may provide tactual information on the location of the mammary clefts and on the production of milk. The extensive contact of mother and calf gives evidence of the strong social bonding between the pair, which may be in part developed and maintained through touch.

Overview

The examples of contactual behavior together with the accumulating material on cutaneous sensitivity suggest that tactual interchanges are an important component of the communication systems of cetaceans. The apparent stereotypy of some tactual behaviors, such as head-rubbing, head-butting, and genital "probing," suggest specializations for a signal function. It would be helpful for further studies to quantify the incidence of contactual behavior and document its context. Puente and Dewsbury (1976) carried out a simple but informative quantitative study of courtship behavior in bottlenosed dolphins, which included comparisons of the frequency of occurrence of selected types of contactual behaviors on observation days on which copulation occurred and on days when it did not. There was a sizeable increase in the frequency of contactual behavior—as well as in other apparent courtship behaviors—on days when copulation occurred. Herman and Forestell (1977) found that the relative incidence of contactual behavior of humpback whales in the Hawaiian breeding grounds increased as the number of animals in the social group (pod) increased.

Psychophysical studies of sensitivity to low-frequency vibrations, touch, pressure, or other cutaneous stimuli would help clarify the somatosensory capabilities of the cetaceans and the potential of the tactual channel for communication. Studies of the type, duration, and location of tactile stimuli that are effective reinforcers could also add to these goals. An interesting question is whether tactile stimulation applied artificially to the dolphin will modify the frequency of occurrence of agonistic or sexual behaviors (cf. Pepper & Beach, 1972).

Visual Capability and Communication

The form and utility of visual communication can vary widely among cetacean species, depending on the photic properties of their habitat, the development of their visual system, and their visual appearance (see Madsen & Herman, this volume). The visual mechanisms, capabilities, and characteristics of selected cetacean species are discussed extensively elsewhere in this volume (Dawson; Madsen & Herman) and need not be repeated in detail here. Briefly, the oceanic dolphins appear to have good photopic and scotopic acuity and to be highly sensitive to those wavelengths dominating the euphotic zone. Some species may be adapted toward the photic characteristics of the deeper layers. All oceanic species have a wide field of vision and, probably, most have excellent movement detection capabilities. Brightness differences are readily detected by bottlenosed dolphins, but form, at least abstract form, is not easily discriminated (Herman, this volume). Color vision is probably weak or absent. Aerial as well as underwater vision is well-developed in some species. However, there seem to be no instances of a cetacean species specialized primarily for aerial viewing, even among those living in turbid waters. Apparently the major ecological pressures have been toward underwater vision, with aerial acuity developing secondarily if underwater vision was biologically adaptive. Mysticete species probably have visual attributes similar to the odontocetes, though less is known about them.

Many cetacean species are conspicuously colored in contrasting shades of black, gray, brown, and white. The display of coloration or the uncovering of hidden coloration under flukes or pectorals may signal movement intentions, identify species, individuals, or classes of individuals, and indicate reproductive state. Body shape, movement, or posture also provide visual signals for these functions and for others.

What appears to be visual signaling has been observed most commonly in sexual and agonistic contexts. The S-shaped posture of a male bottlenosed dolphin during courtship is probably a visual signal, as is the display of the light-colored ventral surface in this species (D. Caldwell & Caldwell, 1972; M. Tavolga, 1966; Puente & Dewsbury, 1976). Facing another animal, lunging toward it, opening and closing of the jaws, and shaking of the head may indicate aggression or threat, while facing away may signal appeasement or subordination (Bateson, 1965; D. Caldwell & Caldwell, 1972). Showing the whites of the eyes may indicate fear (Norris, 1967).

Further long-term observational efforts are needed to corroborate the interpretations of these visual behaviors and to identify others used in social contexts (see Madsen & Herman, this volume). Some forms of visual signaling, such as those involved in simple affiliation, the recognition of associates, or the maintenance of the tightly coordinated group-swimming formations of dolphins, are likely to be subtle and difficult to detect. One elementary approach to studying visual recognition of individuals might be to temporarily apply paint or some similar substance to change body coloration, such as by adding white patches or obliterating white patches. In a well-established cetacean community would there be changes in responses to familiar individuals reintroduced with altered coloration?

In natural settings perhaps the best that can be done at present is to continue to correlate apparent visual cues with behavior and ecology within or across species. For example, do boldly marked animals tend to maintain positions relative to one another that favor the display of coloration? What is the relation of coloration to schooling tendencies (cf. Yablokov, et al., 1972)? Do closely related sympatric species show greater coloration differences than closely related allopatric species (cf. Mitchell, 1970; Perrin, 1972)? What is the relative frequency of occurrence of behaviors having apparent visual components, such as inverted swimming or ventral display, during breeding and nonbreeding seasons?

Further work on visual mechanisms and visual psychophysics is necessary to understand visual capabilities better. Studies of capabilities for detecting real and apparent movement (see Hochberg, 1971) might be especially informative, but at this early stage of knowledge about the cetacean visual system good work on any visual topic would be welcome.

AUDITORY COMMUNICATION IN ODONTOCETES

Receiver Capabilities

The peripheral and central sound production and hearing mechanisms are highly developed among the odontocete cetaceans (see reviews of W. Evans & Maderson, 1973; Jansen & Jansen, 1969; Kinne, 1975; Morgane & Jacobs, 1972; Norris, 1968; Popper, this volume). Audiograms from several odontocete cetaceans indicate sensitivity over as many as 10 octaves (see Popper, this volume), and tests of bottlenosed dolphins reveal excellent pitch discrimination capabilities (Herman & Arbeit, 1972; R. Thompson & Herman, 1975), good resolution of auditory duration (Yunker & Herman, 1974), precise sound localization ability (Renaud & Popper, 1975), and impressive competency in discriminating among different sounds (Herman & Arbeit, 1973) and in remembering their attributes (Herman & Gordon, 1974; R. Thompson & Herman, 1977). These capabilities contribute to orientation and food finding through echolocation or passive listening and also provide the raw material for auditory communication. The exceptional precision of the dolphin in discriminating frequency change may be worth noting if its acoustic communication at all resembles that of many birds and terrestrial mammals that employ frequency change as a key variable in intraspecific calling (J. Brown, 1975).

The extraordinary development of the temporal lobe is probably an expression of the exceptional hearing capabilities of the cetaceans (Jansen & Jansen, 1969). However, the auditory areas have not yet been fully identified (see Russian work cited in Bullock & Gurevich, 1979; also Bullock & Ridgway, 1972). If one compares dolphin with bat brain it is evident that the central structures in dolphins are more than sufficient for support of simple hearing and echolocation and may allow for forms of auditory information processing or communication not available to bats (see Herman, this volume, for further discussion).

Nonvocal Auditory Communication

Nonvocal auditory communication may include the noise of body parts striking the water surface as well as the sounds of percussive jaw claps or of bubble emissions. Norris and Dohl (1980) reported that Hawaiian spinner dolphins reentering the water after a spin, breach, or back-slap, or while head-slapping or tail-slapping, generate omnidirectional noise that propagates over short to intermediate distances. These behaviors seemed constituted to produce noise since they were most common at night, when visual contact is limited, and in the daytime occurred in fully alert but dispersed schools. The amount of leaping may index the level of arousal of individuals, or of the school as a whole. Leaping can be "infectious" and spread rapidly from a few individuals to many, and may help recruit animals for the night's foraging activity or for social activity.

Würsig and Würsig (1979) drew similar conclusions about the functions of leaping in schools of dusky dolphins (*Lagenorhynchus obscurus*). The sounds of the dusky dolphins reentering the water after a leap carried at least 500 m but less than 1 km underwater. In air, the sounds were heard as far away as 3-km distance. Leaps occurred both before and after feeding. The Würsigs hypothesized that prefeeding leaps herded fish or recruited other dolphins to assist in feeding activities (cf., Saayman et al., 1973), while postfeeding leaps established contact with neighboring groups, perhaps conveying a readiness for affiliation.

Leaps may have visual as well as acoustic functions. Madsen and Herman (this volume) suggested that the sight of leaping animals conveys position information to schoolmates and that leaps allow aerial visual inspection for food cues or for environmental features. Würsig and Würsig (1980) have proposed similar hypotheses.

Most observers agree that tail-slaps convey threat or accompany frustration (e.g., D. Caldwell & Caldwell, 1977), in addition to establishing contact. The loud report of a "jaw-clap" has long been regarded as a threat signal (McBride & Hebb, 1948) and an escalation of aggressive behavior may follow if the animal is not appeased or counterthreatened.

Underwater bubbles from the blowhole produce sound and bubbles sometimes accompany the whistles of dolphins (M. Caldwell & Caldwell, 1971), but the significance of these behaviors as communication is untested. Explosive respirations or "snorts" produced in air are associated with disturbance (D. Caldwell & Caldwell, 1977). Altogether, the social functions of nonvocal auditory signals seem mainly limited to affiliation, recruitment, or expressions of excitement, annoyance, or aggression.

Vocal Communication

Types of Vocalizations

The vocalizations of the odontocete cetaceans are of several types and may be very different across species (see reviews in W. Evans, 1967, 1973; Kinne, 1975; Norris, 1969; W. Tavolga, 1968b; Yablokov et al., 1972; also Popper, this volume). The different sounds may be dichotomized broadly into pulsed or unpulsed (continuous) forms. Some pulsed sounds are implicated in echolocation and may be of broad-spectral composition, as in the bottlenosed dolphin (W. Evans, 1973), or of narrow-band composition, as in the narwhal (Ford & Fisher, 1978). Other pulsed sounds, described picturesquely in the literature as barks, squawks, squeaks, blats, groans, moans, and so forth (e.g., D. Caldwell & Caldwell, 1977), may have a social function. The unpulsed or continuous signals are most commonly called whistles. Typically, whistles are unmodulated pure tones or frequency-modulated pure tones limited to the mid- to upper-sonic range of frequency (e.g., 5 to 15 kHz), and of 0.5 to 2 sec in duration. The duration of a whistle can vary from less than 1 sec to several seconds. Also, an animal may repeat a whistle over and over again for long periods of time as well as emit only a short partial whistle.

Research on vocal communication in bottlenosed dolphins and other delphinids has centered on the whistle sounds for largely pragmatic reasons. The sonic range of the whistles is easily recorded and analyzed. Also, whistles are produced by the most common captive species, the bottlenosed dolphin, and appear to have no function other than communication. Such emphasis tends to suggest that whistles are the primary mode of acoustic communication, a supposition that may not be justified. Table 4.1 lists 19 species of odontocete cetaceans, representing five different families, including the family of porpoises Phocoenidae, that do not, or probably do not, whistle. The list might expand further if additional ziphiids or other species of *Cephalorhynchus* eventually proved to be nonwhistlers. The number of nonwhistling species is relatively large and implies that it may be premature to regard whistles as the principal means for vocal communication among odontocetes. For those species without whistles, vocal communication must necessarily take place through pulsed sounds, and for the remaining whistling species it may be that pulsed sounds help carry an important segment of the vocal communication load.

Communication by Pulsed Sound

Other investigators have at times suggested that pulsed sounds may be used in communication (e.g., Busnel & Dziedzic, 1966; Bastian, 1967; D. Caldwell & Caldwell, 1977; M. Caldwell & Caldwell, 1967; Lang & Smith, 1965; Lilly & Miller, 1961a; Norris, 1969). However, there has been little follow-up on the idea. The directional characteristics of many of the pulsed sounds, the relative ease with which they can be localized, their variability, and possibly the power with which they can be produced enhance their potential value as communication signals.

Observers of captive animals have in some cases described contexts in which pulsed sounds appear. A few examples can be given, although the observations are often based on limited data. During precopulatory or copulatory behavior a "yelp" has been heard from male bottlenosed dolphins (M. Tavolga & Essapian, 1957; M. Tavolga, 1966; Puente & Dewsbury, 1976; Wood, 1953) and "squeaking" or "grinding" sounds, as well as a sound resembling a "blast," from the nonwhistling harbor porpoise (Busnel & Dziedzic, 1966). Broadspectrum "alarm or fright" signals were recorded from harbor porpoises newly introduced into a tank and continued to be heard thereafter whenever humans approached (Busnel & Dziedzic, 1966). "Squawks" and "squeaks" were heard from Atlantic spotted dolphins (*Stenella plagiodon*) during periods of excitement associated with training (D. Caldwell & Caldwell, 1971a).

In situations described as alarm, fright, or distress, broadband high-intensity "cracks," "pops," or "squeaks" have been heard from bottlenosed dolphins (M. Caldwell & Caldwell, 1967; M. Caldwell, Haugen & Caldwell, 1962).

Table 4.1. Known or Suspected Nonwhistling Odontocete Species[a]

Family/Species	Source
Platanistidae (river dolphins)	
Inia geoffrensis	Norris et al. (1972)
Platanista gangetica	Andersen & Pilleri (1970)
P. indi	Pilleri, Kraus & Gihr (1971)
?*Lipotes vexillifer*	No data
Physeteridae (sperm whales)	
Physeter catodon	Watkins (1977)
Kogia breviceps	D. Caldwell, Prescott & Caldwell (1966)
?*K. simus*	No data
Phocoenidae (porpoises)[b]	
Phocoena phocoena	Busnel & Dziedzic (1966)
?*P. sinus*	No data
?*P. dioptrica*	No data
?*P. spinipinnus*	No data
Phocoenoides dallii	Evans (1967)
?*Neophocaena phocaenoides*	D. Caldwell & Caldwell (1977)
Delphinidae (dolphins)	
Cephalorhynchus heavisidii	Watkins, Schevill & Best (1977)
?Other *Cephalorhynchus*	Watkins, Schevill & Best (1977)
Orcinus orca	Schevill, Watkins & Ray (1969)
?*Feresa attenuata*[c]	Cf. Evans (1967); Poulter (1968)
Lagenorhynchus australis[d]	Schevill & Watkins (1971)
?*Peponocephala electra*[e]	Unpublished observations (LH)
Ziphiidae (beaked whales)	
Mesoplodon densirostris[f]	D. Caldwell & Caldwell (1971b)
?Other ziphiids[c]	Cf. Poulter (1968); D. Caldwell & Caldwell (1977)

[a]The family Phocoenidae is listed separately, though current systematics favors placing it with the Delphinidae. Question marks indicate uncertainty about vocalizations.

[b]D. Caldwell & Caldwell (1977) say all Phocoenidae are nonwhistlers.

[c]Contradictory statements in literature.

[d]All other species of *Lagenorhynchus* are known to whistle (Schevill & Watkins, 1971).

[e]No underwater sound production to 20 kHz heard in two captive specimens at Sea Life Park, Oahu, but extensive whistling heard from a group of over 100 wild animals identified as probably *Peponocephela* but possible *Feresa,* or both (mixed herd), at the island of Hawaii.

[f]Speculative. Based on in-air recordings of one stranded specimen. Poulter (1968) refers to unpublished recordings of ziphiids which do not include the descriptive category "whistles."

"Squawks" were heard during "play-chase" among bottlenosed dolphins, as well as during agonistic encounters (M. Caldwell & Caldwell, 1967). Squawks have also appeared in agonistic exchanges among captive Amazon bouto. The "buzz" of an echolocation-type sound may occur when a bottlenosed dolphin faces another in apparent threat or during other "emotionally intense" situations (Lilly, 1961; Lilly, 1962).

Among wild animals, Busnel and Dziedzic (1966) recorded "creaking," "blast," or "squeaking" sounds, as well as whistle sounds during pursuit of pilot whales at sea. Watkins (1977) and Watkins and Schevill (1977) recorded temporal sequences of broad-spectrum clicks, or "codas," from sperm whales. Each whale seemed to produce an individually stereotyped and unique coda. This "signature" quality allowed human listeners to easily identify individual emitters and, presumably, the whales could do the same. Codas were heard when a group of whales dove and when whales met underwater. Most often an exchange of codas occurred among closely spaced whales, but among distantly spaced animals an overlap of codas was common. Watkins and Schevill (1974) also observed apparent exchanges of pulse-bursts (not clicks) among wild Hawaiian spinner dolphins, a species which has a whistle sound. In the narwhal, another whistling species, patterned burst-pulse tones have been heard that seem unique to individuals and, like sperm whale codas, may be individual signatures (Ford & Fisher, 1978). Signature characteristics have also been proposed for the whistles of some delphinid species (see later). In general, signature calls are not uncommon among social vertebrates, particularly birds (Brooks & Falls, 1975).

In summary, data from both wild and captive specimens, for both whistling and nonwhistling species, suggest that pulsed sounds may have a significant role in vocal communication. In some cases the same sounds appear in different contexts and in consequence may have several meanings (cf. Smith, 1965, 1977). Overall, however, the data are too sparse to allow for firm conclusions about social context and social function.

Sociality of Whistling vs. Nonwhistling Species

Table 4.1 raises the interesting question of why some species, or even whole families, whistle, while others do not. If the whistle appears only under certain ecological or social conditions these should be educible through careful comparisons of whistlers and nonwhistlers. Although not addressing this question directly, Watkins, Schevill, and Best (1977) pointed to similarities in body shape, habitat, and behavior of the nonwhistling Heaviside's dolphin (*Cephalorhynchus heavisidii*) of the southern hemisphere and the nonwhistling harbor porpoise of the northern hemisphere. Both species have short snouts without a definite beak, a low dorsal fin, and are coastal in habitat. Both were called "timid" animals by Watkins et al. They suggested that the vocal similarities across these species reflected responses to similar ecological niches, though it was unclear how the sounds produced in common were adaptive for that niche.

One feature characterizing many species in Table 4.1 is a relatively low degree of gregariousness. Individuals of these species are often found alone or in small groups of a few animals. This characterization holds for the river dolphins (Kasuya & Aminul Haque, 1972; Layne, 1958), the harbor porpoise (Amundin & Amundin, 1973), probably Heaviside's dolphin given that its behavior is otherwise similar to the harbor porpoise, the finless porpoise (Tomilin, 1967), and the pygmy sperm whale (Mitchell, 1975). Also, Rice (1978) reported that beaked whales (ziphiids) do not aggregate in large schools but form small pods of 3 to 17 animals.

In contrast, many of the whistling species at least occasionally assemble in large herds of dozens, hundreds, or even thousands of animals. Examples are the genera *Stenella, Tursiops,* and some *Lagenorhynchus*, the family Monodontidae (narwhal and belukha), and the common dolphin (Tomilin, 1967; also see summary in Wells et al, this volume). The relationship between whistling and typical aggregation size is imperfect, and there are several overlaps in degree of sociality across whistling and nonwhistling species. For example, killer and sperm whales each occasionally form aggregations of over 100 animals, melon-headed whales have been observed in large groups, and members of the genera *Stenella* and *Tursiops* and of the Monodontidae are at times found in small subgroups. But in general the large-school, communal foraging species seem to be the whistlers and the more solitary species the nonwhistlers. The relationship cannot be pushed too far in the face of incomplete knowledge of school structure and sound production but the apparent correlation merits further study. Comparisons of the social structure of the nonwhistling *L. australis* (Schevill & Watkins, 1971) with others of its genus that do whistle might be especially informative.

Is it reasonable to suppose that whistle sounds have evolved for some special function in the large schools? As one possibility, whistle sounds may have value during communal foraging. Of relatively low frequency, whistles may carry for longer distances in water than do pulsed sounds (Norris & Dohl, 1980). Although less directional than pulsed sounds, whistles nevertheless are probably easily localized by cetaceans, judging by Renaud and Popper's (1975) demonstration that constant-frequency pure tones can be localized by the bottlenosed dolphin with an accuracy of 2° to 4°, depending on frequency. Given these attributes whistles provide a potential vehicle for maintaining vocal communication during food search by echolocation, since bottlenosed dolphins and probably other whistling species can produce whistles and clicks simultaneously (W. Evans & Prescott, 1962; Lilly & Miller, 1961a). Also, whistles have little overlap with the major portion of the echolocation frequency spectrum, minimizing any masking effects. It seems possible, then, that a herd of animals communally foraging by echolocation concurrently transmits supplementary vocal information through whistles. If whistles have species or regional specificity (Graycar,

1976) or individual specificity (M. Caldwell & Caldwell, 1965), this would in the least allow for identification of herd members or familiar associates or aid in the assembly of dispersed animals and in the coordination, spacing, and movements of individuals in rapidly swimming communally foraging herds.

Additional Contexts and Correlates of Whistling

Many different stimuli may elicit whistling or are associated with it. Whistling may appear as a simple phonoreaction, in response to hearing another animal's whistle (Lang & Smith, 1965; Lilly & Miller, 1961b) or a recording of a whistle or other sound (Dreher, 1966; Dreher & Evans, 1964; W. Evans & Prescott, 1962; Morgan, 1971). Mimicry of whistles or of artificial sounds may also occur, revealing the plasticity of the sound-production system (see review of mimicry in Herman, this volume).

Some have attempted to assign specific whistles to particular environmental or social contexts. While many different whistle forms have been described (see later section), the relationship of a particular whistle form to a characteristic circumstance has not been conclusively demonstrated. Commonly, however, changes in rate, duration, or intensity of given whistles, rather than changes in form, accompany changes in context.

Whistles tend to appear or to increase in rate under excited or stressful states. Fully alerted or traveling schools, groups of dolphins meeting, the arrival at a familiar area, the anticipation of feeding or feeding itself, and bow-wave riding are some of the conditions that may induce or increase the amount of whistling. Increased whistling may also occur in unfamiliar surroundings, or when animals are separated from familiar individuals, or stranded, captured, or injured. Other kinds of stressful situations involving confrontations with predators or encounters with strange objects may suppress rather than increase whistling, as discussed in a later section on adaptive silence.

The data in support of the generalizations about excitement- or stress-induced whistling are few but seem consistent. Several examples can be given. Norris and Dohl (1980) observed that whistling rates were high among groups of wild Hawaiian spinner dolphins first entering the shallow bays that were their late morning and afternoon resting place. As the groups settled down to the patterned slow-swimming movements of their resting phase whistle vocalizations ceased, though occasional clicks could still be heard. The end of the rest cycle was signaled by leaps and zigzag swimming and by a resumption of whistling. The groups then moved out of the bays to launch their nightly feeding foray on the scattering-layer organisms in the deeper offshore waters. Norris (1974) reported that on one occasion whistling suddenly increased as a group moving out of the bay encountered another alert, leaping group. According to Norris fully alert traveling schools are generally very noisy.

Dolphins riding the bow wave of a large vessel appear excited and at such times high rates of whistling are commonly heard from Hawaiian spinner dolphins (Norris & Dohl, 1980), common dolphins (Busnel & Dziedzic, 1966), and Pacific bottlenosed and rough-toothed dolphins (LMH, personal observations). Dolphins riding the bow wave of humpback whales in Hawaii are a common sight (unpublished field observations, University of Hawaii), and whistling has been reported in those cases also.

A correlation of whistling with feeding has been noted among wild animals. Dreher and Evans (1964) reported whistling from a feeding group of approximately 30 short-finned pilot whales, and from a subgroup of common dolphins within a larger group of 100 to 150 individuals. Busnel and Dziedzic (1966) also heard whistling from feeding common dolphins.

Captive animals may also show food-related increases in whistling. Powell (1966) noted an early morning peak in whistling of a captive female bottlenosed dolphin that was partially synchronized with a period of feeding. When all feeding was shifted from daytime to night, the locations of the peaks changed from that established earlier. M. Caldwell and Caldwell (1968) observed that the heightened whistling preceding and accompanying feeding decreased markedly in the postfeeding interval. Saayman, Tayler and Bower (1973) reported infrequent whistling from captive Indian Ocean bottlenosed dolphins (*Tursiops aduncus*) in the early morning but greatly increased whistling during the midmorning and mid- and later-afternoon periods when feedings took place. Slow swimming and resting predominated at night, but no sound recordings were made. Russian investigators have also noted the increase in whistling rate of bottlenosed dolphins at feeding times (Titov & Nikolenko, 1975, cited in Bullock & Gurevich, 1979).

Recordings of two young bottlenosed dolphins held together in a tank at the University of Hawaii were made during a 24-hour period one month after their arrival at the facility (D. Richards, unpublished data). The recordings, taken for five minutes each hour, showed little or no vocal activity throughout the night, in keeping with the diminished activity level of the animals. Whistles and echolocation sounds appeared shortly after sunrise and continued variably throughout the day with some of the peaks, but not all, coincident with feeding activity. One year later, the recordings were repeated. At this time, all of the whistling was conicident with feeding or with cues signaling that a training session was in preparation. Training was always accompanied by feeding.

The stressful situations eliciting whistling have at times been extreme. Busnel and Dziedzic (1968) recorded whistles from harpooned bottlenosed dolphins, common dolphins, long-finned pilot whales, and striped dolphins (*Stenella coeruleoalba*). W. Evans (1967) described regular, periodic whistling from specimens of short-finned pilot whales, pantropical spotted dolphins, and rough-toothed dolphins captured at sea and maintained or tethered near the capture vessel.

Less extreme but still stressful circumstances may also elicit whistles. Captive bottlenosed dolphins newly introduced into a tank or temporarily separated from a familiar tank-mate may whistle nearly continuously. A mother separated from her young may also whistle nearly continuously (McBride & Kritzler, 1951), as may an animal placed on a stretcher or stranded in shallow water (M.

Caldwell & Caldwell, 1965, 1968, 1971). Titov and Nikolenko (1975, cited in Bullock & Gurevich, 1979) claimed there was no increase in whistling rate among captive animals caught or being transported. Possibly there animals were well accustomed to transport so that stress was greatly diminished. Generally, captive animals well habituated to their surroundings show decreased whistling rates relative to the early period of their introductin to the setting (M. Caldwell & Caldwell, 1967).

Some of the vocalizations heard in highly stressful contexts have been termed "distress whistles," and are ostensibly species-characteristic. Lilly (1963) described the distress whistle of the bottlenosed dolphin as an upsweep in frequency followed by a downsweep, while Dreher and Evans (1964) characterized it as only a short upsweep whistle. The bottlenosed dolphin harpooned by Busnel and Dziedzic (1968) emitted a series of nearly constant-frequency whistles with brief frequency-descending and frequency-ascending segments attached, respectively, to the beginning and end portions of each whistle. These different whistle forms, together with data considered in later sections, cast doubt on the validity of a species-specific distress whistle.

As a general conclusion, most evidence supports the notion of an increased whistling rate during excitement or stress. Excitement and stress, though psychological concepts, can potentially be measured by accompanying changes in levels of physiological arousal.

Adaptive Vocalization and Adaptive Silence

If threatened by a predator or confronted by new strange stimuli many cetacean species suppress vocalization and fall silent. Whether or not silence greets a given predator seems to reflect the experience of the prey population with that class of predator, implying that silence has a learned component.

Playbacks of killer whale vocalizations, mainly "screams," to Hawaiian spotted dolphins have no apparent effect on their behavior (James Fish, personal communication). Killer whales, inhabitants of temperate and subpolar regions, rarely visit Hawaiian waters. In contrast, playbacks to wild herds of belukhas (Fish & Vania, 1971), gray whales (Cummings & Thompson, 1971b), or southern right whales (Cummings et al., 1972), all known to be occasional prey of killer whales, caused one or more of the following: the cessation of vocalization, fleeing, or the initiation of aerial visual search ("spy-hopping"). On the face of it the vocalizations of the hunting killer whale seem maladaptive if they alert the prey and are not used for echolocation, as seems the case for the scream vocalization. If killer whales hunt largely by passive listening as some suppose (Diercks, 1972; Mate, 1975), their screams might kindle panic in the prey and "flush" it out to the overall advantage of the predator. Mate (1975) postulated this type of predatory strategy for killer whales hunting pinnipeds in near-shore waters. An effective prey counter strategy, then, is to remain still and silent, unless detected. The belukha, dubbed the "sea canary" because of its vociferousness, quickly falls silent when killer whales approach or when hunted by man (Schevill, 1964), though it may also flee the area. Captive bottle-nosed dolphins often become silent when first confronting new, strange objects introduced to their home tank (M. Caldwell & Caldwell, 1967). Typically the dolphins bunch together, warily regarding the object while swimming rapidly past it at some distance. Eventually the interest in the object seems to outweigh fear and the animals approach closer, perhaps briefly interrogating the object through sound and then darting away rapidly. With time, approaches become more frequent, and finally the object may be touched and manipulated. After the initial period of silence whistles may resume gradually (D. Caldwell & Caldwell, 1977), although periods of whistles interspersed with periods of silence may persist for some time (Dreher & Evans, 1964).

There seems a contradiction here between the descriptions of animals falling silent in these fearful predatory situations and other descriptions of intense vocalization during capture or restraint, situations that are also highly fearful. But once captured the animal's camouflage of silence is unmasked and whistling or other vocalizing to attract others of its species becomes the last line of defense. Generally, then, a progression of defensive behaviors may take place in fearful situations: silence (avoidance), clustering, fleeing (escape), possibly fighting if attack is unavoidable, and finally vocalizing in distress.

Portions of this described defensive sequence are illustrated in a provocative report of "scouting" behavior among a group of wild Pacific bottlenosed dolphins (W. Evans & Dreher, 1962). To test for gray whale echolocation, Evans and Dreher strung a barrier of vertically suspended air-filled aluminum poles across the mouth of an inlet to Scammon's lagoon in Baja California, a well-known gray whale winter calving area. The poles were spaced at approximately 15-m intervals. Some time later, five bottlenosed dolphins approached the inlet and apparently sensing the poles at some 350-m distance, stopped and retreated into nearby shoal water, forming a tight school. Their sound activity was described as "very sparse." Approximately five minutes later, a lone animal left the group, made a run on the barrier, echolocating all the while, and returned to the group. A great deal of whistling ensued. This sequence of sonar run, return, and whistling was repeated two to three times before the group finally left the shoal water and proceeded cautiously through the barrier and up the channel.

Rather than concluding that this was a case of "scouting" and "reporting back" through an elaborate whistle code (Dreher & Evans, 1964), the behaviors may instead index the initial defensive sequence of silence when first confronting fearful stimuli and clustering. Two other factors then seem to occur: habituation to fear and the influence of a dominant individual, the "scout." Dominance systems are common among wild and captive bottlenosed dolphin groups (W. Evans & Bastian, 1969; M. Tavolga, 1966; Tayler & Saayman, 1972; also, Wells, et al. this volume), and dominant individuals can greatly influence group behavior. Among wolves, baboons, or rhesus monkeys—

all of which live in groups and form social hierarchies—the movement of the group may await the movement of a dominant animal or animals, especially in times of stress or danger (Chance & Jolly, 1970; Kummer, 1968; Mech, 1970). Dolphin "scouts" may simply be the cetacean analog of dominant individuals exercising control over important group movements.

A brief report of scouting behavoir in dolphins by M. Caldwell et al. (1965) lends support to the interpretation of the behavioral sequence in the Evans and Dreher report as a progression of defense mechanisms, fear habituation, and the influence of dominant individuals. Caldwell et al. described the capture of bottlenosed dolphins in Florida waters by use of nets strung across the mouths of shallow dead-end bays entered by the dolphins. In one case an entrapped group hesitated before the net blocking their exit. Shortly later, the second largest male left the others, scouted the net, and returned to the group. The largest male, followed by a large female, then charged the net, so that the defensive sequence culminated in attack. The animals were sized and sexed after capture, and the individuals described were recognized by markings and fin shapes. This same defensive sequence occurred in other groups as well, and sometimes the animals successfully broke through the net or spilled over its top.

The Signature Character of the Whistle

Disputes about a signature function for the whistle of bottlenosed dolphins or other whistling species of delphinids often seem to confound two issues. One is whether there is in fact a signature component to the whistle—some acoustic character that uniquely identifies one animal to another—and the second is whether the whistle is anything more than a signature. The whistle may indeed contain a signature component, but it may also convey additional information, or assist in such conveyence. For example, the Russian investigators Burdin, Reznik, Skornyakov and Chupakov (1974, cited in Bullock and Gurevich, 1979) reported that an isolated bottlenosed dolphin, linked acoustically through hydrophones and sound projectors to another bottlenosed dolphin in a second tank, simultaneously produced whistles and narrow-band clicks that were harmonically related to the whistles. Burdin et al. postulated that the whistles were identification signals, used for establishing contact between animals, and that the superimposed clicks carried the main supplementary information load. This situation, if valid, is reminiscent of that observed among some songbirds, in which an initial relatively low-frequency song component propagating over long distances establishes contact, while more complex, shorter-range sounds that immediately follow carry the information load (Richards, 1978).

The earlier work of the Caldwells (M. Caldwell & Caldwell, 1965, 1968, 1971; M. Caldwell, Caldwell & Miller, 1973) emphasized the stereotyped character of the whistle of individuals (also see A. G. Tomilin, cited in Bullock & Gurevich, 1979) and tended to suggest that the whistle functioned solely as identification. They have also contended that the so-called distress whistle is nothing more than a variant of the signature call. Later writings (D. Caldwell & Caldwell, 1977) seem to have recognized that variations or elaborations of the individual's basic whistle form may carry information beyond identification. Most of the Caldwells' data on stereotypy have come from in-air recordings of captive animals stranded in low water during tank cleanings or held temporarily on stretchers. In-air recordings bypass the problem of identifying individual emitters from among a group of in-water animals, but at the same time these constrained situations may exaggerate the degree of stereotypy that would obtain under more natural conditions.

The conclusive demonstration of a signature function for whistles or for other sounds of cetaceans ultimately rests on showing an appropriate recognition response by the animals to different whistles. Demonstrating that dolphins can learn to discriminate among recordings of whistles of conspecifics or even of other species (M. Caldwell, Caldwell & Hall, 1973; M. Caldwell, Hall & Caldwell, 1971a, 1971b) is not sufficient evidence, since dolphins are quite facile at learning to discriminate between many different arbitrary sounds (Herman & Arbeit, 1973) or even between the items of a list of as many as six different sounds (R. Thompson & Herman, 1977). A better model may be the experimental demonstration of different responses to sounds of "neighbors" and "strangers" as in the case of territorial songbirds (Brooks & Fall, 1975). The birds ignore the familiar songs of neighbors, but react strongly agonistically to songs of strangers approaching their territory.

A reinterpretation of some findings by Fish and Lingle (1977) provides an interesting demonstration in dolphins of apparent selective responsivity to calls of "neighbor" and "stranger," and comes close to fulfilling the requirements of an experimental test for a signature component in the whistle. A large male pantropical spotted dolphin, apparently a dominant animal in a herd resident in waters off Oahu, Hawaii, was captured and his nearly continuous whistle vocalizations recorded in air aboard the capture vessel. One week later the recordings were played back to the same Oahu herd and consistently produced flight reactions. However, when played back to a herd of the same species resident on the distant island of Hawaii these same sounds elicited approach, not alarm. Individual animals at times even touched the underwater speaker. These results suggest that alarm in spotted dolphins may be conveyed by atypical variations, perhaps prolonged duration or a greatly increased rate of emission (cf. Busnel & Dziedzic, 1968), in the well-known call of familiar and possibly dominant individuals. These same variations in a stranger's call may have no impact because the normal "baseline" characteristics of the stranger's call are not known or because the call itself is not identified with a group member. Alternatively, it may be that different populations have unique alarm calls common to all members of the group, with calls culturally evolved through convergent vocal mimicry (see discussion of mimicry in Herman, this volume). These alternative hypotheses might be distinguished by playing back whistles of different captured herd members, noting which produce alarm, and comparing their acoustic properties. Whichever hypothesis is correct, variations in familiar sounds or local convergent mimicry, learning must play a vital role in interpretation of the sounds.

Size and Characteristics of Whistle Repertoire

SIZE OF REPERTOIRE.   There have been several attempts to inventory the whistle repertoire of wild or captive cetaceans. This can be difficult. Although sophisticated devices such as sonographs or real-time spectrum analyzers can faithfully describe the frequency versus time function ("contour") of a whistle sound, it is left to human judgment to decide when one contour is significantly different from another. There are no data on the reliability of these judgments nor any guarantee that a judged distinction has functional significance for the animals (cf. Busnel, 1966). An additional problem is that it is often difficult to know which animals produced which contours when recording from a group. If there are uniquely individual variants to the whistle, then the total repertoire obtained from a group will include these variants and spuriously extend the estimated size of the species' repertoire.

Efforts to estimate the size of the whistle repertoire have at times produced widely varying results between and even within species. M. Caldwell and Caldwell (1968) obtained only 5 different contours from over 1400 whistles produced by 4 captive common dolphins. Dreher and Evans (1964) identified 17 different contours among 9 captive bottlenosed dolphins, *T. truncatus*; 16 in a group of five wild Pacific bottlenosed dolphins, *T. gilli* (of which 9 overlapped those of *T. truncatus*); 19 in a migrating herd of common dolphins (11 overlapped those of *T. truncatus*); and 7 in a pod of approximately 12 short-finned pilot whales (all overlapping those of *T. truncatus*). Dreher and Evans also recorded some 23 different whistle contours from 3 captive juvenile bottlenosed dolphins but indicated that some of the contours classified as unique may actually have been minor variants of one another. Dreher (1966) listed 20 different contours emitted by 6 captive bottlenosed dolphins in response to objects introduced into their tank. Also, Taruski (1976) described some 30 different contours among wild Atlantic long-finned pilot whales (*Globicephala melaena*), although he grouped the contours into only 7 broad classes.

To circumvent the difficulties in determining which animal in a group produced which sounds some have isolated a pair of animals in separate tanks while allowing acoustic contact between them through hydrophones and sound projectors. Alternating exchanges of vocalizations are common in such cases (Lilly & Miller, 1961b). Lang and Smith (1965), using this "acoustic link" approach, obtained only 6 uniquely different whistle contours during exchanges between 2 bottlenosed dolphins. The earlier-cited work of the Russian team headed by Burdin (cited in Bullock & Gurevich, 1979) identified 8 types of whistles from acoustically linked bottlenosed dolphins. As was noted, pulsatile sounds were heard together with the whistles, but they were also heard apart from them. Other Russian acoustic link work has reported various numbers of different whistles. Bullock and Gurevich, after reviewing these studies, concluded that the size of the vocal repertoire in the bottlenosed dolphin, including both the whistles and the nonecholocation pulsatile sounds, was probably limited to "much less than 40" distinct types. Earlier Busnel (1966) reached a roughly comparable estimate.

While there may be much more to the whistle than its signature function, there seems to be no evidence that the whistle repertoire, if regarded as a set of discrete, functionally different signals, is unusually large. Many of the social primates, for example, have double-digit vocal repertoires. Van Lawick-Goodall (1968) described 24 classes of vocal signals among wild chimpanzees, a species that may not even rely on vocalizations for the majority of its signaling, and Green (1975) differentiated 37 functionally differentiated vocal signals in Japanese macaque monkeys. Busnel (1966), summarizing data of others, listed 15 vocal signals for the gibbon, and among nonprimate mammals noted 23 vocal signals for the pig and 36 for the fox.

The total repertoire of signals of the bottlenosed dolphin or of other delphinid species extends beyond its whistles or even other vocalizations as was illustrated in the earlier discussions of gustatory, tactual, and visual communication. The size of the total repertoire cannot be estimated at this early stage in the analysis of cetacean communication but it is unlikely that it exceeds that of the nonhuman primate or the paralinguistic repertoire (signals other than spoken language) of the human. Wilson (1975, p. 556) estimated the human paralinguistic signal repertory as close to 200 elements. This number, Wilson felt, slightly exceeded the total repertory for rhesus monkey or chimpanzee and greatly exceeded that of the majority of birds and nonprimate mammals.

WHISTLES AS GRADED SIGNALS.   Possibly the whistle sounds comprise a system of graded rather than discrete signals. In a graded system several "basic" types of signals are connected to one another through a series of intermediate forms. The intermediate forms, as well as the basic types, may carry information. The vocalizations of some primate species have been thought to form a graded system—for example, macaque monkey (Rowell & Hinde, 1962; Green, 1975), and chimpanzee and gorilla (Marler, 1973, 1976). Marler (1965, 1973) contended that graded vocal systems characterized those primates, like chimpanzee or rhesus macaque, that spend significant periods of time on the ground, often in open environments where visual contact is easy. In these species variations in vocal signals may be combined with visual signaling or visual context to provide shades of meaning. In contrast, in primarily arboreal dispersed species, such as red-tailed monkeys, visual contact is difficult because of dense foliage, and vocal signals are discrete, without overlap or gradation. Arboreal species often communicate over long distances; the unfavorable signal/noise ratios at these long distances and the unavailability of supportive visual signaling mandate the use of crisp unambiguous auditory signals.

Thus, the constraints of ecology and the spacing of individuals may influence the discrete or graded character of primate vocalizations. The whistles of dolphins might be analyzed from the same perspective. The minor differences in structure that have been referred to and some of the variants in the signature whistles of individuals may index an underlying graded system. Taruski (1976) in fact described the whistles of wild short-finned pilot whales as a graded series. He identified seven main whistle types according to their physical properties but showed that they were connected to one another through a subtle series of intermediate forms. Bastian (1967) also recognized the graded character of the hundreds of whistles produced by a male and a female bottlenosed dolphin maintained in the same tank and arbitrarily divided the whistles into 20 classes.

In those delphinid species that form large herds and that seem to be the "whistlers" (cf. Table 4.1) a graded vocal communication system might prove feasible. In closely schooling herds, supporting contexual information from tactual or visual signals or even from superimposed pulsatile signals is readily available and can form a basis for interpretation of whistle intergrades. Predictions of the degree of gradedness among whistlers might be made from the degree of species sociality, the distance over which communication signals typically travel to a receiver, and the availability of signals other than whistles as supplementary contexual cues. All of this assumes that the analysis of gradedness for primates is correct and that the principles of gradedness can be extended to taxa as divergent from primates as cetaceans.

The Question of Language

The relationship of human language to other forms of animal communication is uncertain and controversial. Questions include the continuity of human language and other forms of animal communication, whether language evolved from primitive nonverbal communication, and whether the capacity for language is unique to humans. The special features of human language, the universality of its major characteristics throughout diverse linguistic groups, its emergence in the individual under fairly prescribed developmental rules, and the special physical and cognitive structures of humans that relate to and support language suggest to most linguists that language is unique to humans (e.g., Lenneberg, 1964, 1967; also see discussions in Nottebohm, 1975; Lyons, 1972). The successes in teaching imposed languages to great apes (Gardner & Gardner, 1971; F. Patterson, 1978; Premack, 1971; Rumbaugh, 1977) appear to demonstrate, however, that some language prerequisites reside in the close relatives of humans. Preliminary work in language comprehension by bottlenosed dolphins, as described by Herman (this volume), suggests that this species may mimic the apes in its ability to learn some language fundamentals. But, like the apes, there is no convincing evidence for true language in the natural communications of dolphins. Over the years, the hypothesis of a natural language in dolphins, vigorously promoted by Lilly (1961, 1967), has failed to receive analytic, experimental, or conceptual support. Lilly's pronouncements are largely behind the popular myth of "talking" dolphins. Partly through uncritical publicity, partly through public gullibility, and partly through poor science, the myth persists today despite many attempts to lay the ghost to rest (Andrew, 1962; Sebeok, 1963; Wilson, 1975; Wood, 1973). Wilson (1975) puts the matter bluntly, "... Lilly's books are misleading to the point of bordering on irresponsibility (p. 473)."

The search for a natural language in dolphins has been both by the cryptographic approach and by experimental behavioral study. The cryptographic method attempts to analyze the information content of dolphin vocalizations, the presumed vehicle for a natural language, if one existed. Even if the language code cannot be discovered, given a natural language the information content of the vocalizations, measured by its diversity or variance, should be exceptionally high (cf. Moles, 1968). The experimental behavioral approach, used by Bastian (1967), places the dolphin in a learning task in which a natural language, with its capability for transmission of symbolic, arbitrary information, seems necessary for problem solution, and then studies performance.

It is hardly worthwhile mentioning a third approach, the serendipitous method. Here, a natural language is assumed and dolphin and human are placed together for long periods of time in the hope that the enforced socialization will somehow break through the language barrier (e.g., see the report of Margaret Howe in Lilly, 1967). Though a strong social relation was developed between dolphin and human, Howe achieved practically no success in teaching human sounds to the dolphin during 2½ months of sharing an enclosure with it or in interpreting any of the dolphin's natural vocalizations.

Information Content of Natural Vocalizations

Earlier sections indicated that no unusual degree of variability or complexity was present in dolphin vocalizations relative to that exhibited by other highly social mammals. The study of Dreher (1966) warrants additional comment, however, because it does claim an exceptionally high level of information content for the whistle vocalizations. Dreher introduced six artificially generated whistles into a tank shared by six bottlenosed dolphins and recorded their responsive whistling. The number of different whistles heard in response to each introduced whistle ranged from five to eight, with some overlap. Altogether 20 different whistles were produced, including six during a pretest period, though some seemed minor variants of one another. Dreher then made a misleading comparison of the information content of the responsive whistles with the statistical information content (Shannon & Weaver, 1949) of letters of the English alphabet. Letters have an average information value of 2.07 bits if one accounts for their relative frequency of occurrence in written English text, while the responsive whistles averaged a somewhat higher 2.17 bits. But it is words or their constituent semantic elements, morphemes—not letters—that are the semantic units of language. Words combined together under the rules of syntax give language an imposing versatility and power. In spoken language strings of words in ordinary discourse convey information at an average rate of 18 bits/sec (unpublished data by Quastler and Wulff cited in Attneave, 1959). In contrast, the dolphins, phonating at a rate of 0.1 whistles per animal per second according to Dreher's data, were transmitting only 0.2 bits/sec.

Human language may be differentiated in part from other forms of animal communication by its flexible syntactical structure, allowing meaning to be modulated through variations in phonemic or word order (Nottebohm, 1975). In this context Busnel (1966) provided an interesting and useful comparison of whistle languages, as used by certain mountain people, and the whistles contours of dolphins. Although an analogy was possible, Busnel emphasized that the level of syntax in the two groups of whistles was so different that the analytic techniques were not interchangeable. It should be added that the differences in origin are even more significant. The human whistle languages are derived syntactically, culturally, and even phonetically from spoken language and, like sign language, are extensions of the characteristic linguistic capacities of humans.

Examination for Properties of Language

Under the controls available in the laboratory one may ask whether an animal is capable of demonstrating important features of language, as identified in analyses of human language. What is "important" is open to interpretation. Many, but not all, of the "design features" of language (Hockett, 1960; Hockett and Altmann, 1968; also see Thorpe, 1972) are found in the natural communication systems of animals. Some design features seem trivial in that they refer to physical rather than conceptual properties of language, such as the use of the vocal-auditory channel or the rapid fading of signals that characterize all auditory communication. Conceptual features differentiate more meaningfully between human language and other forms of animal communication. Examples are the openness of language, which allows new arbitrary words to be added to the vocabulary or new sentences to be devised for conveying new meaning; displacement, or the ability to refer to events remote in space and time; and reflexiveness, or the ability to communicate about the language system itself (metalanguage). It may be argued whether displacement is shown by bees when communicating the locations of distant food resources through their dance, or openness demonstrated by birds when vocally mimicking other species, but these both seem unusual specializations which the animals cannot generalize to other contexts.

Natural communication by dolphins does not seem to have been analyzed from a design-feature perspective, in part because there is so little information to work with. However, Bastian and his colleagues (Bastian, 1967; Bastian, Wall & Anderson, 1968) wondered whether the capacity of language for creating arbitrary symbols to represent new events or objects—the feature of openness—might be revealed in dolphin communication if a situation were structured requiring it.

In the Bastian (1967) study, food reward for each of two bottlenosed dolphins, a male and a female, was made contingent on an apparent cooperative transmission of information between the pair. The two dolphins were housed in the same tank but were physically separated by a net drawn across the center of the tank. An opaque visual screen additionally separated the pair during the final stages of training and testing. To obtain reward each dolphin had to select the correct paddle from a pair located in its half of the tank, and both dolphins had to select correctly for either to receive reward. The training procedure required the female to press her paddle before the male pressed his. With the opaque screen in place, the female could rely on acoustic cues from the male's paddle press to time her response.

The correct paddle was signaled by an out-of-water "cue" light. If flashing, a response on the left-hand paddle of the pair was correct, and if steadily lit a press on the right was required. A cue light was initially present in each half of the tank but as training progressed the male's light was gradually moved into the female's sector and then replaced entirely by the female's cue light. The opaque screen was then gradually drawn across the tank until the cue light on the female's side could no longer be seen by the male. At this stage the male was totally dependent on some form of acoustic information from the female about the state of the cue light or which paddle to press. The transmission of such information would seem to imply an ability to create acoustic symbols to represent something as arbitrary as the state of the light, a feat easily attainable within an open language system but which would seem beyond the closed signal systems of other forms of animal communication.

The provocative findings were that the male responded to the correct paddle nearly flawlessly trial after trial although he could not see the cue light, and that the female was heard to vocalize at the start of most trials. As W. Evans and Bastian (1969) noted, but cautioned against, the results tempt the conclusion that the female was "knowingly" transmitting acoustic information to the male about the state of the cue light or the correct paddle—that is, that in addition to developing symbols for transmission of such information, she understood the need for transmission, its value to her and to the male, and its consequences. In theory, the extensive training procedures used by Bastian might have allowed for such developments.

Additional tests by Bastian et al. (1968) indicated, however, that this conclusion was not warranted. For example, the female's pretrial vocalizations continued when the visual screen was withdrawn even though the male was now able to see the cue light again, and they persisted after the male was removed entirely from the tank. Also, a reversal of the paddle response required for each cue light condition, though initially learned readily by each animal, disrupted joint performance considerably. Finally, an attempt to reverse the roles of the two animals by providing the cue light to the male rather than to the female was successful only after extensive retraining. In a language system senders and receivers ought to be interchanged fairly easily.

A search for the characteristics of the sounds emitted by the female that were controlling the male's responses yielded correlations between the presence and characteristics of echolocation pulses by the female and correct responding by the male, but recordings of these sounds were unsuccessful in controlling the male's responses. A replication of Bastian's (1967) study at Harderwijk dolphinarium in Holland in 1972, while producing results similar to Bastian's, suggested that the position of the female while emitting, whether she faced toward the male's side of the tank or not while vocalizing, keyed the male's responses (W. Dudok van Heel, personal communication). Sound amplitude or source location cues can be altered by the female's position, providing useful auditory information for the male. The study of role reversal by Bastian et al. (1968) also showed that the female, now acting as the receiver, cued acoustically on the male's position in the tank, which changed according to the paddle to be pressed.

W. Evans and Bastian (1969) concluded that the initially high performance level of the animals and the subsequent high performance level after extended practice with the reversed responses to lights were due to adventitious independent learning by each animal. It seems most probable that the "sender" developed different stereotyped postures during the vocalizations that preceded responding to left or right paddle, while the "receiver" learned to key on the different acoustical spatial cues produced.

Boakes and Gaertner (1977) have recently analyzed the Bastian paradigm as a special case of "autoshaping" (Brown & Jenkins, 1968). They concluded that sufficient conditions for successful responding are that the sender develop some "superstitious" behavior pattern in the presence of one signal, a common occurrence in animal learning studies, and that this behavior be detectable by the receiver. The stereotyped behavior of the sender then comes to control the response of the receiver. Boakes and Gaertner further demonstrated that "cooperative" signaling and high performance levels similar to that of the dolphins in the Bastian study was attainable by pairs of pigeons. From these results and analyses it is clear that we cannot conclude that any natural language feature was present in the transactions between the two dolphins.

Summary

The social functions of the various sounds produced by odontocete cetaceans are still in an early state of description. Not all odontocete species produce the narrow-band pure-tone "whistle" sound that has so often been implicated in vocal communication among bottlenosed dolphins. Nonwhistling species are necessarily constrained to the use of pulsatile signals for vocal communication and it may be that even the whistling species use pulsatile signals extensively in vocal communication.

Those species having whistles are typically highly social and often form large herds, while those without tend to be solitary or are found most commonly in small groups. Whistles may have value for maintaining contact or spacing among the members of large herds of dolphins. Whistles propagate well in water, are not masked by echolocation noise, and can be produced contemporaneously with echolocation sounds, so that cooperative foraging by echolocation and social communication through whistles can occur together.

Whistles often appear or increase in repetition rate during periods of excitement or stress. Unique variations in whistle sounds among individuals may be useful for individual identification, though data for this "signature" function are still limited. A signature function has also been suggested for the unique pulsed patterns produced by different sperm whales and for the patterned burst-pulse tones of narwhals.

Attempts to catalog the different whistles produced by a species into discretely different forms have generated lists of frequency versus time "contours" that in many cases seem only minor variants or elaborations of one another. Even at its best the number of different contours cataloged is not unusually large when measured against the vocal repertoires of many other highly social mammals. It was suggested that whistle sounds might be better described as a graded series rather than as discrete signals.

The acoustic link studies and some of the field data described for sperm whales and spotted dolphins demonstrate that exchanges of vocalizations are common between animals, one animal apparently listening while the second is vocalizing. Most workers have properly resisted implying that these exchanges were "dialog," though some have not. Exchanges of vocalizations or of other acoustic signals and simple phonoreactions are common among many animals, including both invertebrate and vertebrate species (Andrew, 1962; Busnel, 1966).

The concept of an extant natural language in dolphins, with properties akin to the more complex "design features" of human language, is unsupported by analytic or experimental evidence, though the idea remains a part of the dolphin popular mystique and is hard to extinguish. There is no solid evidence for any unusual degree of complexity (information content) in the whistle sounds of dolphins, which are presumed to have an important communicative function. Also there was no support in an experimental study for the capability of one dolphin to intentionally transfer information to another about arbitrary environmental events. It was assumed that a true language, with its open, flexible structure, would have allowed for such transfer. Still under study, however, is the degree to which dolphins may be successfully tutored in language fundamentals in the laboratory (see Herman, this volume).

AUDITORY COMMUNICATION IN MYSTICETES

Receiver Capabilities

Whalers have long stood in awe of the sensitivity of their prey to sound. The whine of a motor, the report of a whale-marking gun, or even water disturbance by an oar may put the baleen whale to flight (Gunther, 1949; Norman & Fraser, 1949; Slijper, 1962). In one case, the click of the shutter of an underwater camera caused an orienting response in a humpback whale some 20-m distant (unpublished field observations, University of Hawaii). However, apart from relatively early anatomical descriptions (Fraser & Purves, 1959; Purves, 1966; Reysenbach de Haan, 1957), the hearing system of the mysticetes has received much less attention than that given the odontocetes. The availability of the smaller odontocete species for laboratory study and the great interest in echolocation account for these different emphases. Superficially, many of the mysticete auditory structures resemble those described for odontocetes. Morgane and Jacobs' (1972) estimations of the number of fibers in the cochlear portion of the 8th cranial nerve of the fin whale and the humpback whale greatly exceed the counts for man and also surpass counts for the bottlenosed dolphin and estimations for several other odontocete species. The richness in fiber development suggests excellent hearing sensitivity. Judging by the relatively low upper limits of their vocalization range, the mysticetes probably lack the very high-frequency hearing capabilities of the odontocetes but, on the same basis, may hear well into the low sonic or infrasonic regions.

Nonvocal Auditory Communication

Some mysticete species display aerial behaviors as spectacular as those of the odontocetes. Humpback whales, right whales, and gray whales may leap vigorously into the air (breach), throwing out large plumes of spray and producing airborne and underwater sound on reentry that may carry for several kilometers (Gilmore 1961; Scammon, 1874; Payne, 1978; Payne & McVay, 1971; Saayman & Tayler, 1973). This is a greater distance than the 1-km maximum given for the breaching sound from the very much smaller dusky dolphin (Würsig & Würsig, 1979). Breaching may be a spacing mechanism or may help whales remain in acoustic contact, especially in conditions in which visual contact is difficult. However, since the sounds of a phonating whale can also carry for at least several kilometers, as reviewed in a later section, it seems doubtful that breaching evolved principally for acoustic communication.

   Gilmore (1961) described breaching in gray whales as an exuberant display possibly indicating strength or agility to a competitive male or threat to an intruding male or boat. He noted, however, that calves and their mothers also breach on occasion. Payne (1978) postulated similar functions for breaching (and tail-slapping) in southern right whales. Tomilin (1967) stated that breaching indexed general excitement or arousal deriving from any of several causes, including sexual stimulation, location of food, or a response to injury or irritation. Some breaching, especially among calves, is probably play behavior and some may be social mimicry, breaching begetting breaching without any necessary commitment to communcation (cf. Andrew, 1962). Among humpback whales, the more dramatic forms of twisting breaches seem to be produced by the larger whales (Herman & Forestell, 1977). Calves may breach less elegantly, often failing to rise as completely from the water as larger animals. Head-slapping, forcefully slapping the undersurface of the upper body and head against the water, seems more frequent among calves than among larger animals, and might be an immature form of the breach. Clearly, there can be many immediate causes of breaching and further study is needed to clarify and understand the multiple contexts in which breaching appears.

Tail-slapping, which produces extensive underwater and aerial sound, may occur dozens of times in succession. The loud report of a tail-slap is probably easily localized by other whales and in that context could have some acoustic advantage over vocally produced sounds. Whether tail-slaps among mysticetes have an agonistic component, as they appear to among the odontocetes, is not known. However, on two occasions seen from observer aircraft a humpback whale emerged from well below a vigorously tail-slapping animal and rapidly swam away, as if chased (Herman & Forestell, 1977).

Bubbles are often released from the blowholes of the underwater humpback whale and on occasion great quantities of air may be expelled from the mouth (Herman & Forestell, 1977). "Bubbling" is commonly observed among groups of whales and has a distinct visual component with possible communicative functions (see Madsen & Herman, this volume). The release of bubbles may also produce sounds of social significance. Explosive, wheezing blows that produce sound in air and underwater were reported for humpback whales by Watkins (1967) and P. Thompson, Cummings, and Kennison (ms.). They occur among groups of whales and sometimes when individuals meet. Watkins noted reports in the literature of "unusual" blow sounds for several other species of mysticete whale, but the social significance of any of these sounds remains obscure. Watkins and Schevill (1976) recorded the rattling airborne sounds of the baleen plates of right whales, heard when the head of the animal was partially out of water. No social function was attached to these sounds.

Vocal Communication

Types of Vocalizations

The vocalizations of mysticete whales are, on the whole, very different from those of odontocetes, and more difficult to classify. Generally, they are of lower frequency, narrower bandwidth, and longer duration than odontocete sounds (Norris, 1969). However, there can be wide differences in type and amount of phonation across mysticete species. Some sounds are simple and repetitive, such as the 1-sec duration, narrow-band 20-Hz tonal pulses of fin whales (B. Patterson & Hamilton, 1964). The repetition rate is characteristically slow, that is, 1/12 sec for fin whale pulses in Bermudan waters (B. Patterson & Hamilton, 1964; Schevill, Watkins & Backus, 1964), and 1/1 sec for the humpback whale in its Alaskan feeding grounds (P. Thompson, Cummings & Kennison, ms.). Other sounds are more complex and are best described onomatopoetically, as they appear to the human ear. Some examples are "moans, bubble-type sounds, and knocks" from migrating gray whales (Cummings, Thompson & Cook, 1968), a "metallic sounding pulsed signal" from a captive gray whale

(Fish, Sumich & Lingle, 1974); "belches and moans" from southern right whales (Cummings, Fish & Thompson, 1972); "grunts, yelps, and snorts" from feeding humpback whales (P. Thompson., et al., ms.); and "pings, clicks, and grunt-like, thump-like, and ratchet-like sounds" from minke whales (Winn & Perkins, 1976).

Winn and Perkins (1976) classified the sounds of mysticetes into four acoustic categories: (a) tonal pulses and low-frequency moans between 0.4 to 36-sec duration with fundamental frequencies from 12 to 500 Hz, heard so far from 7 of the 10 mysticete species [minke, sei (Balaenoptera borealis) and pygmy right whale (Caperea marginata) excepted]; (b) gruntlike sounds, generally shorter than moans and with frequencies usually between 40 and 200 Hz, heard from humpback, right, gray, fin, and minke whales; (c) chirps and cries, with frequencies above 1 kHz, heard from the humpback whale; and (d) clicklike sounds, each click a fraction of a second in duration and of either broad-band or pure-tone composition, with frequencies from 3 to 30 kHz, reported for the humpback, gray, blue, minke, and Bryde's whale (Balaenoptera edeni).

The evidence for the use of click sounds for echolocation is controversial (cf. Beamish & Mitchell, 1971; 1973; Beamish, 1977; Winn & Perkins, 1976). In some cases there are doubts as to whether the clicks heard should be ascribed to the whales or to possibly accompanying odontocete species (Fish et al., 1974; Schevill & Watkins, 1972). In other cases whales have failed to avoid obstacles placed in their path (Beamish, 1977; Dreher & Evans, 1964). The low-frequency moans might have an echolocation function limited to rather large targets such as other whales (Winn & Perkins, 1976), dense shoals of prey organisms (Fish et al, 1974; Kinne, 1975), or broad topographic features useful for orientation in space (Norris, 1969; Payne & Webb, 1971). The sea catfish uses 100-Hz pulses for object detection (W. Tavolga, 1976) and some bats can detect objects having diameters of 1 to 3 mm using sound emissions as low as 3 to 8 kHz, demonstrating functionally useful echolocation with low to moderately low emitted sound frequencies. It may also be that the moans of the whales have an aggregative effect on the small schooling fish or planktonic animals favored as food, aiding in their harvest. Certainly, the low-frequency moans of a singing humpback whale can be "felt" by the underwater diver as well as heard (Herman & Forestell, 1977).

Social Functions of Vocalizations

Social functions proposed for the sounds of mysticete whales include long-range contact, assembly calls, sexual advertisement, greeting, spacing, threat, and individual identification (Cummings & Thompson, 1971a; W. Evans, 1967; Payne, 1978; Payne & McVay, 1971; Payne & Webb, 1971; Winn, Perkins & Poulter, 1971; Winn & Winn, 1978). However, only rarely has a specific sound been associated with a given behavioral event. Winn and Winn (1978) observed that a "ratchet" sound regularly preceded the surfacing of singing humpback whales in tropical Atlantic waters. Payne (1978) reported that "screams" were heard from humpback whales charging an observational vessel. As a whale approached the screams intensified and then suddenly stopped as the whale dove under the boat at the last moment. Payne rightfully postulated that this was a threat display.

Low frequency sounds travel long distances underwater and may help dispersed animals maintain contact or attract other whales to an emitter. Winn and Winn (1978) heard humpback whale songs at 32 km from the source, and Cummings and Thompson (1971a) obtained position fixes on 20-Hz fin whale tonal pulses while 160 km from the source. Tomilin (1965, cited in Berzin, 1972) hypothesized that migrating sperm whales might use the deep-water SOFAR (sound fixing and ranging) channel for sound communication over distances of thousands of kilometers. Whale sounds in the 20-Hz range have in fact been recorded at SOFAR depths on many occasions (Northrup, Cummings & Morrison, 1971). Payne and Webb (1971) calculated theoretical detection distances of from 80 to 800 km, depending on background noise, for the high-amplitude 20-Hz fin whale pulses if both source and receiver were near the surface but over a SOFAR channel. With both source and receiver at channel axis depth, which may lie 1000 m or more beneath the surface depending on latitude, the ranges are enormously extended to approximately 900 to 21,000 km. However, there is no evidence that mysticete whales can reach channel-axis depth, although sperm whales may do so (Berzin, 1972).

Payne and Webb (1971) held that a whale herd might be defined as all individuals within acoustic detection range, even though the detection distances might be enormous. This expansive view of a social unit may have limited functional significance for the whales. It is obviously impractical for a whale to approach a sound source some 1000 km away, even assuming that the whale has some means to gauge such distances and that the source will still be there when the listener arrives. Possibly, however, the sounds could serve to synchronize biological or behavioral activities in listeners which promote subsequent feeding or breeding success.

B. Patterson and Hamilton (1964) recorded 20-Hz sounds from two fin whales spaced several kilometers apart. The acoustic tracks of each whale were traced through a multiple hydrophone array, and each whale was uniquely identifiable by a characteristic interval between tonal pulses. Patterson and Hamilton reported that the first whale pulsed alone for approximately 3 hrs. while traveling generally southward. The second whale then began phonating while roughly 5-km easterly of the first whale, which in turn shifted its track in the direction of the second whale. Both whales continued pulsing for the next 52 min, with few instances of sound overlap. Finally, the first whale ceased pulsing when within approximately 2.5 km of the second, which continued pulsing while following a meandering route. These interesting observations are by no means conclusive of a sound-guided approach, let alone of communication, and further observations are needed. Unfortunately, observations of this type require the commitment of specialized equipment and more shore-, ship-, and man-hours than are easily available.

The unique temporal coding of the fin whale tonal pulses are reminiscent of the sperm whale codas described earlier and, like the latter, may allow individuals to identify one another. The simplicity and repetitiveness of the temporally coded pulsed sounds are ideal for long-range communication, which depends greatly on redundancy, crispness, and fidelity for accurate transmission. While frequency information may be greatly distorted by distance, time information is not. The discrete character of the pulsed sound is very different from the gradedness postulated for the whistle sounds of some odontocetes, and is consistent with the idea that gradedness develops for close-contact communication where supplemental context information is available, while discreteness is favored in longer-range acoustic signaling unsupported by other channels of communication (Marler, 1976).

Humpback Whale Vocalizations and the Puzzle of the Mating System

Special comment on humpback whales is appropriate given the many efforts in recent years to study their behaviors. Most work has been done in Hawaii during the winter breeding season. Perhaps 500 or more whales assemble in Hawaii at the mid-February to mid-March seasonal peak (Herman et. al. 1977; Rice, 1977; also see the historical background of the whales in Hawaii in Herman, 1979). The whales are widely distributed throughout the islands but several subregions are most heavily utilized (Herman et al., 1977). Single animals are frequently seen, but groups of two and three animals are also common, with increasingly larger groups progressively rarer (Herman & Antinoja, 1977; Herman et al., 1977). Observations of the University of Hawaii field research team indicate that group structure may be fluid, especially among the larger aggregations. Whales have been seen to come together, temporarily affiliate and then disaffiliate. The overall social structure is still poorly understood, however.

The songs of the humpback whales have been a subject of special interest for many. Intensive singing occurs worldwide in the low-latitude winter assembly areas of the whales and possibly en route to them (Winn et al., 1971; Winn & Winn, 1978; P. Thompson et al., ms.). There have been only a few reports of singing in the summer feeding grounds. Payne and McVay (1971) first characterized the sounds as songs and analyzed them as a hierarchy of progressively more inclusive acoustic elements called units, phrases, themes, songs, and song sessions. A song may be comprised of from four to eight ordered themes, which in turn contain a number of repeated units or phrases (combinations of units). A song session consists of many successive renditions of the song that may itself last for 8 to 20 minutes or longer. Winn and Winn (1978), recording in the shallow banks of the Caribbean Sea favored by North Atlantic humpback whales, heard continuous singing from one whale for 22 hours before terminating observations. The singing was interrupted only when surfacing. As was noted, in the Caribbean a ratchet sound precedes surfacing. In Hawaii, a "chirp" sound and sound attenuation occurs (Peter Tyack, personal communication), and the whale may continue to phonate at low level while at the surface (L. Herman, personal observations). In the Hawaii assembly area, singing takes place throughout the assembly period, from early November to late May or June (Payne, 1978). Near the peak of the season, in areas of heavy concentration, many whales can be heard singing concurrently, day and night, without apparent synchrony or exchanges of vocalizations.

Payne (1978) recorded whales at both Bermuda and Hawaii and found song differences among these groups, as did Winn and Winn (1978). Most interestingly, Payne reported that within a population the song may change from season to season. The changes include additions, deletions, or modifications of themes. At the start of each season, portions of last season's full song can be heard from different whales. As the season progresses, the song becomes longer and more complex and there is a convergence by almost all whales on the full song for that year. A similar type of song plasticity and annual song convergence has been reported in cardueline birds, for example, finches, according to Nottebohm (1970). Like the whales, some cardueline species modify their calls over successive years to match those produced by other flock members or by a mate.

Functions of the whale songs in aggregation, spacing, advertisement, and courtship have been hypothesized (Payne & McVay, 1971; Winn et al., 1971; Winn & Winn, 1978). A signature function for the "cry" vocalization segment of the song has also been proposed (Hafner, Hamilton, Steiner, Thompson & Winn, 1977). Data collected or summarized by Winn and Winn (1978) suggest that singers are always single animals, most probably sexually mature but not old males. Our own observations in Hawaii confirm that almost always a single animal surfaces after attenuation of a song.

Members of the University of Hawaii field team report seeing whales very near a singer on a few occasions. In one case, a mother and calf were positioned immediately above a singer's location, the calf rolling about near the surface while the mother at times remained with it and at times dove deeper, presumably toward the singer. In a second case, again involving a calf rolling at the surface above the singer, the song stopped abruptly when the mother and a second whale rose rapidly from below to join the calf and escort it away from the underwater observer. On a third occasion, an adult-sized whale pressed its head against the side of the singer, near the insertion of the singer's pectoral fin. Both whales were stationary, dorsal surface upward, at approximately 20-m depth. When the singer rolled laterally to observe the diver directly above it near the surface, the singing stopped abruptly.

An immense investment of time and energy must go into the songs and, undoubtedly, more interaction takes place between singers and other whales than is so far apparent. How the songs function in the mating system of the whales is still a matter of conjecture, but some attributes of the singing and singers, and what is known so far of the characteristics of the winter assembly and of the reproductive biology of the humpback whale is reminiscent of a lek mating system, as was proposed by Herman and Forestell (1977). Lekking systems are found among some insects, fish, amphibians, birds, and mammals (Emlen &

Oring, 1977; Wiley, 1978; Wilson, 1975). In a lekking system males assemble on a communal display and mating ground, the lek, and females visit there for breeding. Individual males occupy small territories within the lek. The more dominant males hold those territories more likely to be visited by females and perform the majority of the copulations. All males engage in conspicuous visual behaviors, and, among birds, communal vocalizing occurs as well. The communal displays and vocalizations attract more females than would a lone male, and the well-defined, conspicuous lekking area, generally located away from feeding or nesting sites, is easy to locate. Emlen and Oring (1977) hypothesized that a lek system is typified by a relatively long breeding season, more sexually receptive males than females, and "by the inability of individual males to economically control and monopolize the resources essential for female acquisition (p. 219)." As a form of polygyny, there may be selection for large male size. Loiselle and Barlow (1978) additionally identified five prerequisites to lekking in any species including (*a*) synchronization of reproductive activities of males and females, (*b*) assembly of males at a given area and visits there by females, (*c*) sufficient mobility to reach the assembly area, (*d*) care for the young by only one parent, and (*e*) little or no feeding on the lek.

Some of these factors seem to apply to humpback whales. The concurrent singing of many whales, if they are in fact males, may be thought of as a communal acoustic display. The sex ratio among sexually active animals is probably on the order of two males to one female. Though the sex ratio is equal in the population as a whole, the average two-year reproductive cycle of females removes half of them from the annual reproductive pool (Mackintosh, 1972). The male and female reproductive cycles are seasonally synchronized, according to data on testicular weight and ovarian cycles of humpback whales killed in southern hemisphere waters (Chittleborough, 1965). Females may be seasonally polyoestrus but most of the ovulatory cycles occur during the winter assembly season coincident with the period of maximum testicular weight of the males. Sexually receptive females do not arrive in the breeding areas all at once, but over a period of a month or more (Dawbin, 1966). More generally, the breeding season seems long, almost six months, if one considers not only time spent on the breeding grounds, but travel to and from it. In Hawaii, it was noted that singing whales can be heard from November through to late May or June. Pregnant females are in the vanguard of the return migration to the feeding areas (Chittleborough, 1965; Dawbin, 1966), suggesting that they depart the breeding grounds soon after pregnant, as is typical of a lek mating system.

There is little opportunity in the environment of the breeding ground for a male humpback whale to control individual females. Female arrival would seem to be only grossly predictable, since it apparently is spaced over a protracted period, and it may be difficult to maintain a large defensible area in the expansive three-dimensional world of the whale. Furthermore, there is a stringent problem in energy conservation, since humpback whales and other mysticete whales may not feed on the breeding grounds nor, possibly, enroute (Dawbin, 1966; Tomilin, 1967). It would be clearly wasteful of precious stored resources to engage in costly energetic male-male competition and this may partly account for the typical lack of aggression in social encounters among humpback whales, or among mysticetes in general (Norris, 1967), and the lack of strongly dimorphic male traits. Nevertheless, the evolving image of humpback whale interactions is one of more aggression than has been traditionally supposed, as some of our earlier examples illustrated. Also, the University of Hawaii team has photographed animals with raw freshly bruised dorsal fins on several occasions and two instances of whales with head nodules that appeared red and bruised. Perhaps these sightings are symptomatic of agonistic encounters between whales.

The spacing of singers and the permanency of their occupancy of regions in the breeding areas are poorly understood. If the singers are indeed breeding males, perhaps some quality of the song, or the location of the singer in space, plays a part in a presumptive female choice of mate. If, as in the typical lekking system, only a few males are chosen, then the lack of validated observations of mating in Hawaiian waters may be understood in part as a problem for the human observer in finding the favored few in a whale-sized assembly area covering hundreds of square kilometers.

Trivers (1972) remarked that the sex making the greater investment in reproduction will be a limiting resource for the sex making the lesser investment. There can be little doubt that among humpback whales females invest most heavily in reproduction. After a gestation period of approximately one year (Chittleborough, 1965) the female faces a lactation period almost as long (Yablokov et al., 1972), as well as primary responsibility for rearing the calf. She may remain in the winter assembly area with her new calf for several months (Herman et al., 1977), nursing the calf and apparently sustaining herself on stored bodily reserves alone. Given these costly investment factors, it is not surprising that the female attends so closely to the calf (Herman & Antinoja, 1977; Herman et al., 1977). That female humpback whales and other female mysticete whales are somewhat larger than males may reflect the pressures of these extreme nurturant demands on the mother. As Ralls (1976) noted, bigger mothers are better mothers.

A corollary of Trivers (1972) hypothesis is that the sex making the lesser investment will tend to compete among themselves for access to the opposite sex and may develop extreme epigamic displays—displays for attracting the opposite sex. Perhaps the songs of the whales, if they emanate from males, are the manifestation of the extreme epigamic display of Trivers' hypothesis.

Clearly there is much to learn about the humpback whale mating system. Any proposed mating system will have to account for the various behavioral, social, and reproductive factors discussed above. The lack of validated observations of mating (or even of an erect penis) is puzzling. Mating has been seen among milling surface-active groups of right whales in southern hemisphere waters and is accompanied by ventral postures and much contactual behavior (Payne, 1976; Saayman & Tayler, 1973). While similar groupings and behaviors have been seen among the humpback whales in Hawaiian waters (Herman & Antinoja, 1977) the expected mating episodes were not observed. Sexing of individuals will answer many questions but is difficult in the absence of strongly dimorphic, easily observable, traits. Until such time as more data become available, the lekking hypothesis or any alternative hypothesis must be regarded as highly tentative.

COMMENTS ON DEVELOPMENT AND ON PARENTAL BEHAVIORS IN COMMUNICATION

Tracing the development of communication in the individual and describing the different forms of communication that may occur at different life stages (Burghardt, 1977; Bekoff, 1972) can help greatly in understanding the genetically controlled and learned components of communicative behavior and its functions.

Unfortunately, ontogenetic data on cetaceans are hard to obtain. Developmental processes unfold slowly in most cetaceans: perhaps 4 to 12 years to reach puberty and another 6 to 12 years to attain full physical and social maturity (Bryden, 1972; Harrison & Ridgway, 1971). Tracing any individual's life history is obviously a formidable task, compounded in difficulty by practical problems in observation, availability of animals, or husbandry. In the field one cannot easily keep track of individuals, young or old, for long periods of time, and in the oceanariums there are few infants. Worldwide only 151 bottlenosed dolphin births in captivity were recorded between 1939 and 1977 and many of the newborns did not survive for long (Ridgway & Benirschke, 1977). Furthermore, animals less than 3 years old are highly dependent, limiting the possibilities for manipulating their environment or behavior experimentally.

An alternative strategy to the longitudinal study of individuals is to compare different age groups cross-sectionally to identify similarities and differences in social or communication processes. Many wild cetaceans organize into age-related (or sex-related) subgroups (Leatherwood, 1977; Tomilin, 1967; Wells et al., this volume; Würsig, 1978), offering a natural laboratory for studying the influences of developmental stage on communication. All-juvenile subgroups are found in some dolphin populations (Leatherwood, 1977; Wells et al., this volume; but also see Tayler & Saayman, 1972 for negative evidence), as well as in the sperm whale society (Berzin, 1972) and among humpback whales and perhaps other mysticetes during their migratory cycle (Chittleborough, 1965; Dawbin, 1966). In these juvenile groups with their relative independence from adult care, social skills necessary for survival and reproduction as adults may be practiced and, in the case of dolphins, long-term social affiliations important for later school cohesion may be developed.

So far little advantage seems to have been taken of the opportunities for study of communication in age-segregated groups. Graycar (1977) reported age-related whistle differences among 158 bottlenosed dolphins living in or collected from Florida waters, though how these differences related adaptively or otherwise to age level was not stated.

Behavioral Development of the Calf

Almost all of the descriptions of calf behavior are of the bottlenosed dolphin (e.g., Essapian, 1953; Gurevich, 1977; McBride & Hebb, 1948; McBride & Kritzler, 1951; M. Tavolga & Essapian, 1957; M. Tavolga, 1966). Most emphasize the playfulness of the calf and its tendency to manipulate and investigate objects. Early sexuality is common, expressed sometimes as masturbation but also as direct contact with other individuals of the same or opposite sex. Aggression in calves is minimal or absent.

With respect to communication, play with other animals may allow practice in information transfer or in manipulating others within a nonthreatening context, as in inducing others to chase or be chased. According to Dawkin and Krebs (1978), the essence of communication is the manipulation of others for the advantage of the sender.

The communicative significance, if any, of extreme sexual precocity (Allen, 1977; M. Caldwell & Caldwell, 1967; Gurevich, 1977) within the context of the relatively late development of puberty is not easily understood. The behavior may enhance the later communication of sexual or tactual information, as well as assist in the establishment and maintenance of social bonds among individuals (cf. Saayman & Tayler, 1979). Aggression in dolphins seems to ripen as the animal matures, perhaps in response to increasing biological and social pressures for spacing, for organization of the society hierarchically, or for gaining access to limited resources. Aggression appears to emerge more strongly in males than in females in the apparently polygynous delphinid mating system.

Although almost nothing is known about the progression of sensory development, the precocious behavior of the neonate suggests that the senses develop early. Remaining close to the mother, swimming with her, recognizing her, and finding the nipple, all place early demands on tactual, visual, and perhaps gustatory senses. Recognizing the mother may also depend in part on acoustic cues that she provides. The new bottlenosed dolphin mother may whistle nearly continuously for several days after giving birth (D. Caldwell & Caldwell, 1968), potentially providing a strong acoustic imprinting stimulus. Later, if separated from her calf she may again whistle continuously until the calf returns or is retrieved (McBride & Kritzler, 1951). Reportedly the whistle of the calf is present from the day of birth onward (McBride & Kritzler, 1951). Earlier work of the Caldwells (D. Caldwell & Caldwell, 1972) indicated that the whistle may change in form only slightly as the calf matures, but later work (M. Caldwell & Caldwell, 1979) stresses that the early whistle is relatively unformed and that the "signature" characteristic emerges only later. No one appears to have studied the responsiveness of the mother and calf to playbacks of each other's whistles versus those of others. Such study could help pin down the importance and specificity of acoustic signalling in the mother-calf relationship.

Echolocation sounds are also reportedly present from birth onward (D. Caldwell & Caldwell, 1977), though they may not be wholly functional as sonar signals until much later in life (M. Caldwell & Caldwell, 1967; W. E. Evans, reported in Leatherwood, 1977; also see brief review in Herman, this volume).

Whether the echolocation sounds may have heightened social significance in young dolphins is not known, but the possiblity should be considered.

From what little is known about the behaviors of mysticete calves, it appears that they share some of the traits of young dolphins. Observations of humpback whale calves in Hawaii hint at the inquisitiveness and playfulness of the calf. Calves may roll about vigorously on the surface, leap from the water, and seem much more ready than adults to approach and investigate boats and divers (Herman & Antinoja, 1977; Herman et al., 1977).

Parental and Allomaternal Behavior

The mother-calf relationship among bottlenosed dolphins, and probably among other dolphinids, is one of strong mutual attraction that may last for years (Essapian, 1953; Tavolga & Essapian, 1957; also see review in M. Caldwell and Caldwell, 1966). The factors responsible for the development of these persistent social bonds have not been identified. The one case on record of a dolphin whose development was traced from birth through to her later delivery of several calves provides only a hint of the factors that may be responsible for close bonding. This bottlenosed dolphin, named Spray, was reared in a colony of dolphins, including members of her own species, at Marine Studios in Florida. Her early history is well documented (M. Tavolga & Essapian, 1957; M. Tavolga, 1966). As an adolescent and subadult Spray actively participated in caring for and playing with the newborns of others. She delivered her first calf at the age of seven but was a poor mother. Her negligence in caring for her calf contributed to its death at the age of 15 days (Tavolga & Essapian, 1957). Wood (1977) reported that Spray later gave birth to four additional calves, two at two-year intervals after the first and the third and fourth 11 and 13 years after the first. Two were stillborn and two were apparently normal deliveries. M. Caldwell and Caldwell (1977) stated that Spray showed normal maternal behavior toward all calves after the first, but details are lacking. Though the Caldwells speculated that her poor care for her firstborn may have mainly reflected Spray's young maternal age and attendant hormonal deficiencies, it may also have reflected the constraints of her early rearing in captivity.

Among captive bottlenosed dolphins the birth of a calf may be closely attended by nonpregnant females, or "aunts," exhibiting allomaternal behavior (McBride & Kritzler, 1951; M. Tavolga & Essapian, 1957). Although Dudok van Heel (1977) believes that aunts are not essential to successful parturition and may even be a "nuisance," in the wild they may protect the vulnerable mother from attacks by predators or harassment by mature males. In one observation of a birth in captivity to a dusky dolphin, an attending female of the same species apparently bit the umbilical cord in two (Allen, 1977). In bottlenosed dolphins, breaking of the cord by the mother herself through a sudden "whirling" maneuver is more commonly observed (Essapian, 1953; McBride & Kritzler, 1951).

Aunts show great interest in the calf after birth and typically school closely with the mother-calf pair. Although there are isolated cases of aunts temporarily taking the calf away from its mother, they more typically provide protection for the pair and may "baby-sit" the calf while the mother is on a feeding foray (M. Caldwell & Caldwell, 1972; Gurevich, 1977; McBride & Hebb, 1948; McBride & Kritzler, 1951; Leatherwood, 1977; Tayler & Saayman, 1972).

Unlike the case for dolphins in which school members attend and even assist in the birth process, the humpback whale mother appears to isolate herself from other whales prior to parturition (Herman, Forestell & Antinoja, 1977). The lone large whale is not dependent on a school for protection and is better prepared to defend herself and her calf against predators than is the lone dolphin. The shallow, near-shore areas that are apparently favored for parturition may guard against intrusions from large, deep-water sharks. The isolation period may last for days or possibly weeks and allow time for the formation of strong social bonds between the pair, helping to protect the mother's great investment in her calf.

Aerial surveys reveal that humpback whale calves are widely distributed throughout the waters adjoining the main Hawaiian Islands (Herman & Antinoja, 1977; Herman et al., 1977), indicating that mother-calf pairs eventually wander from the protected near-shore areas. In several years of observation in Hawaii encompassing well over one hundred sightings of calves, two or more mother-calf pairs have never been seen together. However, mother-calf dyads may commonly school with other categories of whales, and are most often seen with a single, usually large, adult called an "escort" by Herman and Antinoja (1977). Field observations show that the escort may play a protective role. It may interpose itself between the pair and approaching vessels or swimmers. If tracked persistently by a vessel the escort may lay a bubble screen in front of the vessel by expelling many small bubbles from its blowhole or large quantities of "gulped" air from its mouth. The bubbles obscure the mother-calf pair traveling ahead from underwater view and would seem to be a useful defensive screen against approaching large sharks or killer whales, occasional predators on humpback whale calves or adults (e.g., see Chittleborough, 1953; K. Evans, 1976). In one isolated and atypical case, observed by members of the University of Hawaii field research team, an escort struck an approaching whale in the flanks with its tail with such force that the struck whale was visibly hurled away a short distance. The escort then continued to interpose itself between that whale and the mother-calf pair.

The sex of the escort is unknown, and it is unclear what resource is being defended—calf, mother, or both. The escort may be an adult female performing allomaternal functions similar to those of the aunt in dolphin groups. Or, it may be a sexually mature male consorting temporarily with a female that is ovulating postpartum. Though postpartum ovulation is relatively rare in humpback whales (Chittleborough, 1958), its occurrence may be frequent enough to warrant the attention of males to females with calves.

As a generalization it appears that among some of the cetaceans, in addition to the obligate social relationship between mother and calf, other adults, mainly females, contribute significantly to the socialization and care of the calf. Allomaternal behavior is common among many group living terrestrial mammals (Spencer-Booth, 1970), affording added protection and nurturance to the young. Allomaternal behavior may also benefit the aunts through giving them practice in the rearing of young which is transferrable to rearing their own future offspring. Aunts may also expect to receive help from other females when they themselves are mothers. In societies in which females share many genes in common, allomaternal behavior may be additionally understood as contributing directly to the likelihood that the aunt's genes will be represented in future generations (inclusive fitness).

Among bottlenosed dolphins, as among mysticete whales, there is no compelling evidence of a significant male role in calf rearing. Large adult males roaming the periphery of wild dolphin schools may give some protection against predators, but it was also noted that attacks on the young by adult male dolphins are common (e.g., McBride & Hebb, 1948; M. Tavolga & Essapian, 1957). Leatherwood (1977) observed that among wild herds of dolphins large solitary animals, presumably males, sometimes made high-speed rushes at a mother-calf pair. The mothers responded by interposing themselves between calf and attacker. Attacks of this type are not easily understood, but some have considered them adaptive spacing mechanisms that reposition the mother-calf pair to the protected center of the school (Norris & Dohl, 1980). An "attack" may also be sexually motivated (Leatherwood, 1977), but its sometimes fatal consequences in captive communities attests to its strongly aggressive component.

In summary, behavioral ontogeny and parenting need much further study. Common patterns that emerge from a number of reports include bonding relationships, allomaternal behavior, the development of peer affiliations, and the subordinate role of the male in calf-rearing. However, these patterns are based on limited data and apply mainly to bottlenosed dolphins. The development of affiliations, social behaviors, and of communication in other cetaceans deserves detailed study. Emerging data should be integrated with materials from broader studies of the social structure of cetacean groups and their mating systems.

DISCUSSION

The description of the communication system of any cetacean species is still at an early stage. Nevertheless, some general adaptations for communication can be seen across species that reflect the common influence of the aquatic habitat. Acoustic communication is favored in that habitat and most cetacean species are highly vocal, although a few, such as the harbor porpoise or other Phocoenidae, seem puzzlingly reserved in vocal output. Possibly, even these animals are highly vocal in ultrasonic portions of the frequency spectrum that are not well sampled by most recording techniques.

The reduced olfactory system of the cetaceans is linked to modifications of the nasal and pharyngeal areas, enabling these animals to live in water and yet breathe air efficiently. Whether or not gustatory communication has been maintained or developed, partially compensating for the loss of olfaction, is still an open question, but seems likely for some species. The laterally placed eyes, the heightened sensitivity to dim light, and the adaptations for improving contrast and resolution found among oceanic cetaceans enhance opportunities for visual detection and presumably visual communication. The extensive innervation of the cetacean integument may be in part a response to the advantages of sensing water movement, but it also allows for communication through tactual interchanges among animals.

Beyond these common trends the form of communication, even within a given channel, can be very different across species or groupings of species, often reflecting local variations in ecology or social structure. Generally little attention has been given to the effects of these latter variables on cetacean communication systems, even in the wake of W. Evans and Bastian's (1969) discussion of the topic. Yet, ecological and social variables may be the "prime movers" (cf. Wilson, 1975) in the evolution of communication systems. It seems likely, for example, that communal foragers, such as many of the oceanic dolphins, have more complex communication systems than the more solitary foragers, such as the harbor porpoise or river dolphins. This type of differential development of communication system has been shown to occur among canids, communal living and hunting species having more complex systems than solitary species (Kleiman, 1967). Some other examples are that feeding ecology can influence body coloration (see Madsen & Herman, this volume), which in turn affects the development of visual displays and visual communication; species-recognition demands may heighten and recognition characters become enhanced with increases in sympatry, as in mixed-species *Stenella* schools [In eastern North Pacific waters, but not in Hawaiian waters, spotted and spinner dolphins school together; adult males of the eastern spinner race develop a large postanal protuberance and a forward-canted dorsal fin not typifying Hawaiian spinner males (Perrin, 1972; Fig. 3.9 in Madsen & Herman, this volume)]; and vocal and possibly chemical alarm signals may develop strongly in species, such as the belukha, victimized by heavy predation, but not in less vulnerable species.

Watkins, Schevill and Best (1977) wondered about the significance of the commonalities in vocal production and in ecology and social organization in Heaviside's dolphin and the harbor porpoise, as was described earlier. We noted that whistling and nonwhistling odontocete cetaceans seemed to be set apart by differences in social structure, and indicated that mating systems can strongly influence the presence or type of ritualized advertising and display, and the degree of agonism exchanged among animals. For example, the extreme sexual dimorphism and apparent male-male competition of sperm whales (D. Caldwell, Caldwell & Rice, 1966) are the expected concomitants of a strongly polygynous mating system, and the extensive singing among humpback whales was hypothesized to be a communal male epigamic display within a lek-type mating system.

Prolonged care of the young is universal among cetaceans, a responsibility that is apparently borne primarily by the female. Allomaternal behavior is probably common, among odontocete cetaceans, but its role among mysticete cetaceans is still largely undefined and uncertain.

The influence of ecology on social structure has been emphasized in many recent papers (e.g., Davies & Krebs, 1978; Eisenberg, Muckenhirn & Rudran, 1972; Wilson, 1975; also, Wells et al., this volume). It is only a small increment in concept to recognize the influence of ecology on communication, and the social forces that shape communication. Marler (1976) emphasized the interaction of communication and society in the premise that "the communication system of a species is a basic component in its social design and must contribute to the organization of its societies (p. 239)." In general, more attention should be given to comparisons of communication systems, or component parts, among cetacean species or populations having different degrees of overlap in social and ecological parameters. One comparison of interest, with a ready-made control on the species variable, is the river and neritic populations of the tucuxi (*Sotalia fluviatilis*). Some populations of this small South American dolphin live entirely within the Amazon and Orinoco River systems, while others are found primarily or exclusively in the coastal marine waters (Mitchell, 1975; Norris, Harvey, Burzell & Krishna Kartha, 1972). Specimens of the river population are on the average smaller in size and lighter in color than the marine forms. Comparisons of the river and neritic populations might be especially revealing of the influences of ecological pressures on social organization and communication systems.

It would also be interesting to compare the tucuxi and the bouto, which are sympatric in the river systems. In this case most portions of the ecology are the same with the probable exception of prey species utilized. The tucuxi produces whistles while the bouto does not (Norris et al., 1972), and the bouto is often found alone while the tucuxi is more often seen in schools of 8 to 20 animals. Is this a case where group size and/or feeding ecology has had a major influence on the development or retention of the whistle sound? Further studies of these two species may help resolve this question and identify other differences in the communication systems of these two species that relate to social factors.

Some long-standing questions about cetacean communication require continued study. The improved acoustic and visual observational techniques available today, together with a healthy dose of observational patience, should eventually yield some firmer correlations between signal and behavior. Improved analytic and playback techniques may help resolve issues about the structure and variability of whistle sounds, and whether they should be regarded as graded or discrete in character. Efforts should be made to identify further the role of pulsatile sounds in acoustic communication. Gustatory and tactual capabilities need to be determined with greater reliability, and the effects of experimentally imposed tactual or chemical stimulation on social behavior merits study. The social role of vision and the visual stimuli that elicit or control social responses of cetaceans should be examined in detail.

The methods of the laboratory offer the easiest approach to the study of mechanisms and of developmental processes, and can be useful for tying down some of the functions of communication. The level or complexity of communication attainable by dolphins, a topic of broad general interest, can be approached through laboratory analyses of language-learning capabilities, as has been done with great-ape species.

Cross-talk between laboratory and field research should be expanded. Laboratory observations arm the field worker with the preliminary kit of behavioral categories with which to describe wild populations and with some knowledge of the general biology and sensory specializations of the species. In the field the full fabric of ecology and social structure are present, but observations of the cetaceans themselves are difficult to come by. The near term prospects for laboratory and field work with cetaceans are favorable, given the heightened interest in and resources available for the study of marine mammals. We may expect, therefore, that knowledge of cetacean communication systems will progress more rapidly in the future.

ACKNOWLEDGMENTS

Preparation of this paper was supported in part by Grants BNS77-24943 and BNS77-16882 from the National Science Foundation to Louis M. Herman and Grant 5 R01 NS13746-02 from the National Institutes of Health: National Institute of Neurological and Communicative Disorders and Stroke to William N. Tavolga. The authors greatly appreciate the comments of George Losey.

REFERENCES

Agarkov, G. B., Khomenko, B. G. & Khadzhinskiy, V. G. (1974). *Morphology of Delphinidae.* Moscow: Naukova-Dumka (Available JPRS 66036, 1975).

Allen, J. F. (1977). Dolphin reproduction in oceanaria in Australasia and Indonesia. In: *Breeding Dolphins: Present Status, Suggestions for the Future* (S. H. Ridgway & K. Benirschke, Eds.), pp. 85–108. Washington, D.C.: U.S. Marine Mammal Commission Report MMC-76/07.

Altmann, S. A. (1962). A field study of the sociobiology of rhesus monkeys, *Macaca mulatta. Ann. N. Y. Acad. Sci.,* **102,** 338–435.

Altmann, S. A. (1967). (Ed.). *Social Communication Among Primates.* Chicago, Ill.: University of Chicago Press.

Amundin, M. & Amundin, B. (1973). On the behaviour and study of the harbour porpoise, *Phocoena phocoena,* in the wild. In: *Investigations on Cetacea,* Vol. V (G. Pilleri, Ed.), pp. 317–328. Berne, Switzerland: Institute of Brain Anatomy, University of Berne.

Andersen, S. & Pilleri, G. (1970). Audible sound production in captive *Platanista gangetica.* In: *Investigations on Cetacea,* Vol. II (G. Pilleri, Ed.), pp. 83–86. Berne, Switzerland: Institute of Brain Anatomy, University of Berne.

Andrew, R. J. (1962). Evolution of intelligence and vocal mimicking. *Science,* **137,** 585-589.

Arvy, L. & Pilleri, G. (1970). The tongue of *Platanista gangetica* and remarks on the cetacean tongue. In: *Investigations on Cetacea,* Vol. II (G. Pilleri, Ed.), pp. 75-77 + 5 Pl. Berne, Switzerland: Brain Anatomy Institute, University of Berne.

Arvy, L. & Pilleri, G. (1972). Comparison of the tongues of some odontocetes: *Pontoporia, Neomeris,* and *Delphinus.* In: *Investigations on Cetacea,* Vol. IV (G. Pilleri, Ed.), pp. 191-200 + 6 Pl. Berne, Switzerland: Brain Anatomy Institute, University of Berne.

Attneave, F. (1959). *Applications of Information Theory to Psychology.* New York: Holt, Rinehart & Winston.

Bastian, J. (1967). The transmission of arbitrary environmental information between bottlenose dolphins. In: *Animal Sonar Systems,* Vol. II (R.-G. Busnel, Ed.), pp. 803-873. Jouy-en-Josas, France: Laboratoire de Physiologie Acoustique.

Bastian, J. & Bermant, G. (1973). Animal communication: An overview and conceptual analysis. In: *Perspectives on Animal Behavior* (G. Bermant, Ed.), pp. 307-357. Glenview, Ill.: Scott, Foresman & Co.

Bastian, J., Wall. C. & Anderson, C. L. (1968). Further investigation of the transmission of arbitrary environmental information between bottlenose dolphins. Naval Undersea Warfare Center, TP 109, San Diego, Calif. pp. 1-40.

Bateson, G. (1965). Porpoise community research: Final report. National Institutes of Health Contract No. N 60530-C-1098.

Beamish, P. (1977). Evidence that baleen whales do not use food-finding sonar. *Proc. 2nd Conf. Biol. Mar. Mammals,* San Diego, Calif., Dec. 1977, p. 72.

Beamish, P. & Mitchell, E. (1971). Ultrasonic sounds recorded in the presence of a blue whale (*Balenoptera musculus*). *Deep-Sea Res.,* **18,** 803-809.

Beamish, P. & Mitchell, E. (1973). Short pulse length audio frequency sounds recorded in the presence of a minke whale (*Balenoptera acutorostrata*). *Deep-Sea Res.,* **20,** 375-386.

Bekoff, M. (1972). The development of social interaction, play and metacommunication in mammals: An ethological perspective. *Q. Rev. Biol.,* **47,** 412-434.

Berzin, A. A. (1972). *The Sperm Whale.* Jerusalem: Israel Program for Scientific Translations.

Best, P. B. (1979). Social organization in sperm whales, *Physeter macrocephalus.* In: *Behavior of Marine Animals: Current Perspectives in Research. Vol. 3. Cetaceans* (H. E. Winn & B. L. Olla, Eds.), pp. 227-289. New York: Plenum.

Boakes, R. A. & Gaertner, I. (1977). The development of a simple form of communication. *Q. J. Exp. Psychol.,* **29,** 561-575.

Braden, I. C., Conklin, C. & Friedman, E. W. (1973). Food and human contact as reinforcers in the Amazon River dolphin (*Inia geoffrensis*). *Proc. 81st Annu. Conv., APA,* **1973,** 871-872.

Brooks, R. J. & Falls, J. B. (1975). Individual recognition by song in white-throated sparrows. I. Discrimination of songs of neighbors and strangers. *Can. J. Zool.,* **53,** 879-888.

Brown, D. H., Caldwell, D. K. & Caldwell, M. C. (1966). Observations on the behavior of wild and captive false killer whales with notes on associated behavior of other genera of captive delphinids. *Los Ang. Cty. Mus. Contrib. Sci.,* **95,** 1-32.

Brown, D. H. & Norris, K. S. (1956). Observations of captive and wild cetaceans. *J. Mammal.,* **37,** 311-326.

Brown, J. L. (1975). *The Evolution of Behavior.* New York: Norton.

Brown, P. L. & Jenkins, H. M. (1968). Auto-shaping of the pigeon's keypeck. *J. Exp. Anal. Behav.,* **11,** 1-8.

Bryden, M. M. (1972). Growth and development of marine mammals. In: *Functional Anatomy of Marine Mammals,* Vol. I (R. J. Harrison, Ed.), pp. 1-79. New York: Academic.

Bullock, T. H. & Gurevich, V. S. (1979). Soviet literature on the nervous system and psychobiology of cetacea. *Int. Rev. Neurobiol.* **21,** 47-127.

Bullock, T. H. & Ridgway, S. H. (1972). Evoked potentials in the central auditory system of alert porpoises to their own and artificial sounds. *J. Neurobiol.*, **3,** 79–99.

Burdin, V. I., Reznik, A. M., Skornyakov, V. M. & Chupakov, A. G. (1974). Study of communicative signals in Black Sea dolphins. "Akusti Cheskiy Zhurnal," **20,** 518–525.

Burghardt, G. M. (1970). Defining communication. In: *Advances in chemoreception*, Vol. I (J. W. Johnston, Jr., D. G. Moulton & A. Turk, Eds.), pp. 5–18. New York: Appleton-Century-Crofts.

Burghardt, G. M. (1977). Ontogeny of communication. In: *How Animals Communicate* (T. A. Sebeok, Ed.), pp. 794–808. Bloomington: Indiana University Press.

Busnel, R.-G. (1966). Information in the human whistled language and sea mammal whistling. In: *Whales, Dolphins, and Porpoises* (K. S. Norris, Ed), pp. 544–568. Berkeley: University of California Press.

Busnel, R.-G. & Dziedzic, A. (1966). Acoustic signals of the pilot whale *Globicephala melaena* and of the porpoises *Delphinus delphis* and *Phocoena phocoena*. In: *Whales, Dolphins, and Porpoises* (K. S. Norris, Ed.). pp. 607–646. Berkeley: University of California Press.

Busnel, R.-G. & Dziedzic. A. (1968). Etude des signaux acoustiques associés à des situations détresse chez certain cétacés odontocètes. *Annls Inst. Océanogr., Monaco*, **46,** 109–144.

Cahn, P. H. (1967). (Ed.) *Lateral Line Detectors*. Bloomington: Indiana University Press.

Caldwell, D. K. & Caldwell, M. C. (1968). The dolphin observed. *Nat. Hist.*, **77,** 58–65.

Caldwell, D. K. & Caldwell, M. C. (1971a). Underwater pulsed sounds of captive spotted dolphins, *Stenella plagiodon*. *Cetology*, **1,** 1–7.

Caldwell, D. K. & Caldwell, M. C. (1971b). Sounds produced by two rare cetaceans stranded in Florida. *Cetology*, **4,** 1–12.

Caldwell, D. K. & Caldwell, M. C. (1972). Senses and communication. In: *Mammals of the Sea* (S. H. Ridgway, Ed.), pp. 466–502. Springfield, Ill.: Thomas.

Caldwell, D. K. & Caldwell, M. C. (1977). Cetaceans. In: *How Animals Communicate* (T. A. Sebeok, Ed.), pp. 794–808. Bloomington: Indiana University Press.

Caldwell, D. K., Caldwell, M. C. & Rice, D. W. (1966). Behavior of the sperm whale, *Physeter catodon* L. In: *Whales, Dolphins, and Porpoises* (K. S. Norris, Ed.), pp. 677–717. Berkeley: University of California Press.

Caldwell, D. K., Prescott, J. H. & Caldwell, M. C. (1966). Production of pulsed sounds by the pygmy sperm whale, *Kogia breviceps*. *Bull. South. Calif. Acad. Sci.*, **65,** 245–248.

Caldwell, M. C. & Caldwell, D. K. (1965). Individualized whistle contours in bottlenosed dolphins, *Tursiops truncatus*. *Nature*, **207,** 434–435.

Caldwell, M. C. & Caldwell, D. K. (1966). Epimeletic (care-giving) behavior in Cetacea. In: *Whales, Dolphins, and Porpoises* (K. S. Norris, Ed.), pp. 755–789. Berkeley: University of California Press.

Caldwell, M. C. & Caldwell, D. K. (1967). Intra-specific transfer of information via the pulsed sound in captive odontocete cetaceans. In: *Animal Sonar Systems*, Vol.II (R.-G. Busnel, Ed.), pp. 879–936. Jouy-en-Josas, France: Labortoire de Physiologie Acoustique.

Caldwell, M. C. & Caldwell, D. K. (1968). Vocalizations of naive captive dolphins in small groups. *Science*, **159,** 1121–1123.

Caldwell, M. C. & Caldwell, D. K. (1971). Statistical evidence for individual signature whistles in Pacific white-sided dolphins, *Lagenorhynchus obliquidens*. *Cetology*, **3,** 1–9.

Caldwell, M. C. & Caldwell, D. K. (1972). Behavior of marine mammals. In: *Mammals of the Sea* (S. H. Ridgway, Ed.), pp. 409–465. Springfield, Ill.: Thomas.

Caldwell, M. C. & Caldwell, D. K. (1977). Social interactions and reproduction in the Atlantic bottlenosed dolphin. In: *Breeding Dolphins: Present Status, Suggestions for the Future* (S. H. Ridgway & K. Benirschke, Eds.), pp. 133–142. Washington, D. C.: U. S. Marine Mammal Commission Report MMC-76/07.

Caldwell, M. C. & Caldwell, D. K. (1979). The whistle of the Atlantic bottlenosed dolphin (*Tursiops truncatus*)—Ontogeny. In: *Behavior of Marine Animals: Current Perspectives in Research. Vol. 3. Cetaceans* (H. E. Winn & B. L. Olla, Eds.), pp. 369-401. New York: Plenum.

Caldwell, M. C., Caldwell, D. K. & Hall, N. R. (1973). Ability of an Atlantic bottlenosed dolphin (*Tursiops truncatus*) to discriminate and potentially identify to individual, the whistles of another species, the common dolphin (*Delphinus delphis*). *Cetology*, **14**, 1-7.

Caldwell, M. C., Caldwell, D. K. & Miller, J. F. (1973). Statistical evidence for individual signature whistles in the spotted dolphin, *Stenella plagiodon*. *Cetology*, **16**, 1-21.

Caldwell, M. C., Caldwell, D. K. & Siebenaler, J. B. (1965). Observations on captive and wild Atlantic bottlenosed dolphins, *Tursiops truncatus*, in the northeastern Gulf of Mexico. *Los Ang. Cty. Mus. Nat. Hist. Contrib. Sci.*, **91**, 1-10.

Caldwell, M. C., Hall, N. R. & Caldwell, D. K. (1971a). Ability of an Atlantic bottlenosed dolphin to discriminate between, and potentially identify to individual, the whistles of another species, the spotted dolphin. *Cetology*, **6**, 1-6.

Caldwell, M. C., Hall, N. R. & Caldwell, D. K. (1971b). Ability of an Atlantic bottlenosed dolphin to discriminate between, and respond differentially to, whistles of eight conspecifics. *Proc. 8th Ann. Conf. Biol. Sonar Diving Mammals*, Menlo Park, Calif.: Stanford Research Institute.

Caldwell, M. C., Haugen, R. & Caldwell, D. K. (1962). High-energy sound associated with fright in the dolphin. *Science*, **138**, 907-908.

Chance, M. R. A. & Jolly, C. J. (1970). *Social Groups of Monkeys, Apes and Man*. New York: Dutton.

Chittleborough, R. G. (1953). Aerial observations of the humpback whale *Megaptera nodosa* (Bonnaterre). *Aust. J. Mar. Freshw. Res.*, **10**, 125-143.

Chittleborough, R. G. (1958). The breeding cycle of the female humpback whale, *Megaptera nodosa* (Bonnaterre). *Aust. J. Mar. Freshw. Res.*, **9**, 1-18.

Chittleborough, R. G. (1965). Dynamics of two populations of the humpback whale *Megaptera novaeangliae* (Borowski). *Aust. J. Mar. Freshw. Res.*, **16**, 33-128.

Cullen, J. M. (1972). Some principles of animal communication. In: *Non-Verbal Communication* (R. A. Hinde, Ed.), pp. 101-122. Cambridge: Cambridge University Press.

Cummings, W. C., Fish, J. F. & Thompson, P. O. (1972). Sound production and other behavior of southern right whales, *Eubalaena glacialis*. *Trans. San Diego Soc. Nat. Hist.*, **17**, 1-13.

Cummings, W. C. & Thompson, P. O. (1971a). Underwater sounds from the blue whale, *Balaenoptera musculus*. *J. Acoust. Soc. Am.*, **50**, 1193-1198.

Cummings, W. C. & Thompson, P. O. (1971b). Gray whales, *Eschrichtius robustus*, avoid the underwater sounds of killer whales, *Orcinus orca*. *Fish. Bull.*, **69**, 525-530.

Cummings, W. C., Thompson, P. O. & Cook, R. D. (1968). Sound production of migrating gray whales, *Eschrichtius gibbosus* Erxleben. *Proc. 74th Meeting Acoust. Soc. Am.*

Davies, N. B. & Krebs, J. R. (1978). Introduction: Ecology, natural selection and social behaviour. In: *Behavioural Ecology* (J. R. Krebs & N. B. Davies, Eds.), pp. 1-18. Oxford: Blackwell Scientific Publications.

Dawbin, W. H. (1966). The seasonal migratory cycle of humpback whales. In: *Whales, Dolphins, and Porpoises* (K. S. Norris, Ed.), pp. 145-170. Berkeley: University of California Press.

Dawkins, R. & Krebs, J. R. (1978). Animal signals: Information or manipulation? In: *Behavioural Ecology* (J. R. Krebs & N. B. Davies, Eds.), pp. 282-309. Oxford: Blackwell Scientific Publications.

Defran, R. H. & Milberg, L. (1973). Tactile reinforcement in the bottlenosed dolphin. *Proc. 10th Annu. Conf. Sonar Diving Mammals*, Palo Alto, Calif., Stanford Research Institute.

Diercks, K. J. (1972). Biological Sonar Systems: A Bionics Survey. University of Texas at Austin: Applied Research Laboratories. Report ARL TR-72-34, 5 Sept. 1972.

Dijkgraaf, S. (1947). Uber die Reizung des Ferntastsinnes bei Fischen und Amphibien. *Experientia,* **3,** 206-216.

Donaldson, B. J. (1977). The tongue of the bottlenosed dolphin (*Tursiops truncatus*). In: *Functional Anatomy of Marine Mammals,* Vol. 3. (R. J. Harrison, Ed.), pp. 175-197. New York: Academic.

Dreher, J. J. (1966). Cetacean communication: Small group experiment. In: *Whales, Dolphins, and Porpoises* (K. S. Norris, Ed.), pp. 529-543. Berkeley: University of California Press.

Dreher, J. J. & Evans, W. E. (1964). Cetacean communication. In: *Marine Bio-Acoustics* (W. N. Tavolga, Ed.), pp. 373-393. Oxford: Pergamon.

Dudok van Heel, W. H. (1977). Dolphin reproduction in western Europe. In: *Breeding Dolphins: Present Status, Suggestions for the Future* (S. H. Ridgway & K. Benirschke, Eds.), pp. 109-112. Washington, D.C.: U.S. Marine Mammal Commission Report MMC-76/07.

Eisenberg, J. F., Muckenhirn, N. A. & Rudran, R. (1972). The relation between ecology and social structure in primates. *Science,* **176,** 863-874.

Emlen, S. T. & Oring, L. W. (1977). Ecology, sexual selection, and the evolution of mating systems. *Science,* **197,** 215-223.

Essapian, F. S. (1953). The birth and growth of a porpoise. *Nat. Hist.,* **58,** 385-392.

Essapian, F. S. (1962). Courtship in captive saddlebacked porpoises, *Delphinus delphis,* L. 1758. *Z. Saugetierkd.,* **27,** 211-217.

Evans, K. (1976). Nightmare at Molokini Island. *Am. Pac.,* **Nov-Dec.,** 21-22.

Evans, W. E. (1967). Vocalization among marine mammals. In: *Marine Bio-Acoustics,* Vol.II (W. N. Tavolga, Ed.), pp. 159-186. Oxford: Pergamon.

Evans, W. E. (1973). Echolocation by marine delphinids and one species of freshwater dolphin. *J. Acoust. Soc. Am.,* **54,** 191-199.

Evans, W. E. & Bastian, J. (1969). Marine mammal communication: Social and ecological factors. In: *The Biology of Marine Mammals* (H. T. Andersen, Ed.), pp. 425-475. New York: Academic.

Evans, W. E. & Dreher, J. J. (1962). Observations on scouting behavior and associated sound production by the Pacific bottlenosed porpoise (*Tursiops gilli* Dall). *Bull. South. Calif. Acad. Sci.,* **61,** 217-226.

Evans, W. E. & Maderson, P. F. A. (1973). Mechanisms of sound production in delphinid cetaceans: A review and some anatomical considerations. *Am. Zool.,* **13,** 1205-1213.

Evans, W. E. & Prescott, J. H. (1962). Observations of the sound production capabilities of the bottlenose porpoise: A study of whistles and clicks. *Zoologica,* **47,** 121-128.

Fine, M. L., Winn, H. E. & Olla, B. L. (1977). Communcation in fishes. In: *How Animals Communicate* (T. A. Sebeok, Ed.), pp. 472-518. Bloomington: Indiana University Press.

Fish, J. F. & Lingle, G. E. (1977). Responses of spotted porpoises, *Stenella attenuata,* to playback of distress (?) sounds of one of their own kind. *Proc. 2nd Conf. Biol. Mar. Mammals,* San Diego, Dec. 1977, p. 34. (Abstract).

Fish, J. F., & Sumich, J. L. & Lingle, G. L. (1974). Sounds produced by the gray whale, *Eschrichtius robustus.* In: *The California Gray Whale* (W. E. Evans, Ed.), pp. 38-45. *Mar. Fish. Rev.,* **36,** 38-45.

Fish, J. F. & Vania, J. S. (1971). Killer whale, *Orcinus orca,* sounds repel white whales, *Delphinapterus leucas. Fish. Bull.,* **69,** 531-535.

Ford, J. K. B. & Fisher, H. D. (1978). Underwater acoustic signals of the narwhal (*Monodon monoceros*). *Can. J. Zool.,* **56,** 552-560.

Fraser, F. C. & Purves, P. E. (1959). Hearing in whales. *Endeavour,* **18,** 93-98.

Frings, H. & Frings, M. (1964). *Animal Communication.* New York: Blaisdell.

Fuller, J. L. (1976). Genetics and communication. In: *Communicative Behavior and Evolution* (M. E. Hahn & E. C. Simmel, Eds.), pp. 23-38. New York: Academic.

Gardner, B. T. & Gardner, R. A. (1971). Two-way communication with an infant chimpanzee. In: *Behavior of Non-human Primates,* Vol. 4 (A. M. Schrier & F. Stollnitz, Eds.), pp. 117-185. New York: Academic.

Gilmore, R. M. (1961). *The Story of the Gray Whale.* San Diego, Calif.: American Cetacean Society.

Graycar, P. J. (1976). Whistle dialects of the Atlantic bottlenosed dolphin, *Tursiops truncatus.* Ph.D. thesis, University of Florida.

Graycar, P. J. (1977). Whistle dialects of the Atlantic bottlenosed dolphin, *Tursiops truncatus. Proc. 2nd Conf. Biol. Mar. Mammals,* San Diego, Calif., Dec. 1977. (Abstract).

Green, S. (1975). Variation of vocal pattern with sound situation in the Japanese macaque (*Macaca fuscata*): A field study. In: *Primate Behavior,* Vol. 4 (L. A. Rosenblum, Ed.), pp. 1-102. New York: Academic.

Gunther, E. R. (1949). The habits of fin whales. *Discovery Rep., 25,* 113-142.

Gurevich. V. S. (1977). Post-natal behavior of an Atlantic bottlenosed calf (*Tursiops truncatus,* Montagu) born at Sea World. In: *Breeding Dolphins: Present Status, Suggestions for the Future* (S. H. Ridgway & K. Benirschke, Eds.), pp. 168-184. Washington, D.C.: U.S. Marine Mammal Commission Report MMC-76/07.

Hafner, G., Hamilton, L., Steiner, W., Thompson, T. & Winn, H. E. (1977). Evidence for signature information in the song of the humpback whale. *Proc. 2nd Conf. Biol. Mar. Mammals,* San Diego, Calif., Dec. 1977, p. 35. (Abstract).

Harrison, R. J. & Ridgway, S. H. (1971). Gonadal activity in some bottlenose dolphins, *Tursiops truncatus. J. Zool. Lond., 165,* 355-366.

Herman, L. M. (1979). Humpback whales in Hawaiian waters: A study in historical ecology. *Pac. Sci, 33,* (in press).

Herman, L. M. & Antinoja, R. C. (1977). Humpback whales in the Hawaiian breeding waters: Population and pod characteristics. *Sci. Rep. Whales Res. Inst., 29,* 59-85.

Herman, L. M. & Arbeit, W. R. (1972). Frequency difference limens in the bottlenosed dolphin: 1-70 KC/S. *J. Aud. Res., 2,* 109-120.

Herman, L. M. & Arbeit, W. R. (1973). Stimulus control and auditory discrimination learning sets in the bottlenose dolphin. *J. Exp. Anal. Behav., 19,* 379-394,

Herman, L. M., Baker, S., Forestell, P. H. & Antinoja, R. C. (1980) Right whale *Balaena glacialis* sightings in Hawaii: A clue to the wintering grounds? *Mar. Ecol. Prog. Ser., 2* (in press).

Herman, L. M. & Forestell, P. F. (1977). The Hawaiian humpback whale: Behaviors. *Proc. 2nd Conf. Biol. Mar. Mammals.* San Diego, Calif., Dec. 1977, p. 29. (Abstract).

Herman, L. M., Forestell, P. H. & Antinoja, R. C. (1977). Study of the 1976/77 migration of humpback whales into Hawaiian waters: Composite description. Report to the National Marine Fisheries Service and U.S. Marine Mammal Commission, October 1977.

Herman, L. M. & Gordon, J. A. (1974). Auditory delayed matching in the bottlenosed dolphin. *J. Exp. Anal. Behav., 21,* 19-26.

Herman, L. M., Peacock, M. F., Yunker, M. P. & Madsen, C. J. (1975). Bottlenosed dolphin: Double-slit pupil yields equivalent aerial and underwater diurnal acuity. *Science, 189,* 650-652.

Hinde, R. A. (1972). (Ed.). *Non-Verbal Communication.* Cambridge: Cambridge University Press.

Hinde, R. A. (1974). *Biological Bases of Human Social Behavior.* New York: McGraw-Hill.

Hinde, R. A. & Rowell, T. E. (1962). Communication by postures and facial expressions in the rhesus monkey *Macaca mulatta. Proc. Zool. Soc. Lond., 138,* 1-21.

Hochberg, J. (1971). Perception. II. Space and movement. In: *Experimental Psychology*, 3rd Ed. (J. W. Kling & L. A. Riggs, Eds.), pp. 475-550. New York: Holt, Rinehart & Winston.

Hockett, C. F. (1960). Logical considerations in the study of animal communication. In: *Animal Sounds and Communication* (W. E. Lanyon & W. N. Tavolga, Eds.), pp. 392-430. Washington, D.C.: American. Institute of Biological Sciences., Pub. 7.

Hockett, C. F. & Altmann, S. A. (1968). A note on design features. In: *Animal Communication* (T. A. Sebeok, Ed.), pp. 61-72. Bloomington: Indiana University Press.

Jansen, J. & Jansen, J. K. S. (1969). The nervous system of Cetacea. In: *The Biology of Marine Mammals* (H. T. Andersen, Ed.), pp. 176-252. New York: Academic.

Kasuyu, T. & Aminul Haque, A. K. M. (1972). Some information on distribution and seasonal movement of the Ganges dolphin. *Sci. Rep. Whales Res. Inst.,* **24,** 109-115.

Kellogg, W. N. (1961). *Porpoises and Sonar.* Chicago, Ill.: University of Chicago Press.

Kinne, O. (1975). Orientation in space: Animals: Mammals. In: *Marine Ecology,* Vol. II (O. Kinne, Ed.), pp. 702-852. London: Wiley.

Kleiman, D.G. (1967). Some aspects of social behavior in the Canidae. *Am. Zool.* **7,** 167-177.

Kritzler, H. (1952). Observations on the pilot whale in captivity. *J. Mammal,* **33,** 321-334.

Kruger, L. (1959). The thalamus of the dolphin, *Tursiops truncatus,* and comparison with other mammals. *J. Comp. Neurol.,* **111,** 133-194.

Kummer, H. (1968). *Social Organization of Hamadrayas Baboons: A Field Study.* Chicago, Ill.: University of Chicago Press.

Kuznetsov, V. B. (1974). A method of studying chemoreception in the Black Sea bottlenose dolphin (*Tursiops truncatus*). In: *Morfologiya, Fisiologiya i Akustika Morskikh Mlekopitayushchikh* (V. Ye. Sokolov, Ed.), pp. 147-153. Moscow: Izdatel'stvo "Nauka." (in Russian).

Kuznetsov, V. B. (1978). Ability to communicate chemically and to transform information about chemical stimuli in the Black Sea *Tursiops. VII-aya Vses. Konf. Morsk. Mlekopitayushchim,* Simpherspol': 178-180. (in Russian).

Lang, T. G. & Smith, H. A. P. (1965). Communication between dolphins in separate tanks by way of an acoustic link. *Science,* **150,** 1839-1843.

Layne, J. N. (1958). Observations on freshwater dolphins in the upper Amazon. *J. Mammal.,* **39,** 1-22.

Layne, J. N. & Caldwell, D. K. (1964). Behavior of the Amazon dolphin, *Inia geoffrensis* Blainville, in captivity. *Zoologica,* **49,** 81-108.

Leatherwood, J. S. (1977). Mother–infant interactions of bottlenosed dolphins in captivity and at sea. In: *Breeding Dolphins: Present Status, Suggestions for the Future* (S. H. Ridgway & K. Benirschke, Eds.), pp. 143-167. Washington, D.C.: U.S. Marine Mammal Commission Report MMC-76/07.

Leatherwood, J. S. & Reeves, R. R. (1978). Porpoises and dolphins. In: *Marine Mammals* (D. Haley, Ed.), pp. 96-111. Seattle, Wash.: Pacific Search.

Lenneberg, E. H. (1964). *New Directions in the Study of Language.* Cambridge, Mass.: M.I.T. Press.

Lenneberg, E. H. (1967). *Biological Foundations of Language.* New York: Wiley.

Lillie, D. G. (1915). Cetacea. The British Antarctic (Terra Nova) expedition. *Nat. Hist. Rep. Zool.,* **1,** 85-124.

Lilly, J. C. (1961). *Man and Dolphin.* New York: Doubleday.

Lilly, J. C. (1962). Vocal behavior of the bottlenose dolphin. *Proc. Am. Philos. Soc.,* **106,** 520-529.

Lilly, J. C. (1963). Distress call of the bottlenosed dolphin: Stimuli and evoked behavioral responses. *Science.* **139,** 116-118.

Lilly, J. C. (1967). *The Mind of the Dolphin.* New York: Doubleday.

Lilly, J. C. & Miller, A. M. (1961a). Sounds emitted by the bottlenose dolphin. *Science,* **133,** 1689-1693.

Lilly, J. C. & Miller, A. M. (1961b). Vocal exchanges between dolphins. *Science,* **134,** 1873-1876.

Ling, J. K. (1974). The integument of marine mammals. In: *Functional Anatomy of Marine Mammals,* Vol. II (R. J. Harrison, Ed.), pp. 1-44. New York: Academic.

Ling, J. K. (1977). Vibrissae of marine mammals. In: *Functional Anatomy of Marine Mammals.* Vol. 3 (R. J. Harrison, Ed.), pp. 387-415. New York: Academic.

Loiselle, P. V. & Barlow, G. B. (1978). Do fishes lek like birds? In: *Contrasts in Behavior: Adaptations in the Aquatic and Terrestrial Environments* (E. S. Reese & F. J. Lighter, Eds.), pp. 31-75. New York: Wiley.

Lyons, J. (1972). Human language. In: *Non-Verbal Communication* (R. A. Hinde, Ed.), pp. 49-85. London: Cambridge University Press.

MacKay, D. M. (1972). Formal analysis of communicative processes. In: *Non-Verbal Communication* (R. A. Hinde, Ed.), pp. 3-25. Cambridge: Cambridge University Press.

Mackintosh, N. A. (1972). Biology of the populations of large whales. *Sci. Prog. Oxf.,* **60,** 449-464.

Marler, P. (1965). Communication in monkeys and apes. In: *Primate Behavior: Field Studies of Monkeys and Apes* (I. DeVore, Ed.), pp. 544-584. New York: Holt, Rinehart & Winston.

Marler, P. (1968). Visual systems. In: *Animal Communication: Techniques of Study and Results of Research* (T. A. Sebeok, Ed.), pp. 103-126. Bloomington: Indiana University Press.

Marler, P. (1973). A comparison of vocalizations of red-tailed monkeys and blue monkeys: *Cercopithecus ascanius* and *C. mitis* in Uganda. *Z. Tierpsychol.,* **33,** 223-247.

Marler, P. (1976). Social organization, communication, and graded signals: The chimpanzee and the gorilla. In: *Growing Points in Ethology* (P. P. G. Bateson & R. A. Hinde, Eds.), pp. 239-280. Cambridge: Cambridge University Press.

Marquette, W. M. (1978). Bowhead whale. In: *Marine Mammals* (D. Haley, Ed.), pp.70-81. Seattle, Wash.: Pacific Search.

Mate, B. R. (1975). An apparent hunting strategy of killer whales using underwater vocalizations. *Proc. Conf. Biol. Conserv. Mar. Mammals,* Santa Cruz, Calif., Dec. 1975, p. 34. (Abstract).

McBride, A. F. & Hebb, D. O. (1948). Behavior of the captive bottlenose dolphin, *Tursiops truncatus. J. Comp. Physiol. Psych.,* **41,** 111-123.

McBride, A. F. & Kritzler, H. (1951). Observations on pregnancy, parturition, and post-natal behavior in the bottlenose dolphin. *J. Mammal.,* **32,** 251-266.

Mech, L. D. (1970). *The Wolf: The Ecology and Behavior of an Endangered Species.* Garden City, N.Y.: Natural History Press.

Menzel, E. W. Jr. (1971). Communication about the environment in a group of young chimpanzees. *Folia Primatologica,* **15,** 220-232.

Mitchell, E. (1970). Pigmentation pattern evolution in delphinid cetaceans: An essay in adaptive coloration. *Can. J. Zool.,* **48,** 717-740.

Mitchell, E. (1975). (Ed.). Report of the Meeting on Smaller Cetaceans. *J. Fish. Res. Board Can.,* **32,** 889-963.

Moles, A. A. (1968). Perspectives for communication theory. In: *Animal Communication: Techniques of Study and Results of Research* (T. A. Sebeok, Ed.), pp. 627-642. Bloomington: Indiana University Press.

Morgan, P. W. (1971). The reactions of belugas to natural sound playbacks. *Proc. 7th Annu. Conf. Biol. Sonar Diving Mammals,* Menlo Park, Calif.: Stanford Research Institute.

Morgane, P. J. & Jacobs, M. S. (1972). Comparative anatomy of the cetacean nervous system. In: *Functional Anatomy of Marine Mammals,* Vol. I (R. J. Harrison, Ed.), pp. 117-244. New York: Academic.

Mountcastle, V. B. (1974). Sensory receptors and neural coding: Introduction to sensory processes. In: *Medical Physiology,* Vol. I (V. B. Mountcastle, Ed.), pp. 285-306. St. Louis: Mosby.

Norman, J. R. & Fraser, F. C. (1949). *Field Book of Giant Fishes.* New York: Putnam.

Norris, K. S. (1967). Aggressive behavior in Cetacea. In: *Aggression and Defense, Neural Mechanisms and Social Patterns* (C. D. Clemente & D. B. Lindsley, Eds.), pp. 225-241. Berkeley: University of California Press.

Norris, K. S. (1968). The evolution of acoustic mechanisms in odontocete cetaceans. In: *Evolution and Environment* (E. T. Drake, Ed.), pp. 297-324. New Haven, Conn.: Yale University Press.

Norris, K. S. (1969). The echolocation of marine mammals. In: *The Biology of Marine Mammals* (H. T. Andersen, Ed.), pp. 391-423. New York: Academic.

Norris, K. S. (1974). *The Porpoise Watcher.* New York: Norton.

Norris, K. S. & Dohl, T. (1980). The behavior of the spinner porpoises *Stenella longirostris* (Schlegel, 1841). *Fish. Bull.,* **77,** 821-849.

Norris, K. S., Goodman, R. M., Villa-Ramirez, B. & Hobbs, L. (1977). Behavior of California gray whale, *Eschrichtius robustus,* in southern Baja California, Mexico. *Fish. Bull.,* **75,** 159-171.

Norris, K. S., Harvey, G. W., Burzell. L. A. & Krishna Kartha, T. D. (1972). Sound production in the freshwater porpoises *Sotalia* cf. *fluviatilis* Gervais and Deville, and *Inia geoffrensis* Blainville in the Rio Negro Brazil. In: *Investigations on Cetacea,* Vol. IV (G. Pilleri, Ed.), pp. 251-262. Berne, Switzerland: Institute of Brain Anatomy, University of Berne.

Norris, K. S. & Prescott, J. H. (1961). Observations on Pacific cetaceans of Californian and Mexican waters. *Univ. Calif. Publ. Zool.,* **63,** 291-402.

Northrup, J., Cummings, W. C. & Morrison, M. F. (1971). Underwater 20-Hz. signals recorded near Midway Island. *J. Acoust. Soc. Am.,* **49,** 1909-1910.

Nottebohm, F. (1970). Ontogeny of bird song. *Science,* **167,** 950-956.

Nottebohm, F. (1975). A zoologist's view of some language phenomena with particular emphasis on vocal learning. In: *Foundations of Language Development: A Multidisciplinary Approach,* Vol.I (E. H. Lenneberg & E. Lenneberg, Eds.), pp. 61-103. New York: Academic.

Palmer, E. & Weddell, G. (1964). The relationship between structure, innervation, and function of the skin of the bottlenose dolphin, *Tursiops truncatus. Proc. Zool. Soc. Lond.,* **143,** 553-568.

Patterson, B. & Hamilton, G. R. (1964). Repetitive 20 cycle per second biological hydroacoustic signals at Bermuda. In: *Marine Bio-Acoustics* (W. N. Tavolga, Ed.), pp. 125-145. New York: Pergamon.

Patterson, F. G. (1978). The gestures of a gorilla: Language in another pongid. *Brain Lang.* **5,** 72-97.

Payne, R. S. (1976). At home with right whales. *Natl. Geogr.,* **149,** 322-339.

Payne, R. S. (1978). Behaviors and vocalizations of humpback whales, *Megaptera* sp. In: *Report on a Workshop on Problems Related to Humpback Whales (Megaptera novaengliae) in Hawaii.* U.S. Marine Mammal Commission Report MMC-77/03.

Payne, R. S. & McVay, S. (1971). Songs of humpback whales. *Science,* **173,** 585-597.

Payne, R. S. & Payne, K. (1971). Underwater sounds of southern right whales. *Zoologica,* **56,** 159-167.

Payne, R. S. & Webb, D. (1971). Orientation by means of long range acoustic signaling in baleen whales. *Ann. N.Y. Acad. Sci.,* **188,** 110-141.

Pepper, R. L. & Beach, F. A. III (1972). Preliminary investigation of tactile reinforcement in the dolphin. *Cetology,* **7,** 1-8.

Perrin, W. F. (1972). Color patterns of spinner porpoises, *Stenella* cf. *S. longirostris,* of the eastern Pacific and Hawaii, with comments on delphinid pigmentation. *Fish. Bull.,* **70,** 983-1003.

Pilleri, G., Kraus, C. & Gihr, M. (1971). The physical analysis of the sounds emitted by

Platanista indi. In: *Investigations on Cetacea,* Vol. III (G. Pilleri, Ed.), pp. 22-30. Berne, Switzerland: Institute of Brain Anatomy, University of Berne.

Poulter, T. C. (1968). Marine mammals. In: *Animal Communication: Techniques of Study and Results of Research* (T. A. Sebeok, Ed.), pp. 405-465. Bloomington: Indiana University Press.

Powell, B. A. (1966). Periodicity of vocal activity of captive Atlantic bottlenose dolphins, *Tursiops truncatus. Bull. South. Calif. Acad. Sci.,* **65,** 237-244.

Premack, D. (1971). On the assessment of language competence in the chimpanzee. In: *Behavior of Non-human Primates,* Vol. 4 (A. M. Schrier & F. Stollnitz, Eds.), pp. 186-228. New York: Academic.

Prescott, J. H. (1977). Comments on captive births of *Tursiops truncatus* at Marineland of the Pacific (1952-1972). In: *Breeding Dolphins: Present Status, Suggestions for the Future* (S. H. Ridgway & K. Benirschke, Eds.), pp. 71-76. Washington, D. C.: U.S. Marine Mammal Commission Report MMC-76/07.

Puente, A. E. & Dewsbury, D. A. (1976). Courtship and copulatory behavior of bottlenosed dolphins, *Tursiops truncatus. Cetology,* **21,** 1-9.

Purves, P. E. (1966). Anatomy and physiology of the outer and middle ear in cetaceans. In: *Whales, Dolphins, and Porpoises* (K. S. Norris, Ed.), pp. 320-380. Berkeley: University of California Press.

Purves, P. E. (1967). Anatomical and experimental observations on the cetacean sonar systems. In: *Animal Sonar Systems: Biology and Bionics* (R.-G. Busnel, Ed.), pp. 197-270. Jouy-en-Josas, France: Laboratoire de Physiologie Acoustique.

Ralls, K. (1976). Mammals in which females are larger than males. *Q. Rev. Biol.,* **51,** 245-276.

Renaud, D. L. & Popper, A. N. (1975). Sound localization by the bottlenose porpoise, *Tursiops truncatus. J. Exp. Biol.,* **63,** 569-585.

Reysenbach de Haan, F. W. (1957). Hearing in whales. *Acta Oto-Laryngol., Suppl.,* **134,** 1-114.

Rice, D. W. (1977). The humpback whale in the North Pacific: distribution, exploitation, and numbers. In: *Report on a Workshop on Problems Related to Humpback Whales (Megaptera novaeangliae) in Hawaii.* (K. S. Norris & R. R. Reeves, Eds.), pp. 29-44. Washington: U.S. Marine Mammal Commission Report MMC-77/03.

Rice, D. W. (1978). Beaked whales. In: *Marine Mammals* (D. Haley, Ed.), pp. 88-95. Seattle, Wash.: Pacific Search.

Richards, D. G. (1978). Environmental acoustics and song communication in passerine birds. Ph.D. thesis, University of North Carolina.

Ridgway, S. H. & Benirschke, K. (1977). (Eds.). *Breeding Dolphins: Present Status, Suggestions for the Future.* Washington, D.C.: U.S. Marine Mammal Commission Report MMC-76/07.

Rowell, T. E. & Hinde, R. A. (1962). Vocal communication by the Rhesus monkey (*Macaca mulatta*). *Proc. Zool. Soc. Lond.,* **8,** 91-96.

Rumbaugh, D. M. (1977). (Ed.). *Language Learning by a Chimpanzee: The Lana Project.* New York: Academic.

Saayman, G. S. & Tayler, C. K. (1973). Some behavior patterns of the southern right whale, *Eubalaena australis. Sonderdruck aus Z. Saugetierkd.,* **3,** 172-183.

Saayman, G. S. & Tayler, C. K. (1979). The socioecology of humpback dolphins (*Sousa* sp.). In: *Behavior of Marine Animals: Current Perspectives in Research. Vol. 3: Cetaceans* (H. E. Winn & B. L. Olla, Eds.), pp. 165-226. New York: Plenum.

Saayman, G. S., Tayler, C. K. & Bower, D. (1973). Diurnal activity cycles in captive and free-ranging Indian Ocean bottlenose dolphins, *Tursiops aduncus* Ehrenburg. *Behaviour,* **44,** 212-233.

Scammon, C. M. (1874). *The Marine Mammals of the Northwestern Coast of North America.* New York: Putnam.

Schevill, W. E. (1964). Underwater sounds of cetaceans. In: *Marine Bio-Acoustics* (W. N. Tavolga, Ed.), pp. 307-316. Oxford: Pergamon.

Schevill, W. E. & Watkins, W. A. (1971). Pulsed sounds of the porpoise *Lagenorhynchus australis. Breviora,* **366,** 1-10.

Schevill, W. E. & Watkins, W. A. (1972). Intense low-frequency sounds from an Antarctic minke whale, *Balenoptera acutorostrata. Breviora,* **388,** 1-8.

Schevill, W. E., Watkins, W. A. & Backus, R. H. (1964). The 20-cycle signals and *Balaenoptera* (fin whales). In: *Marine Bio-Acoustics* (W. N. Tavolga, Ed.), pp. 147-152. Oxford: Pergamon.

Schevill, W. E., Watkins, W. A. & Ray, C. (1969). Click structure in the porpoise, *Phocoena phocoena. J. Mammal.,* **50,** 721-728.

Schmidt, R. F. (1977). Somatovisceral sensibility. In: *Fundamentals of Sensory Physiology* (R. F. Schmidt, Ed.), pp. 81-175. New York: Springer-Verlag.

Scott, J. P. (1968). Observation. In: *Animal Communication: Techniques of Study and Results of Research* (T. A. Sebeok, Ed.), pp. 17-30. Bloomington: Indiana University Press.

Sebeok, T. A. (1963). Aspects of animal communication: The bees and porpoises. *Language,* **39,** 448-466.

Shallenberger, E. W. (1979). The status of Hawaiian cetaceans. Report to the Marine Mammal Commission, Washington, D.C., Feb. 1979.

Shannon, C. E. & Weaver, W. (1949). *The Mathematical Theory of Communication.* Urbana: University of Illinois Press.

Simpson, J. G. & Gardner, M. B. (1972). Comparative microscopic anatomy of selected marine mammals. In: *Mammals of the Sea: Biology and Medicine* (S. H. Ridgway, Ed.), pp. 298-418. Springfield, Ill.: Thomas.

Slijper, E. J. (1962). *Whales.* London: Hutchinson.

Smith, W. J. (1965). Message, meaning, and context in ethology. *Am. Nat.,* **99,** 405-409.

Smith, W. J. (1977). *The Behavior of Communicating.* Cambridge, Mass: Harvard University Press.

Sokolov, V. E. & Volkova, O. V. (1973). Structure of the dolphin's tongue. In: *Morphology and Ecology of Marine Mammals* (K. K. Chapskii & V. E. Sokolov, Eds.), pp. 119-127. New York: Wiley.

Spencer-Booth, Y. (1970). The relationship between mammalian young and conspecifics other than mothers and peers: A review. In: *Advances in the Study of Behavior. Vol. 3* (D. S. Lehrman, R. A. Hinde, E. Shaw, Eds.), pp. 119-194. New York: Academic.

Suchowskaja, L. I. (1972). Morphology of the taste organ in dolphins. In: *Investigations on Cetacea,* Vol. IV (G. Pilleri, Ed.), pp. 201-204. Berne, Switzerland: Institute of Brain Anatomy, University of Berne.

Taruski, A. G. (1976). Sounds and behavior of the pilot whale *Globicephala* spp. Ph.D. Thesis, University of Rhode Island.

Tavolga, M. C. (1966). Behavior of the bottlenose dolphin, *Tursiops truncatus:* Social interactions in a captive colony. In: *Whales, Dolphins, and Porpoises* (K. S. Norris, Ed.), pp. 718-730. Berkeley: University of California Press.

Tavolga, M. C. & Essapian, F. S. (1957). The behavior of the bottlenosed dolphin, *Tursiops truncatus:* Mating, pregnancy, parturition, and mother-infant behavior. *Zoologica,* **42,** 11-31.

Tavolga, W. N. (1968a). Fishes. In: *Animal Communication: Techniques of Study and Results of Research* (T. A. Sebeok, Ed.), pp. 271-288. Bloomington: Indiana University Press.

Tavolga, W. N. (1968b). Marine animal data atlas. Technical Report U.S. Naval Training Device Center No. 1212-1, 50-62.

Tavolga, W. N. (1976). Acoustic obstacle detection in the sea catfish (*Arius felis*). In: *Sound Reception in Fish* (A. Schuijf & A. D. Hawkins, Eds.), pp. 185-204. Amsterdam: Elsevier.

Tayler, C. K. & Saayman, G. S. (1972). The social organization and behavior of dolphins (*Tursiops aduncus*) and baboons (*Papio ursinus*): Some comparisons and assessments. *Ann. Cape Prov. Mus. Nat. Hist. (South Africa),* **9,** 11-49.

Thompson, P. O., Cummings, W. C. & Kennison, S. J. (ms.). Sound production of humpback whales, *Megaptera novaengliae,* in Alaskan waters.

Thompson, R. K. R. & Herman, L. M. (1975). Underwater frequency discrimination in the bottlenosed dolphin (1-140 kHz). *J. Acoust. Soc. Am.,* **57,** 943-948.

Thompson, R. K. R. & Herman, L. M. (1977). Memory for lists of sounds by the bottlenosed dolphin: Convergence of memory processes with humans? *Science,* **195,** 501-503.

Thorpe, W. H. (1972). The comparison of vocal communication in animals and man. In: *Non-Verbal Communication* (R. A. Hinde, Ed.), pp. 27-47. Cambridge: Cambridge University Press.

Titov, A. A. & Nikolenko, G. V. (1975). Quantitative evaluation of the sounds in three species of Black Sea dolphins. *Bionika.* Izdatel'stvo "Nauka Dumka," Kiev, **9,** 115-119. (in Russian).

Tomilin, A. G. (1967). Cetacea. In: *Mammals of the USSR and Adjacent Countries,* Vol. 9. Jerusalem: Israel Program for Scientific Translations.

Trivers, R. L. (1972). Parental investment and sexual selection. In: *Sexual Selection and the Descent of Man, 1871-1971* (B. Campbell, Ed.), pp. 136-179. Chicago: Aldine.

van Lawick-Goodall, J. (1968). The behaviour of free-living chimpanzees in the Gombe Stream Reserve. *Anim. Behav. Monogr.,* **1,** 161-311.

Voisin, J. F. (1972). Notes on the behaviour of the killer whale, *O. orca* (L.). *Norw. J. Zool.,* **20,** 93-96.

Watkins, W. A. (1967). Air-borne sounds of the humpback whale, *Megaptera novaeangliae. J. Mammal.,* **48,** 573-578.

Watkins, W. A. (1977). Acoustic behavior of sperm whales. *Oceanus,* **2,** 50-58.

Watkins, W. A. & Schevill, W. E. (1974). Listening to Hawaiian spinner porpoises, *Stenella* cf. *longirostris,* with a three-dimensional hydrophone array. *J. Mammal.,* **55,** 319-328.

Watkins, W. A. & Schevill, W. E. (1976). Right whale feeding and baleen rattle. *J. Mammal.,* **57,** 58-66.

Watkins, W. A. & Schevill, W. E. (1977). Sperm whale codas. *J. Acoust. Soc. Am.,* **62,** 1485-1490.

Watkins, W. A. & Schevill, W. E. (1979). Aerial observations of feeding behavior in four baleen whales: *Eubalena glacialis, Balenoptera borealis, Megaptera novaeangliae,* and *Balenoptera physalus. J. Mammal.,* **60,** 155-163.

Watkins, W. A., Schevill, W. E. & Best, P. B. (1977). Underwater sounds of *Cephalorhynchus heavisidii* (Mammalia: Cetacea). *J. Mammal.,* **58,** 316-320.

Wiley, R. H. Jr. (1978). The lek mating system of the sage grouse. *Sci. Am.,* **May 1978,** 114-125.

Wilson, E. O. (1968). Chemical systems. In: *Animal Communication: Techniques of Study and Results of Research* (T. A. Sebeok, Ed.), pp. 75-102. Bloomington: Indiana University Press.

Wilson, E. O. (1975). *Sociobiology: The New Synthesis.* Cambridge Mass.: Belknap/Harvard.

Winn, H. E. & Perkins, P. F. (1976). Distribution and sounds of the minke whale, with a review of mysticete sounds. *Cetology,* **19,** 1-12.

Winn, H. E., Perkins, P. F. & Poulter, Y. C. (1971). Sounds of the humpback whale. *Proc. 7th Ann. Conf. Biol. Sonar Diving Mammals,* Menlo Park, Calif.: Stanford Research Institute.

Winn, H. E. & Schneider, J. (1977). Communication in sirenians, sea otters, and pinnipeds. In: *How Animals Communicate* (T. A. Sebeok, Ed.), pp. 809-840. Bloomington: Indiana University Press.

Winn, H. E. & Winn, L. K. (1978). The song of the humpback whale *Megaptera novaeangliae*, in the West Indies. *Mar. Biol.*, **47**, 97-114.

Wood, F. G. Jr. (1953). Underwater sound production and concurrent behavior of captured porpoises, *Tursiops truncatus and Stenella plagiodon*. *Bull. Mar. Sci. Gulf and Caribb.* **3**, 120-133.

Wood, F. G. Jr. (1973). *Marine Mammals and Man*. Washington, D.C.: Luce.

Wood, F. G. Jr. (1977). Births of dolphins at Marineland of Florida, 1939-1969, and comments on problems involved in captive breeding of small cetacea. In: *Breeding Dolphins: Present Status, Suggestions for the Future* (S. H. Ridgway & K. Benirschke, Eds.), pp. 47-60. Washington, D.C.: U.S. Marine Mammal Commission Report MMC-76/07.

Würsig, B. (1978). Occurence and group organization of Atlantic bottlenosed porpoises, *Tursiops truncatus*, in an Argentine bay. *Biol. Bull.*, **154**, 348-359.

Würsig, B. & Würsig, M. (1980). Behavior and ecology of dusky porpoises, *Lagenorhynchus obscurus*, in the South Atlantic. *Fish. Bull.*, **77**, 871-890.

Yablokov, A. V., Bel'kovich, V. M. & Borisov, V. I. (1972). *Whales and Dolphins*. Jerusalem: Israel Program for Scientific Translations.

Yunker, M. P. & Herman, L. M. (1974). Discrimination of auditory temporal differences in the bottlenosed dolphin and by the human. *J. Acoust. Soc. Am.*, **56**, 1870-1875.

CHAPTER 5

The Structure and Functions of Cetacean Schools

Kenneth S. Norris and Thomas P. Dohl

INTRODUCTION

We define a school as any aggregation of aquatic animals that regularly swim together as a unit. This definition is purposely broader than most that have been proposed (see Shaw, 1970, for a review). This breadth is necessary, we believe, because the functions of school formation are themselves the diverse, almost universally occurring needs of free-swimming aquatic animals. In the broadest view, the school provides for social animals the conditions for living in open water. Also, since schools change shape, size, and constituent numbers under differing conditions, it is obvious that these conditions for living are expressed as equilibria between dispersive and attractive forces (see Breder, 1954).

It may be that the evolutionary basis for cetacean schools centers around protection from predation, as is suggested by Williams' (1964) cover-seeking concept or Hamilton's (1971) "selfish herd" model. An alternative view is that the evolutionary mechanisms are food-based. The collective behavior patterns of the school may enhance individual feeding opportunities in the face of unpredictable or patchy food resources. A mechanism of this type has been shown to be the case for the colonial nesting and flocking societies of great blue herons, *Ardea herodias* (Krebs, 1974), and for wagtails, *Motacilla* sp. (Zahavi, 1971). Wagtail groups fluctuate from a territorial system when the food supply is predictable to a flock when it is irregularly distributed. Whatever the ultimate causes of cetacean schooling, during the course of evolution a variety of functions have developed or been transferred to the context of the school.

Most dolphins spend their entire lives in moving schools. Being highly social advanced mammals one can expect their intraschool relationships to be complex, and in fact one can expect that the school is the matrix within which the complexities of these animals are mostly to be understood. This does not imply, however, that the dolphin school is necessarily a permanent unit of a given size, formed of the same individuals from day to day. In some species such as the spinner dolphin (Norris & Dohl, 1980) or the dusky dolphin (Würsig & Würsig, 1980) it clearly is not, and instead school sizes may vary widely and individuals or small groups may move freely from small to large aggregations.

Unraveling the structure of a cetacean school and the functions that the school promotes requires long careful observations of wild aggregations. Unfortunately, however, behavior studies of cetaceans in nature are not easy to do. Only in the last decade has quantitative information on natural behaviors begun to accumulate (for odontocete species see e.g., Evans, 1974, 1975; Evans & Bastian, 1969; Norris & Dohl, 1980; Saayman & Tayler, 1979; Shane, 1977; Würsig, 1976, 1978; Würsig & Würsig, 1979, 1980; also see Russian work cited by Bullock & Gurevich, 1979, and see Wells, Irvine & Scott, this volume). For mysticetes some recent work on humpback whales is reviewed in this volume in the chapter by Herman and Tavolga, while extensive studies of the southern right whale by R. Payne and his colleagues remain unpublished. Rather little, except in Hawaii, has been done underwater, and most of what we know about the details of social interaction has come from captive groups of odontocetes. Unpublished studies of killer whale pods promise to be interesting because sexual dimorphism in *Orcinus* is striking, and pods are small and apparently quite stable. Thus individuals and their histories can be studied.

All of these works must relate eventually to the growing body of knowledge about the substrates of behavior, including sensory and cognitive capacities, which are reviewed in several of the chapters in this volume.

At any rate, to order our current knowledge of schooling in open-water social species of cetaceans we have organized our presentation into six roughly separate functional categories. These are (*a*) food gathering; (*b*) reproduction and growth; (*c*) social integration and communication; (*d*) learning; (*e*) defense;

and (*f*) responses to environmental cycles. We perceive the activities or functions subsumed within these categories to be major determinants of variations in the geometry of cetacean schools.

FOOD GATHERING

Dolphins and whales feed largely on schooling prey of various sorts, and are themselves schooling animals. Between such predators and their prey, food-finding and capture strategies are pitted against concealment, avoidance, or satiation strategies. Brock and Riffenburgh (1960) point out that large prey schools tend to satiate predators with excess numbers and to maximize the cost of finding new prey. On the other hand, schooled predators may optimize capture. Major (1976) has demonstrated that schooled predatory fish are more efficient at food-capture than single animals. This is also true of some large terrestrial mammalian predators (Kruuk, 1972; Schaller, 1972) and almost certainly true of dolphins. In examining these systemic relationships for dolphins we will consider two kinds of information: intraschool optimization of food location and capture by odontocetes, and patterns of school movement that may allow optimal utilization of limited renewable food resources.

Many dolphins and small whale species are known or thought to use special feeding strategies (Würsig & Würsig, 1979, 1980). It seems intuitively obvious that all odontocetes have such strategies and that they will usually vary from species to species, depending on the particular congeries of adaptations possessed by a given form. Sperm whales and spinner dolphins are not likely to use the same strategies, given their widely divergent structures and behavior. The literature contains many passing and a few detailed references to feeding strategy, and many references exist showing that a number of cetacean migrations or seasonal movement patterns are based on seasonally abundant food supplies. [See, for example, Sergeant (1962) for the pothead *Globicephala melaena*; Norris and Prescott (1961) for the northeastern Pacific pilot whale *Globicephala scammoni*; and Neave and Wright (1968) for the harbor porpoise *Phocoena phocoena*.] Information on feeding strategies can be categorized into the following topics: (*a*) food-searching patterns, (*b*) food-capture patterns, and (*c*) learned patterns.

Food-Searching Patterns

Dolphin species such as the common dolphin (*Delphinus delphis*), the pantropical spotted dolphin (*Stenella attenuata*), the spinner dolphin (*Stenella longirostris*), and the dolphins of the genus *Lagenorhynchus*, which travel in schools numbering in the hundreds or even thousands of animals, often gather in schools that are broader than long relative to the direction of school movement. The assumption is that such a school shape allows the group of animals to scan a wide swath of sea for food or enemies, with the aggregate echo-location sounds of many animals. Thus such a school, which may be several kilometers broad, could scan a path of ocean of that width or a little more. Given a 5-km/hr school speed, and 1-km width, 5 km^2 of sea could be scanned per hour, an enormously increased efficiency when compared to the capability of a single animal. Since the food of cetaceans in such schools is typically formed of schooling species, such as surface-dwelling or scattering-layer fishes and squids (Fitch & Brownell, 1968) or anchovies and sprats (Tsalkin, 1938), locating a single school of prey can provide food for a large group of dolphins.

The mode of travel of various species during food-searching varies considerably. Some, like pilot whales, move deliberately in broad ranks (Norris & Prescott, 1961), while spinner or spotted dolphins may move more rapidly, engaging in spectacular aerial behavior along the way. The right whale dolphin (*Lissodelphis* sp.) may at times move very rapidly and spend much time in the air (Tomilin, 1967). Its extremely attenuate body may allow it to take three or four rapid propulsive strokes of its flukes while its anterior body is in the air, thus reducing drag. At any rate this species may be seen leaping in dense ranks, traveling considerable distances over the surface with each leap.

The most clearly defined case of lateral spreading of schools during food-searching is probably found in the pilot whales, in which rather modest numbers of animals (usually 100 or less) travel as a rank essentially abreast during food-finding. The lateral breadth of such schools may cover 2 km or more at times. Similar behavior is often seen in the smaller killer whale (*Orcinus orca*) schools.

It is interesting that the shape of fleeing schools is similar to schools searching for food, though generally much more tightly packed.

Searching may rely upon sea-bottom topography as a guidepost. The best-documented case of this sort has been reported for the species *Delphinus delphis* and *Stenella longirostris* by Evans (1974). *Delphinus* schools off southern California and Baja California were found to congregate with considerable predictability over escarpments and seamounts which might be as much as 2000 m below the surface (Fig. 5.1). It is surmised that the dolphins may be able to "hear" such topography by passive listening, even though it is too deep to reach by diving, since there is an increase in ambient sea noise in such areas. Increased density of scattering-layer organisms exist over such sea-bottom topography as compared to sea bottom with less relief, and the stomach contents of dolphins feeding over areas of high relief contain fishes and squids of the vertical migrant layers, as well as surface anchovies. Whether the echolocation systems of the dolphins could be used to discriminate such deep-lying topography is a moot point. A similar association of large toothed whales such as the sperm whale and Baird's beaked whale (*Physeter catodon* and *Berardius bairdi*) with subsurface seamounts has been reported to us by whalers at the Richmond Whaling Station, California, and was used by them as a guide to capture. Also it is thought that large baleen whales such as the fin whale

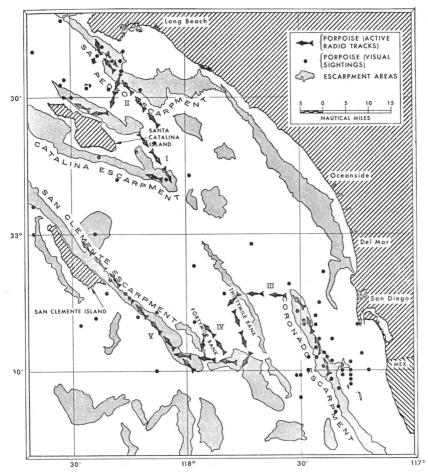

Fig. 5.1. Association of feeding common dolphin schools (*Delphinus delphis*) with subsurface topography in the southern California bight area. Note that most sightings are over escarpments and seamounts, with relatively fewer sightings in the intervening areas of low bottom relief. From W. E. Evans, *Ann. N. Y. Acad. Sci.*, **188**, 1971, 142-160. Reprinted by permission.

(*Balaenoptera physalus*) and the blue whale (*B. musculus*) migrate in the general region of the continental slope (Mackintosh, 1946), though this has not been correlated with feeding strategy.

In addition smaller odontocetes frequent these same areas and must feed there. Included are the common dolphin, *Delphinus delphis,* and members of the genera *Lissodelphis*, *Grampus*, and *Stenella*. Studies of otoliths and squid beaks from stomach contents (Fitch & Brownell, 1968) show feeding on vertical migrant or mesopelagic forms, some of which require dives in excess of 200-m depth. Even though many of these food organisms are small, individual capture rather than batch feeding is required.

Baleen whales generally do not form the tightly cohesive schools seen in odontocetes. They are much more often seen widely dispersed or alone, though occasional large assemblages of baleen whales are encountered on feeding, breeding, or calving grounds (Tomilin, 1967; Norris, 1967a). Considering their capability at long-distance sound transmission (Payne & Webb, 1971) it may be that the widely spread baleen whales should be considered as highly dispersed schools, perhaps covering many kilometers of sea.

An interesting sidelight on food search patterns is what has been called "social parasitism" by Norris and Prescott (1961). This behavior consists of small groups of dolphins of various species associating themselves with other species that are presumed to have more efficient food-finding capability. A typical example is the very common association of bottlenosed dolphins (*Tursiops* sp.) with northeastern Pacific pilot whale schools. Often small groups (20 or less) of bottlenosed dolphins swim at the ends of pilot whale food-searching ranks. Two observations by Norris and Prescott (1961) showed that this sort of association cannot be called casual. When the pilot whales sounded the bottlenosed dolphins normally sounded with them but usually surfaced well before the whales, presumably because of lesser breath-holding capabilities. However, the position of the dolphins on the surface was predictive of the place where the whales ultimately surfaced. That is to say, the dolphins at the surface seemed to travel over the whales beneath. The second observation was of a group of right whale dolphins swimming with a pilot whale school. Observations were made simultaneously from an aircraft circling above the schools and from a capture vessel harrying the whale school. The aircraft reported that the right whale dolphins followed each evasive maneuver of the whales, changing course as they did, even though they were swimming an estimated kilometer away. Presumably contact was maintained acoustically.

A striking association of a somewhat different sort is that between tropical oceanic dolphins (principally *Stenella* and *Delphinus*) and yellowfin tuna. The fish typically follow behind the dolphins and are presumably benefited by the dolphins' superior food-finding abilities (Green, Perrin & Petrich, 1971). Even when the dolphins are herded by power boats, the tuna generally remain with them and can be netted. This behavior forms the basis for much of the purse-seine tuna industry, and will be discussed in greater detail under "Optimization of Energy Supplies" later in this chapter.

Seasonal movements of many and perhaps most cetacean schools seem regulated by variable food resources. For instance, Clarke (1957) and Sergeant (1962) report seasonal migration by pilot whale schools to be regulated by seasonal abundance of squid and schooling fish. Wilke, Taniwaki, and Kuroda (1953) report such migrations for Dall's porpoise (*Phocoenoides dalli*), the Pacific white-sided dolphin (*Lagenorhynchus obliquidens*), and the northern right whale dolphin (*Lissodelphis borealis*) along the shores of Japan. Other observations of food-related migrations of many cetacean species may be found in summary volumes such as that of Tomilin (1967). Würsig and Würsig (1980) describe the movements of dusky dolphins (*Lagenorhynchus obscurus*) in relation to both seasonal and diurnal movements of their anchovy prey, concluding that the major movements of this species within the region they studied were probably food related.

While most cetaceans seem to move widely in relation to prey aggregations, often on a seasonal basis, some species have more restricted ranges and may be seen from day to day in certain regions. The latter group includes those living in rivers and lakes, such as the boutu (*Inia geoffrensis*), the platanistid dolphin of the Amazon–Orinoco Basins (see Layne, 1958), the susu (*Platanista gangetica*) of the Ganges–Brahmaputra Rivers (Pilleri, 1970), or the belukha (*Delphinapterus leucas*) that sometimes swims long distances up rivers during the downstream migration of salmon smolts. All of these forms, nevertheless, exhibit seasonal or other movements. It seems that topography simply channels or restricts movements like those normally seen in the open sea. Only those dolphins associating with oceanic islands, or perhaps certain features of the shoreline such as river mouths or bays, seem wholly restricted to specific geographic locations in part of their range. Some populations of the spinner dolphin, for example, habitually retreat to atoll lagoons during the day, presumably to rest (Norris & Dohl, 1980), and historic records suggest that this behavior has long continuity through time. Only these latter forms therefore fit the definition of refuging forms as developed by Hamilton and Watt (1970) and may be expected to have developed feeding relations such as core areas, trampling zones, and feeding arenas. An example is the school of spinner dolphins that each morning enters Bigej pass into the lagoon at Kwajalein Atoll, Marshall Islands, and leaves late in the day, presumably to feed offshore (personal observation of K. Norris). The problem of attempting to describe the more usual cetacean feeding patterns in Hamilton and Watt's (1970) less restrictive category of the "radial pack" (that is "a pack moving within a restricted area") lies in the almost universal lack of information about the total range of movement of any species. Do schools move in definable areas, or are they wholly opportunistic? It seems that the answer for oceanic forms lies somewhere between these two extremes. Along the California coast, for example, one can roughly recognize an inshore group of species, a continental-shelf-margin group, and a central-Pacific group. There are, however, records of all classes in near-shore waters on occasion. For instance, the oceanic false killer whale, *Pseudorca crassidens*, strands occasionally on San Nicolas Island (Mitchell, 1965), which is located on the continental shelf, and has been seen in a large school off near-shore Santa Catalina Island (Norris & Prescott, 1961), and the continental-slope-dwelling Pacific right whale dolphin is seen occasionally within a few kilometers of the California coast. These occurrences indicate movements of 160 km or more inshore of normal areas of occurrence, but even so the general faunal separations can still be recognized, and hence restriction of a very fluid sort to a general region of the sea occurs.

Food-Capture Patterns

Food-capture by odontocetes has been described in a number of sources (see Mitchell, 1978, for a recent summary of feeding preferences and strategies of the mysticete whales). The behavior of dolphin schools during food-capture typically differs strikingly from the searching mode. We classify the capture patterns as (*a*) spread-school formations, and (*b*) cooperative-capture methods.

It seems typical of oceanic schooling genera, such as *Stenella* and *Delphinus*, that once concentrations of food are located the form of the dolphin school changes. The frontally moving or relatively tightly bunched school enters a spread phase in which members may separate many times as far from their neighbors as during searching, and the entire school may become discoidal or essentially shapeless instead of ranked (see Saayman, Bower & Tayler, 1972). Evans (1974), Norris and Dohl (1980), and Tsalkin (1938) describe feeding schools in oceanic dolphins such as *Delphinus* and *Stenella*. Both genera when feeding spread over large areas of sea, reducing school cohesion and breaking into small subgroups. Single animals may sometimes occur within the boundaries of such a school. For some species diving patterns change markedly during the change from searching to feeding. During feeding the dives of the spinner dolphin become longer in duration and may become regular. Even though spread over a wide area of sea, *Stenella* schools may still continue to dive more or less synchronously, except in very large schools. In these very large schools diving or surfacing synchrony is not perfect and sections of a school a mile or more in largest dimension may dive or reach the surface with some independence. Nevertheless, it is impressive in such cases that considerable segments of these schools, even though widely spread, may dive together. Thus some sort of signal system, probably acoustic, must be involved in diving synchrony. It may also be that diving synchrony is part of the strategy of food-capture. Many animals attacking a prey school may have a greater percentage of success than an animal fishing alone. Such feeding strategies are known for a variety of animals such as cormorant flocks (Bartholomew, 1942), shearwaters (Shallenberger, 1973), pinnipeds (Roger Gentry, personal communication), and fish (Major, 1976).

Spread formation begins at dusk and persists in darkness in spinner schools (Norris & Dohl, 1980) but occurs in daytime in common dolphin schools (Evans, 1974). Thus these dolphin school formations may not have the same sensory basis as Radakov (1967) ascribes for prey fish schools, which disperse at night. He speculates that darkness protects scattered fish because the visual sense will fail to locate most individuals. (Radakov maintains that day-time schooling is protective because it mathematically reduces the likelihood of a predator locating prey). In contrast, using much the same argument as that presented by Brock and Riffenburgh (1960), we assume that the spread formation in dolphin schools is a response to the needs of food-getting, and that echolocation allows effective use of the spread formation at night even on rather dispersed prey.

Belukhas are reported to enter river mouths and rivers regularly in search of salmon smolts and other fish migrating seaward. Sometimes the aggregations at sea may be very numerous. For example, Tomilin (1967) reports about 10,000 belukhas in the Straits of Novaya Zemlya in pursuit of schooling fish migrating there in autumn. Belukhas are extremely sensitive to tidal changes during such feeding movements (Fish & Vania, 1971).

Lengthy detailed observations have been made of Black Sea common dolphin (*Delphinus delphis*) schools, and their schooling and feeding patterns are probably as well known as for any small cetacean (Tsalkin, 1938; Barabash-Nikiforov, 1940; Golenchenko, 1949). The dolphins reportedly concentrate over abundant food in essentially stationary spread schools of great extent, numbering, according to Tsalkin, as many as 250,000 to 300,000 individuals, numbers far in excess of any we can report. When food organisms become sparse or disperse, the large dolphin schools fragment or move away. Irregular migrations are made following the two major food species, anchovies and sprats, which have different thermal requirements and hence move to varying parts of the sea. Mating groups were also noted from the air, with groups of six to eight males and a single female reported, though parenthetically one wonders how accurate sex determination can be when carried out from the air, even though they report seeing males with erections (Tomilin, 1967).

Tomilin (1967, p. 498) describes feeding schools of *Delphinus* as follows:

Feeding individuals . . . dive more steeply (than in moving schools). When they feed on a stationary fish school they dive vertically. If the school stays at a depth of 60 to 70 m, the dolphins chase it to the surface, where they can catch fish with greater facility. Fishermen are convinced that this animal cannot dive deeper than about 70 m because it never escapes under the foot rope of the "alaman" (or purse seine) which is about 70 m deep.

Tarasevich (1957) suggests that if two species feed on closely similar or identical food supplies they will form mixed schools. Mixed schools are indeed evident features in dolphin schools in areas where single prey species predominate, or where a single phenomenon such as the rising of the deep scattering layer determines the food sources. For instance, in southern California waters, where either anchovy or squid schools are common prey, mixed feeding schools are common.

Würsig and Würsig (1980) describe dusky dolphins diving to round up segments of anchovy schools, which they then encircle and force against the surface. Individual dolphins may then transect the trapped fish and feed, taking as many as five fish in a single mouthful.

Dolphin species that form small or moderate-size schools have rather often been reported to use more specialized kinds of food-capture patterns. For example, killer whales have been reported to work cooperatively in food-capture. Andrews (1931) and Tomilin (1967) reported seeing gray whales (*Eschrichtius robustus*) whose tongues and flukes had been eaten or frayed by killers. Cooperative behavior in which a group of killers harried right and humpback whales being harpooned by hand whalers near shore in southern Australia has been chronicled by Dakin (1934). Some members of the killer school reportedly hung on the lips of the stricken whale while others pressed over its nostrils and still others hung on its flukes.

Killers are well known to aggregate around pinniped rookeries during pupping season (Norris & Prescott, 1961; Tomilin, 1967). The ability to find the small, isolated rookeries suggests that killer whales have considerable navigational ability and are sensitive to the seasonal variations indicating a rookery season is in progress.

One common defensive response of prey species to killer whales is to remain motionless or, in the case of the gray whale, often to turn belly up with flippers extended. The lack of motion presumably minimizes directional acoustic cues from the prey. The belly-up position of the gray whale may also allow free use of the tail, its major defensive weapon (Norris & Gentry, 1974). On first consideration this behavior might seem to represent a submissive posture like that of a subordinate wolf exposing its throat to a dominant animal. But such subordinate behavior would seem to be of little value in staving off an attacking animal of another species such as the killer whale. Evasive behavior is also evident, as for example by belukhas who changed river channels or avoided a river mouth altogether when recorded killer whale sounds were played in the river (Fish & Vania, 1971), or when gray whales were turned in their migration course or were caused to swim inside kelp beds by underwater projection of killer whale sounds (Cummings & Thompson, 1971).

Killer whales seem to use encirclement of large prey such as sea lions or walruses, with some regularity. Once the prey is bunched, a member or members of the killer whale group may cut through the compressed and often nearly immobile prey school, feeding as they go (Zenkovich, 1938; Norris & Prescott, 1961). Both Norris and Prescott (1961) and Leatherwood and Samaris (1974) report herding and capture of California sea lions (*Zalophus californianus*) by killer whales in which the whales formed a crescent behind the fleeing sea lions and raced in to make kills. Similar behavior has also been reported by M. Caldwell and Caldwell (1972) and Leatherwood (1975) in which *Tursiops* circled schools of prey fish, and individuals darted through the fish schools periodically in attempts at capture.

Saayman, Tayler, and Bower (1973) described the herding and crisscrossing maneuvers used by bottlenosed dolphins in trapping fish schools. Busnel (1973) described an unusual relationship between native fishermen and dolphins on the west coast of Africa in which mullet (*Mugil* sp.) during a near-shore migration, were herded into shallow water by Atlantic humpbacked dolphins (*Sousa teuszi*) and bottlenosed dolphin (*Tursiops* sp.) schools. The presence of the dolphins offshore of the fish schools allowed the fishermen to place gill nets among the fish while the dolphins also fed among the milling fish. The dolphins are reported to come to the fishing site in response to slaps of sticks against the water surface. Fink (1959) reports somewhat similar cooperative feeding of harbor porpoises in Monterey Bay in conjunction with California sea lion schools.

Prey-capture seems more often successful when the predatory cetacean forces the prey fish against a barrier of some kind, such as the surface, the shoreline, or underwater formations, producing, we suppose, predictable behavior on the part of the prey. This advantage has been shown by Major (1976) to exist for predatory fish forcing prey fish up against the water surface. One of us (Norris) has noted similar behavior in the Pacific bottlenosed dolphins, *Tursiops gilli*. In upper Magdalena Bay, Baja California Sur, Mexico, a lone dolphin was seen working along the steeply sloping sandy shore, so close to the beach that the dolphin was turned partly on its side. Small schooling fish were periodically forced into the wedge of water along the beach, whereupon the dolphin attempted capture, primarily with movements of its neck and head. A small stingray flapped completely out of the water at the observer's feet as the dolphin passed.

The humpback whale, *Megaptera novaeangliae,* encircles herrings and perhaps plankton organisms in a cylinder of bubbles streamed from its blowhole. The whale surfaces inside the cylinder with mouth open to receive entrapped prey that will not transgress the bubble barrier (Ingebrightsen, 1929; Jurasz & Jurasz, 1978).

Schools must also be the place where the young learn to feed on their own. McBride (1940), discussing captive *Tursiops truncatus*, reports that young animals would grasp fish after three weeks of nursing but that four weeks elapsed before any ingestion took place, and 11 weeks before it became routine. Presumably in the natural environment a similar time sequence would be followed by young animals.

Optimization of Energy Supplies

Both flocking birds and herding plains-dwelling ungulates seem to follow feeding strategies that optimize food availability to the group and reduce energy expenditures needed to acquire food. Specifically, by randomization of flock or herd movements over a feeding area and by changes in density of feeding animals relative to seasonal changes in food abundance, the group is able to feed almost constantly on essentially unexploited resources and allow the resources just utilized enough time to reconstitute themselves (seed populations build up or grass grows to optimum height for grazing). In this way the entire food resource is never reduced en masse but only randomly in part, and the difficulty of food-finding is accordingly reduced to a minimum. In the sea a human parallel is the new technique used primarily by Russian fishermen called "pulse fishing" in which a given resource area is very heavily utilized and then left to reconstitute itself while other areas are heavily fished.

The feeding patterns of some dolphins are suggestive that such a pulse-fishing mechanism may be at work, especially among common dolphins and spotted and spinner dolphin schools. We will return to the circumstances in these animals after discussing one case in terrestrial animals. Cody (1971) has studied feeding in mixed desert finch flocks. He found that flocks followed random paths ("a random walk") through a feeding area and that this behavior allowed much of the seed crop to reach full growth in the area being fed upon. The flocks were complexes of species, each member of which had a different feeding strategy and thus cropped a different part of the food supply. Differing rainfall conditions caused changed responses in the bird flocks (spreading or contraction of feeding grounds), once again tending to optimize instantaneous food availability from a finite renewable resource.

Perrin, Warner, Fiscus, and Holts (1973) have discussed the multispecies feeding aggregations of pelagic predators in the eastern tropical Pacific, including schools of the spotted dolphin, the spinner dolphin, and the yellow-fin tuna (*Thunnus albacares*). They found that the tuna and the spotted dolphin fed on fishes, squids, and crabs, largely from the epipelagic or pelagic faunas, while the spinner dolphins fed mostly on mesopelagic species. Further, the spinner dolphins in their sample showed a preponderance of empty stomachs, indicating a separation of the time of day during which spotted and spinner dolphins feed. As noted elsewhere in this chapter, the Hawaiian spinner dolphin is felt to be a nocturnal feeder, while the spotted dolphin feeds predominantly during the day.

This association of tuna and dolphin is a complicated one. The tuna apparently follow the spotted dolphin schools, or as the fishermen put it: "the spotters carry fish." The bond is strong enough that fishermen can herd the dolphins with speedboats, knowing that a net set around the dolphin school is likely to catch the fish swimming behind and below. If a mixed school of spotted and spinner dolphins is fragmented during the chase the fish are reported to follow the spotted dolphins. Because of this behavior, and because the food of the spotted dolphin and the tuna are very similar, sometimes identical, it is felt that the association is probably food-based.

Additional species are involved in this association. Fishermen locate both dolphins and fishes by first sighting flocks of seabirds, which include frigate-birds, boobies, terns, shearwaters, storm petrels, and jaegers. The frigate-birds, which are predominant, are primarily kleptoparasitic on the food of other birds, but are also reported to us by Hawaiian fishermen to take flying fish flushed from the water by swimming predators. Jaegers, too, are klepto-parasitic. The other birds feed on aquatic prey. Additional associates are known to assemble in the water, such as pelagic sharks and billfish (W. Perrin, personal communication).

Whether this complicated multispecies aggregation optimizes energy availability for some of its participants is not uniformly clear. A Hawaiian fisherman reported to one of us (Norris) that he could generally estimate the size of spotted dolphin schools by the number of frigatebirds flying over them; many birds meant a large school, while a few birds were less indicative. From subsequent observations we can confirm that this numerical relationship seems generally to be true.

The pantropical spotted dolphin in its association with tuna seems nuclear (W. Perrin, personal communication) and may assist the tuna in daytime food-finding. Both fish and "spotters" are apparently feeding within 30 m of the surface, while the spinner feeds "several hundred feet down" (Fitch & Brownell, 1968). Thus the spinner dolphins in mixed schools are apparently not feeding in these aggregations and may simply orient to spotted dolphin schools during rest periods and swim passively with the other species of the aggregation, thereby achieving a degree of "cover-seeking" in the open sea. Insular populations seem to achieve the same degree of cover by retreating to sheltered coves or atoll lagoons.

The Hawaiian spinner dolphin feeds on scattering layer organisms off the Hawaiian Islands, and does so in a manner suggestive of randomized search. Most feeding is performed by diving in deep water (400–2000 m) over the steep slopes of the islands where they merge with the deep sea. After daytime rest in protected coves, schools set out to sea in late afternoon or evening, and do so in variable directions from night to night. Thus nighttime feeding apparently occurs along different parts of the slope over time. Scattering layers apparently are denser in the areas of feeding than elsewhere near the islands, and one assumes that some sort of concentrating effect upon the scattering layer is at work there, perhaps based on subsurface currents. If so one can think in terms of replenishment rate and regard the use of variable feeding grounds as a mechanism allowing replenishment to occur and also, on the average, to bring the dolphins into areas of maximum food abundance along the slope.

William E. Evans (personal communication) reports a similar circumstance in common dolphins preferentially feeding over submarine escarpments in the southern California bight.

REPRODUCTION AND GROWTH

Reproduction

It seems probable that the proximity of individuals in a school allows reproductive synchrony to take place at less energetic cost than if animals swim alone separated by long stretches of open water, though proximity may also increase the chances of interference by sexual rivals. Seemingly the need for mechanisms allowing location of sexually ready animals over long distances is reduced in the close confines of dolphin schools. Yet schools may sometimes be large and diffuse, and the need for short-range mechanisms is to be expected. The larger whales whose "schools" may cover many miles of sea may require longer-range mechanisms. For instance, Payne and Webb (1971) have suggested that baleen whale schools may be spread very widely with animals often far out of visual contact, and that low-frequency sounds may serve to integrate such "schools." Certainly some species such as the humpback whale (*Megaptera novaeangliae*) seem to produce complex calls largely or wholly on the breeding grounds and are mainly silent when at other places. However, the right whale (*Eubalaena glacialis*) produces complex "stanza calls" far from calving areas (Cummings, Fish & Thompson, 1972). The calls of baleen whales may travel as far as 200 km or more at sea and clearly could serve as a means for localization of mates. In the more cohesive schools of dolphins and toothed whales such mechanisms are doubtless functional over much shorter ranges. It is possible that such special reproductive calls as that described by Dreher (1966) are effective in organizing reproductive activity within schools, though M. Caldwell and Caldwell (1972) regard such signals as primarily related to individual identification and not specifically assignable to the circumstances of reproduction (also see discussion of the "signature" function in Herman and Tavolga, this volume).

In the first few days of life many newborn cetaceans, especially delphinids, may engage in echelon swimming (Norris & Prescott, 1961), a swimming posture in which the baby animal positions itself above the midline forward of the dorsal fin of its mother. It typically lies against her side without swimming movements as she moves through the water. Dohl, Norris, and Kang (1974) report for a rough-toothed dolphin-bottlenosed dolphin (*Steno bredanensis* × *Tursiops truncatus*) hybrid that the young animal performed this maneuver only when it actively pressed its inboard pectoral flipper against the mother's side (Fig. 5.2). Such behavior may be the basis for reports that female belukhas carry their babies on their backs (Chapskii, 1941).

While location of potential mates represents the first level of discrimination needed for successful reproduction, a more refined level concerns detection of

Fig. 5.2. Echelon swimming, or assisted locomotion, between a rough-toothed dolphin mother (*Steno bredanensis*) and her hybrid baby (*Steno* × *Tursiops truncatus*) shortly after birth. Note the contact of the baby's inboard pectoral fin with the mother's side and the cupped flaccid flukes extended directly to the rear. In this position the young is "dragged" along without locomotory movements of its own as the mother swims.

sexual readiness. Such detection can reduce the energetic cost of the reproductive process and contribute to the precise timing of reproductive events. In migratory species at least, such timing would seem of crucial importance and is probably generally assisted by a photoperiodically mediated timing mechanism (Dawbin, 1966; Clarke, 1957; Norris, 1967a) (Fig. 5.3).

In general, as with terrestrial animals and birds, sexual timing seems most precise in migrators and in species or populations living in seasonal climates where both light levels and temperature may undergo strong seasonal fluctuations. This is not to say, however, that many tropical terrestrial animals may not have precisely timed reproduction. Tropical and subtropical cetacean species generally seem to exhibit very broad modes of reproduction spread over many months (W. F. Perrin, 1970, 1972a,b, and personal communication, for *Stenella attenuata* and *S. longirostris*; Evans, 1975, for *Delphinus*; Tomilin, 1967, for *Delphinus delphis*, in the Black Sea). Tarasevich (1957) suggests that seasonality in reproduction ensures segregation of pregnant females and young within dolphin schools, since sexually mature males do not participate in training of or caring for the young.

Gilmore (1961) has suggested that gray whales might locate lagoon mouths or follow one another by taste during migrations. It seems an important possibility that schooling animals, such as most porpoise species, may indicate sexual readiness by specific tastes imparted to the water through which the school swims (see M. Caldwell and Caldwell, 1967b; Herman and Tavolga, this volume). This point was brought home with some force to us (Norris and Dohl, 1980) when observing spinner dolphins from an underwater viewing vehicle. We found that these animals possessed a well-defined defecation period during which the water was literally filled with olive-colored trails of

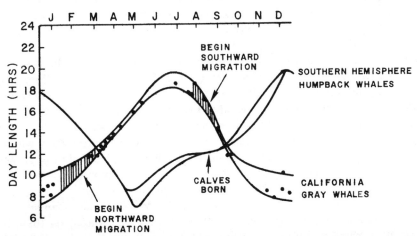

Fig. 5.3. A comparison of day length regimes of migrating California gray whales and southern hemisphere humpback whales. (Gray whale data mostly from Pike, 1962. Humpback whale data from Dawbin, 1966.)

finely divided fecal material. This period occurred for a few hours in the morning, after a night's feeding, and before the usual midday rest period. An impressive feature of this activity was that most school members swam through such clouds of material. Urination, which is probably a nearly constant process in cetaceans, may provide chemical cues about or to sexually ready animals. If indeed such a mechanism exists the school is clearly useful in promoting successful mating.

Sexual readiness is otherwise evident in dolphins. For instance, changes in the shape and color of the genital area in porpoises entering sexual readiness have been reported (Tavolga & Essapian, 1957; Tayler & Saayman, 1972). Behavioral cueing systems are obviously important too, in refining timing of mating (Tavolga & Essapian, 1957).

Some odontocetes, such as killer whales, pilot whales, sperm whales, and other strongly dimorphic forms, seem to be uniformly polygynous, with relatively few adult males or even a single male in schools having many females or immature animals. Sergeant (1962) regards this as the inevitable consequence of higher mortality in males, resulting in sex ratios greatly different than 1:1. Pelagic dolphin species that are not highly dimorphic, such as the common dolphin of the Black Sea are reported near to 1:1, or actually slightly biased in favor of males. Among 82,843 specimens examined, 53.6% were males and 46.8% were females, with very similar figures being recorded for embryos taken (Tsalkin, 1938).

Sexual segregation occurs commonly in dolphin schools. Norris and Prescott (1961) report it for northeastern Pacific pilot whale schools in southern California waters. True (1890) reports it for *Tursiops truncatus* in samples taken by the Cape Hatteras seine fishery, and many other examples could be cited for other genera (see Wells, Irvine, and Scott, this volume).

Intraspecific scarring, often sexually related, is widespread in odontocete schools. Dolphins and toothed whales commonly rake one another with their teeth, and from spacing and number of such marks it is often possible to assign the marks to schoolmates (Norris, 1967b). A subadult female Cuvier's beaked whale (*Ziphius cavirostris*), for instance, bore marks of the two teeth of an adult male of the same species (Norris & Prescott, 1961) (Fig. 5.4). Gunter (1943) reported a high incidence of scarring in males of *Tursiops truncatus* in Texas bays. Norris and Prescott (1961) cite observations of eastern north Pacific pilot whales in which juvenile males were usually covered with scar marks from larger animals, while adult males were largely without scars from conspecifics, presumably having healed completely during growth. Sex- and age-segregated sperm whale schools are commonly reported (see, for example, Clarke, 1957). Nursery herds of mothers and young numbering 100 to 150 animals have been reported in the Mozambique Channel, with such schools staying in water warmer than about 17°C. Males, on the other hand, may penetrate into polar regions (D. Caldwell, Caldwell & Rice, 1966).

At any rate sexuality seems likely to develop extremely early in schooling dolphins and certainly plays a central part in play and in social ordering of

Fig. 5.4. Scarring of a female Cuvier's beaked whale (*Ziphius cavirostris*). The separation and depth of the scars suggest that they were made by an adult male of the same species. Specimen found stranded on February 21, 1956, Pebbly Beach, Santa Catalina Island, California.

the school (Bateson, 1965). Erections and sexual play begin very early in dolphin life. M. Caldwell and Caldwell (1967b) report a newborn *Tursiops* having an erection at the age of two days when brushed by its mother.

Pregnancy and parturition are also involved in social ordering within the school. McBride and Kritzler (1951) present evidence that when captive pregnant female Atlantic bottlenosed dolphins are about four months from term they may segregate themselves from their normal position in the social order of the school. Swimming activity tends to slow, and during contractions a peculiar "yawning behavior" was seen and a unique bark was heard. After birth, the females and their young were often subject to savage attack by the adult males present, behavior which in wild schools would clearly result in structuring of the school. In fact, since such behavior is confined by tank limits in captivity, what seems to be destructive behavior might in fact be adaptive, perhaps resulting in the locations of females and young within the center of the school, as has often been observed. The attacks may represent herding. Structuring would be enhanced further by the circling behavior and whistling (McBride & Kritzler, 1951) which allows the mother to locate young animals that have strayed, even though widely separated. In the same captive school it was noted that when danger threatened "auntie dolphins" (consorts of the mother-young pair) rushed to them, and the school as a whole tended to coalesce around them.

A commonly observed feature of dimorphic odontocete schools is the positioning of adult males on school wings, with mothers and young more centrally located. This feature is sometimes strikingly evident in killer whale and pilot whale schools. On December 7, 1971, while the authors were on board the research vessel *Alpha Helix* on a cruise under the scientific leadership of Dr. Gerald Kooyman, a school of 15 killer whales was encountered and followed for 6½ hours. The school was continually harassed by the vessel, yet it consistently maintained its structure—the adult males on the wings of the rank-shaped school with what were presumed to be females and young in the central position or sometimes as a subgroup behind the middle region of the school. The school as a whole maneuvered constantly to elude us, and yet this form remained. It was a strking exhibition of the cohesiveness of this aspect of school formation and was made clearly evident to the observers by the extreme dimorphism of the animals. It is interesting to consider this structure in relation to the selfish herd hypothesis of Hamilton (1971) which shows that animals on the margins of schools are more vulnerable to predation than interior animals. In the very dimorphic species discussed above it seems obvious that the very large males should be less subject to predation than the smaller females and young, and hence dimorphism should, to some degree, compensate for exposed position in males.

A unique roseate formation has been described for sperm whales by Nishiwaki (1962) in which several whales lie at the surface facing each other, positioned like the spokes of a wheel. This behavior has also been noted by E. Barham and W. E. Evans (personal communication) during an aerial survey in the eastern tropical Pacific. Its function is unknown.

A number of reports describe sweeping changes in school structure related to sex or reproduction. Unisexual schools have been reported to occur under certain circumstances by a number of authors, usually during migration. A few examples will suffice. Pregnant female common dolphins in the Black Sea may migrate offshore to calve but yet remain within the general confines of the very large and spread schools (Tsalkin, 1938). The belukha regularly segregates into a unisexual school during migration, adult males generally forming a vanguard, while females and young may travel behind (Arsen'ev, 1939). The narwhal (*Monodon monoceros*) likewise forms unisexual schools at times, including migratory groups in northern Canada in which adult males may go on entirely different migratory paths than remaining portions of the school (Newman, 1968).

Ohsumi (1971) has reported on an interesting experiment concerning school composition of sperm whales. An attempt was made to catch all members of nursery schools which occur in certain localized areas, and three such schools were taken. Examination revealed that average school size was 27.1 animals, males were all calves and represented 22.9% of all these schools, with adult females and female calves comprising the remainder. Males are reported to leave these schools at puberty to form "bachelor schools." Bulls are thought to struggle with one another to assume a position in "harem schools" during the breeding season. Ohsumi regards the harem as temporary, and males are often seen as solitary animals at other seasons. In nursery schools Ohsumi feels leadership is exercised by old females, though all-male groups apparently lack a leader.

Periodic coalescence of smaller schools into larger aggregations during parturition has been reported for the common dolphin in the Black Sea by Tsalkin (1938). Similar behavior was noted by the authors for the Hawaiian spinner dolphin. An aggregation of all or most schools of spinner dolphins living on the lee coast of the Island of Hawaii was noted in early fall at a time when births were especially common, though birth was far from restricted to this period but instead spread rather widely through the year.

Such aggregations may provide more than protection. For instance, D. Brown, Caldwell, and Caldwell (1966) refer to schoolmates assisting in the removal of the afterbirth from a newly parturient mother, and Dohl et al. (1974) report similar though incomplete behavior in the rough-toothed dolphin (*Steno bredanensis*). Aggregation on breeding grounds is probably both best known and most restricted geographically for the gray whale, the black right whales (*Eubalaena glacialis*), and the humpback whale (*Megaptera novae-angliae*) which gather for calving in inland lagoons or along island shores. Clarke (1957) suggests that such aggregation enhances reproductive success, and that it played an important part in allowing such populations to recover from near extinction, at least in the case of the gray whale.

Growth

Often the growth stage of cetaceans seems to regulate their location within a school. As has been discussed above, mother–young pairs are prominent features of cetacean groupings (Fig. 5.5), and such pairs may persist on an intermittent basis for remarkably long periods. For example, in the Atlantic bottlenosed dolphin the members of a mother–young pair may return to each other in time of stress for years after the young has reached adulthood (Tavolga & Essapian, 1957). Infants often remain near their mothers during capture (Gunter, 1943). For example, whalers sometimes caught young sperm whales because the mother, and sometimes the other schoolmates, were attracted to the captive and could be killed (D. Caldwell et al., 1966). Mal'm (1946) reports that female common dolphins whose infants have been killed in nets may swim about them in spite of nearby fishing activities.

Yet it is also a common observation that juvenile animals may be found throughout a school and that very small young may sometimes actually swim in the lead of spinner dolphin schools (Norris & Dohl, 1980). This observation, we believe, does not indicate actual leadership in the school falling upon the young, but instead the lack of any such defined leader in the school. Half-grown spinner dolphins have been noted performing repeated spins at the

Fig. 5.5. Nursing posture between rough-toothed dolphin mother (*Steno bredanensis*) and her hybrid baby (*Steno* × *Tursiops truncatus*).

advancing front of a school traveling to the feeding grounds after completion of a rest period, yet when the same school changed direction 180° these animals were at once part of the trailing group.

That dispersal, especially of juvenile whales, may occur over wide areas and thus influence genetic intermixture is indicated by tagging results from Antarctic baleen whales (S. Brown, 1954) which showed as much as 50° longitudinal dispersal in fin and blue whales. Such dispersal occurred rapidly according to tag returns. Whether such dispersal movements are features in other species, as they are in some terrestrial and avian species, is unknown.

Dominance hierarchies are known from captive odontocete schools and reflect, to some degree, age structure within the school (see, for example, Tavolga & Essapian, 1957 and McBride & Hebb, 1948 for *T. truncatus*; Norris & Prescott, 1961 for *Globicephala scammoni*; Bateson, 1965 for *Stenella longirostris* and *S. attenuata*) and also affect the overall structure of schools. For instance, McBride (1940) reported a well-defined size-related order in captive *Tursiops truncatus* schools. New arrivals were first bitten by established males and females. After a few weeks such new-comers were assimilated into the established schools, though some seemingly permanent intimidation seemed to exist, signaled by jaw-claps and a peculiar body posture "like that taken by a dolphin preparing to get underway."

In natural schools at sea dominance may be expressed largely by spacing and placement of subgroups and by sexual segregation within the school as a whole, while in captivity confinement may convert such dispersive pressure into the establishment of hierarchies. Thus one should be careful in interpretation from captive observations to the conditions which might be supposed to occur in nature.

SOCIAL INTEGRATION AND COMMUNICATION

A considerable body of anecdotal evidence can be mustered showing that odontocete schools function to allow social integration, and some of it has been mentioned in earlier contexts. In fact one can conceive of the school as an equilibrium system ordered by two sorts of social phenomena: those factors tending to promote cohesion of the school and hence to bring school members together, and those factors that tend to spread the school apart. These two composite phenomena have been called "centrifugal" and "centripetal" factors by Breder (1954). Indeed, it is notable that dolphin schools in general may change drastically in size or in their spacing of animals and subgroups under differing conditions. For instance, spinner dolphins are highly aggregated during daytime rest periods; in fact the animals swim so closely packed that some members may be near to touching (Bateson, 1965, reports that sleeping animals do not actually touch in mixed *Stenella* schools), and groups of as many as 50 animals may occupy an area no larger than 40 to 60 m in diameter. The same school may assume a spread formation later in the day and evening, during feeding, and may spread over a kilometer of sea, with subgroups and even individuals widely spaced.

The parts of such a social integration system are obviously diverse. We can, however, list some phenomena or patterns that tend toward cohesion and some that tend to dispersion. Patterns tending toward cohesion seem to involve protection, fright, sleep or rest, as well as familial or habitual associations. Patterns tending toward dispersion seem to involve alertness, aggression, feeding, and a lack of familial or habitual association.

Such social integration is obviously mediated by various sensory and signaling systems. It seems clear that vision and hearing and perhaps taste are the main senses for most cetaceans (see Popper; Dawson; Herman & Tavolga, this volume). Hearing may be particularly important. Suffice it to say that ample evidence exists that sounds, either produced by the animals themselves or from other sources in their environment, serve many vital functions. It is probable that most unified actions of schools when spread over considerable expanses of water are acoustically mediated. For example, flight, synchronous diving in conditions of low visibility, epimeletic behavior in which school members may come from considerable distances to assist distressed members, feeding behavior in which cetaceans may swim long distances to specific feeding situations, synchronized intraspecific school movements, and probably migration and avoidance of stranding in areas of great tidal fluctuations, are apt to be assisted by phonation, passive listening, or active echolocation. McBride and Hebb (1948) presented some of the first well-documented examples of these sorts, and Tomilin (1955, cited in Tomilin, 1967) expanded these examples further.

Epimeletic behavior, mentioned above, is especially afforded female and young animals, adult males often being left untended though in distress (Hubbs, 1953). Such behavior is very widely known among odontocetes and is also known from some mysticetes (M. Caldwell and Caldwell, 1966) (Fig. 5.6).

It is likely that protection of individual cetaceans, especially small ones, is greatly enhanced by being in a school, just as has been shown to be the case for fish (Breder, 1967). In fact for an oceanic dolphin the school represents the focus of all living activity, and lone animals at sea tend to be severely frightened. This was shown graphically in early tests using single trained dolphins to work in the open sea (Norris, 1965).

Fright induces bunching of dolphin schools, often of a very extreme sort. Animals may come in to close proximity or perhaps actually touch at times in a dense fleeing school. If a school is disturbed and pursued and if great disturbance is caused, as by a shot fired into the water, the animals will come together and usually will begin rapid synchronized swimming away from the disturbance, often accompanied by frequent leaps. Roughly, the greater the fright, the tighter the school may become and the more time it may spend leaping.

MacAskie (1966) reported on the behavior of a school of killer whales which had been "cornered" against a sea cliff by three vessels. All five animals reared bolt upright in the water with their heads facing offshore toward the encircling vessels. They then drove and appeared on the sea side of the vessels and swam away.

Perhaps no other group or school behavior in cetaceans has excited such controversy as mass stranding. Many species of cetaceans have stranded as single individuals, and these seem uniformly ill, but whole schools of certain

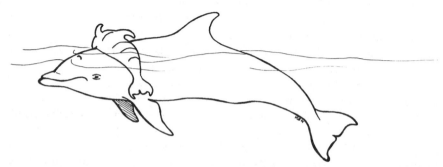

Fig. 5.6. Epimeletic behavior in a mother-young dolphin pair. A stillborn baby is carried on the snout of its mother, sometimes until decomposition is far advanced (see text).

species strand en masse, especially globicephaline whales, killer whales, false killer whales and their near relatives the electra or melon-headed whale (*Peponocephala electra*) and the pygmy killer whale (*Feresa attenuata*) (Fig. 5.7). Some mass strandings seem to include nonmoribund individuals, since members of such schools have sometimes been kept alive in oceanaria, while it is rare for single strandings to recover. A number of theories designed to explain this perplexing behavior have been advanced. In most cases the most logical cause seems to be that for a variety of reasons, such as faulty navigation, parasitism, or disease (especially of the nervous system), a single animal or the majority of the school may strand, and the remainder may come onto the beach because of factors relating to the tight social integrity of the school, probably mediated by acoustic and other signals (see Norris, 1967a). Most of the major stranding species are members of groups in which pronounced sexual dimorphism and organized school structure are known (though other nondimorphic genera such as *Lagenorhynchus* also strand occasionally). In fact, pronounced sexual dimorphism is almost confined to odontocete species that regularly strand as schools.

Sleep or rest in natural schools has seldom been observed. It has been noted in the spinner dolphin (Norris & Dohl, 1980), the common dolphin (Evans, 1974), and the northeastern Pacific pilot whale (Norris & Prescott, 1961). Sleep groups are reported for captive Hawaiian spinner and spotted dolphins (Bateson, 1965). Occasional records by whalers of quietly floating animals that could easily be approached, or which were run into by the whaling vessel, have sometimes been described as sleep. For example, the bowhead or Greenland right whale (*Balaena mysticetus*) has been encountered in this fashion (Gray, 1887), as has the fin whale. McBride (1940) describes similar nearly motionless sleep in captive Atlantic bottlenosed dolphins, in

Fig. 5.7. A stranded school of Pacific pilot whales (*Globicephala* sp.) at 4.8 km N. La Paz, Baja California, Mexico, on May 12, 1959.

which animals sculled slowly with eyelids closed, except for a brief opening every 15 seconds or so.

Social order may be altered between sleep and active states, with subgroup size changing and shifts in social grouping also taking place (Bateson, 1965; M. Caldwell & Caldwell, 1964).

The typical sleep pattern seems to consist of marked aggregation of school members, slowing of swimming, changes in diving rhythms, and a suppression or alteration of normal familial and habitual association patterns (see Slijper & van Utrecht, 1959). As mentioned earlier, in the Hawaiian spinner dolphin rest seems usually to occur near midday and to occur in insular populations in sheltered coves or atoll lagoons. Rest behavior in oceanic spinner dolphin populations remains unobserved but, as noted elsewhere, may take place with other *Stenella* species. In insular spinner dolphins a school will typically come inshore in rank formation from the deep-water feeding grounds. Subgroup structure will still be in evidence and much activity such as spinning, leaping, and head-slapping will be seen. Once in the rest area the school tends to slow and become discoidal, very tightly packed (to or near the point of actual contact of members), and aerial behavior soon ceases altogether. Diving, which is shallow in the traveling school, becomes predominant, with surface times reduced until only a few breaths are taken before resubmergence. Phonation changes from a varied mixture of squeals, barks, whistles, and clicks in the traveling school to desultory click emission in the resting school. Powell (1966) has described a similar diurnal pattern of phonation in *Tursiops,* with peaks in the morning and afternoon (Fig. 5.8). Arousal in the resting spinner dolphin school may first be noticed either by a resurgence of phonation or aerial behavior (Norris & Dohl, 1980).

From captive studies familial or habitual associations are known to be prominent features of bottlenosed dolphin schools (McBride & Hebb, 1948;

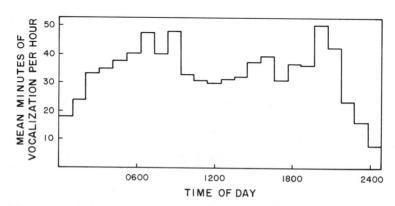

Fig. 5.8. Mean levels of vocal activity in relation to time of day for five Atlantic bottlenosed dolphins (*Tursiops truncatus*). From B. A. Powell, *Bull. South. Calif. Acad. Sci.,* **65,** 1967, 237–244. Reprinted by permission.

Tavolga & Essapian, 1957; M. Caldwell & Caldwell, 1964; Tavolga, 1966). While familial associations are essentially unknown for other odontocete species because captive schools have not been available for study over long enough periods, anecdotal evidence suggests that the general pattern of familial or habitual groupings found in the bottlenosed dolphin may be common throughout the Odontoceti. The rather solitary Ganges susu (*Platanista gangetica*), which is usually seen alone or in pairs (Kasuya & Aminul Haque, 1972), needs further study in this context. The presence within most odontocete schools of associations such as mother–young pairs and age and sex groups is of unknown permanence.

The spinner dolphin study of Norris and Dohl (1980) revealed fluidity, not permanence, of all groupings that could be tested. Nonetheless, on a day-to-day basis aggregations of stable membership were seen. It is simply unknown what is of long-term and what is of short-term permanence in the groups found within such schools. Würsig (1976), in one of the most successful radiotracking studies of dolphins to date, demonstrated that two radiotagged individual dusky dolphins (*Lagenorhynchus obscurus*) taken from separate subgroups of the same school went on widely divergent paths. At times on days following the tagging the pair was separated by 50 km, indicating that the school from which they came was probably casual.

Würsig and Würsig (1977) and Würsig (1978), from a study of bottlenosed dolphins in Argentina, show the same mixture of group fluidity and permanence. Marked animals could be divided into transient individuals located as far as 300 km from the initial recording site and those that were always present whenever dolphins were sighted. Wells, Irvine, and Scott (this volume) in a study of visual and radiotagged bottlenosed dolphins in Florida waters found an overall population range for a group of dolphins, as well as seasonal sexual segregation within the population.

Casual groups of various species are often assembled in oceanaria, and when studied are found to have developed social relationships usually involving all captive species. Bateson (1965), observing such an assemblage of spinner and spotted dolphins (Fig. 5.9), noted that groups of two species were organized into dyadic and triadic associations, with most sexual behavior occurring between species (Fig. 5.10). A complex hierarchy was also noted, involving both species. D. Brown and Norris (1956) reported that interspecific matings or mating attempts occurred nearly daily between a male bottlenosed dolphin and a North Pacific striped dolphin. D. Brown et al. (1966) reported on behavioral interactions between a false killer whale (*Pseudorca crassidens*) and a Pacific pilot whale (*Globicephala scammoni*) and between North Pacific striped dolphin–bottlenosed dolphin pairs. This intergeneric activity included sexual behavior, interaction during parturition, and strong social attachments. In fact, broadly speaking, odontocetes in captivity seem to be able to construct social groups from species which may be widely divergent phylogenetically and morphologically. An extreme example of this sort was the remarkable imitative social and sexual behavior noted by Tayler and Saayman (1973) between an In-

Fig. 5.9. A mixed captive school of spinner dolphins (*Stenella longirostris*) and Pacific spotted dolphins (*Stenella attenuata*) at Sea Life Park, Hawaii. Note the black beak tips of spinner dolphins above, and the white beak tips of spotted dolphins below. These may be species-specific school identification marks to animals at sea. Note also the pectoral contact between the animals.

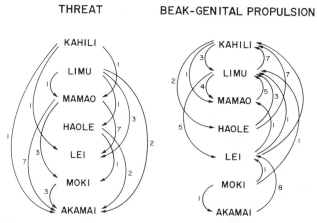

Fig. 5.10. Systems of ascendancy of threat and beak–genital propulsion in a captive mixed Pacific spotted dolphin (*Stenella attenuata*)-spinner dolphin (*Stenella longirostris*) school, at Sea Life Park, Hawaii (see Bateson, 1965). Kahili: *S. attenuata* male; Lei: *S. attenuata* female; Moki: *S. longirostris* male; Akamai: *S. longirostris* male; Haole: *S. longirostris* male; Mamao: *S. longirostris* female; Limu: *S. longirostris* male.

236

dian Ocean bottlenosed dolphin (*Tursiops aduncus*) and a Cape fur seal (*Arctocephalus pusillus*).

A similar association between widely divergent marine mammal taxa has been noted by Norris, Goodman, Villa-Ramirez, and Hobbs (1977) in which California sea lions were seen associating closely with courting and mating California gray whales at the entrance to a calving lagoon. The sea lions, all subadult animals, swam closely among the whale groups, and followed them as they moved. The basis for this association is unclear.

Most interesting and challenging, it seems to us, is the potential contribution of learning the contingency patterns of communication to school behavior and structure. Such second-order learning (Reusch & Bateson, 1951) opens the way for much subtlety of behavior. With it communication may become a conscious process for both sender and receiver, rather than the perhaps more usual mammalian pattern in which an emotional state in one animal is perceived by another and reacted to without the sender necessarily knowing it had transmitted signals in the first place. The way becomes open to the use of ideas, however simple. Ideas are, after all, second-order abstractions themselves. Was the case of the killer whale and the Pacific bottlenosed dolphin discussed by Pryor (1973) of this sort? These animals, kept in adjoining tanks, learned to join each other, the more agile bottlenosed dolphin jumping the barrier between the two tanks in which the animals were held. When trainers sought to contain them by increasing the width and height of the barrier the dolphin was no longer able to hurdle it, but the whale, with its superior strength, repeatedly and apparently covertly pushed the barrier aside enough for the dolphin to swim in. As with most anecdotes we are left unsatisfied, but together with experimental evidence cited earlier it seems a fertile area for study. With such contextual capability as part of the behavioral armament of the dolphin in a school we might expect to find a number of advanced patterns involving behavior based on ideas. One might find evidence of altruism resulting from behavioral transactions in which the context might be widely known. Assistance or support behavior of cetaceans has commonly been observed and is clearly altruistic (M. Caldwell & Caldwell, 1966). Examples have been observed both in captive and wild animals, and are sometimes interspecific. For instance, a North Pacific striped dolphin died in an oceanarium tank, and as it sank to the bottom an adult Atlantic bottlenosed dolphin attempted to raise it in typical supporting behavior (D. Brown & Norris, 1956). Two additional examples are provided by Norris and Prescott (1961) and Caldwell, Brown, and Caldwell (1963). In the first example a distressed Dall's porpoise, which had just been introduced to an oceanarium tank, was supported by two North Pacific striped dolphins, and in the latter, an adult male Northeastern Pacific pilot whale carried a dead Pacific striped dolphin for many hours in an oceanarium and resolutely prevented the staff from removing it from the tank. An example of intergeneric "standing-by" behavior was noted during the capture of a Northeastern Pacific pilot whale, in which the whale was accompanied throughout capture by a group of Pacific striped dolphins, which left the collecting vessel only after the whale was hoisted on board (Norris, 1974). The smaller dolphins swam very close to the tethered whale during the entire half hour of capture. One wonders if the rather frequent reports of dolphins supporting people at sea are of this sort. At any rate the question arises from such examples whether the animals were responding by means of a fixed action pattern to the "distress signals" produced by the stricken animal, or whether the animal involved in epimeletic behavior had perceived the context of the situation (i.e., distress and the imminent possibility of death) and was able to react to it.

Because of the experimental evidence for higher-order learning in dolphins (see Herman, this volume) we do not think the latter explanation can any longer be cast aside on grounds of parsimony. As far as we know, no wild animal other than man will come to the assistance of a distressed member of another taxon.

Altruism and its relation to ideas of group selection has received much attention recently by students of evolution (see, for example, Lewontin, 1970) because true altruism between noninterbreeding forms may be difficult to explain without resort to selection at levels of relation more distant than kin groups. This we feel may sometimes be a misleading simplification. Cetaceans seem clearly to exhibit intergeneric or even interfamilial altruism. The basic pattern of epimeletic behavior, however, must be used largely intraspecifically, and the examples given most likely represent "spillover" of an intraspecific pattern into the domain of more distant relationships. Selection operates by compromise and balance, in a systemic sense, with other traits of the evolving species. Even though the trait of supporting schoolmates in distress is valuable to the species, the complex of signals, cognitions, and response that it entails may allow it to be expressed by one cetacean to another, even though they may be reproductively isolated. The more interesting question, it seems to us, is whether or not it represents a contextual response to a common circumstance.

Evidence of distress is typical of the giver of epimeletic behavior. The pilot whale mentioned above, during its support of the stricken Pacific striped dolphin was highly agitated, as could be seen by the widely opened eyes with whites showing (M. Caldwell et al., 1963). In another example, an Atlantic bottlenosed dolphin carried a dead leopard shark for 8 days, avoiding all attempts to take the shark from her, and was nearly inappetant the entire time. Once the shark was finally removed the dolphin returned quickly to a normal feeding pattern (Norris & Prescott, 1961).

That individual or species recognition is of great importance in the social integration of dolphin schools is probably indicated by the observation of Perrin (1972b) on the dimorphism of the eastern tropical Pacific spinner dolphin. He found that bizarre dimorphism of the dorsal fin and postanal protuberance exists in the species only in the tropical eastern Pacific, even though the species is probably pantropical (see Fig. 3.9 in Madsen & Herman, this volume). Only in this area of the Pacific in which extreme dimorphism is found is the species known to travel regularly with other genera or species, such as *Stenella attenuata* and *Delphinus delphis*. Thus the supposition is that heightened intraspecific recognition systems are necessary.

Sensory Integration

Sensory integration, we believe, is an extremely important function of the dolphin school, just as it seems to be for fish schools (Shaw, 1970). By this we mean that the school serves to integrate the sensory inputs of its many members to provide important environmental information to part or the entire school. For example, an incoming sensation of danger by one or a few animals can quickly transmit through the school and result in appropriate evasive or aggressive action. This is clearly true in dolphin schools, in which harassment of one part of a school will often cause the school as a whole to react. That is not to say that the school may not fragment, especially if it is large, and come to form two or more subschools. It is a common observation that such is the case. The parts may or may not rejoin after the disturbance has ceased.

Sensory integration thus provides to the individual dolphin the joint sensory capabilities of many animals sensing in many different loci and directions. Presumably much sensory input of this sort is acoustic, either from active echolocation or passive listening, though other sources of data such as vision, taste, temperature sensation, tactile sensation, and possibly other modalities are likely to be involved. For instance, the commonly observed echelon formation of the animals in cetacean school subgroups may be a function of turbulence patterns generated by adjacent animals (Breder, 1965). Not only does sensory integration provide individuals with increased environmental information, but as a result it frees them for other activities within the school. A dolphin in a school engages in many kinds of social activity, such as sexual behavior (Fig. 5.11), play, familial interchange of various sorts, in addition to sensing the surrounding world. In fact the level of intricacy of social activity is likely directly dependent on the reliance the animal may put on its security within the sensory-integration system of the school. Thus sensory integration is basic to other school activities; in essence it provides a high level of environmental awareness at modest individual expense. The school thus becomes a moving enclave of safety within which the animal is free to engage in other activities. In this sense it serves some of the functions generally ascribed to "home range" and "territory" for terrestrial animals.

Examples of such sensory integration have seldom been mentioned in the literature. Norris and Dohl (1980) describe the movement patterns of resting spinner dolphins in what seems to provide a clear example.

The resting school moved slowly back and forth ranging from a few meters off the cliff face to perhaps 250 meters offshore. It traveled slowly, mostly underwater, rising in a tight discoidal group for four or five quiet breaths before submerging again. The animals rose rather steeply to the surface, not as a single tightly integrated group, but seriatum, as a column of animals. Often after rising the animals spread outward from this rising column a short distance before turning to define the compact confines of the school, like the petals of a flower opening. Such schools seem to be without striking subgroup structure in that animals seem much more equidistantly spaced than in alert schools. To be sure, some bunching could be discerned, but it was much less evident than in feeding or traveling groups, and done with deliberateness. Such groups, though seemingly less alert than moving schools, were found to be exquisitely aware of any strange objects, such as buoys, boats, lines, or swimmers in their environment. It was striking that they reacted to such foreign objects in much the same way as we have come to expect from groups of schooling fishes, and not with typical dolphin individuality. For instance, when such a school cruised along the shoreside of the point where we waited in a quietly rocking skiff some 150 meters offshore, the school approached slowly as a discoidal group, thinned as it reached a point directly inshore of us, streamed between the skiff and the cliff as a long line of quietly moving animals and reformed its discoidal shape once past us. We found that our skiff or our anchored workboat could deform such discoidal groups from some distance, causing the side nearest the skiff to become dented or malformed as the entire school reacted to our presence. When a four hydrophone array capable of triangulation by sound was placed near the path of such resting schools it was assiduously avoided and no animals passed through it (Watkins & Schevill, 1972). A line stretched across the water was capable of deflecting such schools. In all of these cases, even though the animals moved slowly and other evidences of alertness such as complex phonation or aerial behavior were suppressed, the school as a whole remained alert, through sensory integration we suppose, and was even more aware of strange circumstances in its environment than obviously alert schools. For instance, it is usually possible to cruise among alert schools and many school members will station at the bow within a few feet of the observer, but resting animals never allow this privilege but move away in the manner of a fish school.

Fig. 5.11. Beak-genital propulsion in captive spinner dolphins at the Oceanic Institute, Oahu, Hawaii. Male above; female below.

Shaw (1970) cites an interesting correlative observation in fish in which a frightened school with erratic schooling behavior was tranquilized en masse, causing schooling to become at once "more precise and classical." Thus it may be that these observations indicate that dolphins, at least, may normally exist in their schools at a higher level of social integration and complexity and descend to a simpler level of school integration during sleep or rest. In the case of frightened fish, however, what superficially looks like the less regular alert dolphin school may, in fact, represent a breakdown of the normal constraints producing tightly integrated schooling.

A number of other possible examples of sensory integration mechanisms are known for cetaceans. Layne and Caldwell (1964) mention the existence of rheotaxis (orientation to current) in the boutu of the Amazon River, and one wonders if this orientational method might not in part regulate swimming patterns for entire schools of this riverine species.

Vocal signals are, of course, common in dolphins and whales, and within the framework of the school may serve to integrate school activities as a whole (see, for example, D. Brown et al. 1966, who report incessant vocalization in traveling false killer whale schools). McBride (1940) first reported that variations in whistle types were related to emotional state, an observation which has often been repeated. What are termed "distress calls" has been described by Lilly (1963) as a two-part whistle of rising and falling inflection. However, it has proved difficult to assign such a context to the call with security as F. G. Wood (in Norris, 1966, p. 542) points out, since distressed animals may not always emit it, and since similar calls may be given when no distress is evident (see also Herman & Tavolga, this volume.) The stereotypy of dolphin whistles to individual animals (signature whistle) is especially emphasized by M. Caldwell and Caldwell (1972) who report signature forms being emitted under a variety of emotional states, including distress, by many animals. A repetitive sterotyped call is often heard during capture sequences from the captive animal (other schoolmates are typically silent at such times) or in stressed animals in captivity.

The spin of spinner dolphins consists of a leap from the water accompanied by a rapid rotation along the longitudinal axis causing a cavity to be scooped out of the surface upon reentry (Fig. 5.12). Hester, Hunter, and Whitney (1963) have described the spins as most prevalent in slowly moving schools, and to be performed by both adults and young. We (Norris & Dohl, 1980) concur and report that they are most common when animals are in spread formation. Such conditions occur especially during nighttime feeding. We have also been able to record the sound produced by such spins and to plot the frequency of spins throughout daylight hours, finding they were most prevalent at dusk. We propose that such spins may provide a short-range omnidirectional sound signal (in comparison with the highly directional and probably longer-range forwardly directed biosonar beams) useful in defining the positions of schoolmates when vision is ineffective. Spins are very common

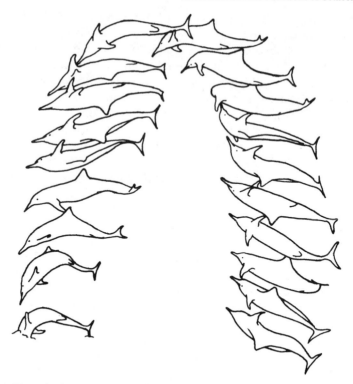

Fig. 5.12. Frame-by-frame composite of a spinning sequence in a young male eastern tropical Pacific spinner dolphin, *Stenella longirostris*. (From Hester, Hunter, and Whitney, 1963.)

at night, as we know from captive observations, but we have not been able to quantify nighttime frequency in nature. Würsig and Würsig (1980) describe all the kinds of aerial behavior shown by spinner dolphins in the dusky dolphin and record the sound propagated by such behavior 0.5 km away. They likewise conclude that a major function of such behavior may be sound production and its consequent signaling value to the dolphin school.

Another possible sensory integration system may be found in the visual assessment of the coloration of one animal by another. It is known for birds, for example, that the remarkably precise formation flight of shorebirds such as sandpipers may be assisted by mutual observation of the color patterns of adjacent birds (Diana Tombach, personal communication). Both birds and dolphins typically show a dorsoventral partitioning of coloration. Both are generally dark above but often with important contrasting pattern components such as wing specula or patterned dorsal fins. Both typically show a differentiated lateral pattern, often with dark lines or blotches, and both typically show ventral light patterns often demarcated from the pattern above by longitudinal dark markings. Also both forms often show pectoral contrasting markings that may be differentially shown when a limb is moved (Fig. 5.13).

Fig. 5.13. Ventral pattern of the Chilean dolphin or tonina, *Cephalorhynchus eutropia*, at Corrál, Chile. Note sharp demarcation of ventral and lateral patterns, dark marking on genital area and white mark in axilla.

That is, in the preparation for banking flight or during maneuvers when swimming, an animal may lift a limb, showing the contrasting pattern mark in greater or lesser degree. Distinctive axillary marks are known in dolphins, as, for example, in the tonina or Chilean dolphin (*Cephalorhynchus eutropia*), and the llampa or Peale's dolphin (*Lagenorhynchus australis*).

In either the dolphin or the bird, as school- or flockmates commence changes in direction such as dives, wheeling movements, or leaping, the intention is probably very rapidly communicated to adjacent schoolmates by visual changes in these pattern components. In the case of birds high-speed photography has shown that such directional changes, even though appearing simultaneous to a human observer, are indeed progressive changes throughout the flock, indicating that animals across the width or length of a flock are reacting seriatim to changes in direction initiated at some locus in the flock. The same system may well operate in dolphin schools to integrate school movements during feeding, flight, or at other times when movements co-ordinated between individuals are of special value.

Doubtless schools act upon information from the physical as well as the biological environment. Passive listening to waves or rheotaxis may give excellent cues for orientation in tidal situations such as are regularly encountered by all estuarine or shore-dwelling species. In general these species are obviously fully aware of tidal changes and the complexities of changing channels in shoaling water. True (1885) mentions movements of schools of Atlantic bottlenosed dolphins in and out with the tide, as does Gunter (1943). Gray whale mother and young pairs must constantly deal with such problems in their shallow calving lagoons, and many young seem to be lost to tidal stranding (Eberhardt & Norris, 1964; Norris et al., 1977), indicating that learning is involved. Adults seldom strand and seem to have learned to cope for themselves as well as their young when the two remain together. Changing light levels and spectral quality related to shoaling water around oceanic islands may give rather precise indications of depths to marine animals (Norris, 1974).

The fin whale, which is usually an open-water species, strands with some regularity. In the Sea of Okhotsk, Middendorf (1869, cited in Tomilin, 1967) reports fin whales being commonly trapped on certain muddy shoals where large tides create bars during tidal recession.

Pilleri and Knuckey (1967) speculate that sun navigation is the basis of regular movements of common dolphin schools in the Mediterranean Sea. The schools move into the sun in the morning and reverse direction in the afternoon, once again moving into the sun.

LEARNING

The highly developed learning capabilities of odontocetes are well known from work with captive animals (Kellogg & Rice, 1966; Pryor, Haag & O'Reilly, 1969; Herman & Arbeit, 1973; and see Herman, this volume). As is the case with other higher mammals, this seems to translate into an heightened importance of learning in the social patterns of odontocetes. The school is the social unit within which such learning must find its meaning. Enough examples of intraschool learning are known to surmise that it may be of major importance in the integration and expression of school functions. Further, learning during maturation must depend to a considerable extent on feedback relations within the school. Young animals must learn the right things at the right time (Eibl-Eibesfeldt, 1967) to fit such learning into the proper context of dolphin life and growth patterns as a whole.

Examples of learned behavior in dolphin schools are well known. It is a common observation that dolphins will learn to recognize vessels that harass them and will selectively avoid such craft (Norris, 1974). For example, harassment of a small (ca. 30) spinner dolphin school on the coast of Oahu by collectors resulted in avoidance of the collecting craft. The avoidance lasted for more than a year. Its inception took approximately the following pattern: the first and second animals were rather easily netted from the school. However, when a try was made for a third, avoidance took place. The collector, suspecting his motor noise was alerting the animals, came in upon them at widely different engine speeds than he had customarily used and was able to net a third. A fourth animal was taken by masking engine noise with an auxiliary generator engine, but no further animals could be taken from that ship. The avoidance was probably reinforced from time to time by the collector who periodically attempted to approach experienced animals. After several attempts to net bottlenosed dolphins in San Diego Bay the Marineland of the Pacific collecting vessel *Geronimo* was avoided, even though the same school was noted riding the bow of passing vessels such as harbor tugs and merchant craft (Norris, 1974). In large schools avoidance of collecting vessels may not come to involve the entire school, even though repeated collections are made from it. This was the case with a spotted dolphin school that ranges along the Oahu Island shore lines on both sides of Kaena Point. Over five years more than a dozen animals have been collected from this school. It appears in this case that a given capture increased the skittishness of some animals in the school but that other parts of the school remained approachable by the collecting vessel. This school numbers an estimated 250 animals. It may simply be that the fear of individual experienced animals translates to school avoidance in small schools, while in larger ones it does not.

School behavior seems rather often to be modified by learning in special feeding situations. In some cases where relatively transitory human patterns are also involved one can be sure that learning is important in the dolphin's behavior. Two examples of this sort are presented by Norris and Prescott (1961) for the Pacific bottlenosed dolphin. The first example concerned the rather regular attendance of a school of bottlenosed dolphins at the Navy garbage station located off Imperial Beach, California. On a regular schedule the Navy gathered garbage from Navy ships at anchor in San Diego Harbor and transported it out of the bay and some kilometers down the coast to the dumping location. A bottlenosed dolphin school, normally seen inside or near the bay, was often in attendance presumably feeding on fish attracted to the garbage. This school's movements and direction could be predicted roughly by the schedule of the garbage barge, though the porpoises and the barge did not normally travel together.

A similar habitual pattern was observed in relation to the shrimp trawling industry of the northern Gulf of California, Mexico. In this case otter trawls were dragged at slow speed across the sea bottom for a number of hours, usually with no dolphins in attendance. However, once the winch began to retrieve the net, dolphins were usually noted rapidly converging on the vessel, often leaping free of the water en route. As they arrived at the vessel they began to mill around her close aboard, as the catch was sorted on the after-deck. When trash fish and invertebrates were shoveled overboard the dolphins picked off and ate individual prey items. They were often so close aboard that they could be touched with oars (Norris & Prescott, 1961). Similar behavior in relation to trawling vessels has been noted by Kleinenberg (1958) for *Tursiops truncatus* and by Leatherwood (1975). In the latter case, groups of *Tursiops* were observed to follow trawls, presumably to feed on organisms stirred up by the net.

Habituation and play may both be involved in such dolphin–human inter-actions as the Opononi incidents (see Alpers, 1961), in which a large bottle-nosed dolphin swam with bathers and allowed swimmers to grasp its fin and flippers and take rides on its back. The behavior lasted for several months before the animal was found dead in a tidal pool. Similar interactions between humans and a single bottlenosed dolphin have recently been noted off the Cornwall Coast of England (Webb, 1978). Saayman and Tayler (1971) described a similar incident at Fish Hoek Bay, in the Cape region of South

Africa. Two cow bottlenosed dolphins established contact with bathers and allowed swimmers to ride by holding fins or straddling the animals' backs. The animals shied away at once if a swimmer brought along strange objects. This seems to indicate that the dolphins knew when extraneous objects were involved. A number of other examples of this general sort could be cited, including the original "boy and the dolphin" story by Pliny the Elder (see a reprint of Pliny in Turner, 1962). Usually these are difficult to interpret in terms of learning or motivation on the part of the cetaceans.

Fairholm (1856) describes South Pacific natives being assisted by dolphins in mullet fishing. He reports that the familiarity was so great that the individual animals were recognized and named. The dolphins came in, driving the fish before them, and when the fish were concentrated natives threw nets over the fish. In the melee dolphins and men gathered fish together. Only in the precise area of Amity Bay was this known to occur, and it was reported to have happened as long as men could remember. Busnel's (1973) report is of a very similar circumstance along the African coast and was described earlier in this chapter.

Play seems inextricably linked to learning, and its functions are to be found at least in part in the establishment of finely tuned patterns such as leaping, bow-riding, familial interaction patterns, and so on. It is a prominent part of dolphin school behavior as it is in the behavior of most higher mammals (Eibl-Eibesfeldt, 1967) and in fact could only be fully expressed in the relative safety of the school. It is common to observe young animals repeatedly "practicing" aerial patterns such as spins, leaps, and fin- and head-slaps. For example, such repeated leaps in young animals have been reported by Norris and Prescott (1961) for North Pacific striped dolphins. D. Caldwell and Fields (1959) report surf-riding in Atlantic bottlenosed dolphins in which about 12 animals split off from a larger school and repeatedly cruised inshore to play in the surf, riding waves in and out. Similar behavior, including surf-riding by young animals, has been observed near Puerto Peñasco, Baja California, Mexico for the Pacific bottlenosed dolphin by the senior author. Gunter (1943) describes Atlantic bottlenosed dolphins "playing" with food fish by throwing them repeatedly into the air, and similar observations, both in nature and in captivity, were reported for the Pacific bottlenosed dolphin by Norris and Prescott (1961), though in the latter case the behavior seemed to be related to food preparation rather than play. The field observation was made in Estero de Punta Banda, Mexico and was much like Gunter's. Detailed observations of the behavior in captivity were made by D. Brown and Norris (1956), and the behavior was attributed to the need to soften and break up large prey before swallowing. It not only included tossing the fish in the air (which randomized the end of the fish grasped by the dolphins) but included rubbing the fish against the bottom so firmly that rather prominent furrows were left in the bottom gravel. Ultimately the heads of the 1- to 1.5-kg yellowtail (*Seriola dorsalis*) were broken free and their headless flaccid bodies were swallowed.

What is especially interesting for our discussion here was that this behavior was carried out with great dispatch by animals which had been taken as adults in the wild, while two captive-born animals who had never been forced to deal with large food objects before attempted to imitate the others, and then spent considerable time beyond that required by the wild animals in manipulating the fish, ultimately without success. Another revealing feature was that obvious ownership of a given fish existed. No animal touched another's fish, even though it drifted to the bottom and was left for several minutes. It is not unlikely therefore that dolphins, like other higher mammals, must practice, even as adults, in order to perform various difficult finely integrated feats that are part of their lives.

DEFENSE

An important function of the school clearly is the protection of the individuals that make it up, as was described in the section on sensory integration. In fact Williams (1964) regards schooling behavior in fishes as a form of cover-seeking, and we suggest protection from predation as probably causal in the formation of cetacean schools. Further, school shape and internal structure seem related to protective mechanisms, varying according to circumstances. Whether there is any socially based division of defense within schools is not clear.

In spinner, bottlenosed, and other schools sharks are important predators (Norris & Dohl, 1980), as is clearly evident from the prevalence of scarring from shark bites, including both large species capable of killing adult dolphins and the smaller *Isistius* that apparently bites circular chunks from the blubber coat of tropical and subtropical dolphins and whales. The killer whale (*Orcinus orca*) is also known to prey on dolphins (Eschricht, 1866).

Dolphins in schools seem exquisitely aware of their surroundings and a typical reaction to strange objects in the environment is avoidance and tightening of the school. Townsend (1914) was one of the first to record such avoidance when he noted that Atlantic bottlenosed dolphins seined at Cape Hatteras, North Carolina, refused to cross seine ropes, and even though the net had not yet come ashore the catch was accordingly considered secure. Similar experiments were run by us in Hawaii in which entire spinner dolphin schools were encircled and held by a single line at the surface from which 20-m lines hung at 3-m intervals. Schools were held for nearly 4 hours inside such "nets" even though ample room existed for the dolphins to swim untouched into the open sea beneath or through the vertical lines (Norris & Dohl, 1980). Norris, Stuntz, and Rogers (1978) discuss the reactions of dolphins captured in yellowfin tuna seines. The major species caught, the Pacific spotted dolphin (*Stenella attenuata*) typically gathered into a tight discoidal school that sought a position that was as far as possible from all sources of disturbance—motorboat noise, winch noise, net, swimmers, and so forth. Within this group was a central rafting group in which mothers with young and juveniles were typically found. The peripheral animals engaged in much aggressive activity, which served to contain the rafting animals centrally. Underwater the peripheral animals were found to dive continuously under the rafters and to surface about them on all sides. Thus the younger and more vulnerable school members were contained within a moving cup of adults that was protected in all directions by moving alert adults. Spinner dolphins (*Stenella longirostris*) typically were seen outside this fluid group, sometimes under it, and occasionally penetrated briefly into it.

Sometimes the "sending of scouts" to investigate the source of disturbance or danger had been suggested, though we do not believe such purposive behavior to be established (M. Caldwell & Caldwell, 1964; Evans & Dreher, 1962). It may be akin to commonly observed behavior of a few animals leaving a large school to run the bow of a passing ship or simple curiosity on the part of a few school members. Whether or not such scouting behavior represents true division of labor within the school is wholly unclear, and whether animals are in any way "sent" on a scouting mission is also unknown, and we think unlikely (also see discussion in Herman & Tavolga, this volume).

Aggregation may occur without the presence of strange objects in the environment. For example, Uda and Nasu (1956) report bunching of whale schools during hurricanes, though one cannot be sure that such weather conditions induce fear in the animals concerned. It may be, for instance, that the noises of breaking waves cause acoustic interference, requiring closer order. Or it may be that storm waves simply entrain dolphin schools in their moving water mass. In fact we have watched a school of pilot whales "surfing" on the breakers of a very rough, gale-swept sea without any indication that conditions were disturbing for the animals.

The freshwater platanistid dolphin, *Inia*, seems to form only small schools (Layne, 1958), even though in the same river system the delphinid genus *Sotalia* forms typical, though modest-sized, schools (Norris, Harvey, Burzell & Krishna Kartha, 1972). In the case of *Platanista gangetica* single animals or pairs seem the rule, though larger but scattered groups are sometimes found (Kasuya & Aminul Haque, 1972).

If one views the entire cetacean group from whales to the smallest dolphins some regularity is evident. The largest whales hardly school at all in the normal sense of the word. They are usually widely spaced, solitary, or found in small groups related to reproduction. On the feeding grounds aggregations may occur, but these may well be related to a mutually available mass of food rather than to social cohesion. River dolphins form small groups or are solitary; coastal dolphin species form small to moderate-sized schools, while pelagic forms may at times form huge schools of several thousand animals (the numbers of animals in such large schools cannot, as yet, be estimated with any accuracy).

The variation in school size is exactly what would be predicted if predation were the basis for the degree of schooling tendency (Williams, 1964; Hamilton, 1971). Large whales seem singularly free of predation as adults, except for the occasional depredations of killer whales (Andrews, 1931). River dolphins in the Amazon–Orinoco system may suffer minor predation from large fish or crocodilians and schooling seems poorly developed. In the Ganges-Indus- Brahmaputra system schooling appears even less developed. In these latter locations predators would also seem all but absent, especially in the upper river systems (see Pilleri, 1970; Kasuya & Aminul Haque, 1972).

RESPONSES TO ENVIRONMENTAL CYCLES

Repetitive behavior patterns are common features of dolphin schools. Some, such as the presence of reproductive seasons in many cetaceans, may result in cyclic phenomena in their schools. Sexual and age groupings may come and go, and a variety of cyclic, apparently sexually related migrations take place. The physiologic needs of the animals, such as energy conservation, may also be determinative, as, for example, in the reproductive migrations of harbor porpoises from icy northern waters to somewhat warmer southern ones (Dudok van Heel, 1962). Thermal balance seems crucial to these small cetaceans (the smallest cetaceans in northern waters, 1.8 to 2.0 m total length), and they bear enormously large young in proportion to adult size (Kanwisher & Sundnes, 1966). The case with larger whales seems of a somewhat different sort. Their long migrations to polar feeding areas (Mackintosh, 1946) seem more complex and are likely to be controlled by the presence of seasonal food in polar seas, and perhaps by very low winter air temperatures (which would have to be breathed by overwintering animals) in polar latitudes.

Aside from such annual patterns it is becoming increasingly evident, as more detailed natural history accumulates for wild cetaceans, that daily patterns may be highly structured. The daily fluctuations in vocalizations of bottlenosed dolphins found by Powell (1966) is a case in point. Norris and Dohl (1980) found, for the Hawaiian spinner dolphin, that similar schooling patterns were repeated over and over on a daily basis, and that after some experience the observer could predict the sequence. They found that nighttime was devoted to feeding along the submerged island slope usually about 200 to 2000 m depth, where the island pitched sharply into the abyssal sea. During this period the animals dove in widely spread formation. Once morning came the school tightened, assuming a rank formation and movement toward shore occurred, with much aerial behavior. This was terminated during a period in which defecation was widespread throughout the school, presumably occasioned by the digestion of the night's catches. Dolphins caught in the afternoon had empty stomachs. To what extent food is stored during the night in the esophageal stomach is unknown.

Once the spinner school came near shore it moved into one of a few specific coves or lagoons (usually the same places are occupied rather regularly even within a given cove or lagoon). The rest period followed, discussed in detail earlier, in which rank formation changed to discoidal shape, with animals tightly packed. During this period the diving pattern changed dramatically from short, shallow dives, with the animals staying mostly at the surface in the ranked school, to a pattern of deeper dives, with only brief periods for respiration being spent on the surface. Spinning and vocalization, except for desultory clicks, ceased, and subgroup structure became obscure. The school became a reactive system much like a fish school. No school leaders could be discerned except within subgroups where dominance hierarchies exist. Once rest periods were complete, individual alertness and activity increased, spinning and vocalization became strongly evident, as did play and reproductive behavior. The school reassumed a rough rank or linear formation and began movement out to sea toward the evening feeding grounds.

Such diurnal patterns have been alluded to in observations of captive animals. Saayman et al. (1973) report diurnal changes in captive Indian Ocean bottlenosed dolphin schools involving school shapes, activity patterns, signaling and aggression (Fig. 5.14). Bateson (1965) reported shifts in the social structure of a bispecies school (*Stenella longirostris* and *S. attenuata*) in which partners were changed when shifting from rest to play periods.

Morrow (1948) and Hunter (1968) discussed the "breakdown" of fish schooling during darkness. They assumed that schooling was maintained principally by vision. Beyond the question of the sensory modality involved may lie an adaptive question. It may be that in the absence of visually oriented

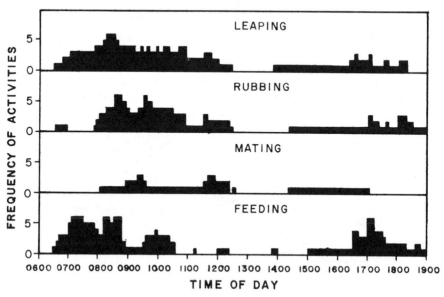

Fig. 5.14.   The duration of feeding and courtship activities of bottlenosed dolphins (*Tursiops*) in Plettenberg Bay, South Africa, in relation to time of day. From G. S. Saayman, C. K. Tayler, and D. Bower, *Behavior*, **44**, 1973. Reproduced by permission.

predators tight schooling is no longer the optimum strategy for survival. Also, since schools are presumably essential for daytime survival, such dispersed fish must reassemble in the morning. Thus nighttime fish schools must maintain some degree of cohesion, as Breder (1954) perceived. This represents a close parallel to the condition observed in *S. longirostris*. That is, since the school state represents an equilibrium condition between opposing sets of dispersive and aggregative forces, different degrees of aggregation or dispersion occur at different times of day, varying from the tight aggregation to resting schools to the highly dispersed aggregation of diving and feeding schools. The dolphins, however, maintain their principal long-distance sensory system, echolocation, regardless of light level.

DISCUSSION

When one compares the structures of fish and dolphin schools it is clear that there are important differences between them. Large pelagic fish schools, for the most part, seem to be without cohesive subschool units, while several kinds of subschool units can be discerned in dolphin schools. Even during rest periods in dolphin schools one can discern subschool units, though resting animals behave in a much more unified fashion than when wholly alert. Resting dolphin schools are reminiscent of fish schools in terms of total movement patterns. At such times they tend to act as a group in a fashion that suggests that the main determinants of movements are the integrated sensory inputs of many school members, as is the case with daytime fish schools.

Both fish and dolphin schools maintain clear discrete boundaries to their schools; school density does not trail off, in either case, toward the school margins, but changes abruptly. The differences one sees between fish and dolphin schools are in internal structure within the envelope of the school, such as spacing, segregation of various classes of animals, and of individualized or group movement. What might account for these differences?

The most obvious causal difference lies in the complicated sociality of dolphins (e.g., Tayler & Saayman, 1972, compare the complexities of dolphin social life to that of the baboon and Saayman & Tayler, 1979, draw analogies with chimpanzee social structure), and its general simplicity in schooling fish, and in a broad sense this is certainly a phylogenetic difference. That is, the general pattern of difference seems derivative from the piscene level of neural organization on the one hand, and the mammalian level on the other. Typically, schooling fish do not show such social features as family or play groups. Well-developed parental care and familial bonds characterize schooling dolphins but generally not schooling fish. While schooling fish often show courtship behavior, both courtship and spawning tend to be strongly potentiated by similar simultaneous activity of other fish throughout the school, and spawning is typically compressed into a short space of time. On the other hand, courtship and mating patterns are widely spread over the year in many dolphin species and for mysticetes in at least the California gray whale (see Sauer, 1963; Norris et al., 1977). One often cannot determine from simply watching the behavior of dolphin schools when true mating takes place because reproductive patterns are so widely used in other social contexts. However, the timing of births gives a clue. In tropical and warm temperate dolphins the birth period tends to be broadly modal, often spanning several months or even the entire year. Field observation usually reveals a mixture of growth stages of young in a single dolphin school. In contrast, the timing of births in the migratory whales seems to be much more precise, coinciding with the stay at the warm water end of long migrations (see, for example, Dawbin, 1966).

In common, neither fish nor dolphin schools have been shown to have true leaders for short-term directional movements such as are used in avoidance of strange objects in the environment. In fact such leadership would probably defeat the responsiveness of school guidance as mediated by sensory integration. It seems that whether it be in open air or water, the third dimension makes it profitable for social animals as divergent as crustaceans, fish, dolphins, or birds to use the integrated sensory inputs of many individuals to determine immediate group movements. Where leadership exists in the form of dominance, it results in social ordering within and between the subgroups of a school, but there seems to be no evidence that the school as a whole receives direction directly from this source.

Long-term school movements, such as to seasonal feeding grounds, are more difficult to interpret. Some movements, e.g., those synchronized with human fishing activities, clearly involve learning by some school members. It is probable that such learning may then be transmitted "culturally" to other school members. In migrating bird flocks, for example, the learning of older flock members guides the migrating group (Perdeck, 1958). How learning by individual birds or dolphins is transformed into directional signals for the entire group is unclear, but one suspects that eventually a large fraction of the group learns the directions culturally. When events dictate migration or movement the "consensus" arrived at by such subtleties as the average precision of orientation and the average persistence of movement toward a given direction works in much the same fashion as immediate sensory integration to guide the animal group. Both dominance hierarchies and school or flock cohesion may then further potentiate such unified movement. That is, once a large number of a group project common signals about the direction of movement, the factors which determine school or flock structure act to ensure its unified application.

Whether the large-scale movements of fish, such as the albacore schools that circle the entire North Pacific Ocean (Shomura & Otsu, 1956) involve learning in this way is unclear, but it seems unlikely that they do.

None of this discussion requires that dolphin schools maintain complete structural constancy through time. As was noted, Norris and Dohl (1980) found that spinner schools are very fluid with respect to numbers at different locations and times. Recognizable individuals and subgroups could be found in groups ranging from a few animals to hundreds of individuals. Feeding schools often represented the coalescence of a number of smaller resting schools. Schools found at certain locations were regularly larger than those found in other areas, even though individual dolphins moved back and forth between the two areas.

The differences in expression of schooling tendency and resultant school structure between different widely separated phylogenetic groups seems to us to be a fertile field of investigation, albeit difficult. The dolphin school is especially worthy of investigation because of the flexibility of dolphin behavior and the forms it may take in the context of the school.

What might be the behavioral bases for the putative levels of complexity in dolphin schools? Two features of dolphin behavior impress us as providing a possible substrate for school complexity. They are: (a) the capacity for complex learning (see Herman, this volume), and (b) the capacity to understand contingency patterns within which learning occurs ("deuterolearning"—Ruesch & Bateson, 1951; Bateson, 1966). For example, the experiment of Pryor et al. (1969) showed that a dolphin could grasp generalized rules; namely, that for each reward the dolphin was required to produce a novel piece of behavior. The grasping of this general rule happened quickly. Many other examples of rule-governed behaviors in dolphins are given in the chapter by Herman.

Let us look at these two capabilities and see if it is possible to discern how they might affect school dynamics in dolphins. There is much evidence that dolphins learn many things about their environment and that this learning is translated into school behavior. Clearly the response of dolphins to trawlers, described earlier, is of this sort, as is the association of dolphin schools with seine fisherman working from the shore. The bow-riding of dolphins is another case (see Lang, 1966). Around the world dolphins of several species come to ships to ride almost effortlessly in the pressure and flow field around the bow or along the waist. Ships have plied the seas only briefly, thinking in evolutionary time, and one must look to a natural antecedent for an explanation. It is probably found in dolphins riding the waves at the heads of swimming whales, which has occasionally been observed. The translation to ships must be based on repeated learning at various places in the world. One can recount other examples, but these will suffice. It seems apparent that dolphin learning capabilities provide for a high level of behavioral flexibility in nature, and that this is translated into local variations in group behavior that one might call culture.

The translation from individual learned patterns to school behavior may occur by imitation, a feature well known in dolphin behavior (see review in Herman, this volume). Dolphins often learn from each other in oceanarium shows (Pryor, 1973), and sometimes it seems probable that simple observation by the animal is enough for learning, without actual practice of the pattern involved. For instance, not all members of captive schools will be asked to perform certain behavior patterns, but many may learn them to some degree. Such an example is given by Norris (1974) in which socially subordinate bottlenosed dolphins were shipped from one oceanarium to another, and when established in their new home without dominant animals, they began to perform the old patterns of the former dominant animals. The expression of such capacity in wild schools seems imperfect. The learned refusal of dolphins to travel with a collecting vessel is a common observation. We noted that whole schools of Hawaiian spinner dolphins (10 to 60 animals approximately) may refuse to come to certain ships that have harassed them. But in much larger schools (200 to 500 individuals), such as those of the Hawaiian spotted dolphin, collective wariness is less evident. Clearly, learned patterns do not always penetrate entire schools, or at least the wary part of such schools is not able to transmit its instantaneous fear to other parts of the school. Thus it is uncertain whether the animals that learn fear are able to teach some or all of their fellows, or whether when the collecting vessel arrives their "instantaneous fear" transmits a warning to others in the school. In large schools such transmission may be imperfect, and because such aggregations may be transitory groupings a schoolwide behavior pattern may be impossible to establish.

Vocal imitation is well known in odontocetes, both experimentally (Lilly, Miller & Truby, 1968, for the bottlenosed dolphin; also see review in Herman, this volume) and naturally (Backus & Schevill, 1966, for the sperm whale). While there is no clear evidence how this capability might be used in nature, it will be interesting to see, as our knowledge of wild schools grows, if local dialects exist (cf. Wilson, 1975).

At any rate when we attempt to conceptualize how dolphin schools may function we must allow into our thinking the demonstrated complex learning abilities of dolphins. Such an admission will allow us to pose new and important questions about school structure that may then be tested.

On a simpler plane a great many basic aspects of cetacean schooling important to contemporary ideas about group behavior are still unknown or incompletely understood. For example, if one attempts to apply to cetaceans the theoretical constructs related to spacing of mobile animals discussed by J. Brown & Orians (1970) many gaps in knowledge are at once apparent. What, for example, might home ranges or territories mean in cetacean groupings? Do schools occupy exclusive territories that they defend from other schools? How does predation affect group structure and composition? Does intraschool aggression influence school structure and intraschool affiliations? How is resource partitioning and utilization related to school structure? Most of what we know of these topics is based on anecdotal evidence, so that the emerging quantitative studies of cetacean schooling (e.g., see Wells, Irvine & Scott, this volume) are especially welcome and should help us to better answer these fundamental questions.

ACKNOWLEDGMENTS

Our work on dolphin schools has long been supported by the Biological Branch of the Office of Naval Research, and we are most grateful for this long-continued assistance. This chapter was prepared largely under support from the National Marine Fisheries Service.

Many individuals have helped with our field studies and we gratefully acknowledge their assistance. In preparing this review we have been especially aided by Editor Louis Herman, Professor Arthur Popper, of Georgetown University, Professor George Barlow of the University of California, Berkeley, and Professor Gregory Bateson of the University of California, Santa Cruz.

REFERENCES

Alpers, A. (1961). *Dolphins, the Myth and the Mammal.* Boston, Mass.: Houghton Mifflin.

Andrews, R. C. (1931). *Whale Hunting with Gun and Camera. A Naturalist's Account of the Modern Shore-whaling Industry, of Whales and Their Habits and of Hunting Experiences in Various Parts of the World.* New York: Appleton-Century-Crofts.

Arsen'ev, V. A. (1939). Distributions and migrations of the beluga in the far east. *Izvest. Tikhook. (TINRO),* 1–108. [In Russian.]

Backus, R. H. & Schevill, W. E. (1966). *Physeter* clicks. In: *Whales, Dolphins and Porpoises* (K. S. Norris, Ed.), pp. 510–528. Berkeley: University of California Press.

Barabash-Nikiforov, I. C. (1940). Cetacean fauna of the Black Sea—Its origins and composition. *Izd. Voronezhskogo Universiteta, 12.* [Cited in Tomilin, 1967.]

Bartholomew, G. A. (1942). The fishing activities of double-crested cormorants of San Francisco Bay. *Condor, 44,* 13–21.

Bateson, G. (1965). Porpoise Community Research: Final Report. National Institutes of Health Contract No. N 60530-C-1098.

Bateson, G. (1966). Problems in cetacean and other mammalian communication. In: *Whales, Dolphins, and Porpoises* (K. S. Norris, Ed.), pp. 569–582. Berkeley: University of California Press.

Breder, C. M., Jr. (1954). Equations descriptive of fish schools, and other animal aggregations. *Ecology, 35,* 361–370.

Breder, C. M., Jr. (1965). Vortices and fish schools. *Zoologica, 50,* 97–114.

Breder, C. M., Jr. (1967). On the survival value of fish schools. *Zoologica, 52,* 25–40.

Brock, V. E. & Riffenburgh, R. H. (1960). Fish schooling: A possible factor in reducing predation. *J. Cons. Perm. Int. Explor. Mer., 25,* 307–317.

Brown, D. H., Caldwell, D. K. & Caldwell, M. C. (1966). Observations on the behavior of wild and captive false killer whales, with notes on associated behavior of other genera of captive delphinids. *Los Ang. Cty. Mus. Nat. Histo. Contrib. Sci., 95,* 1–32.

Brown, D. H. & Norris, K. S. (1956). Observations of captive and wild cetaceans. *J. Mammal., 37,* 311–326.

Brown, J. L. & Orians, G. H. (1970). Spacing patterns in mobile animals. *Ann. Rev. Ecol. Syst., 1,* 239–262.

Brown, S. G. (1954). The movements of fin and blue whales in Antarctic waters. *Norsk Hvalfangsttid., 43,* 301–309.

Bullock, T. H. & Gurevich, V. S. (1979). Soviet literature on the nervous system and psychobiology of cetacea, 1960-1977. *Int. Rev. Neurobiol., 21,* 47–127.

Busnel, R.-G. (1973). Symbiotic relationship between man and dolphins. *Trans. N.Y. Acad. Sci., Ser. 11, 35,* 112–131.

Caldwell, D. K. & Fields, H. M. (1959). Surf-riding by Atlantic bottlenosed dolphins. *J. Mammal., 40,* 454–455.

Caldwell, D. K., Caldwell, M. C. & Rice, D. W. (1966). Behavior of the sperm whale *Physeter catodon* L. In: *Whales, Dolphins, and Porpoises* (K. S. Norris, Ed.), pp. 677-717. Berkeley: University of California Press.

Caldwell, M. C., Brown, D. H. & Caldwell, D. K. (1963). Intergeneric behavior by a captive Pacific pilot whale. *Los. Ang. Cty. Mus. Nat. Hist. Contrib. Sci.,* **70,** 1-12.

Caldwell, M. C. & Caldwell, D. K. (1964). Experimental studies of factors involved in care-giving behavior in three species of the cetacean family Delphinidae. *Bull. South. Calif. Acad. Sci.,* **63,** 1-20.

Caldwell, M. C. & Caldwell, D. K. (1966). Epimeletic (care-giving) behavior in cetacea. In: *Whales, Dolphins, and Porpoises* (K. S. Norris, Ed.), pp. 755-789. Berkeley: University of California Press.

Caldwell, M. C. & Caldwell, D. K. (1967a). Intraspecific transfer of information via the pulsed sound in captive cetaceans. In: *Animal Sonar Systems: Biology and Bionics,* Vol. II (R.-G. Busnel, Ed.), pp. 879-936. Jouy-en-Josas, France: Laboratoire de Physiologie Acoustique.

Caldwell, M. C. & Caldwell, D. K. (1967b). Dolphin community life. *Los Ang. Cty. Mus. Nat. Hist. Contrib. Sci.,* **5,** 12-15.

Caldwell, M. C. & Caldwell, D. K. (1972). Behavior of marine mammals. In: *Mammals of the Sea. Biology and Medicine* (S. H. Ridgway, Ed.), pp. 419-465. Springfield, Ill.: Thomas.

Chapskii, K. K. (1941). Marine mammals of the Soviet Arctic. *Izdat. Glavsev. Leningrad-Moscow.* [Cited in Tomilin, 1967.]

Clarke, R. (1957). Migrations of marine mammals. *Norsk Hvalfangsttid.,* **46,** 609-630.

Cody, M. L. (1971). Finch flocks in the Mohave Desert. *Theor. Popul. Biol.,* **2,** 142-158.

Cummings, W. C., Fish, J. F. & Thompson, P. O. (1972). Sound production and other behavior of southern right whales, *Eubalaena glacialis. Trans. San Diego Soc. Nat. Hist.,* **17,** 1-14.

Cummings, W. C. & Thompson, P. O. (1971). Gray whales, *Eschrichtius robustus,* avoid the underwater sounds of killer whales, *Orcinus orca. Fish. Bull.,* **69,** 525-530.

Dakin, W. J. (1934). *Whalemen Adventurers, the Story of Whaling in Australian Waters and Other Southern Seas Related Thereto, from the Days of Sails to Modern Times.* Sydney: Angus & Robertson.

Dawbin, W. H. (1966). The seasonal migratory cycle of humpback whales. In: *Whales, Dolphins, and Porpoises* (K. S. Norris, Ed.), pp. 145-170. Berkeley: University of California Press.

Dohl, T. P., Norris, K. S. & Kang, I. (1974). A porpoise hybrid *Tursiops* × *Steno. J. Mammal.,* **55,** 217-221.

Dreher, J. (1966). Cetacean communication: Small group experiment. In: *Whales, Dolphins, and Porpoises* (K. S. Norris, Ed.), pp. 529-543. Berkeley: University of California Press.

Dudok van Heel, W. H. (1962). Sound and cetacea. *Neth. J. Sea Res.,* **1,** 407-507.

Eberhardt, R. L. & Norris, K. S. (1964). Observations of newborn Pacific gray whales on Mexican calving grounds. *J. Mammal.,* **45,** 88-95.

Eibl-Eibesfeldt, I. (1967). Concepts of ethology and their significance in the study of human behavior. In: *Early Behavior: Comparative and Development Approaches* (H. W. Stevenson, Ed.), pp. 127-146. New York: Wiley.

Eschricht, D. F. (1866). On the species of the genus *Orca* inhabiting the northern seas. In: *Recent Memoranda on the Cetacea* (W. H. Flower, Ed.), pp. 151-188. London: Ray Society.

Evans, W. E. (1974). Radio telemetric studies of two species of small odontocete cetaceans. In: *The Whale Problem: A Status Report* (W. E. Schevill, Ed.), pp. 385-394. Cambridge, Mass.: Harvard University Press.

Evans, W. E. (1975). The biology of the common dolphin, *Delphinus delphis* Linnaeus. Ph.D. thesis, University of California, Los Angeles.

Evans, W. E. & Bastian, J. (1969). Marine mammal communications: Social and ecological factors. In: *The Biology of Marine Mammals* (H. T. Andersen, Ed.), pp. 425-476. New York: Academic.

Evans, W. E. & Dreher, J. J. (1962). Observations on scouting behavior and associated sound production by the Pacific bottlenose porpoise (*Tursiops gilli* Dall). *Bull. South. Calif. Acad. Sci.*, **61**, 217-226.

Fairholme. (1856). The blacks of Moreton Bay and the porpoises. *Proc. Zool. Soc. Lond.*, **24**, 353-354.

Fink, B. D. (1959). Observations of porpoise predation on a school of Pacific sardines. *Calif. Fish Game*, **45**, 216-217.

Fish, J. F. & Vania, J. S. (1971). Killer whale, *Orcinus orca*, sounds repel white whales, *Delphinapterus leucas*. *Fish. Bull.*, **69**, 531-535.

Fitch, J. E. & Brownell, R. L., Jr. (1968). Fish otoliths in cetacean stomachs and their importance in interpreting feeding habits. *J. Fish. Res. Board Can.*, **25**, 2561-2574.

Gilmore, R. M. (1961). *The Story of the Gray Whale*, 2nd ed. San Diego, Calif.: privately printed.

Golenchenko, P. (1949). [Cited in Tomilin, 1967.]

Gray, R. (1887). Notes on a voyage to Greenland seas in 1886. *Zoologist*, **XI**, 122-124.

Green, R. E., Perrin, W. R. & Petrich, B. P. (1971). The American Tuna Purse Seine Fishery. In: *Modern Fishing Gear of the World*, **3**, 182-194. London: Fishing News (Books), Ltd.

Gunter, G. (1943). Contributions to the natural history of the bottlenose dolphin *Tursiops truncatus* (Montagu) on the Texas coast, with particular reference to food habits. *J. Mammal.*, **23**, 267-275.

Hamilton, W. D. (1971). Geometry for the selfish herd. *J. Theor. Biol.*, **31**, 295-311.

Hamilton, W. J. & Watt, K. E. F. (1970). Refuging. Paper 4010, *Ann. Rev. Ecol. System.*, **1**, 263-286.

Herman, L. M. & Arbeit, W. R. (1973). Stimulus control and auditory discrimination learning sets in the bottlenose dolphin. *J. Exp. Anal. Behav.*, **19**, 379-394.

Hester, F. J., Hunter, J. R. & Whitney, R. R. (1963). Jumping and spinning behavior in the spinning porpoise. *J. Mammal.*, **44**, 586-588.

Hubbs, C. L. (1953). Dolphins protecting dead young. *J. Mammal.*, **34**, 498.

Hunter, J. (1968). Effect of light on schooling and feeding of jack mackerel *Trachurus symmetricus*. *J. Fish. Res. Board Can.*, **25**, 393-407.

Ingebrightsen, A. (1929). Whales caught in the North Atlantic and other seas. *Rapp. Cons. Explor. Mer.* **56**, 1-26.

Jurasz, C. & Jurasz, V. (1978). Humpback whales in southeastern Alaska. *Alaska Geogr.*, **5**(4), 117-127.

Kanwisher, J. & Sundnes, G. (1966). Thermal regulation in cetaceans. In: *Whales, Dolphins, and Porpoises* (K. S. Norris, Ed.), pp. 397-409. Berkeley: University California Press.

Kasuya, T. & Aminul Haque, A. K. M. (1972). Some information on distribution and seasonal movement of the Ganges dolphin. *Sci. Rep. Whales Res. Inst.*, **24**, 109-115.

Kellogg, W. N. & Rice, C. E. (1966). Visual discrimination and problem solving in a bottlenose porpoise. In: *Whales, Dolphins, and Porpoises* (K. S. Norris, Ed.), pp. 731-754. Berkeley: University of California Press.

Kleinenberg, S. E. (1958). Quelques données sur l'alimentation de *Tursiops tursio*. Fabr. dans la mer noire. *Bull. Soc. Nat. Muscov*, **47**, 406-413.

Krebs, J. R. (1974). Colonial nesting and social feeding as strategies for exploiting food resources in the great Blue Heron (*Ardea herodias*). *Behaviour*, **51**, 99-134.

Kruuk, H. (1972). *The Spotted Hyena*. Chicago, Ill.: University of Chicago Press.

Lang, T. G. (1966). Hydrodynamic analysis of cetacean performance. In: *Whales, Dolphins, and Porpoises* (K. S. Norris, Ed.), pp. 410-434. Berkeley: University of California Press.

Layne, J. N. (1958). Observations on freshwater dolphins in the upper Amazon. *J. Mammal.*, **39**, 1–22.

Layne, J. N. & Caldwell, D. K. (1964). Behavior of the Amazon dolphin, *Inia geoffrensis* (Blainville) in captivity. *Zoologica,* **49,** 81–801.

Leatherwood, J. S. (1975). Some observations of feeding behavior of bottlenosed (*Tursiops truncatus*) in the northern Gulf of Mexico and (*Tursiops* cf. *T. gilli*) off southern California, Baja California, and Nayarit, Mexico. *Mar. Fish. Rev.*, **37**, 10–16.

Leatherwood, J. S. & Samaris, W. F. (1974). Some observations of killer whales, *Orcinus orca,* attacking other marine animals. *Proc. South. Calif. Acad. Sci. Meet.*, Fullerton, Calif.

Lewontin, R. C. (1970). The units of selection. *Ann. Rev. Ecol. System.*, **1970**, 1–18.

Lilly, J. C. (1963). Distress call of the bottlenose dolphin: Stimuli and evoked behavioral responses. *Science,* **139**, 116–118.

Lilly, J. C., Miller, A. M. & Truby, H. M. (1968). Reprogramming of the sonic output of the dolphin: Sonic burst count matching. *J. Acoust. Soc. Am.*, **43**, 1412–1424.

MacAskie, I. V. (1966). Unusual example of group behavior by killer whales (*Orcinus rectipinna*). *Murrelet,* **47**, 38.

Mackintosh, N. (1946). The natural history of whalebone whales. *Biol. Rev.* **21,** 60–74.

Major, P. (1976). The behavioral ecology of predator–prey interactions in schooling fish. Ph.D. thesis, University of California, Santa Cruz.

Mal'm, E. N. (1946). Dolphins of the Black Sea. *Izdatel'stvo Akademii Nauk SSSR Leningrad,* 1–23. [Cited in Tomilin, 1967.]

McBride, A. F. (1940). Meet Mister Porpoise. *Nat. Hist. Mag.*, **45**, 16–29.

McBride, A. F. & Hebb, D. O. (1948). Behavior of the captive bottlenose dolphin, *Tursiops truncatus. J. Comp. Physiol. Psychol.*, **41**, 111–123.

McBride, A. F. & Kritzler, H. (1951). Observations on pregnancy, parturition and post-natal behavior in the bottlenose dolphin. *J. Mammal.*, **32**, 251–266.

Mitchell, E. (1965). Evidence for mass stranding of the false killer whale (*Pseudorca crassidens*) in the eastern North Pacific Ocean. *Norsk Hvalfangsttid.*, **8**, 172–177.

Mitchell, E. (1978). Finner whales. In: *Marine Mammals of Eastern North Pacific and Arctic Waters* (D. Haley, Ed.), pp. 36–45. Seattle, Wash.: Pacific Search.

Morrow, J. E., Jr. (1948). Schooling behavior in fishes. *Q. Rev. Biol.*, **23**, 27–38.

Neave, D. J. & Wright, B. S. (1968). Seasonal migrations of the harbor porpoise (*Phocoena phocoena*) and other cetacea in the Bay of Fundy. *J. Mammal.*, **49**, 259–264.

Newman, M. A. (1968). Narwhals: Sea going unicorns of the north Pacific. *Search,* 17–18.

Nishiwaki, M. (1962). Aerial photographs show sperm whales' interesting habits. *Norsk Hvalfangsttid.*, **5**, 395–398.

Norris, K. S. (1965). Trained porpoise released in the open sea. *Science,* **147**, 1048–1050.

Norris, K. S. (1966). (Ed.). *Whales, Dolphins, and Porpoises.* Berkeley: University of California Press.

Norris, K. S. (1967a). Some observations on the migration and orientation of marine mammals. In: *Animal Orientation and Navigation* (R. S. Storm, Ed.), pp. 101–125. Corvallis: Oregon State University Press.

Norris, K. S. (1967b). Aggressive behavior in cetacea. In: *Aggression and Defense: Neural Mechanisms and Social Patterns* (C. D. Clemente & D. B. Lindsley, Eds.), pp. 225–241. Berkeley: University of California Press.

Norris, K. S. (1974). *The Porpoise Watcher.* New York: Norton.

Norris, K. S. & Dohl, T. P. (1980). The behavior of the Hawaiian spinner porpoise, *Stenella longirostris. Fish. Bull.,* **77,** 821–847.

Norris, K. S. & Gentry, R. L. (1974). Capture and harnessing of young California gray whales, *Eschrichtius robustus. Mar. Fish. Rev.,* **36**, 58-64.

Norris, K. S., Goodman, R., Villa-Ramirez, B. & Hobbs, L. (1977). The behavior of California gray whales (*Eschrichtius robustus*) in southern Baja California, Mexico. *Fish. Bull.,* **75**, 159-171.

Norris, K. S., Harvey, G. W., Burzell, L. A. & Krishna Kartha, T. D. (1972). Sound production in the freshwater porpoise *Sotalia* cf. *fluviatilis* Gervais and Deville and *Inia geoffrensis* Blainville in the Rio Negro, Brazil. In: *Investigation on Cetaceans,* Vol. IV (G. Pilleri, Ed.), pp. 251-260. Berne, Switzerland: Institute of Brain Anatomy, University of Berne.

Norris, K. S. & Prescott, J. H. (1961). Observations on Pacific cetaceans of Californian and Mexican waters. *Univ. Calif. Publ. Zool.,* **63**, 291-402.

Norris, K. S., Stuntz, W. E. & Rogers, W. (1978). The behavior of porpoises and tuna in the eastern tropical Pacific yellowfin tuna fishery: Preliminary studies. Final Report, U.S. Marine Mammal Commission Contract No. MMC6AC022.

Ohsumi, S. (1971). Some investigations on the school structure of sperm whales. *Sci. Rep. Whales Res. Inst.,* **23**, 1-25.

Payne, R. & Webb, D. (1971). Orientation by means of long range acoustic signalling in baleen whales. *Ann. N.Y. Acad. Sci.,* **188**, 110-141.

Perdeck, A. C. (1958). Two types of orientation in migrating starlings *Sturnus vulgaris* L. and Chaffinches *Fringilla coelebs* L., as revealed by displacement experiments. *Ardea,* **46**, 1-37.

Perrin, W. F. (1970). Color pattern of the eastern Pacific spotted porpoise *Stenella graffmanni* Lonnberg (Cetacean Delphinidae). *Zoologica,* **54**, 135-142.

Perrin, W. F. (1972a). Variation and taxonomy of spotted and spinner porpoises (genus *Stenella*) of the eastern tropical Pacific and Hawaii. Ph.D. thesis, University of California, Los Angeles.

Perrin, W. F. (1972b). Color patterns of spinner porpoise (*Stenella* cf. *S. longirostris*) of the Eastern Pacific and Hawaii, with comments on delphinid pigmentation. *Fish. Bull.,* **70**, 983-1003.

Perrin, W. F., Warner, R. R., Fiscus, C. H. & Holts, D. B. (1973). Stomach contents of porpoise, *Stenella* spp., and yellowfin tuna, *Thunnus albacares,* in mixed species aggregations. *Fish. Bull.,* **71**, 1077-1092.

Pike, G. (1962). Migration and feeding of the gray whale (*Eschrichtius gibbosus*). *J. Fish. Res. Board Can.,* **19**, 815-838.

Pilleri, G. (1970). Observations on the behavior of *Platanista gangetica* in the Indus and Brahamputra Rivers. In: *Investigations on Cetacea,* Vol. II (G. Pilleri, Ed.), pp. 27-60. Berne, Switzerland: Institute of Brain Anatomy, University of Berne.

Pilleri, G. & Knuckey, J. (1967). The distribution, navigation and orientation by the sun of *Delphinus delphis* L. in the western Mediterranean. *Experientia,* **24**, 394-396.

Powell, B. A. (1966). Periodicity of vocal activity of captive Atlantic bottlenose dolphins: *Tursiops truncatus. Bull. South. Calif. Acad. Sci.,* **65**, 237-244.

Pryor, K. W. (1973). Behavior and learning in porpoises and whales. *Naturwissenschaften,* **60**, 412-420.

Pryor, K. W., Haag, R. & O'Reilly, J. (1969). The creative porpoise: Training for novel behavior. *J. Exp. Anal. Behav.,* **12**, 653-661.

Radakov, D. V. (1967). The biological and practical importance of schooling behavior in fishes. *Proc. FAO Conf. Fish Behavior Relation Fishing Techniques Tactics,* Bergen, Norway.

Ruesch, J. & Bateson, G. (1951). *Communication: The Social Matrix of Psychiatry.* New York: Norton.

Saayman, G. S., Bower, D. & Tayler, C. K. (1972). Observations on inshore and pelagic dolphins on the southeastern Cape coast of South Africa. *Koedoe,* **15**, 1-24.

Saayman, G. S. & Tayler, C. K. (1971). Responses to man of captive and free-ranging cetaceans. *Baralogia: Proc. 1st 2nd South African Symp. Underwater Sci.*, University of Pretoria, pp. 1–9.

Saayman, G. S. & Tayler, C. K. (1979). The socioecology of humpback dolphins (*Sousa* sp.). In: *Behavior of Marine Animals*, Vol. 3. *Cetaceans* (H. E. Winn & B. L. Olla, Eds.), pp. 165–226. New York: Plenum.

Saayman, G. S., Tayler, C. K. & Bower, D. (1973). Diurnal activity cycles in captive and free-ranging Indian Ocean bottlenose dolphins (*Tursiops aduncus* Ehrenburg). *Behaviour*, **44**, 212–233.

Sauer, E. G. F. (1963). Courtship and copulation of the gray whale in the Bering Sea at St. Lawrence Island, Alaska. *Psychol. Forsch.*, **27**, 157–174.

Schaller, G. (1972). *The Serengeti Lion.* Chicago, Ill.: The University of Chicago Press.

Sergeant, D. E. (1962). The biology of the pilot or pothead whale *Globicephala melaena* (Traill) in Newfoundland waters. *Bull. Fish. Res. Board Can.*, **132**, 1–84.

Shallenberger, R. J. (1973). Breeding biology, homing behavior, and communication patterns of the wedge-tailed shearwater *Puffinus pacificus chlororhynchus.* Ph.D. thesis, University of California, Los Angeles.

Shane, S. H. (1977). The population biology of the Atlantic bottlenose dolphin, *Tursiops truncatus,* in the Aransas Pass areas of Texas. Master's thesis, Texas A. & M. University.

Shaw, E. (1963). Some new thoughts on the schooling of fishes. FAO Fisheries Rpt. 62, Vol. 2. FAO Conf. Fish Behavior in Relation to Fishing Techniques and Tactics. Bergen, Norway.

Shaw, E. (1970). Schooling in fishes: Critique and review. In: *Development and Evolution of Behavior* (L. R. Aronson, E. Tobach, D. S. Lehrman, & J. S. Rosenblatt, Eds.), pp. 452–480. San Francisco, Calif.: W. H. Freeman.

Shomura, R. A., and Otsu, T. (1956). Central North Pacific albacore surveys, January 1954–February 1955. *U.S. Fish. Wildl. Spec. Sci. Rep. Fish,* **173**, 1–29.

Slijper, E. J. & van Utrecht, W. L. (1959). Observing whales. *Norsk Hvalfangsttid.,* **3**, 101–117.

Tarasevich, M. N. (1957). Comparison of the composition of herds of aquatic and amphibious mammals. *VNIRO,* **33**, 199–218. [In Russian].

Tavolga, M. C. (1966). Behavior of the bottlenose dolphin (*Tursiops truncatus*): Social interactions in a captive colony. In: *Whales, Dolphins, and Porpoises* (K. S. Norris, Ed.), pp. 718–730. Berkeley: University of California Press.

Tavolga, M. C. & Essapian, F. S. (1957). The behavior of the bottlenosed dolphin (*Tursiops truncatus*): Mating, pregnancy, parturition, and mother–infant behavior. *Zoologica,* **42**, 11–31.

Tayler, C. K. & Saayman, G. S. (1972). The social organization and behavior of dolphins (*Tursiops aduncus*) and baboons (*Papio ursinus*): Some comparisons and assessments. *Ann. Cape Prov. Mus. Nat. Hist.,* **9**, 11–49.

Tayler, C. K. & Saayman, G. S. (1973). Imitative behaviour by Indian Ocean bottlenose dolphins (*Tursiops aduncus*) in captivity. *Behaviour,* **44**, 286–298.

Tomilin, A. G. (1967). *Mammals of the U.S.S.R. and Adjacent Countries,* Vol. IX. *Cetacea.* Jerusalem: Israel Program Scientific Translations.

Townsend, C. H. (1914). The porpoise in captivity. *Zoologica,* **1**, 289–299.

True, F. W. (1885). The bottle-nose dolphin *Tursiops tursiops* as seen at Cape May, New Jersey. *Science,* **5**, 336–339.

True, F. W. (1890). Observations on the life history of the bottlenose porpoise. *Proc. USNM,* **13**, 195–203.

Tsalkin, V. I. (1938). Respredelinie obyknovennogo chemomorskogo del'fina (*D. delphis*) letne-osennii period. *Trudy Azova-Chernomomorskogo Nauchno-Issledovatel'skogo Instituea Rybnaga Khozyaistva No. 2.* (Cited in Tomilin, 1967.)

Turner, P. (1962). (Ed.). Pliny the Elder, mermaids and mermen. In: *Natural History,* 9th Book, Centaur Press.

Uda, M. & Nasu, K. (1956). Studies of the whaling grounds in the northern sea region of the Pacific Ocean in relation to the meteorological and oceanographic conditions (Pt. 1). *Sci. Rep. Whales Res. Inst.,* **11**, 163, 179.

Watkins, W. A. & Schevill, W. E. (1972). Sound source location by arrival times on a non-rigid three-dimensional hydrophone array. *Deep-Sea Res.,* **19**, 691–706.

Webb, N. G. (1978). Boat towing by a bottlenose dolphin. *Carnivore,* **1**, 122–130.

Wilke, F., Taniwaki, T. & Kuroda, N. (1953). *Phocoenoides* and *Lagenorhynchus* in Japan, with notes on hunting. *J. Mammal.,* **34**, 488–497.

Wilson, E. O. (1975). *Sociobiology: The New Synthesis.* Cambridge Mass.: Belknap/Harvard.

Williams, G. C. (1964). Measurements of consociation among l. fishes and comments on the evolution of schooling. *Mich. State Mus. Publ. Biol. Sci.,* **2**, 351–383.

Würsig, B. (1978). Occurrence and group organization of Atlantic bottlenose porpoises (*Tursiops truncatus*) in and Argentine Bay. *Biol. Bull.,* **154**, 348–359.

Würsig, B. (1976). Radiotracking dusky porpoises (*Lagenorhynchus obscurus*) in the South Atlantic: A preliminary analysis. United Nations Fisheries Series (Food and Agricultural Organization) No. 5.

Würsig, B. & Würsig, M. (1977). The photographic determination of group size, composition, and stability of coastal porpoises (*Tursiops truncatus*). *Science,* **198**, 755–756.

Würsig, B. & Würsig, M. (1979). Behavior and ecology of bottlenose dolphins, *Tursiops truncatus,* in the south Atlantic. *Fish. Bull.,* **77**, 399–442.

Würsig, B. & Würsig, M. (1980). Behavior and ecology of dusky dolphins, *Lagenorhynchus obscurus,* in the south Atlantic. *Fish. Bull.,* **77**, 871–890.

Zahavi, A. (1971). The function of pre-roost gatherings and communal roosts. *Mich. State Mus. Publ. Biol. Sci.,* **113**, 106–109.

Zenkovich, B. A. (1938). Vokrug sucta za Kitami. (Around the world after whales.) [Cited in Tomilin, 1967.]

CHAPTER 6

The Social Ecology of Inshore Odontocetes

Randall S. Wells, A. Blair Irvine, and Michael D. Scott

INTRODUCTION

Cetacean groups, especially among the odontocetes (toothed whales), may vary greatly from species to species or even among populations of a species. Variations may occur in typical group size, age and sex composition, nature and extent of interactions between individuals, and duration of associations. For example, freshwater dolphins are often seen alone or in small groups, while some pelagic species may swim in schools of several thousand individuals. Separate all-male and nursery groups are prevalent during parts of the migratory cycle for such animals as sperm whales (*Physeter catodon*) and narwhals (*Monodon monoceros*). Age and sex segregation of subgroups has been reported for a number of species, including bottlenosed dolphins (*Tursiops* spp.) and Hawaiian spinner dolphins (*Stenella longirostris*), while groups of mixed age and sex seem to be the rule in many other cases. Interactions and associations between individuals may range from simply swimming in the same migratory group or the same feeding aggregation to being a member of a permanent family unit, as in the case of killer whales (*Orcinus orca*). Among *Tursiops truncatus*, group composition may change frequently within a stable herd, but associations within subgroups may be of greater duration.

Why do we find such diversity of sociality in such a relatively small taxonomic group? Also, through a description of this diversity, can we begin to obtain some insight into the selective forces responsible for social development in the cetaceans? A comparable degree of social diversity can be found among some groups of terrestrial mammals, including macropods (Kaufmann, 1974), ungulates (Estes, 1974; Geist, 1974; Jarman, 1974; Owen-Smith, 1977), and primates (Crook & Gartlan, 1966; Clutton-Brock, 1974; Eisenberg et al., 1972). Some major advances in the understanding of sociality have recently come from the relating of differences in social organization of relatively closely related species to differences in ecology (see Crook et al., 1976; Krebs & Davies, 1978, for reviews). Studies of this kind have found that the availability and distribution of food resources, the degree and type of predation pressure, and the physical characteristics of the habitat are especially important sources of ecological pressure acting on sociality. Jarman (1974), for example, found that the smaller antelopes are forced by their diet to disperse widely in less-open habitats. Larger antelopes are less selective in their diet, need larger quantities of food, and are found in open grasslands. The dense biomass of the larger antelopes and the need for communal defense from predators in the exposed grasslands promote group formation, and one finds larger, more stable, and more socially complex groups than in the case of the smaller antelopes.

The pervasiveness of the influence of these ecological variables on sociality urges the study of their influence on cetacean groups. To date most field studies of cetacean social organization have focused on the description of the forms or functions of the social structure, but have not yet defined in detail the ecological forces that may shape the structure. This chapter attempts to examine the relationships between social organization and ecology for selected inshore odontocete species—those dwelling primarily in rivers or coastal regions or making frequent excursions into these waters from more pelagic habitats. After a brief discussion of methods for the study of wild cetacean groups a broad description of available data on selected odontocete species is given to describe some correlations between social and ecological variables. The social variables considered are principally group size and composition and strategies for acquiring food and avoiding predation. The ecological variables are those noted earlier—habitat physical characteristics, food resources, and predation pressures. The chapter then provides a detailed "case study" of a population of bottlenosed dolphins (*Tursiops truncatus*) inhabiting the coastal and inland waters along Florida's west coast. Finally, some general remarks are given, summarizing the relationships of ecology to sociality as observed in these cetacean groups, and some parallels are drawn with what is known about such relationships among terrestrial mammals.

STUDYING FREE-RANGING CETACEANS

Field studies of cetaceans lag far behind those of terrestrial mammals because of the inherent difficulties of observing highly mobile aquatic mammals. Until recently, most available information on free-ranging odontocetes was based upon opportunistic field observations (e.g., Brown, Caldwell & Caldwell, 1966; Brown & Norris, 1956; Gunter, 1942; Norris & Prescott, 1961; Norris, 1974). However, because group structure in many cetacean species is fluid (Norris & Dohl, this volume) a single opportunistic observation of a group may provide only limited insight into its structure and dynamics. Some additional data on group structure have come from mass strandings (see Norris & Dohl, this volume, for review), the collection of individuals (Gunter, 1942), and the occasional harvest of entire groups (Mitchell, 1975; Miyazaki, 1977; Ohsumi, 1966, 1971; Sergeant, 1962).

Long-term studies conducted with several odontocete populations within the last decade have begun to provide information on the influence of ecology on cetacean behavior. Observational tools being employed include observations from cliffs, aircraft, boats and underwater viewing vessels, and repeated identification and observation of individuals through use of natural marks, tags, and telemetry.

Observations from cliffs and airplanes facilitate determination or description of numbers, swimming formations, and some behaviors. Clifftop vantage points have been used for detailed observations of several odontocete species (e.g., Norris, 1974; Norris & Dohl, 1980; Saayman & Tayler, 1979; Tayler & Saayman, 1972; Würsig and Würsig, 1979, 1980). The use of surveyor's transits or theodolites on these cliff tops has been increasing and allows for tracing the movements of individuals with high precision and reliability.

Observers in aircraft may potentially examine both nearshore and offshore groups, but observation time and sighting probability are limited by flight speed and altitude. Studies from aircraft have provided information on group size and distribution, feeding behavior, and behavior of groups containing calves (Barham et al., 1979; Leatherwood, 1975, 1977; Leatherwood et al., 1978; Odell, 1976; Odell & Reynolds, 1977). Studies from clifftop vantage points and aircraft are additionally limited because identification of individuals is usually difficult (but see Würsig & Würsig, 1977).

The use of boats permits close-range observations and may allow for following of individuals or groups. Observers in vessels with underwater viewing capabilities have reported three-dimensional deployment of subgroups and segregation of dolphins by age and sex classes (Evans & Bastian, 1969; Norris, 1974). Visibility from vessels may be limited, however, and the effect of the craft on behavior is difficult to assess.

Recognizable markings or characteristics such as dorsal fin shapes can identify individuals; repeated sightings of these individuals may provide information about home-range limits and group composition (Caldwell, 1955; Essapian, 1962; Irvine, Scott, Wells, Kaufmann & Evans, 1979; Irvine & Wells, 1972; Norris & Dohl, 1980; Norris & Prescott, 1961; Saayman & Tayler, 1979; Shane, 1977; Tayler & Saayman, 1972; Würsig & Würsig, 1977). Analyses of cetacean photographs can often reveal distinctive markings not readily observable in the field (Balcomb & Goebel, 1976; Würsig & Würsig, 1977). While natural markings may identify individuals that are present, information about the sex and age of the animals is usually not available, except for highly sexually dimorphic species such as the killer whale or strongly age-dimorphic species, such as pantropical spotted dolphins (*Stenella attenuata*), which exhibit age-related color variations (Perrin, 1975b).

The capture of animals for tagging allows the sex, size, and age of individuals to be determined. Tags or marks placed on the dorsal fin or body of the temporary captives (e.g., Evans, Hall, Irvine & Leatherwood, 1972; Irvine, et al., 1979; Irvine & Wells, 1972; Norris & Pryor, 1970; Perrin, 1975a; Sergeant & Brodie, 1969) facilitate the identification of more individuals than is generally possible by natural marks alone. Artificial tags are generally more visible and are more easily recognized than scars or natural markings.

A promising technique for studying individual cetaceans is biotelemetry, which has been used to monitor the movements and activities of several odontocete species (Erickson, 1978; Evans, 1971, 1974; Evans et al., 1972; Gaskin, Smith & Watson, 1975; Irvine et al., 1979; Jennings, 1979; Leatherwood & Evans, 1979; Martin, Evans & Bowers, 1971; Norris, Evans & Ray, 1974; Würsig, 1976). Radiotelemetry can provide information about an animal's location, dive time and depth, heart rate and body temperature, but instrumentation costs are high and attachment problems are always present (Irvine et al., 1979). The success of radio- or visual tagging programs will be influenced by the species being tagged, the habitat in which captures are planned, the number and type of tags used, and the effort made to resight tagged animals. Tag designs, which vary in durability and potential effects on the animals (Irvine, et al., 1979; Nishiwaki, Nakajima & Tobayama, 1966), greatly influence sighting success. Possible injury to the animals either during capture or from the tags must also be guarded against. Local habitat is an important criterion for deciding what tagging techniques are to be used and usually determines the accessibility of the animal for observation and study.

SOME ECOLOGICAL CORRELATES AND SOCIAL CHARACTERISTICS OF ODONTOCETE SPECIES

Summary Data

Table 6.1 summarizes available data on selected ecological and social characteristics for 24 odontocete species. Included are typical habitats, movement patterns, group size, food and feeding behavior, predation pressure, and the composition, stability, and ordering of social groups. Habitats can be roughly classified into three categories on the basis of physical and biological characteristics: riverine, including river systems, lakes, and the upper reaches of estuaries; coastal, ranging from estuaries to open coastlines and including relatively shallow waters up to several miles offshore; and pelagic or open ocean. While some species or populations may occupy only a single habitat type, others may range over several classes of habitats. Generally habitats decrease in physiographic complexity as one proceeds offshore into areas free of reefs or abutting land masses that tend to constrain cetacean movements. Because of greater accessibility, most long-term field studies of small cetaceans have involved inshore animals ranging through riverine or coastal habitats.

Movement patterns are given in Table 6.1 when it is possible from the literature to categorize the animals as migratory or resident. Group size reflects the reported range of numbers of animals per group as well as the general consensus of typical group size. Food and feeding behavior includes, within the limits of available data, the kind and distribution of food resources exploited, and the type of feeding strategy used. Predation is infrequently documented for cetaceans, but when such cases have been reported or when authors have reasonably assumed that predation may occur the nature of the pressure is included in the table. Published data on group composition are rarely based on confirmed identifications of age or sex classes as might be obtained through capture, for example, but are more commonly derived from observations of secondary and often subtle sexual or age characteristics.

Intensively Studied Species

Though most of the information in Table 6.1 is compiled from bits and pieces available in the literature, detailed studies of populations of six species included in the table have recently been completed. These studies have incorporated combinations of the techniques discussed earlier to investigate the behavioral ecology of humpback dolphins (*Sousa* sp.) and Indian Ocean bottlenosed dolphins (*Tursiops aduncus*) in nearshore waters off South Africa, dusky dolphins (*Lagenorhynchus obscurus*) and Atlantic bottlenosed dolphins (*Tursiops truncatus*) in Argentine coastal waters, Atlantic bottlenosed dolphins in waters adjoining the southeastern United States, killer

Table 6.1 Behavioral and Ecological Characteristics of Some Inshore Odontocetes

Species	Adult Length (cm)	Typical Habitat and Movements	Group Size	Food and Feeding Behavior	Predators	Group Characteristics	References
Ganges River dolphin (*Platanista gangetica*)	230	Riverine	Usually alone, occasionally in pairs; groups of up to 10	Demersal fish & crustaceans; individual feeding		Paris frequently are mother/calf; possibly territorial	Aminul Haque et al., 1977; Kasuya & Aminul Haque, 1972; Nishiwaki, 1972; Pilleri, 1970; Rice, 1977; Yablokov et al., 1972
Amazon River dolphin (*Inia goeffrensis*)	250–300	Riverine	Usually alone, but occasionally in dispersed groups of 3–4	Fish; individual feeding	Apparently none	Relatively stable	M. Caldwell et al.; 1966; Layne, 1958; Layne & Caldwell, 1964; Nishiwaki, 1972; Rice, 1977; Yablokov et al., 1972
White fin dolphin (*Lipotes vexillifer*)	220	Riverine	Usually 3–4 occasionally 10 or more	Demersal fish & crustaceans			Nishiwaki, 1972; Rice, 1977; Yablokov et al., 1972
Tucuxi (*Sotalia fluviatilis*)	120	Riverine, shallow coastal waters	Usually alone, in pairs, or trios	Fish & crustaceans		Tight knit, relatively stable	Layne, 1958; Nishiwaki, 1972; Rice, 1977; Yablokov et al., 1972
Finless porpoise (*Neophocaena phocaenoides*)	150–190	Shallow coastal waters, riverine	Alone or groups of 2–4	Small fish & squid			Nishiwaki, 1972; Rice, 1977; Tomilin, 1967
Whitefront dolphin (*Cephalorhynchus hectori*)	180	Coastal	Up to 6				Marcuzzi & Pilleri, 1971; Rice, 1977
Irrawaddy dolphin (*Orcaella brevirostris*)	220	Coastal, riverine	Rarely alone, but only form small groups	Fish			Nishiwaki, 1972; Rice, 1977; Yablokov et al., 1972

Species	Length (cm)	Habitat	Group size	Feeding	Predators	Group composition	References
Harbor porpoise (*Phocoena phocoena*)	150–180	Shallow coastal waters, riverine; migratory	2–10; larger feeding aggregations (seldom more than 50)	Fish, squid, crustaceans; cooperative feeding	Killer whales, sharks	1. mixed ♀♀, juvenile groups 2. adult ♂♂ groups	Arnold, 1972; Evans & Bastian, 1969; Gaskin et al., 1975; Leatherwood & Reeves, 1978; Leatherwood et al., 1976; Neave & Wright, 1968; Nishiwaki, 1972; Rice, 1977; Tomilin, 1967; Yablokov et al., 1972
Humpback dolphin (*Sousa* sp.)	240	Coastal, riverine; year-round residents (follow habitual inshore routes)	Often alone or in pairs; usually less than 10	Opportunistic; fish (especially reef fish); Cooperative hunting, individual feeding	Sharks, killer whales	Very labile; larger groups (>10) of mixed composition	Nishiwaki, 1972; Rice, 1977; Rice & Scheffer, 1968; Saayman & Tayler, 1973, 1979
Beluga (*Delphinapterus leucas*)	400–500	Coastal, riverine; migratory	Usually 5–10, but larger migratory groups	Fish, crustaceans, molluscs	Killer whales, sharks, Polar bears	1. mixed ♀♀, juvenile groups 2. adult ♂♂ groups 3. ♂ & ♀ pairs only during mating 4. Large males lead small traveling groups	Fay, 1978; Leatherwood et al., 1976; Rice, 1977; Sergeant, 1962, 1973; Sergeant & Brodie, 1969; Tomilin, 1967; Yablokov et al., 1972
Narwhal (*Monodon monoceros*)	500	Deep coastal waters, riverine; migratory	Usually 3–15; larger migratory groups	Squid & fish	Walrus, Killer whales, sharks	Groups of mixed age and sex; also of same sex	Leatherwood et al., 1976; Mansfield et al., 1975; Newman, 1978; Rice, 1977; Tomilin, 1967; Walker, 1968; Yablokov et al., 1972
Dusky dolphin (*Lagenorhynchus obscurus*)	200	Coastal; ~1500 km² range (in Argentina)	Usually 6–15; occasional feeding aggregations of 300	Schooling fish and squid; individual & cooperative feeding	Killer whales	Small groups stable over at least several days	Rice, 1977; Würsig, 1976; Würsig & Würsig, 1980; Yablokov et al., 1972

Table 6.1 *(Continued)*

Species	Adult Length (cm)	Typical Habitat and Movements	Group Size	Food and Feeding Behavior	Predators	Group Characteristics	References
Whitebeaked dolphin (*Lagenorhynchus albirostris*)	270–310	Coastal; migratory	Usually small groups; feeding aggregations of 1500	Schooling fish, cephalopods, crustaceans			Leatherwood et al., 1976; Nishiwaki, 1972; Rice, 1977; Tomilin, 1967
Atlantic bottlenosed dolphin (*Tursiops truncatus*)	240–370	Coastal, riverine, pelagic, distinct inshore & offshore populations; inshore residents	Usually <20, but occasional aggregations of several hundred	Opportunistic; mostly fish; individual & cooperative feeding	Sharks, killer whales	1. Adult ♀ ♂ and offspring 2. Subadult males 3. Subadult ♂ ♂, mixed ♀ ♀ and offspring; fluid associations within a herd	D. Caldwell, 1955; M. Caldwell & Caldwell, 1972; Essapian, 1962; Irvine & Wells, 1972; Irvine et al., 1979; Leatherwood, 1975; Leatherwood et al., 1976; Shane, 1977; Tomilin, 1967; Wells, 1978; Würsig, 1978; Würsig & Würsig, 1977, 1979; Yablokov et al., 1972
Pacific bottlenosed dolphin (*Tursiops gillii*)	300	Coastal, pelagic; distinct inshore & offshore populations; some migratory, some resident	Usually 25 or less inshore; occasionally groups of several hundred	Fish & squid; individual & cooperative feeding		Age-sex segregation of subgroups	Evans & Bastian, 1969; Leatherwood, 1975; Leatherwood et al., 1972; Leatherwood & Reeves, 1978; Nishiwaki, 1972; Norris & Prescott, 1961; Rice, 1960
Dalls porpoise (*Phocoenoides dalli*)	210	Coastal, pelagic; migratory	Usually 3–15, occasional aggregations of several hundred	Schooling fish and squid	Killer whales	Mixed age classes; dominance by larger individuals	Leatherwood et al., 1972; Morejohn, 1979; Nishiwaki, 1972; Norris & Prescott, 1961; Rice, 1977; Tomilin, 1967

Species	Length (cm)	Distribution	Group size	Food	Predators	Social organization	References
Killer whales (*Orcinus orca*)	700–1000	Coastal, riverine, pelagic; resident, ranging & transient pods; residents range over ~320–480 km	Usually 1–40, sometimes over 100	Fish, squid, marine mammals; coordinated feeding	None	Frequently segregated by age and sex into subgroups; permanent family units	Balcomb & Goebel, 1976; Balcomb et al., 1979; Bigg, pers. comm.; Condy et al., 1978; Erickson, 1978; Leatherwood et al., 1976; Martinez & Klinghammer, 1970; Nishiwaki, 1972; Norris & Prescott, 1961; Rice, 1977; Scheffer, 1978; Tomilin, 1967
Atlantic white-sided dolphin (*Lagenorhynchus acutus*)	230–270	Coastal	Usually 10–50, feeding aggregations of 1000–1500	Fish			Leatherwood et al., 1976; Rice, 1977; Tomilin, 1967
Longfin pilot whale (*Globicephala melaena*)	620	Coastal, pelagic; migratory	Inshore = 85 Offshore = 20	Squid and schooling fish	Killer whales	1. Adult ♂ groups 2. Mixed adult groups 3. Mixed adults & young Disperse for feeding, otherwise tight knit	Leatherwood et al., 1976; Rice, 1977; Sergeant, 1962; Taruski, 1976; Tomilin, 1967; Yablokov et al., 1972
Shortfin pilot whale (*Globicephala macrorhynchus*)	470–600	Coastal, pelagic; migratory—follows squid inshore during spring	Usually 5–60, occasional aggregations of several hundred	Squid and schooling fish; coordinated hunting	Killer whales	1. Groups of mixed age & sex, tight knit 2. Adult males may separate from females and juveniles in winter 3. Haremlike subgroups	Brown & Norris, 1956; Leatherwood et al., 1972, 1976; Reilly, 1978; Rice, 1977; Yablokov et al., 1972

Table 6.1 *(Continued)*

Species	Adult Length (cm)	Typical Habitat and Movements	Group Size	Food and Feeding Behavior	Predators	Group Characteristics	References
Pacific pilot whale (*Globicephala scammoni*)	630	Coastal, pelagic; migratory	As many as several hundred	Squid, group foraging and dispersed feeding	Killer whales	Segregation of old males noted	Nishiwaki, 1972. Norris, 1967; Norris & Prescott, 1961
Pacific white-sided dolphin (*Lagenorhynchus obliquidens*)	240	Pelagic—summer & fall/coastal—winter & spring; migratory	Usually 2–100 but occasionally thousands	Schooling fish & squid	Killer whales	Groups of mixed age/sex with segregated subgroups	Evans & Bastian, 1969; Leatherwood & Reeves, 1978; Norris & Prescott, 1961; Rice, 1977; Tomilin, 1967; Yablokov et al., 1972
Indian Ocean bottlenosed dolphin (*Tursiops aduncus*)	260	Coastal, pelagic; some apparently resident	Usually 100–200; ranges 3–1000	Individual feeding on benthic & reef-dwelling fish; cooperative feeding on schooling fish	Sharks possibly killer whales	Dominance hierarchy (from captive studies)	Ross, 1977; Saayman & Tayler, 1973; Tayler & Saayman, 1972
Hawaiian spinner dolphin (*Stenella longirostris*)	210	Coastal, pelagic; "resident" to some degree	6–250	Schooling fish and squid	Sharks	Age-related subgrouping in larger herds; dominance hierarchy (from captive studies)	Bateson, 1974; Leatherwood et al., 1972; Norris & Dohl, this volume, 1980; Perrin et al., 1973; Rice, 1977

272

whales (*Orcinus orca*) resident in waters along the northeast Pacific coast and in the Indian Ocean, and spinner dolphins (*Stenella longirostris*) in Hawaiian waters. Available data on movements, group characteristics, predation pressure, and feeding behavior are briefly summarized for each of these species in the following sections.

Humpback Dolphin

Clifftops immediately adjoining a large bay in South Africa were used to observe humpback dolphins (Saayman & Tayler, 1973, 1979; Saayman et al., 1972). Resightings of naturally marked individuals over a three-year period suggested that a group was permanently resident in the bay. The sheltered bay was used for resting and short periods of social behavior, with the animals remaining within 1 km of shore. Feeding was done on an individual basis in more open coastal areas. The bay was shared at times with Indian Ocean bottlenosed dolphins, but these tended to range further offshore than the humpback dolphins.

Sharks and killer whales were considered potential predators on the humpback dolphins. The dolphins were seen to enter shallows and to cross over reefs in apparent response to approaching killer whales.

Adult dolphins often traveled between areas of the bay singly or in pairs, while subadults almost always traveled in groups. A local group of about 25 individuals that regularly dispersed into dynamic subunits with fluctuating membership may have been the basic social unit or herd.

Indian Ocean Bottlenosed Dolphin

The Indian Ocean bottlenosed dolphins inhabiting the shallow bay together with the humpback dolphins or roaming into adjoining coastal areas were also studied extensively by Saayman and colleagues (Saayman, Bower & Tayler, 1972; Saayman & Tayler, 1973; Saayman, Tayler & Bower, 1973; Tayler & Saayman, 1972). Repeated sightings of several distinctly marked individuals indicated that, like the humpback dolphins, at least some of the bottlenosed dolphins were residents of the areas observed, including some 32 km of open, rocky coastline outside the bay. Some groups of bottlenosed dolphins ranged from coastal shallows to well offshore.

The bottlenosed dolphins were at least occasionally attacked by sharks, as evidenced by shark bite scars. Ross (1977) found that 2 of 58 (3.4%) dolphins examined bore sharkbite scars, and Tayler and Saayman (1972) observed dolphins actively avoiding approaching sharks.

Feeding style varied depending on the resource being exploited. Dispersed, individual feeding was observed when benthic and reef-dwelling fish were hunted, while highly coordinated, cooperative herding and feeding methods were employed with schooling fish.

The mean group size was 140 animals. Apparent age and sexual segregation was observed within these groups. Additionally, more than one reproductively mature male was reportedly observed in the same group.

Dusky Dolphin

Dusky dolphins have been studied intensively in the Golfo San Jose region of Argentina (Würsig, 1976; Würsig & Würsig, 1980). Radiotracking data and sightings of spaghetti-tagged individuals suggest that at least some individuals were permanent residents, although total abundance varied seasonally. The animals ranged over an area approximately 1500 km², including a large bay and an unsheltered coastline. Resting occurred in shallow water, while feeding took place in deeper areas. Feeding was generally associated with cooperative herding of large schools of small fish by aggregations of up to 300 dolphins. Killer whales approached groups of dolphins on several occasions; in response the dolphins moved rapidly into very shallow water.

A number of subgroups, totaling about 300 animals, were sometimes dispersed over a 10-km diameter area in the Golfo San Jose region. The large group may form the primary breeding herd in the area. Subgroup composition was stable for at least several days. Mixed sex groups of 6 to 15 individuals were typical while traveling.

Atlantic Bottlenosed Dolphin

Bottlenosed dolphins studied by the Würsigs from clifftops in southern Argentina typically inhabited coastal waters less than 10 m deep and made only brief excursions into deeper water (Würsig, 1978; Würsig & Würsig, 1977). During the summer the dolphins were usually observed in water less than 6 m deep. Their movements appeared to be influenced by tides and submerged obstructions. Recognizable individuals were photographically identified in the area throughout the year. Home range limits were not determined, but several dolphins from the study group were recognized 300 km from the study site on one occasion, and were later identified back in the study area.

Shane (1977) observed bottlenosed dolphins from a small boat in the channels and passes of the Gulf of Mexico near Port Aransas, Texas. She was able to define home ranges for a number of dolphins that were maintained during much of the year, but some of the animals apparently left the area during the winter.

Little information is available about predation pressure on these bottlenosed dolphins. Würsig and Würsig (1979) reported two apparently unsuccessful attacks by killer whales, as well as apparent killer whale tooth scars on another bottlenosed dolphin, all in Argentine waters. Wood et al. (1970) noted that of 22 bottlenosed dolphins captured off the southeastern United States, 4 had definite shark attack scars, and 7 others had scars that were probably from shark bites.

Feeding strategies may vary with the nature of the food source or habitat (Leatherwood, 1975; Hamilton & Nishimoto, 1977; Hoese, 1971). Würsig and Würsig (1979) reported that the dolphins increased their swimming speed, used a "wider-than-long" food-searching formation, and engaged in group feeding, apparently on small schooling fish, when in deeper water. In contrast, when close to shore individuals usually fed alone. Shane (1977) noted that relatively small groups of bottlenosed dolphins were involved in feeding and that peaks in feeding activity occurred during early morning and late afternoon.

In general group size may vary greatly from region to region for the Atlantic bottlenosed dolphin (Table 6.2) and appears related to such factors as season and physical characteristics of the habitat. Odell and Reynolds (1977) reported an increase in group size during the winter for bottlenosed dolphins in the offshore waters along Florida's west coast, but Shane (1977) noted a slight decrease in group size in waters off Texas during the same period.

Shane (1977) also noted that groups were, on the average, larger when in open waters than when in the constrained regions of the channels or passes. Odell (1976) found a similar relationship in comparing group size in the open waters of Biscayne Bay with that in the complex aquatic habitat adjoining the Florida Everglades. Similarly, Leatherwood and Platter (1975) reported that groups were larger in the open sounds of the northern Gulf of Mexico than in the shallow, coastal marshlands. Within the marshland area, the largest groups were in the deep channels connecting shallower feeding areas. However, in a later study Leatherwood et al. (1978) reported no such correlation for similar regions.

Group composition was dynamic for the bottlenosed dolphins in Argentine waters (Würsig, 1978; Würsig & Würsig, 1977) and Texas (Shane, 1977). Würsig (1978) suggested that food availability may be a primary determinant of subgroup size and stability. Subgroups, composed of combinations of the 53 recognizable dolphins that passed by the shore station, generally consisted of 8 to 22 animals. Although a small core unit of individuals were consistently observed together, subgroup composition generally varied from day to day.

Killer Whale

Killer whales are widely distributed through coastal, pelagic, and some riverine habitats (Tomilin, 1967). Recent work by Bigg (1979, personal communication) and Balcomb and others (Balcomb & Goebel, 1976; Balcomb et al., 1979) has involved the photographic identification of killer whales in Puget Sound, Washington. Bigg has classified groups (pods) in this area as (*a*) resident, if inhabiting a relatively limited area; (*b*) ranging, if pod movement included several resident ranges; or (*c*) transient, if the pod was seen only

Table 6.2. Group Sizes for the Genus _Tursiops_

Location	Habitat Type	Mean Dolphins/ Group	Std. Dev.	Range	No. of Groups	Source
Gulf of Mexico, offshore Sarasota, Florida	Coastal-pelagic	2.87			31	Odell & Reynolds, 1977
Everglades National Park	Coastal (shallow bays, many small islands)	2.95		1–25		Odell, 1976
Aransas Pass, Texas	Coastal (bays, channels, passes)	3–4				Shane, 1977
Florida, West Coast	Pelagic-coastal	4.66		1–50	326	Odell & Reynolds, 1977
Sarasota, Florida	Coastal (bays, flats, channels, passes)	4.84	4.31	1–40	688	This chapter
Texas	Coastal (bays)	5–10				Gunter, 1942
Indian River, Florida	Riverine	7.8	2.12	1–35	64	Leatherwood, 1979
Biscayne Bay, Florida	Coastal (open bay)	8		3–13		Odell, 1976

Location	Habitat				Reference
Florida	Coastal	12			D. Caldwell & Caldwell, 1972
Mediterranean Sea	Pelagic	10–15			Pilleri & Knuckey, 1969
Mississippi Sound (1975)	Coastal	10.5			Leatherwood & Platter, 1975
Argentina	Coastal	15	3.28	8–22	Würsig & Würsig, 1977; Würsig, 1978
Mississippi Sound (1974)	Coastal (open bay)	15.9	2.66		Leatherwood et al., 1978
California	Coastal	20			Evans & Bastian, 1969
Atchafalaya Bay, Louisiana	Coastal (open bay)	23.2	4.37		Leatherwood et al., 1978
N. Gulf of Mexico	Coastal (sounds and islands)	28.9	8.7	1–175	Leatherwood & Platter, 1975
Hawaii	Coastal (within 1 km of reefs)	1–50			Rice, 1960
Cape Hatteras, N. Carolina	Coastal	12 to several hundred			Townsend, 1914
South Africa	Coastal (open bay)	140.3	21.4		Saayman et al., 1973

briefly or occasionally in Puget Sound. Some resident pods inhabited ranges, or "community territories," that extended for 320 to 480 km.

Condy et al. (1978) studied killer whales off Marion Island in the south Indian Ocean. The animals approached to within 100 m of the coast, an irregular shoreline characterized by large open bays, smaller coves, deep inlets, and offshore rocks. The abundance of killer whales in these waters varied seasonally, and was synchronized with the seasonal haul-out of southern elephant seals (*Mirounga leonina*). Hunting occurred close to shore, and resting took place in sheltered coves. One recognizable whale was observed during three consecutive summers.

Killer whales have no known predators. The prey of killer whales includes fish, squid, other cetaceans, and pinnipeds. Highly coordinated, cooperative hunting and feeding of "packs" of killer whales is well documented (Condy et al., 1978; Martinez & Klinghammer, 1970; Norris & Prescott, 1961).

Killer whales were infrequently found alone in Puget Sound (Balcomb & Goebel, 1976; Balcomb et al., 1979; Bigg, 1979). They were usually sighted in what appear to be permanent family groups of up to 40 individuals. These groups were often segregated into subgroups by age and sex. While traveling, subgroups often maintained a consistent position relative to other subgroups (Norris & Dohl, this volume).

Condy et al. (1978) reported seeing solitary animals as well as groups of 2 to 25 animals. The solitary animals were generally adult males. Within the larger groups subgroups might range from 2 to 11 individuals; single animals were also seen accompanying the group. Subgroups of two animals were either females with calves or two subadults, while subgroups of three consisted of an adult male with a female and calf or an adult male with subadults.

Hawaiian Spinner Dolphin

Spinner dolphins are primarily pelagic, but in Hawaiian waters some populations regularly move near shore. At the island of Hawaii spinners have been studied from cliffs and from aircraft and boats (Norris, 1974; Norris & Dohl, 1980; Norris & Dohl, this volume). Radio transmitters have been attached to some animals, and together with distinctive natural markings have provided a means for study of individuals.

During the day, groups of dolphins were seen resting in relatively shallow protected bays. The location of the daylight rest areas varied for given animals. Late in the day the animals moved offshore, where they fed nightly on organisms associated with the deep scattering layer.

Observations of sharks attacking pelagic spinner dolphins have been reported in the eastern Pacific Ocean (Leatherwood et al., 1971), and shark bite scars have also been seen on spinner dolphins in Hawaii (Norris & Dohl, 1980).

Group size for Hawaiian spinner dolphins ranged from 6 to 250 individuals. Composition was fluid, and only the composition of small subgroups remained stable for more than a few days. Small groups of dolphins joined into larger groups as they moved offshore for the evening feed. Diving by group members was relatively synchronized while feeding, suggesting coordinated and possibly cooperative feeding. However, because feeding is at night in deep water, the feeding strategies could not be documented.

Overview

Several trends become evident from examination of Table 6.1 and from the summary data on the more intensively studied groups. Among the smaller odontocetes, group size varies relatively predictably from habitat to habitat. In general species occupying riverine habitats are found alone or in small groups of 2 to 10 individuals. Coastal species are intermediate in group size, generally 50 or fewer individuals comprise a group, though occasionally much larger aggregations occur, usually associated with feeding or migration. Finally, species utilizing both coastal and pelagic waters are often found in large groups containing hundreds or thousands of individuals.

Migrations are generally limited to the more pelagic species. The major exceptions involve species in colder waters such as harbor porpoises (*Phocoena phocoena*), belugas (*Delphinapterus leucas*), and narwhals (*Monodon monoceros*), where temperature and/or ice flows apparently strongly influence movements. Movements of migratory species are often correlated with the movements of their prey. In species such as bottlenosed dolphins, Pacific white-sided dolphins (*Lagenorhynchus obliquidens*), spinner dolphins, or killer whales, inshore populations are often resident to a given area, while offshore populations may range over a much larger area or be migratory.

The food resources utilized by odontocetes vary with the habitat, with a concomitant variation in feeding strategies. Riverine species consume demersal fish and crustaceans primarily. Feeding is generally accomplished on an individual basis. Coastal species are opportunistic feeders, taking advantage of bottom fish, schooling fish, and cephalopods, to name but a few. Both individual feeding and coordinated group foraging and feeding strategies are used. The more pelagic species tend to concentrate their feeding efforts on clumped, patchily distributed resources such as schooling fish and squid and often rely on coordinated foraging and feeding.

Predation pressure varies from habitat to habitat. Riverine species are apparently subjected to little or no predation. Sharks and killer whales are implicated as at least occasional predators on most of the coastal and pelagic species.

Trends for group characteristics are more difficult to identify, because of the paucity of information on group composition and stability. Age-related or sexual segregation appears to be common in many of the more social species. This may take the form of separate groups of a single sex, or a more functional organization, such as nursery groups, or segregated subgroups within a larger group. Variations across both of these last two possibilities may occur. For example, among bottlenosed dolphins and spinner dolphins segregated subgroups of a larger resident population unit may be dispersed over the "home range" of the group, with occasional encounters between subgroups sometimes resulting in changes in their composition.

Groups of the larger inshore odontocetes such as pilot whales (*Globicephala* sp.) or killer whales involve very tight social bonds, as evidenced by the frequent mass strandings of the pilot whales (Fehring & Wells, 1976; Irvine, Scott, Wells & Mead, 1979) and the permanent family units of the killer whales. Territoriality has been suggested as the possible dominance system used by the Ganges River dolphin (*Platanista gangetica*) (Pilleri, 1970), based on observations of captives. Dominance hierarchies within multimale, mixed sex groups of captive bottlenosed dolphins or spinner dolphins have also been observed (Bateson, 1965; Tavolga, 1966; Tayler & Saayman, 1972). More data from observations of groups of known age and sex composition are needed, however, to identify social systems in the wild.

These general trends are not perfect—much variation occurs not only across species but within species as well. The noted trends do, however, provide a perspective from which to examine the social ecology of individual species in greater detail, as in the following case study, and they facilitate speculation about selective forces that may influence cetacean social organization in general.

CASE STUDY: BOTTLENOSED DOLPHINS IN FLORIDA WATERS

During 1970 and 1971 (Irvine & Wells, 1972) and then again from 1975 through 1978 (Irvine et al., 1979) boat surveys and tagging and observational studies were carried out on a population of Atlantic bottlenosed dolphins in waters near Sarasota on the central west coast of Florida (Fig. 6.1). Radio- and visual tagging techniques (Fig. 6.2) were used extensively during the latter period. Of an estimated population of 102 animals 47 dolphins were captured at least once and tagged (Table 6.3). Also 12 additional dolphins, not captured, were individually identified by distinctive dorsal fins or natural markings (Fig. 6.3). During the course of the study there were more than 3000 dolphins sighted. Of these 781 were tagged animals that were identified, 129 were tagged but the tags could not be read, and 87 were naturally marked.

Most of the tags were installed between December 1975 and July 1976, and the majority of the tag sightings were made within this period. Captures and boat surveys were conducted more frequently in the northern part of the study area, resulting in a higher number of tag sightings in that region than in the southern areas. Consequently, the data may in part reflect seasonal or geographic sampling biases.

Habitat Subareas

The study area extended 40 km south from Tampa Bay and included waters up to 3 km offshore and the inshore bays and channels of the north–south

Table 6.3 Summary of Capture and Sighting Data for *Tursiops truncatus* Tagged near Sarasota Florida January 1975 to July 1976

Dolphin Identification Number	Length (cm)	Males Initial Sighting	Number of Sightings	Number of Captures		Dolphin Identification Number	Length (cm)	Females Initial Sighting	Number of Sightings	Number of Captures
03	210	Apr 75	68	3	S	31	207	May 76	19	1
40	215	Jun 76	11	1	U	11	224	Aug 75	20	5
21*	219	Feb 76	42	2	B	10	230	Aug 75	1	1
26	221	Apr 76	56	2	A	28	233	Apr 76	33	3
19	224	Dec 75	1	1	D					
14	227	Dec 75	27	4	U					
27	237	Apr 76	20	1	L					
20	237	Feb 76	30	1	T					
01	246	Jan 75	3	1		17	240	Dec 75	31	3
06*	249	June 75	12	1		45	248	Jul 76	1	1
02	254	Jan 75	50	4	A	44	249	Jul 76	1	1
08*	262	Jan 75	12	1	D	09	250	Aug 75	79	2
07*	262	Jun 75	6	1	U	43	252	Jul 76	1	1
35*	264	May 76	10	1	L	33	261	May 76	15	2
30*	276	May 76	1	1	T	32*	269	Apr 76	24	2
34*	286	May 76	19	3						

281

Table 6.3 (*Continued*)

Dolphin Identification Number	Males				Dolphin Identification Number	Females			
	Length (cm)	Initial Sighting	Number of Sightings	Number of Captures		Length (cm)	Initial Sighting	Number of Sightings	Number of Captures
	Calves					Presumed Mothers			
42m	136	Jul 76	1	1	41	239	Jul 76	1	1
46m**	146	Jun 76	7	1	18*	242	Dec 75	14	4
47f**	138	Jun 76	7	1	29	243	Apr 76	20	2
16m	192	Dec 75	2	1	15	247	Dec 75	19	2
23m	149	Mar 76	1	1	22	248	Mar 76	1	1
37f	198	May 76	9	2	36	250	May 76	9	2
25m	172	Mar 76	1	1	24*	253	Mar 76	1	1
05m	182	Apr 75	34	6	04	253	Apr 75	34	4
39f	212	May 76	17	2	38	256	May 76	17	2
13m	190	Oct 75	34	2	12*	257	Oct 75	34	3

* = Also tagged during 1970–71; ** = known to have been born during study; m = male; f = female.

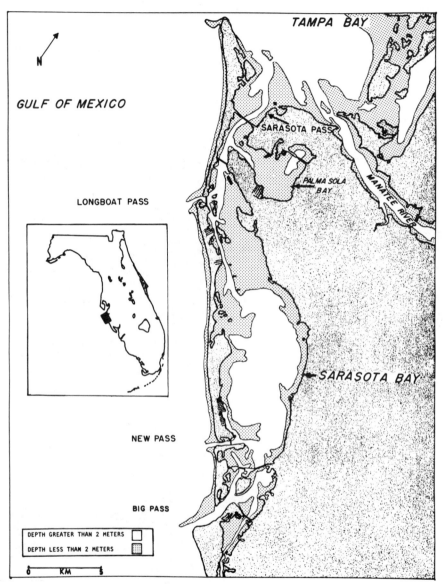

Fig. 6.1. Location of the 1975–1976 study area (shaded portion of inset) on the west coast of Florida.

Fig. 6.2. Examples of tags and marks on free-ranging small cetaceans. Dolphins are positioned from the foreground; (1) Radio transmitter package mounted on the dorsal fin, with a freezebrand below and a yellow plastic Roto tag attached to the trailing edge of the fin; (2) yellow fiberglass visual tag (#22) bolted through the fin, with a freezebrand on the body below and a yellow Roto tag attached to the trailing edge of the dorsal fin; (3) two yellow Roto tags attached to the trailing edge of the dorsal fin; (4) unmarked. (Photograph by M. D. Scott near Sarasota, Florida, 1976.)

Fig. 6.3. Naturally marked dolphin "Moonfin" in Tampa Bay, Florida. Note also the synchronous surfacing and close associations typical of primary groups in the Sarasota, Florida area. (Photograph by R. S. Wells.)

284

barrier island chain (Fig. 6.1). Six physiographic subdivisions characterize the area: (a) Channels—narrow waterways 2 to 3 m deep and often formed or maintained by dredging; (b) Flats—extensive seagrass beds in waters less than 2 m deep; (c) Bays—large areas of open water between the barrier islands and the mainland, 2 to 5 m deep; (d) Passes—channels between barrier islands, 2 to 11 m deep; (e) Inshore Gulf—Gulf of Mexico waters less than 2 m deep; and (f) Offshore Gulf—Gulf of Mexico waters deeper than 2 m. All depths are given as mean low tide.

Range

The largest population unit within the study area was called a "herd," and referred to all those dolphins that regularly occupied the study area and interacted socially with each other to a much greater degree than with dolphins in nearby areas. This definition of herd is comparable to Wilson's (1975) demographic society as defined for chimpanzees. Individual dolphins were categorized into adult or subadult classes on the basis of body length, using the length vs maturity curve of Sergeant et al. (1973). Small subadults were classified as calves if consistently seen in association with a significantly larger female.

"Home range" has been defined as the area over which an animal normally travels during its routine activities (Burt, 1943; Jewell, 1966). Available evidence suggested the existence of a herd with a home range of approximately 85 km^2 and included the entire study area south from Tampa Bay to approximately Big Pass, and the inshore and offshore waters previously noted (Fig. 6.1). The herd home range contained a number of distinctive individual home ranges that differed according to age-sex class. Females with calves had the largest home ranges and were most often observed in the northern part of the study area, though some were sighted throughout the study area (Fig. 6.4). The home ranges of adult females without calves were somewhat more concentrated in the northern part of the study area (Fig. 6.5). Subadult female (Fig. 6.6) and adult male (Fig. 6.7) home ranges were located almost entirely in the northern part of the study area. Five of seven subadult males, however, had home ranges centered in Sarasota Bay and were not sighted in the northernmost areas of the herd home range (Fig. 6.8). The two smallest subadult males were exceptions and had home ranges resembling those of females with calves.

Average sizes for the home range of each age-sex class are shown in Table 6.4. Home range sizes were determined by measuring the area enclosed by a polygon that included all sightings of an individual, excluding land masses. In general, home range size increased markedly with each additional sighting of an individual, up to about 15 sightings. Consequently, only those dolphins seen at least 15 times were included in calculations of the mean home-range sizes. Significantly larger home ranges were observed for subadult males

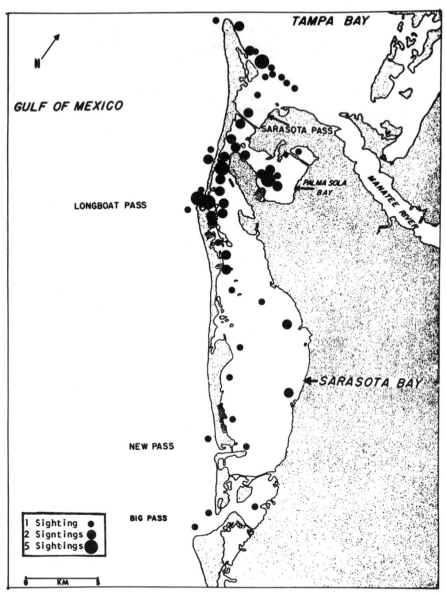

Fig. 6.4. Cumulative sightings of females with calves. Dots indicate the number of sightings of female–calf pairs at a given location during 1975–1976.

286

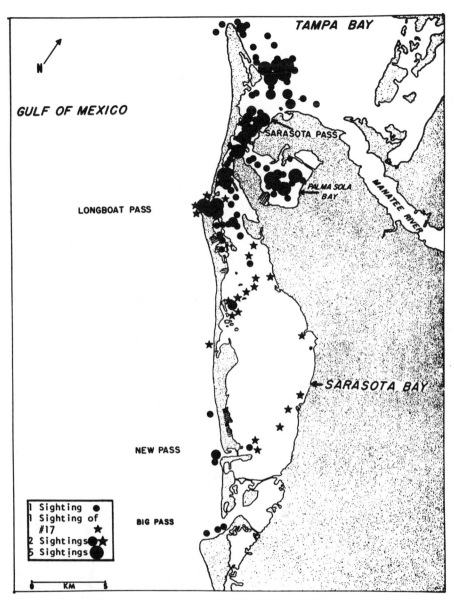

Fig. 6.5. Cumulative sightings of adult females without calves. Dots indicate the number of sightings of individuals by location during 1975–1976. Stars indicate sightings of dolphin #17, a female with an unusual home range. See text for discussion.

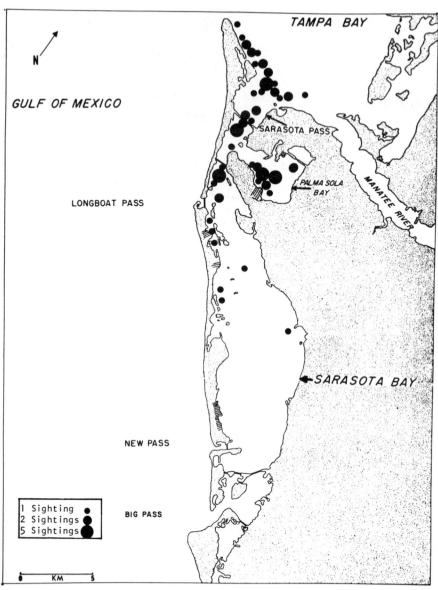

Fig. 6.6. Cumulative sightings of subadult females. Dots indicate the number of sightings of individuals by location during 1975–1976.

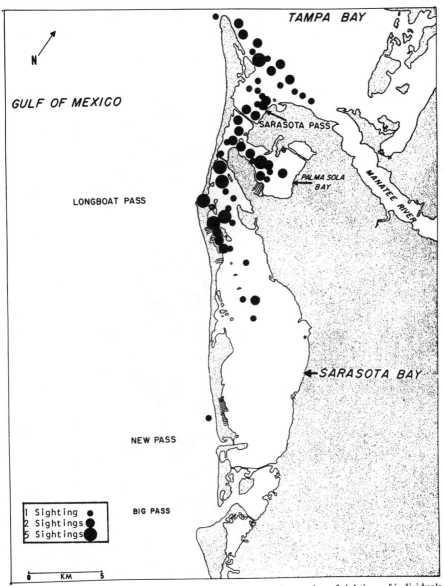

Fig. 6.7. Cumulative sightings of adult males. Dots indicate the number of sightings of individuals by location during 1975–1976.

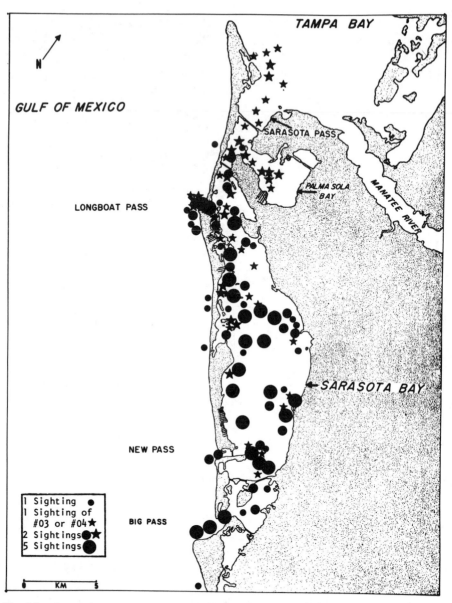

Fig. 6.8. Cumulative sightings of subadult males. Dots indicate the number of sightings of individuals by location during 1975–1976. Stars indicate sightings of the two smallest subadult males, dolphins #03 and #04, that had unusual home ranges for this class. See text for discussion.

Table 6.4 **Mean Home Range Sizes**[a] **for Individuals of Given Age-Sex Classes of** *T. truncatus* **near Sarasota, Florida (1975–1976)**

Age (Size)/Sex Class	Mean (km^2)	Std. Dev. (km^2)	No. of Dolphins
Female with calf	41.43	20.90	3[b]
Subadult male	40.03	5.94	6
Adult male	20.84	10.52	2
Adult female	18.40	13.29	5
Subadult female	14.84	1.92	3

[a] Minimum of 15 sightings per animal.
[b] Female–calf pairs.

and females with calves than for other age-sex classes (Mann-Whitney U test, $p < 0.05$).

Observations made while following groups of dolphins for extended periods indicated that the use of certain areas of the herd home range was emphasized over others. Typically, a group of dolphins was found in one part of the herd home range for several days, milling over flats or in passes, often while apparently feeding. This pattern was usually then shifted to another part of the home range and repeated.

Use of particular regions of the herd home range appeared to vary seasonally. The number of dolphins found to the east of the barrier islands decreased during the winter but increased around the passes, along the beaches, and offshore in the Gulf. Sightings of females with calves suggested that certain areas were used seasonally as nurseries, especially shallow flats, sheltered bays, and river mouths.

Predation Pressure

Sharks are probably the only predator of bottlenosed dolphins in the Sarasota area, although dolphins probably do not form the major portion of the sharks' diet (AIBS report, 1967). More than 6% of all captured dolphins displayed healed shark bite scars, and two other captured dolphins were lacking large portions of their flukes, possibly as a result of shark attack. Within the Sarasota area stranded dolphins, both live and dead, displayed wounds that appeared to be from shark bites, but it is impossible to determine if or how the attacks contributed to the strandings.

The three shark species reported by Wood, Caldwell, and Caldwell (1970) to most commonly have dolphin remains in their stomachs are present in the Sarasota area. The tiger shark (*Galeocerdo cuvier*) and dusky shark (*Carcharhinus obscurus*) are generally less prevalent than the bull shark (*Carcharhinus leucas*) (Clark & von Schmidt, 1965). Bull sharks are found both inshore and offshore (Green, 1968; Springer & Woodburn, 1960) with abundance peaking during the summer (Clark & von Schmidt, 1965). Figure 6.9 shows percentages of dolphin sightings each month in various physiographic regions as compared with the catch frequencies of bull sharks in the Sarasota area (Clark & von Schmidt, 1965). When bull sharks were most prevalent most dolphin sightings were in the shallower, more complex habitats; and those dolphins observed in the deeper waters were generally in large groups.

Food Resources

Atlantic bottlenosed dolphins have been described as opportunistic feeders (D. Caldwell & Caldwell, 1972; Leatherwood, 1975). Mullet (*Mugil* sp.) is reportedly its primary prey in the southern United States (Gunter, 1942, 1954; McBride, 1940), although local fishermen claim the dolphins will eat almost any available fish. Mullet are common in the Sarasota area, particularly in areas of flats.

In the study area, the northern part of the herd range was characterized primarily by flats. From spring to fall dolphins were often seen feeding over the flats occasionally catching leaping mullet in midair. Commercial fishermen reported that in November mullet move from the shallow inshore waters to the passes and into the Gulf of Mexico to spawn and remain for the winter. Dolphins were not usually sighted on the flats during winter but were instead in the passes and offshore, presumably drawn to the deeper areas by their food source. Both individual feeding and apparent cooperative feeding on fish schools were observed.

Group Size

During field observations all dolphins sighted within an approximately 100 m radius were classified as belonging to the same group. This generally accounted for all dolphins visible from the boat at any given time. Further distinctions were made to reflect the fluidity of groups. Primary groups were defined as the smallest number of dolphins observed to be closely associating and engaging in similar activities. Two or more primary groups often combined to form secondary groups for periods ranging from minutes to hours. Secondary groups were rarely observed intact on consecutive days, but primary groups occasionally stayed together for days or weeks. The term group can therefore be considered to be equivalent to Wilson's (1975) casual group as defined for chimpanzees.

Dolphins were infrequently sighted alone. The average group consisted of 4 or 5 dolphins, but ranged at times up to approximately 40 individuals. Group size varied during the day. Significantly smaller groups (comparison of 95% confidence limits) occurred during early morning and late afternoon than during midday and early afternoon (Fig. 6.10). Group size was not significantly affected by ebb and flood tides (Mann-Whitney U test, $p < 0.05$).

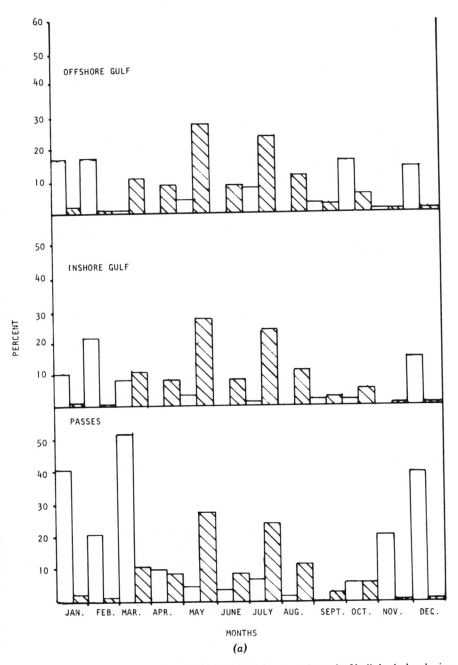

Fig. 6.9. Monthly percentages of total dolphin sightings and yearly catch of bull sharks by physiographic region in the Sarasota area. Dolphin data (Wells, 1978) are represented by open bars, shark data (Clark & von Schmidt, 1965) are represented by hatched bars.

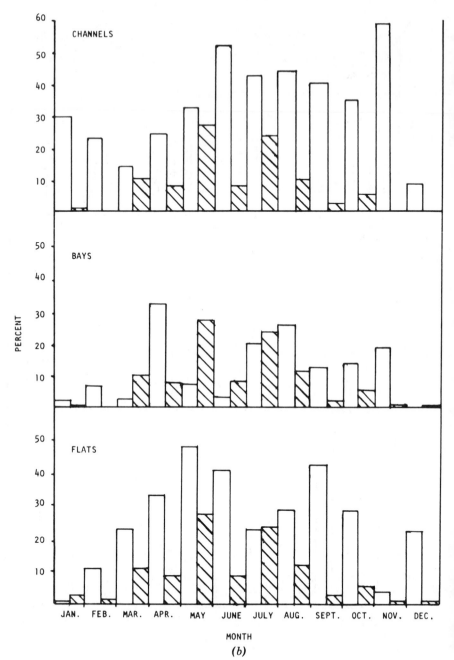

Fig. 6.9 (*Continued*)

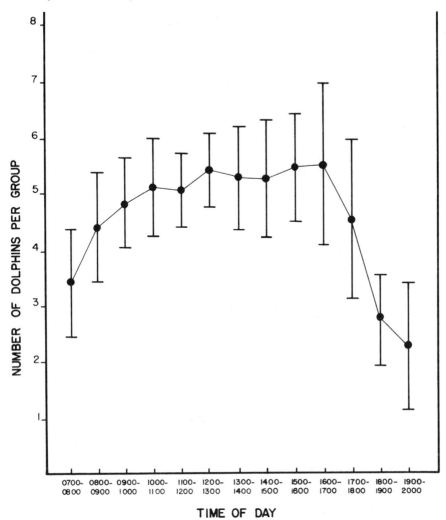

Fig. 6.10. Group size in the Sarasota area as a function of time of day with 95% confidence limits indicated for each mean.

Group size varied with the physiography of the study area. Frequency distributions of group size are given in Figure 6.11, for each of the six physiographic regions. Significantly larger groups (Mann-Whitney U test, $p < 0.05$) were found in the passes and offshore Gulf and in general group size seemed to increase with water depth.

When the average size for each age–sex class in each of the 91 groups of known composition was calculated, group size appeared to be related to the age and sex composition. Adult females, including those with young, were found in significantly smaller groups than any other classes (Mann-Whitney U test, $p <$

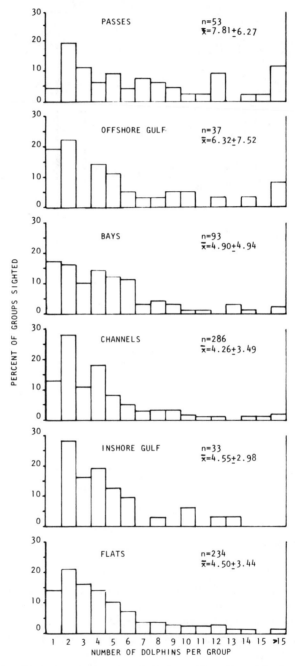

Fig. 6.11. Group size frequencies for selected physiographic regions of the Sarasota study area. Mean, standard deviation, and number of samples (*n*) are included for each region.

0.05). Adult male group size was similar to the average for the entire study herd, and subadult animals tended to be found in larger groups (Table 6.5).

No consistent arrangement or formations of individuals or classes within groups were observed. Individuals in groups tended to disperse over several hundred meters when engaged in apparent feeding activities over flats, but group members were often within 1 to 3 m of each other when groups traveled or were sighted offshore or in passes. Subgroups frequently changed relative position in a moving group but usually remained discrete. Synchronized breathing by all members of a subgroup within a 5 to 10 sec period, followed by a 30 sec to 1 min dive, was common.

Group Composition

The composition of groups within the study herd was variable from day to day, but certain patterns of association between age and sex classes were apparent. Some age–sex classes were not represented in groups containing other classes or particular combinations of classes (Table 6.5). Females were associated with a larger variety of classes than males. Of a total of 28 successful capture attempts adult and subadult males were not once found together (Table 6.6). Observations without captures revealed adult and subadult males together on only six occasions, with the two smallest tagged subadult males always involved in such associations.

Association patterns between identifiable individuals were quantified and represented in computer-produced dendrograms constructed for each month (Wells, 1978). Figure 6.12 is a representative example of one of the dendrograms. A 100% similarity between two dolphins indicates that both were always observed together. However, the clusters of associations at 15 to 25% similarity levels were the most typical (Fig. 6.12) and there was little variation in association patterns from month to month.

Table 6.5 Mean Group Size and Number of Class Association Combinations for Each Age/Sex Class of *Tursiops truncatus* near Sarasota, Florida (1975-1976)

Age/Sex Class	Mean Group Size	Std. Dev.	No. of Groups in Which Class Was Represented	No. of Class Association Combinations (see text)
Female with calf	4.16	2.78	56	11
Adult female	4.68	3.21	28	9
Adult male	5.30	3.77	10	7
Subadult male	5.44	4.49	32	7
Subadult female	5.81	3.73	16	9

Table 6.6 Size and Composition of *Tursiops truncatus* Groups Captured near Sarasota, Florida (1975–1976)

Date	Location	Group Size Before Capture	Number Captured[a]	Calves ♂	♀	Subadult ♂	♀	Adult ♂	♀
29 Jan 75	Sarasota Pass	2	2	0	0	0	0	2	0
28 Apr 75	Palma Sola Bay	4	3	1	0	1	0	0	1
15 Jun 75	N. Sarasota Bay Flats	4	3	0	0	0	0	3	0
1 Aug 75	N. Sarasota Bay Flats	18	4	1	0	0	2	0	1
2 Oct 75	Sarasota Pass	10	2	1	0	0	0	0	1
15 Dec 75	Gulf of Mexico	2	1	0	0	1	0	0	0
	Gulf of Mexico	8	4	1	0	1	0	0	2
	Gulf of Mexico	3	3	0	0	2	0	0	1
16 Dec 75	Palma Sola Bay	11	4	1	0	0	1	0	2
	Cortez	7	7	1	0	2	1	0	3
13 Feb 76	N. Sarasota Bay Flats	4	4	1	0	0	1	0	2
	Sarasota Bay	2	2	1	0	0	0	0	1
14 Feb 76	New Pass	3	2	0	0	2	0	0	0
19 Mar 76	Palma Sola Bay	4	4	2	0	0	0	0	2
15 Apr 76	Sarasota Bay	6	4	0	0	4	0	0	0
16 Apr 76	Sarasota Pass	3	2	0	0	0	0	0	2
	Palma Sola Bay	2	1	0	0	0	1	0	0
	N. Sarasota Bay Flats	10	3	0	0	2	0	0	1
7 May 76	Sarasota Bay	6	1	0	0	1	0	0	0
	Palma Sola Bay	8	6	0	0	0	2	2	2
8 May 76	Palma Sola Bay	9	9	1	2	0	0	2	4
8 Jun 76	N. Sarasota Bay Flats	11	4	1	0	1	1	0	1
	Cortez	4	4	1	1	0	0	0	2
9 Jun 76	Sarasota Bay	6	2	1	0	0	0	0	1
	Bulkhead	13	5	0	0	0	1	2	2
24 Jul 76	Bulkhead	9	7	1	0	0	0	2	4
	N. Sarasota Bay Flats	4	1	0	0	0	0	0	1
25 Jul 76	Cortez	6	6	1	2	0	0	0	3

[a] Includes identifiable escapees.

Adult males (bulls) were often found together and regularly associated with all classes except subadult males. Dendrogram similarity levels were greater between bulls and adult or subadult females than between bulls and females with calves. Adult and subadult females were frequently found together. Though occasionally seen in the company of females with calves, subadult males were found primarily in bachelor groups or with one or two adult females.

The overall low levels of similarity in the dendrograms (Wells, 1978) were an indication of the dynamic nature of the groups within the study herd. Particular

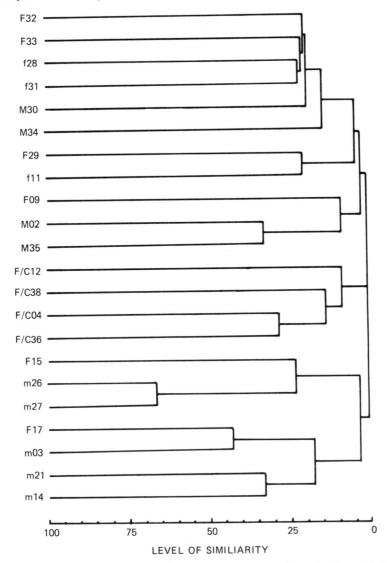

Fig. 6.12. Dendrogram analysis of May 1976 tag sightings. M = adult male, F = adult female, F/C = adult female with a calf, m = subadult male, f = subadult female. Numbers refer to dolphin identification numbers (Table 6.1). See text for discussion.

groups of several dolphins were sighted on consecutive days in the study area, after which the members of the group were found in several new groups in different areas (Fig. 6.13). After several weeks the original group was sometimes once again encountered intact.

Some long-term associations were noted. Female–young pairs were together for lengthy periods. One pair was observed repeatedly over a period of 15

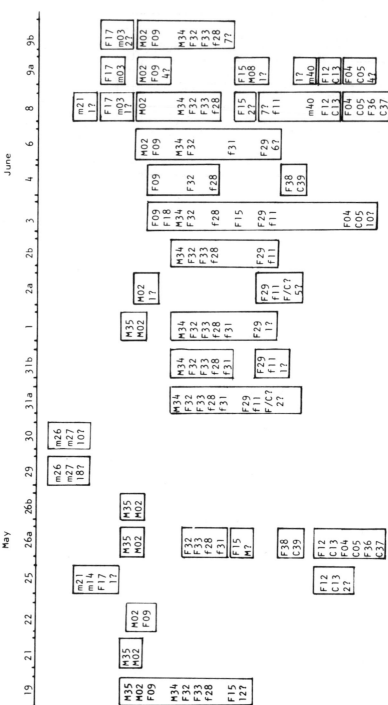

Fig. 6.13. Tagged dolphin interactions, May 19, 1976 through June 9, 1976. Each horizontal line represents a tagged dolphin. Sightings of the individual are indicated by its identification number (from Table 6.1) and its age-sex class designation. Boxes enclose groups. Letters following dates indicate more than one sighting of members of a group during a given day. M = adult male, F = adult female, C = calf, m = subadult male, f = subadult female, and a number followed by "?", within a box, indicates the number of unidentified dolphins in the group. M? indicates an unidentified adult mal

300

months. At the first capture of this pair the calf was already the size of a 6 to 12 month old (Sergeant et al., 1973), suggesting that by the time of the final observation the pair had been together for at least 21 months.

Prolonged or repeated associations were noted for other dolphins, in addition to mother–young pairs. Some subadult males remained together for several weeks. Recurrent associations following separation were observed, as shown in Figure 6.13. Several adult males (#06, #07, and #08) that were repeatedly captured together during the 1970–1971 study were captured together again in 1975 and were resighted separately and together several times over the next 13 months.

Behavioral Mechanisms

Behavioral mechanisms that may have influenced group composition and structure were not observed. Some insight into these mechanisms may be gained, however, through comparisons with information from other studies of captive and free-ranging groups. Studies of captive groups of odontocetes can be valuable because they make possible prolonged, detailed observations of the structure and maintenance of social organization. Such detail is not easily obtained from studies of cetaceans in the field. Captive groups also potentially provide opportunities to study and manipulate variables that may influence social structure.

Behavioral studies of captive groups have been conducted on relatively few species, primarily those that are easily captured and that survive well in oceanariums (see Defran & Pryor, this volume). Coastal species seem better adapted to the shallow waters found in most captive environments than most pelagic species, and members of the genus *Tursiops* are the most commonly held and most often studied captives.

A colony of captive dolphins was established at Marine Studios (now Marineland of Florida in St. Augustine) in the late 1930s. During the next 40 years the colony usually consisted of at least one mature male, five to six mature females, several subadults (some born in captivity), and one or two adult spotted dolphins. These dolphins were the objects of numerous studies on behavior and social organization (Essapian, 1953, 1963; McBride, 1940; McBride & Hebb, 1948; McBride & Kritzler, 1951; Tavolga, 1966; Tavolga & Essapian, 1957). The group was socially structured as a hierarchy with the adult male dominant over all of the other male and female tank residents. The females were also organized into a less rigid dominance hierarchy, with the largest and presumably oldest females dominant over younger animals (McBride & Hebb, 1948; Tavolga, 1966). Subsequent studies in Florida and elsewhere support a hypothesis that dominance hierarchies may be the "normal" social system for captive Atlantic bottlenosed dolphins, although the hierarchy may not always be rigidly maintained (Brown & Norris, 1956; Caldwell et al., 1965; M. Caldwell & Caldwell, 1967, 1972; D. Caldwell & Caldwell, 1972; Norris, 1967; Tavolga, 1966).

Hierarchies similar to those described for Atlantic bottlenosed dolphins were also reported for captive Indian Ocean bottlenosed dolphins in South Africa (Saayman, Tayler & Bower, 1973). Two adult males, two adult females, and later their young were in the group studied. The largest male was dominant over all other residents, and generally the largest female was dominant over the other females and subadults Bateson (1965) studied a mixed group of captive Hawaiian spinner dolphins and tropical spotted dolphins (*Stenella attenuata*) consisting of five males and two females. He described a well-defined dominance hierarchy with the largest adult male apparently dominant over other tank residents.

When adult male bottlenosed dolphins were captured from the same groups and maintained together in captivity, the larger male was dominant, maintaining priority of access to food and females with little obvious aggression (McBride, 1940). McBride and Kritzler (1951), however, reported that aggressive interactions occurred between three adult males from different capture localities. During sexual activity these bulls fought with such viciousness that two had to be removed, and the normal composition of the colony thereafter included only a single adult male (Wood, 1977).

Dominance by captive bottlenosed dolphins was displayed through jaw claps, chasing, biting, ramming, or tail slaps against subordinates (D. Caldwell & Caldwell, 1972; M. Caldwell, Caldwell & Townsend, 1968; Essapian, 1953; Lawrence & Schevill, 1954; McBride & Hebb, 1948; Norris, 1967; Tavolga, 1966). Dominance of males over females was usually expressed for mate selection (Essapian, 1953; Tavolga, 1966). Adult males usually swam alone, or for brief periods (minutes or hours) with an adult female, throughout most of the year (M. Caldwell & Caldwell, 1972, 1977; Tavolga, 1966; Tavolga & Essapian, 1957). During the apparent breeding season, however, dominant bulls maintained longer associations (days or weeks) with a given female, and this courtship period was terminated at the option of the bull (Tavolga & Essapian, 1957).

Relationships between captive adult males and females with calves may vary. The Caldwells (D. Caldwell & Caldwell, 1972; M. Caldwell & Caldwell, 1972) reported that bulls seldom associated with calves and Tavolga and Essapian (1957) suggested that females without calves might be preferred partners for bulls. There are also many examples in the literature of violent aggression by captive bulls toward calves. Responses by the mothers included either reciprocated aggression or prolonged avoidance of the bulls (D. Caldwell & Caldwell, 1972; M. Caldwell & Caldwell, 1972; Essapian, 1953, 1963; McBride & Hebb, 1948; Tavolga & Essapian, 1957).

Captive bulls were frequently aggressive toward subadult males attempting to copulate with females or toward subadult males that were recent additions to the tank (D. Caldwell & Caldwell, 1972; McBride, 1940; Norris, 1967; Tavolga, 1966). The response of captive subadults to aggression by adult males was usually to remain away from the bulls, and Tavolga (1966) noted that subadult males spent much of their time together.

While the above observations of captive cetaceans suggest the existence of a hierarchial social organization, captive studies must be viewed with caution. Generalizations about the social behavior of free-ranging cetaceans cannot necessarily be based on captive studies (Bateson, 1965; Evans & Bastian, 1969; Norris, 1967; Norris & Dohl, this volume; Saayman & Tayler, 1979; Herman & Tavolga, this volume; Yablokov, Belkovich & Borisov, 1972). Confinement in an unnatural habitat and artificial combinations of individuals (often of different species) must certainly affect behavior, but the extent of these effects is impossible to determine in the absence of comparative data from free-ranging animals. Because of the potential for detailed observations of social behavior, however, captive studies are valuable for complementing information gained from field studies.

Recent field studies have begun to provide some of the requisite information on social organization for several species, but data are still insufficient to define adequately the social system of any free-ranging odontocete. The following synthesis of captive and field observations of Atlantic bottlenosed dolphin social behavior, however, may provide some insight into the mechanisms involved in the social organization of free-ranging dolphins.

McBride and Kritzler (1951) suggested that the basic social unit for Atlantic bottlenosed dolphins was a family unit consisting of a single adult male and three to five adult females with either first- or second-year young. Recent field observations do not support this suggestion. Group composition of free-ranging animals is far more dynamic than captive observations suggest (Irvine et al., 1979; Würsig, 1978; Würsig & Würsig, 1977, 1979). Observations of wild groups with more than one sexually mature male (Wells, 1978) contradict the "family unit" concept. Even during the apparent mating season multiadult male groups were evident in the northern part of the Sarasota study area, both with and without females. Adult male groups of terrestrial mammals such as ungulates (Jarman, 1974; Owen-Smith, 1977) and primates (Eisenberg et al., 1972) are usually indicative of a dominance hierarchy, rather than one of the common forms of territoriality (Eisenberg, 1966; Kaufmann, 1971). No aggressive interactions were observed between adult males in the Sarasota area. If a dominance order existed among the bulls it may have been established and not frequently contested.

While dominance may be expressed in a hierarchy in captivity, it may be expressed by spacing, placement of subgroups, and sexual segregation in the wild (Norris & Dohl, this volume). Leatherwood (1977) reported that in the northern Gulf of Mexico bottlenosed dolphin subgroups containing small calves and presumed mothers were located near the center of groups. This may have been a protective mechanism to avoid nearby sharks. Similar formations were not obvious in the Sarasota area. Norris and Dohl (this volume) suggest that aggression by adult male dolphins might also serve as a herding mechanism, keeping mothers and calves in the center of the group. The different distributions of adult males and females with calves observed in the Sarasota area potentially could be due to female avoidance to lessen aggression by adult males toward the calves.

Segregation of wild groups of subadult male bottlenosed dolphins as noted in the Sarasota area has previously been reported from the Hatteras region of North Carolina (Townsend, 1914; True, 1891) and might be related to the dominance of adults over the younger animals. Bite marks and scars, which are usually considered testimony of past intraspecific aggression (McBride, 1940; Norris, 1967), were observed on a number of subadult dolphins near Sarasota and Texas (Gunter, 1942). In the northern Gulf of Mexico juvenile subgroups of unknown sexual composition were frequently observed on the perimeter of a herd or separated from it by several hundred meters (Leatherwood, 1977) and may also be examples of sexual segregation.

Summary and Conclusions

The data presented for bottlenosed dolphins in Florida waters are consistent with the generalized characteristics of *Tursiops truncatus* in Table 6.1. As an inshore population of a primarily coastal-riverine species, they are apparently resident and found in small to medium-sized groups. Group size increases in deeper waters. They are opportunistic feeders, subsisting on a variety of schooling and nonschooling fishes, and utilize both individual and apparently cooperative feeding techniques. Group structure may be influenced by shark predation in this area. Age-related and sexually segregated subgroups within a large, stable, dispersed herd occur. Multiadult male groups, with or without females, are evident throughout much of the year, suggesting the existence of a dominance system other than territoriality. Studies of captives suggest the existence of an individual dominance hierarchy system.

DISCUSSION

Ecological Pressures on Sociality

The similarities in sociality of odontocetes in similar habitats suggest that ecological forces act strongly on the social behavior of these cetaceans. Ecological pressures may take the form of either attractive or dispersive forces for sociality, as noted by Norris and Dohl (this volume). If predation pressure and food resources serve as selective forces for cetaceans, as they do for terrestrial mammals, then predation pressure would be generally an attractive force, and food resources, depending on abundance and distribution, either an attractive or dispersive force. Similarly, the physical nature of the habitat, whether open or complex, can be either an attractive or dispersive force.

Predation Pressure

Sharks and killer whales are the most commonly reported predators of odontocetes (see Table 6.1), but the influence of predation on cetacean group structure is not completely understood. Most of the available information on cetaceans as prey comes from anecdotal field sightings or reports of shark stomach contents. Wood et al. (1970) stated that "... the number of reports that sharks have dined on porpoises, as well as the fact that such reports encompass far-flung geographical areas, lends credence to the assumption that porpoises do indeed constitute a not infrequent article of diet for some of the larger sharks."

Wood et al. reported a higher frequency of shark bite scars on inshore dolphins than on dolphins frequenting offshore waters. Leatherwood et al. (1971) reported few shark bite scars on pelagic dolphins and suggested that either the fast-swimming dolphins might outmaneuver most sharks in open water or attacks that did occur were generally fatal. It may also be that the larger groups formed by the pelagic species are a relatively effective deterrent to attack. Bite scars on healthy individuals might provide a relative indication of predation pressure, but, as noted by Wood et al. (1970), healed bite scars represent only unsuccessful predation attempts at some unknown time during the animal's life. Bite scars are therefore of uncertain value as an indicator of the level of predation.

Odontocete responses to the presence of potential predators vary in relation to species and size of the predator, the physical environment, and the activities and size of the cetacean group. Observed cetacean responses to the proximity of potential predators include tolerance, active avoidance, and active aggression.

Captive studies have indicated that Atlantic bottlenosed dolphins may recognize certain shark species as potential predators. McBride and Hebb (1948) noted a marked reaction by dolphins to tiger sharks, known dolphin predators, compared with other shark species. In another case a study to condition a bottlenosed dolphin to repel various species of sharks was relatively successful with three shark species. When tests included bull sharks (*C. leucas*), a known predator of the bottlenosed dolphin, the dolphin became agitated and refused to respond to commands (Irvine et al., 1973).

Mutual tolerance between cetaceans and sharks has been reported both in captivity and in the field. No agonistic interactions between pairs of bottlenosed dolphins and a variety of shark species were observed during captive experiments in the Bahamas (AIBS, 1967; Gilbert et al., 1971) and Florida (Irvine et al., 1973). Similar observations have been reported from several oceanariums where dolphins and sharks have been maintained in the same tanks (Essapian, 1953; Wood et al., 1970). Nonaggressive interactions, including feeding on the same fish schools, have been reported for free-ranging bottlenosed dolphins (AIBS, 1967; Caldwell, 1960; Leatherwood, 1977; Saayman & Tayler, 1979; Wood et al., 1970).

Increased group cohesion and active avoidance have also been reported as responses to predators. Upon introduction of sandbar sharks (*Carcharhinus milberti*) captive bottlenosed dolphins and spotted dolphins (*Stenella plagiodon*) increased their swimming speed and formed tighter groups (McBride & Hebb, 1948; Wood et al., 1970). In another instance all the bottlenosed dolphins in a tank gathered around a female giving birth and herded approaching sharks away (McBride & Hebb, 1948). In the wild active avoidance has been reported for Indian Ocean bottlenosed dolphins upon encountering hammerhead sharks (*Sphyrna zygaena*) or a great white shark (*Carcharodon carcharias*) (Tayler & Saayman, 1972). Humpback dolphins, Atlantic bottlenosed dolphins, dusky dolphins, and common dolphins (*Delphinus delphis*) have been observed to flee rapidly in tight groups from approaching killer whales (Saayman & Tayler, 1979; Würsig & Würsig, 1979, 1980).

A number of reports of aggression by dolphins toward sharks are available. Brown and Norris (1956), Norris and Prescott (1961), McBride and Hebb (1948), and Essapian (1953) observed bottlenosed dolphins attacking sharks in captivity. Coordinated agonistic responses by dolphin groups toward sharks in the field have also been reported. Gunter (1942) gave an account of bottlenosed dolphins killing a shark off Texas. Saayman and Tayler (1979) reported that half of a group of 10 humpback dolphins chased a 4- to 5-m long great white shark into the open sea from a sheltered bay. Springer (AIBS, 1967) observed members of a group of common dolphins apparently cooperating to protect their young by repeatedly rushing at approaching sharks.

Habitats and Responses to Predators

The physical characteristics of the habitat appear to influence cetacean responses to predation pressure. Norris and Dohl (1980) considered the tendency of spinner dolphins to rest in shallow bays to be an adaptation for reducing the chances of shark attack. The increased use of the shallow waters of the Sarasota area by bottlenosed dolphins during the spring–summer peak abundance of bull sharks could also be a protective adaptation by reducing the volume of water that must be kept under surveillance for potential predators.

The use of shallow waters appears to be important to some odontocetes for avoidance of killer whales as well. On three occasions a group of dusky dolphins moved into water less than 1 m deep and swam rapidly away in single file when a group of killer whales approached (Würsig & Würsig, 1980). The authors believed that the dolphins' movements near the biologically noisy shoreline might have hidden the animals from the echolocation or hearing of the killer whales. Saayman and Tayler (1979) observed two groups of humpback dolphins remaining within several meters of shore, over a barely submerged reef, while killer whales approached. As the killer whales slowly withdrew, the dolphins swam away in small groups but still remained very close to shore. Saayman and Tayler attributed the survival of humpback dolphins in that area to their ability to utilize wave action to cross shallows when threatened by predators.

Norris and Dohl (this volume) suggest that the variation in cetacean group size is what would be predicted if predation were the primary influence on schooling or the tendency toward group formation. Smaller groups are generally found in riverine habitats where few predators are found, and larger groups are found in the coastal and pelagic habitats where more predation pressure might be expected (see Table 6.1). Norris and Dohl emphasized the importance of sensory integration to groups. Larger groups potentially provide more sensory input, allowing more efficient surveillance.

Larger groups also provide increased cover for the individual (Hamilton, 1971) and are potentially better suited for cooperative, active defense of the group. In general it seems that as the physical habitat provides less protection for a cetacean the importance of the group increases as a means for protection from predation.

Food Resources and Group Characteristics

The increase in group size from inshore to offshore habitats may also be influenced by food sources. While large numbers of animals in a group may be advantageous for cooperative foraging and feeding on patchy, rich food resources, the advantages are greatly diminished given more evenly distributed food resources such as bottom fish or reef fish. Riverine species of odontocetes, which feed on the relatively evenly distributed bottom fish or crustaceans, are typically found in small groups and usually forage alone. Coastal and more pelagic species consume primarily schooling fish or squid (Table 6.1)—a rich but potentially patchy resource—and are found in larger groups and may utilize coordinated, cooperative hunting and feeding strategies (Leatherwood, 1975; Norris & Dohl, this volume). Fish tend to school in open habitat (Williams, 1964), and the advantage of sensory integration to an individual odontocete for finding scattered fish schools is obvious. Small groups of both spinner dolphins (Norris & Dohl, 1980) and dusky dolphins (Würsig & Würsig, 1980) have each been observed merging to form large feeding aggregations. Cooperative feeding by cetacean groups has been reported frequently (Busnel, 1973; Hamilton & Nishimoto, 1977; Hoese, 1971; Leatherwood, 1975; Morozov, 1970; Saayman & Tayler, 1973).

Odontocete species with relatively limited coastal home ranges may be able to rely on the resident fish resources but also utilize seasonally available fish schools. Humpback dolphins generally feed individually on reef fishes but also feed on pelagic fish schools moving into the dolphins' home range with changing water temperatures (Saayman & Tayler, 1979). Atlantic bottlenosed dolphins in Argentine waters follow a similar pattern, usually feeding on reef or rock-dwelling fish but also preying on anchovy schools when available (Würsig & Würsig, 1979). In Florida bottlenosed dolphins apparently prefer mullet but additionally feed on numerous other available fish (Irvine et al., 1979). The diurnal and seasonal variations in group size seen in the coastal species may reflect responses to the changing food resources and the relative advantages of dispersing or aggregating.

Habitat and Reproduction

Though Norris and Dohl (this volume) attribute many functions to cetacean schools, one of the most important is to bring males and females into contact with each other for reproduction. In the open habitat of pelagic species this is accomplished by maintaining large mixed groups throughout the year or by synchronized migrations of different age and sex classes. In some coastal species it

may not be necessary or desirable to be part of a large group at all times. Instead, a comparatively limited home range for a herd of mixed age and sex
classes, often with segregated subgroup structure, ensures that encounters with
the opposite sex can be made at appropriate intervals. In the riverine species
movements of cetaceans are highly constrained by the physical environment.
This potentially increases the likelihood of encounters between individuals sharing a given river system and reduces the need to maintain large social groups.
Additionally, large groups would be difficult to maintain and coordinate in a
shallow, complex habitat.

Evaluation of Selective Mechanisms

In summary pelagic dolphins are highly adapted to life in the open sea. They
may be relatively small in size but streamlined for prolonged high-speed search
and pursuit of rapidly moving schools of small prey fish, which they catch with
elongated jaws containing numerous small teeth. The primarily coastal species
of small odontocetes are characterized by shorter beaks and fewer teeth for
catching more diverse prey, both in the water column and on the bottom.
Riverine species are highly maneuverable for swimming in waters containing
numerous obstructions. Their long, narrow jaws are equipped with numerous
teeth that are well designed to catch individual prey on the bottom (Myrick,
1979; Yablokov et al., 1972).

The more solitary riverine species occupy regions where movements are
restricted by land masses and other obstructions, obtain their prey individually
primarily from the bottom, and are infrequently subjected to predation. Coastal
species generally form small-sized subgroups that may be part of a larger socially interacting resident population unit, feed individually or cooperatively on
a variety of prey species, and suffer from predation to some degree. The more
pelagic species or populations are frequently found in large groups, often are
migratory in apparent response to the movements of patchy but rich food
resources, and apparently are subjected to predation pressure.

To say that cetacean group size, composition, or stability is the result of any
single environmental factor would be an oversimplification. While some factors
may be more important than others in determining group characteristics, the
resultant group is the product of complex interactions between environmental
factors and inherent species characteristics. Though experimentation with cetacean groups is difficult, natural experimental situations exist that may be used
to elucidate the relative importance of different habitat characteristics.
Especially interesting in this regard are situations where the same species may
be found in different habitats in different parts of its range (cf. Herman &
Tavolga, this volume). Examples include *Sotalia,* which is found both in the
riverine and coastal habitats; *Tursiops*, which can be found in all three major
habitat types; and *Stenella longirostris,* which is primarily pelagic but becomes
coastal around islands in parts of its range. In these situations one can control
for habitat characteristics and look for variations in group characteristics.

   Comparisons with other social mammals can also clarify some of the selective forces acting on group characteristics and provide models for understanding cetacean social structure. Eisenberg (1966) and Estes (1974) noted a parallel in the tendency to form groups in the openness of the pelagic marine environment and the plains and savanna regions inhabited by African ungulates. In general African ungulates inhabiting the less-open habitats having natural cover are found singly or in small groups, while the plains-dwellers are more gregarious (Estes, 1974; Jarman, 1974). "Edge" species living in the forest–plains interface may form groups only when venturing away from cover (Estes, 1974). It has been suggested by Jarman (1974) and others that predation pressure in open habitats induces ungulate herd formation and determines minimum group size, while the distribution and availability of food limits the maximum group size.

   To a somewhat lesser degree group formation of macropods is similar to that of ungulates. The most social kangaroo and wallaby species are found in the open forest or plains of Australia, while solitary and less social species are found in denser forest (Kaufmann, 1974). Similarly Eisenberg et al. (1972) noted that semiterrestrial primate species tended to be found in larger groups than strictly arboreal species.

   In these various taxa group-living is a recent evolutionary development and can be associated with increased use of open habitat. Among cetaceans the primitive Platanistidae (Kellogg, 1928, Kulu, 1972; Lipps & Mitchell, 1976; Miller, 1923) inhabit only the relatively complex riverine and coastal habitat and are usually found in the smallest groups of any of the small odontocetes. Conversely, the more specialized small Delphinidae inhabiting the coastal and open offshore waters are often found in large groups with apparently complex social structure.

   The ecological factors influencing the social evolution of the larger delphinids, such as killer whales and pilot whales, are presumably somewhat different than those noted for the smaller delphinids or platanistids. In particular the influence of predation pressure on group size is less important to larger animals, while factors such as prey density, prey size, and predator mobility may be more important, as has been suggested for canids and felids (Kleiman & Eisenberg, 1973).

   Many aspects of inshore odontocete social behavior and social organization seem convergent with those of some terrestrial mammals, but only a sampling will be mentioned here. Eisenberg (1966) and Wilson (1975) noted that both ungulates and captive dolphins display dominance hierarchies, group defense, care-giving behavior, and allomaternal behavior, such as the association of escort females with pregnant females near term.

   Similarities can be found between aspects of the social systems of bottlenosed dolphins and the most socially complex of the bovids, the African buffalo (*Syncerus caffer*). In open habitat the buffalo form herds. Each herd has a fixed home range that only slightly overlaps that of other herds. Within the herd the buffalo are found in mixed groups of fluid composition. Group defense has been reported. The young are precocial, but may associate with their mothers for 2 or 3 years. Resources are evenly distributed rather than clumped. A dominance hierarchy among adult males and females, with females subordinate, exists within the herd. Bachelor groups are occasionally observed (Estes, 1974; Jarman, 1974).

Differential use of the herd home range by adult and subadult male buffalo has not been reported, as was the case for the Sarasota group of dolphins, but some ungulates, particularly territorial species, do exhibit this behavior. Bachelor male wildebeests and gazelles are relegated to substandard habitat from the time they are expelled from nursery herds until they mature and become territorial. Adult male impalas select the more productive habitat to defend as their territories, and other males attempting to enter a territory are driven off. Female impalas may move at will through several male territories and are possibly attracted more by the available resources than by the territorial male (Jarman, 1974).

A number of authors have compared the social systems of primates and bottlenosed dolphins. Tayler and Saayman (1972) made a detailed comparison between the social organization of the baboon (*Papio ursinus*) and the Indian Ocean bottlenosed dolphin. They noted that both groups formed dominance hierarchies and that the mating system consisted of rotating consort relationships without permanent sexual pair bonds. Coordinated group defense against predators occurred in both species. Many similarities have been found between bottlenosed dolphins and chimpanzees (*Pan* sp.) as well. Demographic societies of chimpanzees maintain well-defined, stable home ranges. The composition of each casual group within a society is quite fluid, but a highly developed dominance hierarchy exists. Cooperative hunting and defense have been reported for chimpanzees as for the dolphins (Goodall, 1965; Saayman & Tayler, 1979; Wells, 1978; Wilson, 1975; Würsig, 1978).

Unfortunately the volume of data available on inshore odontocete social ecology does not begin to approach that collected for terrestrial mammals. The extensive data for the terrestrial species have allowed the construction of comprehensive hypotheses on the evolution and ecology of sociality. Whether the ecological pressures that shape cetacean social behavior, from the riverine to the deep-ocean habitat, eventually can be synthesized into a single unified organizational scheme remains to be determined and awaits the collection of much more data on individual species. The development of new study techniques and the recent surge in interest in comprehensive field studies should add to those data, answer many questions, and create directions for further studies.

ACKNOWLEDGMENTS

The *Tursiops* case study reported here was supported by U.S. Marine Mammal Commission contracts MM4AC004 and MM5AC0018. The valuable assistance and dedication of numerous volunteers from the University of Florida and New

College Campus of the University of South Florida is greatly appreciated.
Logistical assistance in the field was provided by Carol Blanton, Mary Moore,
John Morrill, Fran and Jack Wells, and Fred Worl. Stephen A. Bloom directed
the computer analyses. Dianna Foster, Donna Gillis, Jean Shufro, Dawn
Stephens, and Estella Duell helped prepare the manuscripts. John H. Kauf-
mann, Bernd Würsig, Louis M. Herman, Steve Leatherwood, Albert C.
Myrick, Jr., and Paul Forestell were especially helpful with their reviews of the
manuscripts.

REFERENCES

AIBS (1967). Conference on the shark–porpoise relationship. American Institute of Biological Science, Washington, D.C.

Aminul Haque, A. K. M., Nishiwaki, M., Kasuya, T. & Tobayama, T. (1977). Observations on the behaviour and other biological aspects of the Ganges susu, *Platanista gangetica. Sci. Rep. Whales Res. Inst.*, **29**, 87-94.

Arnold, P. W. (1972). Predation on harbour porpoise, *Phocoena phocoena,* by a white shark, *Carcharodon carcharias. J. Fish. Res. Board Can.,* **29**, 1213-1214.

Balcomb, K., Boran, J., & Osborne, R. (1979). Killer whales, *Orcinus orca,* in Greater Puget Sound. Presented at Third Biennial Conference of the Biology of Marine Mammals, October 7-11, 1979, Seattle, Washington.

Balcomb, K. C., III & Goebel, C. A. (1976). A killer whale study in Puget Sound. Final Report National Marine Fisheries, Contract NASO-6-35330.

Barham, E. G., Sweeney, J. C., Leatherwood, S., Beggs, R. K. & Barham, C. L. (1980). Aerial census of bottlenose dolphins (*Tursiops truncatus*) in a region of the Texas coast. *Fish. Bull.,* **77**, 585-595.

Bateson, G. (1965). Porpoise community research: Final report. U.S. Navy Undersea Research & Development Center, Contract No. N60530-C-1098.

Bigg, M. A. (1979). Interaction between pods of killer whales off British Columbia and Washington. Presented at Third Biennial Conference of the Biology of Marine Mammals, October 7-11, 1979, Seattle, Washington.

Brown, D. H. & Norris, K. S. (1956). Observations of captive and wild cetaceans. *J. Mammal.,* **37**, 311-326.

Brown, D. H., Caldwell, D. K. & Caldwell, M. C. (1966). Observations on the behavior of wild and captive false killer whales, with notes on associated behavior of other genera of captive delphinids. *Los Ang. Cty. Mus. Nat. Hist. Contrib. Sci.,* **95**, 1-32.

Burt, W. H. (1943). Territoriality and home range concepts as applied to mammals. *J. Mammal.,* **30**, 25-27.

Busnel, R.-G. (1973). Symbiotic relationship between man and dolphins. *Trans. N.Y. Acad. Sci.,* **35**, 112-131.

Caldwell, D. K. (1955). Evidence of home range of an Atlantic bottlenose dolphin. *J. Mammal.,* **36**, 304-305.

Caldwell, D. K. (1960). Notes on the spotted dolphin in the Gulf of Mexico. *J. Mammal.,* **41**, 134-136.

Caldwell, D. K. & Caldwell, M. C. (1972). *The World of the Bottlenosed Dolphin.* New York: Lippincott.

Caldwell, M. C. & Caldwell, D. K. (1967). Dolphin community life. *Los Ang. Cty. Mus. Nat. Hist. Contrib. Sci.,* **5**, 12-15.

Caldwell, M. C. & Caldwell, D. K. (1972). Behavior of marine mammals. In: *Mammals of the Sea: Biology and Medicine* (S. H. Ridgway, Ed.), pp. 419-465. Springfield, Ill.: Thomas.

Caldwell, M. C. & Caldwell, D. K. (1977). Social interactions and reproduction in the Atlantic bottlenosed dolphin. In: *Breeding Dolphins: Present Status, Suggestions for the Future* (S. H. Ridgway & K. W. Benirschke, Eds.), pp. 133-142. Washington, D.C.: Marine Mammal Commission Report MMC-76/07.

Caldwell, M. C., Caldwell, D. K. & Evans, W. E. (1966). Sounds and behavior of captive Amazon freshwater dolphins, *Inia geoffrensis. Los Ang. Cty. Mus. Nat. Hist. Contrib. Sci.,* **108,** 1-24.

Caldwell, M. C., Caldwell, D. K. & Siebenaler, J. B. (1965). Observations on captive and wild Atlantic bottlenosed dolphins, *Tursiops truncatus,* in the Northeastern Gulf of Mexico. *Los Ang. Cty. Mus. Nat. Hist. Contrib. Sci.,* **91,** 1-10.

Caldwell, M. C., Caldwell, D. K. & Townsend, B. G., Jr. (1968). Social behavior as a husbandry factory. In: *Symposium of Disease and Husbandry of Aquatic Mammals* (D. K. Caldwell & M. C. Caldwell, Eds.), pp. 1-9. (Mimeograph.)

Clark, E. & von Schmidt, K. (1965). Sharks of the central Gulf Coast of Florida. *Bull. Mar. Sci.,* **15,** 13-83.

Clutton-Brock, T. H. (1974). Primate social organization and ecology. *Nature,* **250,** 539-542.

Condy, P. R., van Aarde, R. J. & Bester, M. N. (1978). The seasonal occurrence and behaviour of killer whales *Orcinus orca,* at Marion Island. *J. Zool. Lond.,* **184,** 449-464.

Crook, J. H., Ellis, J. E. & Goss-Custard, J. D. (1976). Mammalian social systems: Structure and function. *Anim. Behav.,* **24,** 261-274.

Crook, J. H. & Gartlan, J. S. (1966). Evolution of primate societies. *Nature,* **210,** 1200-1203.

Eisenberg, J. F. (1966). The social organizations of mammals. *Handbuch Zool.,* **10,** 1-92.

Eisenberg, J. F., Muckenhirn, N. A. & Rudran, R. (1972). The relation between ecology and social structure in primates. *Science,* **1976,** 863-874.

Erickson, A. W. (1978). Population studies of killer whales (*Orcinus orca*) in the Pacific Northwest: A radio-marking and tracking study of killer whales. Washington, D.C.: Marine Mammal Commission Report, MMC-75/10.

Essapian, F. S. (1953). The birth and growth of a porpoise. *Nat. Hist.,* **62,** 392-399.

Essapian, F. S. (1962). An albino bottle-nosed dolphin, *Tursiops truncatus,* captured in the U.S. *Norsk Hvalfangst-tid.,* **9,** 341-344.

Essapian, F. S. (1963). Observations on abnormalities of parturition in captive bottlenosed dolphins, *Tursiops truncatus,* and concurrent behavior of other porpoises. *J. Mammal.,* **44,** 405-414.

Estes, R. D. (1974). Social organization of the African Bovidae. In: *The Behavior of Ungulates and Its Relation to Management* (V. Geist & F. Walther, Eds.), pp. 166-205. Morges, Switzerland: International Union for Conservation of Nature and Natural Resources.

Evans, W. E. (1971). Orientation behavior of delphinids: Radio telemetric studies. *Ann. N.Y. Acad. Sci.,* **188,** 142-160.

Evans, W. E. (1974). Radio-telemetric studies of two species of small odontocete cetaceans. In: *The Whale Problem: A Status Report* (W. E. Schevill, Ed.), pp. 385-394. Cambridge, Mass.: Harvard University Press.

Evans, W. E. & Bastian, J. (1969). Marine mammal communication: Social and ecological factors. In: *The Biology of Marine Mammals* (H. T. Andersen, Ed.), pp. 425-476. New York: Academic.

Evans, W. E., Hall, J. D., Irvine, A. B. & Leatherwood, J. S. (1972). Methods for tagging small cetaceans. *Fish. Bull.,* **70,** 61-65.

Fay, F. H. (1978). Beluhka whale. In: *Marine Mammals* (D. Haley, Ed.), pp. 132-137. Seattle, Washington: Pacific Search Press.

Fehring, W. K. & Wells, R. S. (1976). A series of strandings by a single herd of pilot whales on the west coast of Florida. *J. Mammal.,* **57,** 191-194.

Gaskin, D. E., Smith, G. J. D. & Watson, A. P. (1975). Preliminary study of movements of harbor porpoises (*Phocoena phocoena*) in the Bay of Fundy using radio telemetry. *Can. J. Zool.,* 53, 1466-1471.

Geist, V. (1974). On the relationship of ecology and behaviour in the evolution of ungulates: Theoretical considerations. In: *The Behaviour of Ungulates and Its Relation to Management* (V. Geist & F. Walther, Eds.), pp. 235-246. Morges, Switzerland: International Union for Conservation of Nature and Natural Resources.

Gilbert, P. W., Irvine, B. & Martini, F. H. (1971). Shark-porpoise behavioral interactions. *Am. Zool.,* **11,** 636.

Goodall, J. (1965). Chimpanzees of the Gombe Stream Reserve. In: *Primate Behavior* (I. DeVore, Ed.), pp. 425-473. New York: Holt, Rinehart & Winston.

Green, J. (1968). *The Biology of Estuarine Animals.* London: University of Washington Press.

Gunter, G. (1942). Contributions to the natural history of the bottlenose dolphin, *Tursiops truncatus* (Montagu), on the Texas coast, with particular reference to food habits. *J. Mammal.,* **23,** 267-276.

Gunter, G. (1954). Mammals of the Gulf of Mexico. *Fish. Bull.,* **55,** 543-551.

Hamilton, P. V. & Nishimoto, R. T. (1977). Dolphin predation on mullet. *Fla. Sci.,* **40,** 251-252.

Hamilton, W. D. (1971). Geometry for the selfish herd. *J. Theor. Biol.,* **3,** 295-311.

Hoese, H. D. (1971). Dolphin feeding out of water in a salt marsh. *J. Mammal.,* **52,** 222-223.

Irvine, A. B., Scott, M. D., Wells, R. S., Kaufmann, J. H. & Evans, W. E. (1979). A study of the activities and movements of the Atlantic bottlenosed dolphin, *Tursiops truncatus,* including an evaluation of tagging techniques. National Technical Information Service PB-298 042, 54 pp.

Irvine, A. B., Scott, M. D., Wells, R. S. & Mead, J. G. (1979). Stranding of the pilot whale, *Globicephala macrorhynchus,* in Florida and South Carolina. *Fish Bull.* **77,** 511-513.

Irvine, B., Wells, R. S. & Gilbert, P. W. (1973). Conditioning an Atlantic bottlenosed dolphin, *Tursiops truncatus,* to repel various species of sharks. *J. Mammal,* **54,** 503-505.

Irvine, B. & Wells, R. S. (1972). Results of attempts to tag Atlantic bottlenosed dolphins, *Tursiops truncatus. Cetology,* **13,** 1-5.

Jarman, P. J. (1974). The social organisation of antelope in relation to their ecology. *Behaviour,* **58,** 215-267.

Jennings, J. G. (1979). Range and movements of dolphins with information obtained from a satellite-linked transmitter. Presented at Third Biennial Conference of the Biology of Marine Mammals, October 7-11, 1979, Seattle, Washington.

Jewell, P. A. (1966). The concept of home range in mammals. *Symp. Zool. Soc. Lond.,* **18,** 85-109.

Kasuya, T. & Aminul Haque, A. K. M. (1972). Some information on distribution and seasonal movement of the Ganges dolphin. *Sci. Rep. Whales Res. Inst.,* **24,** 109-115.

Kaufmann, J. H. (1971). Is territoriality definable? In: *Behavior and Environment* (A. H. Esser, Ed.), pp. 36-40. New York: Plenum.

Kaufmann, J. H. (1974). The ecology and social organization in the kangaroo family (Macropodidae). *Am. Zool.,* **14,** 51-62.

Kellogg, R. (1928). The history of whales—Their adaptation to life in the water. *Q. Rev. Biol.,* **3,** 29-76 & 174-208.

Kleiman, D. G. & Eisenberg, J. F. (1973). Comparisons of canid and felid social systems from an evolutionary perspecitve. *Anim. Behav.,* **21,** 637-659.

Krebs, J. R. & Davies, N. B. (1978). *Behavioural Ecology: An Evolutionary Approach.* Oxford: Blackwell Scientific Publications.

Kulu, D. D. (1972). Evolution and cytogenetics. In: *Mammals of the Sea: Biology and Medicine* (S. H. Ridgway, Ed.), pp. 503-527. Springfield, Ill.: Thomas.

Lawrence, B. & Schevill, W. E. (1954). *Tursiops* as an experimental subject. *J. Mammal.,* **35,** 225-232.

Layne, J. N. (1958). Observations on freshwater dolphins in the upper Amazon. *J. Mammal.,* **39,** 1-22.

Layne, J. N. & Caldwell, D. K. (1964). Behavior of the Amazon dolphin *Inia geoffrensis* (Blainville), in captivity. *Zoologica,* **49,** 81-112.

Leatherwood, S. (1975). Some observations of feeding behavior of bottlenosed dolphins (*Tursiops truncatus*) in the northern Gulf of Mexico and (*Tursiops* cf. *T. gilli*) off southern California, Baja California, and Nayarit, Mexico. *Mar. Fish. Rev.,* **37,** 10-16.

Leatherwood, J. S. (1977). Some preliminary impressions on the numbers and social behavior of free-swimming, bottlenosed dolphin calves (*Tursiops truncatus*) in the northern Gulf of Mexico. In: *Breeding Dolphins: Present Status, Suggestions for the Future* (S. H. Ridgway & K. W. Benirschke, Eds.), pp. 143-167. Washington, D.C.: Marine Mammal Commission Report MMC-76/07.

Leatherwood, S. (1979). Aerial survey of the bottlenose dolphin, *Tursiops truncatus,* and the West Indian Manatee, *Trichechus manatus,* in the Indian and Banana Rivers, Florida. *Fish. Bull.,* **77,** 47-59.

Leatherwood, J. S., Caldwell, D. K. & Winn, H. E. (1976). Whales, dolphins, and porpoises of the western North Atlantic: A guide to their identification in water. Seattle, Wash.: NOAA Technical Report, National Marine Fisheries Service, CIRC-396.

Leatherwood, S. & Evans, W. E. (1979). Some recent uses and potentials of radiotelemetry in field studies of cetaceans. In: *Behavior of Marine Animals* (H. E. Winn & B. L. Olla, Eds.) pp. 1-32. New York: Plenum.

Leatherwood, J. S., Evans, W. E. & Rice, D. W. (1972). The whales, dolphins, and porpoises of the eastern North Pacific: A guide to their identification in the water. San Diego, Calif.: Naval Undersea Center Technical Publication No. 282.

Leatherwood, J. S., Gilbert, J. W. & Chapman, D. G. (1978). An evaluation of some techniques for aerial censuses of bottlenosed dolphins. *J. Wildl. Mgmt.,* **42,** 239-250.

Leatherwood, J. S., Perrin, W. F., Garvie, R. L. & LaGrange, J. C. (1971). Observations of sharks attacking porpoises (*Stenella* spp. and Delphinus cf. *D. delphis*). San Diego, Calif.: Naval Undersea Center Technical Publication No. 908.

Leatherwood, J. S. & Platter, M. F. (1975). Aerial assessment of bottlenosed dolphins off Alabama, Mississippi, and Louisiana. In: *Tursiops truncatus Assessment Workshop* (D. K. Odell, D. B. Siniff, & G. H. Waring, Eds.), pp. 49-86. Final Report to U.S. Marine Mammal Commission.

Leatherwood, J. S. & Reeves, R. R. (1978). Porpoises and dolphins. In: *Marine Mammals of the Eastern North Pacific and Arctic Waters* (D. Haley, Ed.), pp. 97-111. Seattle, Wash.: Pacific Search.

Lipps, J. H. & Mitchell, E. (1976). Trophic model for the adaptive radiations and extinctions of pelagic marine mammals. *Paleobiology,* **2,** 147-155.

Mansfield, A. W., Smith, T. G. & Beck, B. (1975). The narwhal, *Monodon monoceros,* in eastern Canadian waters. *J. Fish. Res. Board Can.,* **32,** 1041-1046.

Marcuzzi, G. & Pilleri, G. (1971). On the zoogeography of cetacea. In: *Investigations on Cetacea* (G. Pilleri, Ed.), pp. 101-170. Berne, Switzerland: Institute for Brain Anatomy, University of Berne.

Martin, H., Evans, W. E. & Bowers, C. A. (1971). Methods for radio tracking marine mammals in the open sea. *Proc. IEEE '71 Conf. Engineering Ocean Environ.,* pp. 44-49.

Martinez, D. R. & Klinghammer, E. (1970). The behavior of the whale *Orcinus orca*: A review of the literature. *Z. Tierpsychol.,* **27,** 828-839.

References

McBride, A. F. (1940). Meet Mr. Porpoise. *Nat. Hist.,* **45,** 16-29.

McBride, A. F. & Hebb, D. O. (1948). Behavior of the captive bottlenose dolphin, *Tursiops truncatus. J. Comp. Physiol. Psychol.,* **41,** 111-123.

McBride, A. F. & Kritzler, H. (1951). Observations on pregnancy, parturition, and postnatal behavior in the bottlenose dolphin. *J. Mammal.,* **32,** 251-266.

Miller, G. S., Jr. (1923). The telescoping of the cetacean skull. *Smithson. Misc. Coll.,* **76,** 1-71.

Mitchell, E. (1975). Porpoise, dolphin, and small whale fisheries of the world: Status and problems. IUCN Monograph No. 3. Morges, Switzerland: International Union for Conservation of Nature and Natural Resources.

Miyazaki, N. (1977). School structure of *Stenella coeruleoalba. Rep. Int. Whale Comm.,* **27,** 298-299.

Morejohn, G. V. (1979). The natural history of Dall's porpoise in the North Pacific Ocean. In: *Behavior of Marine Animals* (H. E. Winn & B. L. Olla, Eds.), pp. 45-84. New York: Plenum Press.

Morozov, D. A. (1970). Dolphin hunting. *Rybnoe Khoziaistvo,* **46,** 16-17.

Myrick, A. C., Jr. (1979). Qualitative adaptive models for pelagic and river dolphins. Presented at Third Biennial Conference of the Biology of Marine Mammals, October 7-11, 1979, Seattle, Washington.

Neave, D. J. & Wright, B. S. (1968). Seasonal migrations of the harbor porpoise (*Phocoena phocoena*) and other cetacea in the Bay of Fundy. *J. Mammal.,* **45,** 259-264.

Newman, M. A. (1978). Narwhal. In: *Marine Mammals* (D. Haley, Ed.), pp. 138-144. Seattle, Washington: Pacific Search Press.

Nishiwaki, M. (1972). General biology. In: *Mammals of the Sea: Biology and Medicine* (S. H. Ridgway, Ed.), pp. 3-204. Springfield, Ill.: Thomas.

Nishiwaki, M., Nakajima, M. & Tobayama, T. (1966). Preliminary experiments for dolphin marking. *Sci. Rep. Whale Res. Inst.,* **20,** 101-107.

Norris, K. S. (1967). Aggressive behavior in cetacea. In: *Aggression and Defense* (C. D. Clemente & D. B. Lindsley, Eds.), pp. 225-241. Berkeley: University of California Press.

Norris, K. S. (1974). *The Porpoise Watcher.* New York: Norton.

Norris, K. S. & Dohl, T. P. (1980). The behavior of the Hawaiian spinner porpoise, *Stenella longirostris. Fish. Bull.,* **77,** 821-849.

Norris, K. S., Evans, W. E. & Ray, G. C. (1974). New tagging and tracking methods for the study of marine mammal biology and migration. In: *The Whale Problem: A Status Report* (W. E. Schevill, Ed.), pp. 395-408. Cambridge: Harvard University Press.

Norris, K. S. & Prescott, J. H. (1961). Observations on Pacific cetaceans of Californian and Mexican waters. *Univ. Calif. Publ. Zool.,* **63,** 291-402.

Norris, K. S. & Pryor, K. W. (1970). A tagging method for small cetaceans. *J. Mammal.,* **51,** 609-610.

Odell, D. K. (1976). Distribution and abundance of marine mammals in south Florida: Preliminary results. *Univ. Miami Sea Grant Special Rep.,* **5,** 203-212.

Odell, D. K. & Reynolds, J. E., III (1977). Distribution and abundance of the bottlenose dolphin, *Tursiops truncatus,* on the west coast of Florida. Draft Final Report to U.S. Marine Mammal Commission.

Ohsumi, S. (1966). Sexual segregation of the sperm whale in the North Pacific. *Sci. Rep. Whales Res. Inst.,* **20,** 1-16.

Ohsumi, S. (1971). Some investigations on the school structure of sperm whale. *Sci. Rep. Whales Res. Inst.,* **23,** 1-25.

Owen-Smith, N. (1977). On territoriality in ungulates and an evolutionary model. *Q. Rev. Biol.,* **52,** 1-38.

Perrin, W. F. (1975a). Distribution and differentiation of populations of dolphins of the genus *Stenella* in the Eastern tropical Pacific, *J. Fish. Res. Bd. Can.*, **32**, 1059-1067.

Perrin, W. F. (1975b). Variation of spotted and spinner porpoise (Genus *Stenella*) in the eastern tropical Pacific and Hawaii. *Bull. Scripps Inst. Oceangr.*, **21**.

Perrin, W. F., Warner, R. R., Fiscus, C. H. & Holts, D. B. (1973). Stomach contents of porpoises, *Stenella* spp. and yellow fin tuna, *Thunnus albacares*, in mixed-species aggregation. *Fish. Bull.*, **71**, 1077-1092.

Pilleri, G. (1970). Observations on the behavior of *Platanista gangetica* in the Indus and Brahmaptura Rivers. In: *Investigations on Cetacea* (G. Pilleri, Ed.), pp. 27-60. Berne, Switzerland: Brain Anatomy Institute, University of Berne.

Pilleri, G. & Knuckey, J. (1969). Behaviour patterns of some delphinidae observed in the western Mediterranean. *Z. Tierpsychol.*, **26**, 48-72.

Reilly, S. B. (1978). Pilot whale. In: *Marine Mammals* (D. Haley, Ed.), pp. 112-119. Seattle, Washington: Pacific Search Press.

Rice, D. W. (1960). Distribution of the bottlenosed dolphin in the leeward Hawaiian Islands. *J. Mammal.*, **41**, 407-408.

Rice, D. W. (1977). A list of the marine mammals of the world. NOAA Technical Report NMFS, SSRF-711.

Rice, D. W. & Scheffer, V. B. (1968). A list of the marine mammals of the world. U.S. Fish & Wildlife Service Special Scientific Report—Fisheries No. 579.

Ross, G. J. B. (1977). The taxonomy of bottlenosed dolphins *Tursiops* sp. in South African waters, with notes on their biology. *Ann. Cape. Prov. Mus. Nat. Hist.*, **11**, 135-194.

Saayman, G. S., Bower, D. & Tayler, C. K. (1972). Observations on inshore and pelagic dolphins on the south-eastern cape coast of South Africa. *Koedoe*, **15**, 1-24.

Saayman, G. S. & Tayler, C. K. (1973). Social organization of inshore dolphins (*Tursiops aduncus* and *Sousa*) in the Indian Ocean. *J. Mammal.*, **54**, 993-996.

Saayman, G. S. & Tayler, C. K. (1979). The socioecology of humpback dolphins (*Sousa* sp.). In: *The Behavior of Marine Animals*, Vol. III (H. E. Winn & B. L. Olla, Eds.), pp. 165-226. New York: Plenum.

Saayman, G. S., Tayler, C. K. & Bower, D. (1973). Diurnal activity cycles in captive and free-ranging Indian Ocean bottlenose dolphins (*Tursiops aduncus* Ehrenburg). *Behaviour*, **44**, 212-233.

Scheffer, V. B. (1978). Killer whale. In: *Marine Mammals* (D. Haley, Ed.), pp. 120-127. Seattle Washington: Pacific Search Press.

Sergeant, D. E. (1962). The biology of the pilot or pothead whale, *Globicephala melaena* (Traill), in Newfoundland waters. *Fish. Res. Board Can. Bull.*, **132**, 84 pp.

Sergeant, D. E. (1973). Biology of white whales (*Delphinapterus leucas*) in western Hudson Bay. *J. Fish. Res. Board Can.*, **30**, 1065-1090.

Sergeant, D. E. & Brodie, P. F. (1969). Tagging white whales in the Canadian Arctic. *J. Fish. Res. Board Can.*, **25**, 2201-2205.

Sergeant, D. E., Caldwell, D. K. & Caldwell, M. C. (1973). Age, growth and maturity of bottle-nosed dolphins (*Tursiops truncatus*) from northeast Florida. *J. Fish. Res. Board Can.*, **30**, 1009-1011.

Shane, S. H. (1977). The population biology of the Atlantic bottlenose dolphin, *Tursiops truncatus*, in the Aransas Pass area of Texas. M.Sc. thesis, Texas A. & M. University.

Springer, V. G. & Woodburn, K. D. (1960). An ecological study of the fishes of the Tampa Bay area. *Fla. Sta. Board Conserv. Prof. Pap. Ser.*, **1**, 104 pp.

Taruski, A. G. (1976). Whistles of the pilot whales (*Globicephala* spp.): Variations in whistling related to behavioral/environmental context, broadcast of underwater sounds, and geographic location. Ph.D. thesis, University of Rhode Island.

Tavolga, M. C. (1966). Behavior of the bottlenose dolphin (*Tursiops truncatus*): Social interactions in a captive colony. In: *Whales, Dolphins, and Porpoises* (K. S. Norris, Ed.), pp. 718–730. Berkeley: University of California Press.

Tavolga, M. C. & Essapian, F. S. (1957). The behavior of the bottlenosed dolphin (*Tursiops truncatus*): Mating, pregnancy, parturition and mother–infant behavior. *Zoologica,* **42,** 11–31.

Tayler, C. K. & Saayman, G. S. (1972). The social organization and behaviour of dolphins (*Tursiops aduncus*) and baboons (*Papio ursinus*): Some comparisons and assessments. *Ann. Cape Prov. Mus. Nat. Hist.,* **9,** 11–49.

Tomilin, A. G. (1967). *Mammals of the U.S.S.R. and Adjacent Countries*. IX. Cetacea. Jerusalem: Israel Program for Scientific Translations.

Townsend, C. H. (1914). The porpoise in captivity. *Zoologica,* **1,** 289–299.

True, F. W. (1891). Observations on the life history of the bottlenose porpoise. *Proc. Nat. Mus.,* **13,** 197–203.

Walker, E. P. (1968). *Mammals of the World,* Vol. 2. Baltimore, Md.: Johns Hopkins.

Wells, R. S. (1978). Home range characteristics and group composition of Atlantic bottlenosed dolphins, *Tursiops truncatus* on the west coast of Florida. M.Sc. thesis, University of Florida.

Williams, G. C. (1964). Measurement of consocation among fishes and comments on the evolution of schooling. *Mich. State Mus. Publ. Biol. Sci.,* **2,** 351–383.

Wilson, E. O. (1975). *Sociobiology: The New Synthesis*. Cambridge, Mass.: Harvard University Press.

Wood, F. G. (1977). Births of dolphins at Marineland of Florida, 1939–1969, and comments on problems involved in breeding of small cetacea. In: *Breeding Dolphins: Present Status, Suggestions for the Future* (S. H. Ridgway & K. Benirschke, Eds.), pp. 47–60. Washington, D.C.: U.S. Marine Mammal Commission Report MMC-76/07.

Wood, F. G. Jr., Caldwell, D. K. & Caldwell, M. C. (1970). Behavioral interactions between porpoises and sharks. In: *Investigations on Cetacea,* Vol. II (G. Pilleri, Ed.), pp. 264–277. Berne, Switzerland: Brain Anatomy Institute, University of Berne.

Würsig, B. (1976). Radio tracking of dusky porpoises (*Lagenorhynchus obscurus*) in the South Atlantic, a preliminary analysis. Advisory Committee on Marine Resource Research, Scientific Consultation on Marine Mammals, Bergen, Norway, pp. 1–21.

Würsig, B. (1978). Occurrence and group organization of Atlantic bottlenose porpoises (*Tursiops truncatus*) in an Argentine bay. *Biol. Bull.,* **154,** 348–359.

Würsig, B. & Würsig, M. (1977). The photographic determination of group size, composition, and stability of coastal porpoises (*Tursiops truncatus*). *Science,* **198,** 755–756.

Würsig, B. & Würsig, M. (1979). Behavior and ecology of bottlenose porpoises, *Tursiops truncatus,* in the South Atlantic. *Fish. Bull.,* **77,** 399–442.

Würsig, B. & Würsig, M. (1980). Behavior and ecology of dusky porpoises, *Lagenorhynchus obscurus,* in the south Atlantic. *Fish. Bull.,* **77,** 871–890.

Yablokov, A. V., Bel'kovich, V. M. & Borisov, V. I. (1972). *Whales and Dolphins*. Jerusalem: Israel Program for Scientific Translations.

CHAPTER 7

The Behavior and Training
of Cetaceans in Captivity

R. H. Defran and Karen Pryor

This chapter reviews behavioral material on various species of delphinids and other cetaceans maintained in captivity. Emphasis is on the behaviors characterizing and differentiating the most commonly maintained species, and on the trainability of these species. New data are provided based on surveys of cetacean trainers and researchers.

PUBLISHED BEHAVIOR DESCRIPTIONS

History and Scope of Public Display of Cetaceans

The first species of cetaceans to be maintained in captivity were apparently specimens of the beluga whale, *Delphinapterus leucas*, and the Atlantic bottlenosed dolphin, *Tursiops truncatus*, in the early 1860s. These animals were displayed at The Aquarial Gardens in New York as part of P. T. Barnum's museum (Wyman, 1853). At about the same time a harbor porpoise, *Phocoena phocoena*, was kept for several months in the Brighton aquarium in England (Matthews, 1971). During the late 1870s, two different shipments of belugas were sent to England and displayed at the aquarium in Westminster and at shows in Manchester and Blackpool. At about the same time a solitary specimen of the Ganges susu, *Platanista gangetica*, was briefly maintained by Anderson (1878) in Calcutta.

The beluga from The Aquarial Gardens may have also been the first cetacean to be trained in captivity (Wood, 1973). Reportedly the animal's repertoire included hand-feeding, harness-wearing, and towing a car ridden by a young lady around its tank.

The first extensive published observations on captive cetaceans were by Townsend (1914), the director of the New York Aquarium. His early and colorful description of a group of five bottlenosed dolphins begins as follows:

They are the same jolly porpoises that make high-speed dashes under the bows of ships. No more popular exhibition of wildlife has ever been made anywhere. After seven months . . . they remain in good condition, feeding, leaping, and otherwise disporting themselves after the manner of porpoises on the high seas [p. 289].

Townsend's report, though brief, was comprehensive and included descriptions of swimming, feeding, and play behaviors, aerial displays, social behavior, and comments on their visual acuity. Townsend stressed the importance to the animal's health of proper techniques of transport and water treatment, but was not able to implement these measures himself. He reported sadly, in June 1916, that all his cetaceans had died (Devine & Clark, 1967). At about that same time, the first cetacean birth in captivity was recorded at Brighton, England. The harbor porpoise calf was, unfortunately, stillborn. It was not until some 30 years later, in 1947 at Marine Studios, that a live cetacean birth was recorded. The mother was a bottlenosed dolphin who was pregnant at the time of her collection (Alpers, 1960).

The modern era of maintaining cetaceans in captivity began in 1938 with the establishment of the Marine Studios aquarium at Marineland in Florida. Originally established as an aquatic movie set, the facility eventually became involved in the display and training of captive cetaceans for public viewing. Early curators and aquarium staff pioneered modern techniques of cetacean collection, transport, husbandry, and medicine. These techniques have been continuously refined and improved over the years by many workers, and are reviewed in a number of sources (Asper, 1975; Bigg & Wolman, 1975; Goforth, 1973; Griffin & Goldsberry, 1968; Newman & McGeer, 1966; Norris, 1974; Pryor, 1975; Ridgway, 1972; Ridgway & Van Dyke, 1975; Pilleri, 1970; Walker, 1975).

The early population of cetaceans at Marine Studios consisted mainly of bottlenosed dolphins. Reports by McBride (1940) and McBride and Hebb (1948) gave detailed behavioral descriptions of this early colony of cetaceans. Later, McBride and Kritzler (1951) gave the first extensive description of pregnancy, birth, and parturition in a captive cetacean (*T. truncatus*). Additional early material on these topics was provided a few years later by Tavolga and Essapian (1957). Wood (1977) has recently summarized the history of 40 births at Marineland from 1959 through 1969. Since the establishment of Marine Studios, a wide variety of cetaceans have been maintained in public oceanaria and research facilities around the world. In a number of these locales, breeding programs have begun. The history, current status, and success of these programs is outlined in contributions to the volume by Ridgway and Benirschke (1977).

Table 7.1 presents a complete list, to the best of our knowledge, of species maintained in captivity, currently or in the past, at various locales. This list follows the taxonomic designations suggested by the Marine Mammal Commission report on marine mammal names (1976), and by Rice (1977). The list comprises seven families and 35 species of cetaceans, including two representatives

Table 7.1 Species of Cetaceans Maintained in Captivity

Scientific Name	Common Name	Location
Family Balaenopteridae		
Balaenoptera acuotro-strata	Minke Whale	Japan
Family Eschrichtiidae		
Eschrichtius gibbosus	Gray Whale	West Coast USA
Family Delphinidae		
Cephalorhynchus hectori	Hector's dolphin	New Zealand, South Africa
Delphinus delphis[a]	Common dolphin	Japan, New Zealand, Rumania, Spain, USSR, West Coast USA
Feresa attenuata	Pygmy killer whale	Hawaii, Japan
Globicephala sp.[a]	Pilot whale	England, Hawaii, Japan, Mainland USA
Grampus griseus[b]	Risso's dolphin	Japan, USA
Lagenorhynchus obliquidens[a]	Pacific white-sided dolphin	Japan, New Zealand, West Coast USA
Lagenorhynchus obscurus[b]	Dusky dolphin	New Zealand, South Africa
Lagenorhynchus acutus	Atlantic white-sided dolphin	New England
Lissodelphis borealis	Northern right whale dolphin	West Coast USA
Orcaella brevirostris	Irawady dolphin	Indonesia, New Zealand
Orcinus orca[a]	Killer whale	Canada, England, France, Holland, Japan, Mainland USA

Table 7.1 (*Continued*)

Scientific Name	Common Name	Location
Peponocephala electra	Melon-headed whale	Japan, Hawaii
Pseudorca crassidens[a]	False killer whale	Florida, Hawaii, Japan, West Coast USA
Sotalia fluviatilis	Tucuxi	Australia, USA
Sousa chinensis[b]	Indo-Pacific humpbacked dolphin	New Zealand, South Africa
Stenella attenuata[b]	Pacific spotted dolphin	Hawaii, Japan
Stenella coeruleoalba	Striped dolphin	Japan
Stenella longirostris[a]	Hawaiian spinner dolphin	Hawaii
Stenella plagiodon	Atlantic spotted dolphin	East Coast USA
Steno bredanensis[a]	Rough-toothed dolphin	Hawaii, Japan
Tursiops aduncas[b,c]	Indian Ocean bottlenosed dolphin	Indonesia, New Zealand, South Africa
Tursiops gilli[a,c]	Pacific bottlenosed dolphin	Hawaii, Indonesia, Japan, West Coast USA
Tursiops truncatus[a]	Atlantic bottlenosed dolphin	Worldwide
Family Monodontidae		
Delphinapterus leucas[a]	Beluga whale	Canada, England, Germany, Japan, Mainland USA
Monodon monoceros	Narwhal	West Coast Canada
Family Phocoenidae[d]		
Neophocaena phocaenoides	Finless porpoise	Japan
Phocoena phocoena[b]	Harbor porpoise	East Coast USA, England, France, Holland, Japan, USSR
Phocoenoides dalli	Dall's porpoise	Japan, West Coast USA
Family Physeteridae		
Kogia breviceps	Pygmy sperm whale	Florida, New Zealand, New England
Physeter catodon	Sperm whale	New England
Family Platanistidae		
Inia geoffrensis[a]	Bouto	Hawaii, Japan, Mainland USA
Platanista gangetica	Ganges susu	Japan, Switzerland, West Coast USA, India
Platanista minor	Indus susu	Switzerland

[a] Most frequently maintained.
[b] Frequently maintained.
[c] Referred to in this chapter as separate species, but sometimes regarded as separate populations of *T. truncatus* (e.g., Mitchell, 1975; Rice, 1977).
[d] This family is recognized in MMC (1976) but not by Rice (1977), who places species listed here in the Delphinidae.

of the mysticete whales and, among the odontocetes, representatives of the families of true dolphins, river dolphins, sperm whales, and the narwhal/beluga family of monodonts.

Not all species have been kept routinely or with equal success. Such factors as availability, ease of collection and transport, and state-of-the-art medicine, husbandry, and training technology have favored the maintenance of some species over others. Additionally, the ease of maintaining a species in tank settings seems to reflect in part the ecological characteristics of the natural habitat of the species. The shallow, coastal water favored by the bottlenosed dolphin in the northern areas of the Gulf of Mexico apparently preadapt it well for tank living. In contrast, open-ocean pelagic species such as Dall's porpoise, *Phocoena dalli*, seem to have much greater difficulty in adapting to the tank environment. The sociobiology of the species, especially the degree of dependence on conspecific tank mates, also plays a part in the adjustment process.

Observations of Captive Specimens

Several authors have reviewed the literature on behavior of captive cetaceans (e.g., M. Caldwell & Caldwell, 1972; Evans & Bastian, 1969; Kinne, 1975; Yablokov, Bel'kovich & Borisov, 1972). Given the number and diversity of species that have been maintained in captivity (Table 7.1) one would expect to find a rich literature on species-typical behaviors and on the comparative behaviors of captive cetaceans. Such is far from the case. Relatively little published information is available for captive species, and what does exist is heavily weighted toward the bottlenosed dolphin. Even for this ubiquitous captive species, little systematic behavioral work has been done since the work of McBride and Hebb (1948), McBride and Kritzler (1951), Tavolga and Essapian (1957), Essapian (1963), and Tavolga (1966). Saayman and his colleagues in South Africa, in their recent descriptions of the Indian Ocean bottlenosed dolphin (Saayman, Tayler, & Bower, 1973; Tayler & Saayman, 1972, 1973), are one exception, but most of the emphasis on behavior in recent times seems to have shifted toward the natural setting (e.g., see chapters by Norris and Dohl and by Wells, Irvine, and Scott in this volume). The result is that, together with some of the older monographs on natural history and behavior of cetacea (e.g., Kleinenberg, Yablokov, Bel'kovich & Tarasevich, 1964, for the beluga, and see the compendium provided by Tomilin, 1967), the total behavioral literature for most cetacean species is weighted toward descriptions in the natural setting. This seems a reasonable and necessary trend but, at the same time, the unique advantages and opportunities that can be gained from studying animals in the laboratory and the oceanarium should not be bypassed. In this section, we briefly summarize the available scientific literature on the behavior of various cetaceans maintained in captivity. Some of this material provided the basis for the selection of behavior items used in the later-described questionnaire surveys given to cetacean trainers and researchers.

Bottlenosed Dolphin

Historically and currently the species of cetacean most frequently maintained in captivity has been the bottlenosed dolphin. In 1975 the population of captive bottlenosed dolphins worldwide was estimated as over 450 animals (Ridgway & Benirschke, 1977).

General descriptive accounts by Townsend (1914), McBride (1940), McBride and Hebb (1948), Norris and Prescott (1961), M. Caldwell and D. Caldwell (1967, 1972), Bel'kovich, Krushinskaya & Gurevich (1970), Wood (1953), are in good agreement on the broader aspects of bottlenosed dolphin behavior in captivity. Playful and gregarious, these animals may engage in a wide variety of daily activities. Individuals may spend a significant amount of time manipulating objects in their tank or carrying about some toy, leaf, or other chance debris on their rostrum. Spontaneous diversions such as balancing a ball between the pectoral fins may arise suddenly and disappear just as quickly. Play with other cetaceans is common, including chases or even "keep-away" games. Bottlenosed dolphins housed with other aquatic animals such as sea lions, turtles, eels, and fish are frequently reported to manipulate, or even harass these other species by pushing, towing, threatening, or biting them. Aerial displays including breaching and porpoising and aquatic behaviors such as fast swimming in a circular or irregular pattern are frequently seen. These activities may be carried out alone or in unison with tankmates.

Typically bottlenosed dolphins show strong interest in humans who approach their enclosure or enter their tank. The dolphins generally draw close and make contact (Defran & Milberg, 1973). Solicitation of human contact, while rare in newly collected specimens, is more common among trained or well-adapted animals. Individual dolphins may make persistent attempts to be rubbed and petted and will frequently retrieve balls or other objects and bring them to the human. For many captive animals these activities have been specifically trained, while for others they arise from the social contact with humans who feed them daily. Thus it would be expected that the mere presence of humans acquires a generalized reinforcing function. Observations of the solicitation of human contact by dolphins by those unfamiliar with an animal's history in captivity or with naive specimens has probably given rise to the popular belief that dolphins, particularly the bottlenosed dolphin, have a "natural" orientation to humans. Contact with humans may be at least as strongly motivated by tactile stimulation as by social factors. When Defran and Milberg (1973) substituted a water jet as a reinforcer in place of human tactile stimulation, responding continued at the previously high levels. Further evidence of the tactile sensitivity of this species comes from the report of a subadult female bottlenosed dolphin which lost visual sensitivity shortly after collection. This animal eventually acquired a repertoire of over 50 separate behaviors under the control of tactile cues (Kathy Krieger, personal communication). Although atypical, this case provides insight into the tactile sensitivity of the species. This sensitivity to tactile stimulation is also supported by references in the literature (see later) to gentle fluke and pectoral stroking which form a component of precopulatory activities.

Aggressive behavior by bottlenosed dolphins toward humans is not uncommon (Norris, 1967; Pryor, 1973; Shurepova, 1973). Aggressive acts such as threats, biting, and butting are well known to dolphin handlers and trainers.

Many of the behavioral characterizations of bottlenosed dolphins in captivity are derived from well-adapted specimens. However, much individual variation has been reported in the degree and frequency with which different types of behavior are seen. Many variables may contribute to the behavior observed. Most obvious are the conditions of collection and transport as well as initial adaptation.

Many observers have reported highly stereotyped behavior during the days and weeks following collection. Tightly-bunched swimming in narrow circles, frequent whistling in unison, and rapid flight from novel stimuli are examples. The presence of well-adapted tankmates as well as their relative degree of aggressiveness may affect the speed and success of adaptation.

Beyond the initial period of adaptation, other conditions of captivity may contribute to behavior. Feeding schedules, general husbandry practices, training history, and social grouping may all have important effects (D. Caldwell & Caldwell, 1972).

Hierarchical arrangements, especially in the context of sexual activity, have been described in a number of reports (M. Caldwell & Caldwell, 1966, 1972, 1977; McBride & Hebb, 1948; McBride & Kritzler, 1951; Prescott, 1977; Puente & Dewsbury, 1976; Tavolga, 1966; Tavolga & Essapian, 1957; Tayler & Saayman, 1972; Wood, 1977). There is general agreement that social hierarchies develop within the captive colony, but the organization seems to be fluid (nonlinear) rather than strict. Some animals may be dominant in sexual situations, while still others may have the "right" to perform a trained routine first.

Copulation typically is preceded by a variety of courtship activities (M. Caldwell & Caldwell, 1977; Puente & Dewsbury, 1976; Saayman & Tayler, 1977; Tavolga & Essapian, 1957), including posturing, tactile stimulation with fins, rostrum, and teeth, displays such as "jaw-clapping," vocalizations, as well as more aggressive interactions such as "head-butting." Again, there are wide individual differences in the degree, sequencing, and significance of these precopulatory activities.

Developmental aspects of bottlenosed dolphin behavior in captivity have been described by M. Caldwell and Caldwell (1966), McBride and Hebb (1948), McBride and Kritzler (1951), Tavolga and Essapian (1957), Gurevich (1977), Leatherwood (1977), Essapian (1953, 1963), and others. Close affiliation between the newborn and mother continues for an extended period of time and dependency may persist even into adulthood. Occasional care for the young may be actively or passively (e.g., tolerance of nearby swimming) assumed by tankmates while the mother is feeding. In the case history of a male calf described by Gurevich (1977) most forms of general activity described for adult specimens, including manipulation of objects and aquatic displays were seen within the first six months. Immature sexual behavior with both male and female tank-mates was also described.

Reports of care-giving among adult cetaceans were reviewed by M. Caldwell and Caldwell (1966). This behavior has been reported in captivity for the bottle-nosed dolphin as well as other species such as the pilot whale (*Globicephala* sp.), the common dolphin (*Delphinus delphis*), and the Pacific white-sided dolphin (*Lagenorhynchus obliquidens*). While such care-giving does occur, Caldwell and Caldwell concluded that it is extremely rare.

Diurnal activity cycles have been given only limited attention among captive bottlenosed dolphins. Powell (1966) found that vocalization periodicity surrounded the periods of twice-daily feeding.

More complete data collected by Saayman et al. (1973) on the Indian Ocean bottlenosed dolphin (*Tursiops aduncas*) suggest that the frequency of social interactions is significantly affected by a diurnal cycle. Social behavior among the animals they observed was infrequent at night and in the early morning but increased significantly to a peak at midday and in the early afternoon.

Bouto, Ganges Susu, and Indus Susu

Published observations on captive bouto (*Inia geoffrensis*) have been restricted thus far to four isolated pairs (three male–male and one male–female) and several isolated male specimens (M. Caldwell, Caldwell, & Evans, 1966; Curtis, 1962; Layne & Caldwell, 1964; Penner & Murchison, 1970). The primary source for observations on the bouto's general activity and social behavior is the report by Layne and Caldwell (1964) who studied two male specimens, one mature and the other a subadult maintained in the same enclosure. Some of the more conspicuous behavior previously reported for the bottlenosed dolphin has also been reported for the bouto, including play, orientation to humans, sexual activity (mostly masturbation) and various solitary swimming and surface activities. While quantitative comparisons are not possible, it appears that both the frequency and range of activities displayed by bouto are attenuated with respect to those seen in the bottlenosed dolphin. An exceptional lack of timidity was observed by M. Caldwell et al. (1966). In none of the 10 bouto they tested was there a fright or flight reation to the introduction of sound measurement apparatus into their enclosures. The authors noted that this pattern was in sharp contrast to the initial fright reactions of most bottlenosed dolphins they had tested with the same apparatus.

Other contrasts include the lack of aerial behavior or fast swimming in the bouto and the rarity of coordinated swimming or breathing behavior among pairs of animals. Perhaps the most conspicuous activity reported for the bouto is its slow, articulated swimming style. The flexible neck and large pectoral flippers apparently enable it to convolute itself and reverse directions in a short space. The even more diverse and articulated side-swimming style of the Ganges susu has been described by Pilleri, Gihr, and Kraus (1971). Additional work on the Ganges susu has described its activity cycles, feeding behavior, the apparent establishment of exclusive individual territories among pairs of animals (male-female and male–male pairs), and sound production (Aminul Haque, Nishi-waki, Kasuya, & Tobayama, 1977; Andersen & Pilleri, 1970; Pilleri, Gihr, & Kraus, 1970).

The behavior in captivity of four Indus susu (*Platanista minor*), two young females and a subadult pair, were recently described by Pilleri, Gihr, Purves, Zbinden, and Kraus (1976). This species displayed many of the behavioral traits of the Ganges susu, including side-swimming and continuous emission of echolocation signals during the 24-hour cycle. Like the Ganges susu, the visual system is greatly regressed. Complex swimming patterns developed as the animals adapted to the tank environment and included chase behavior among pairs of animals. Chase behavior by the male was directed toward the subadult female in the spring mating period. Unison swimming was common throughout the year as was contactual behavior between the animals and manipulation of objects in their tank. One of the two young female specimens in the tank died, but a second specimen maintained with it appeared to offer no assistance. The only social interaction between dolphins and humans noted by the authors was some reluctance on the part of the animals to accept food offered by hand.

Harbor Porpoise

Andersen and Dziedzic (1964) described the general behavior of a mixed colony of eight harbor porpoises observed over a year's time. Many aspects of this colony's initial adaptation to captivity were similar to those previously described for the bottlenosed dolphin. Observations on courtship and other sexual behavior were made on a group of three sexually mature females and two males, one of which was sexually mature. Stereotyped courtship displays included swimming in pairs, chasing, body contact (stroking, rubbing, mouthing, and swimming over another), posturing (ventral exposure, S-shaped posture, head pointing) and occasional erections. Other forms of sexual behavior, reported for bottlenosed dolphins, such as nuzzling, head- and fluke-slaps, as well as intromission and pelvic thrusts, were not observed in the harbor porpoise. Although the harbor porpoises were reported to be inventive in the construction of "games" and to manipulate and play with starfish or with food in their tank, they were also noted to frighten easily when presented with novel stimuli.

Andersen (1976) stated that of approximately 35 animals of this species he observed in captivity, aggression toward humans was never seen. This is unlike the bottlenosed dolphin and a number of other cetaceans, which on occasion may directly threaten or even attack human handlers. Also in contrast to the bottlenosed dolphin, the harbor porpoise displays little aerial behavior (leaping) in captivity unless specifically trained for it (Amundin & Amundin, 1973). High leaps have been observed occasionally in the wild when two or more animals meet, but typically wild harbor porpoises are not known as leaping animals. The wild animals observed by Amundin and Amundin were mostly alone or in pairs, and large groups are apparently rare in this species.

Andersen (1969) reported that care-giving was not common among captive harbor porpoises he observed. Although there were numerous instances of sickness among the animals maintained in Anderson's laboratory over many years, only a few cases of care-giving were observed: A juvenile female was supported for two days by a male of the same age and a sick calf was supported by a juvenile female not its mother. There appear to be no data on care-giving interactions among wild harbor porpoises against which these captive data may be compared.

Pacific White-Sided Dolphin and Common Dolphin

Brown and Norris (1956) described the mating and play activities of a small colony of Pacific white-sided dolphins maintained at Marineland of the Pacific. In spite of the presence of opposite-sexed conspecifics, both males and females directed most of their courting and mating activities toward bottlenosed dolphins housed in the same tank. These interspecific matings took place on a near daily basis, most often in the early morning hours.

Other observations on Pacific white-sided dolphins included epimeletic behavior directed to a wounded female by a male tankmate. Also members of this species spent some time playing with objects such as a feather or piece of kelp but did so less often than bottlenosed dolphins.

Most descriptions of common dolphins in captivity suggest a shy, easily frightened animal that adapts poorly to the confinement of captivity and, once there, has little inclination to play and is reluctant to socialize with other tankmates (Brown & Norris, 1956). In contrast, however, the isolated male–female pair of common dolphins observed by Essapian (1962) spent considerable time together engaged in courtship and some mating activities. Following the sudden death of the male, the female slowly developed an attachment to a male bottlenosed dolphin. This relationship continued for over a month until the death of the female.

Yablokov et al. (1972) described a male common dolphin that often played with a live piked dogfish in its tank, holding it carefully in its jaws and dragging it about the tank, sometimes stroking it and sometimes throwing it in the air. This animal also played with dead fish in a variety of ways. The playfulness of this specimen is at variance with the report of Brown and Norris (1956).

Spotted Dolphin and Spinner Dolphin

Bateson (1974) observed a colony of two spotted and five spinner dolphins for three weeks at Sea Life Park in Hawaii. Bateson's report focused on the dominance hierarchies which were evidenced in sexual and aggressive behavior. Specific behaviors included threats, beak–genital propulsion, chasing and various forms of body contact. Only in the case of threats were the interactions between animals not reversible; that is, no animal was ever threatened by an animal it had previously threatened. Other findings in Bateson's study were that highly active behaviors such as fast swimming and/or chasing occurred only in solitary animals or pairs of animals. Triads, when formed, were more likely to engage in slow unison swimming.

Pryor (1973) reported on eight years of observations by herself and others at Sea Life Park, Hawaii, on the same group studied by Bateson (1974). Some additional individuals of *Stenella* sp. that joined the original colony were also described. Pryor characterized *Stenella* sp. as timid, easily frightened, fearful of objects, highly dependent on the presence of other dolphins, and as reacting adversely to novelty. Other characteristics of spinner or spotted dolphins included a tendency to mimic each other's actions, swimming and leaping in unison and engaging in unison aerial displays. Aerial displays in spinner dolphins include rapid rotation of the body while in air. Aerial displays in spotted dolphins consist of a high leap in which the animal travels horizontally above the water with the body in an S-shape. Individuals of both species also display a tail over head flip in midair. Pryor (1975) gave a general account of the response to training in both species as well as additional descriptions of the social interactions described by Bateson (1974).

Rough-Toothed Dolphin

Norris, Baldwin, and Sampson (1965), in an early study, successfully trained a rough-toothed dolphin (*Steno bredanensis*) to perform in the open sea. They reported the animal to have retrieved small floating objects when not at work, to have voluntarily followed boats, and to have carried bits of seaweed on a pectoral or dorsal fin while following the boat. Norris et al. also reported that the animal slapped the water repeatedly with both flukes and pectoral fins when alarmed. Pryor, Hagg, and O'Reilly (1969) reported on the establishment of novel behavior as a criterion for reinforcement in two individuals of this species. Norris et al. described the rough-toothed dolphin as being heavily scarred in nature and in captivity as being unafraid of objects or barriers and highly manipulative. Pryor (1973) described the rough-toothed dolphin as being bolder and more investigative that the bottlenosed dolphin, highly trainable but volatile and aggressive. Pryor (1975) reported rough-toothed dolphins to have removed objects from each others' bodies, to have manipulated strange objects placed in their enclosure, and, in one case, to have manipulated a bell leading to a food reward within 24 hours of collection. Newly collected rough-toothed dolphins were also reported to gut and dehead all food fish, no matter how small, by smashing them against the water's surface. This practice gradually diminished over the initial period of adaptation.

Pilot Whale

Kritzler (1952) provided extensive notes on the behavior of a young male pilot whale that survived a mass stranding at St. Augustine Beach, Florida. The animal was judged to be less than a year old when brought into captivity. It required adaptation to the diurnal regime of the aquarium. Once adapted, it was noticeably active when feeding and in play, often imitating some of the behavior exhibited by its bottlenosed dolphin tankmates. Frequently the animal was a participant in play. It was often the target of threat or true aggression by the bottlenosed dolphins. After nine successful months in captivity, the animal succumbed to a fatal injury received from a bottlenosed dolphin that rammed it during a feeding period.

Kritzler's (1952) observations were expanded in later reports by Brown (1960, 1962), M. Caldwell, Brown, and Caldwell (1963), and Brown, Caldwell, and Caldwell (1966) on pilot whales maintained by Marineland of the Pacific. Sexual behavior among the two females and one male specimen was reported to be sporadic, initiated by the females, and mostly limited to precopulatory activities. Many of the courtship behaviors previously described for bottlenosed dolphins, including vigorous head-butting (Tavolga & Essapian, 1957), were also seen in these captive whales. Other characteristics of pilot whale behavior included limited play activity, little fear reaction to novel stimuli introduced into the tank, occasional tail-slapping, and aggressive harassment of fish and turtle tankmates, as well as progressively more violent attacks on humans entering the tank. Two separate instances were reported of the male animal carrying and copulating with a dead female conspecific and carrying and protecting a dead Pacific white-sided dolphin tankmate. Later, this same male engaged in a series of aggressive attacks upon its female conspecific tankmates, eventually inflicting a fatal injury on one of them. During the same period the male's aggressive behavior alternated with sporadic inappetance and depression. These aberrant behaviors were reported to cease following simultaneous administration of an antidepressant drug and removal of the animal from participation in the trained whale show.

Several specimens of the pilot whale have been trained for open ocean work, as have specimens of the killer whale, bottlenosed dolphin, Pacific white-sided dolphin, and rough-toothed dolphin (see Wood, 1973). In "Project Deep-Ops" a pilot whale carried a special recovery device to depths of over 500 m and placed it about a dummy warhead to float it automatically to the surface (Bowers & Henderson, 1972). These results reveal the tractability of this and other similarly trained cetaceans.

False Killer Whale

Brown et al. (1966) reported on the general and interspecific behavior of a subadult female false killer whale. In contrast to other species of cetaceans observed by Brown et al., this animal adapted rapidly to the conditions of captivity and within a short time had acquired a repertoire of trained show behaviors. The acquisition of show behaviors was apparently facilitated by an adeptness at observational learning. Several instances of interspecific sexual behavior were observed. The first involved a male pilot whale, which engaged the false killer whale in precopulatory activities such as rubbing his melon against her tail flukes, followed by intromission. In a second case, the false killer whale established a strong social relationship with a female bottlenosed dolphin. Eventually the two females exchanged sexual stimulation characterized by various forms of nuzzling each other's urogenital regions. Later, this behavior was repeated by the false killer whale with a small pilot whale. Other observations on this specimen included a protective relationship developed with a female common dolphin. On several occasions the false killer whale intervened when caretakers attempted to handle and medicate the common dolphin. Intervention several times took the form of grasping a trainer's leg until he moved away from the small dolphin. Pryor (1973) described a young false killer whale cooperating with a bottlenosed dolphin to remove a partition between them. The same false killer whale also pinned a diver, who had been teasing it, to the tank wall for several minutes.

Killer Whale

The first behavioral account of a killer whale (*Orcinus orca*) in captivity involved a young male specimen wounded at the time of its collection (Newman & McGeer, 1966). It recovered slowly from its wounds and fasted for the first 54 days of its captivity. Once it began feeding the whale exhibited a moderate level of surface activity such as slapping its pectoral flipper and tail sharply against the water, some leaping and playfulness, and reduced shyness toward its human handlers. After a total of 83 days in captivity the whale died of unknown causes.

Burgess (1968) reported an early and successful attempt to maintain and train a killer whale at Sea World, San Diego. The small female began accepting hand-held food almost immediately and was conspicuously nonaggressive. Among the other activities reported for the small whale were solicitation of rubbing and petting from its handlers, vigorous pectoral fin slaps against the water, and limited aerial activity.

Martinez (1973) and Martinez and Klinghammer (1978) carried out an observational study on the solitary and social activities of seven captive and approximately 40 wild killer whales. An ethogram consisting of over 50 behavioral events was then constructed from their observations. The specific items noted by Martinez and Klinghammer coincide extensively with the items used by the present authors in their survey of the behaviors of captive cetaceans, as described in the later section on Behavioral Ratings. No attempt was made by Martinez and Klinghammer to quantify the relative frequency of individual behavior items, but the authors mentioned that most were seen in all captive specimens and in many of the wild animals they observed. In addition to describing individual behavioral events, Martinez and Klinghammer provided a statement of the context in which the behavior occurred and occasionally an inference as to the functional significance of the behavior.

Pygmy Killer Whale

Pryor, Pryor, and Norris (1965) reported their observations of an apparently mature pygmy killer whale (*Feresa attenuata*) male which survived for 20 days in captivity. The animal was unusually aggressive compared to other species of cetaceans observed by the authors. It snapped its jaws and "growled" (by emitting air through the blowhole) when approached. Unlike most newly collected cetaceans, the pygmy killer whale made no attempt to avoid obstacles or observers, often making an aggressive response instead. When undisturbed the isolated animal seemed listless. Introduction of two companion pilot whales brought about increased activity and some aggression directed at the smaller of the two whales. One morning the smaller whale was found dead, the apparent victim of an attack by the pygmy killer whale. Other aggressive displays were later directed at spinner dolphins briefly housed with this individual.

Beluga Whale

The beluga has been one of the most frequently maintained cetacean species, yet its behavior in captivity has been given only limited attention in the published literature.

Bartman (1974) gave a brief behavioral description of two female beluga whales maintained at the Duisberg Zoo for over four years. The animals required a comparatively long four months in captivity before they would approach a feeding station. Once adapted, the animals displayed a marked preference for close contact with each other and for tactile stimulation by their handlers. Both animals exhibited cautious curiosity followed by aggressive playfulness toward divers entering their enclosure. The devising of games that involved manipulation of objects in their tank was a frequently observed activity.

BEHAVIORAL RATINGS

Experienced trainers and researchers familiar with cetacean behavior constitute a valuable information resource which deserves more serious scientific attention. Few people have spent more time observing or attempting to understand cetacean behavior than those training cetaceans in oceanariums or research facilities. A trainer's work is facilitated by the ability to recognize and to interpret correctly the nature and functions of various forms of behavior. Sometimes the trainer's own well-being can be at stake: for example, a trainer who does not recognize a threat display may be in danger of injury if the threat is followed by active aggression.

In the research reported here the common body of behavioral and training knowledge of widely experienced trainers and laboratory investigators of cetaceans was sampled with a questionnaire. Respondents were asked to estimate the frequency of occurrence of a provided list of behaviors for each species of captive cetacean with which a respondent was familiar.

A similar approach was later taken for the analysis of species-typical aspects of cetacean training. However, here respondents were asked to compare various aspects of a species' trainability with those of the most familiar captive species, the bottlenosed dolphin.

Respondents and Species

Questionnaires on behavior and on training were sent to a selected list of respondents having a minimum of five years of experience working with cetaceans and who were usually in supervisory positions. A wide geographic distribution of trainers was sought and efforts were made to obtain representation from different institutions. The observational and training experience of the 22 respondents is summarized in Table 7.2; the name and institutional affiliation of each respondent is given in the acknowledgment section.

Table 7.2 also lists the 11 species for which behavior and training data were obtained. In the behavior survey those species for which less than four respondents returned questionnaires were omitted from the final analysis. The exceptions were *D. leucas* and *D. delphis*. With these latter species, the two respondents had had extensive experience with a sizable sample of animals, at Sea World in one case, and in several laboratories in Russia in the other case. Species omitted due to insufficient numbers of respondents were *S. plagiodon*, the Atlantic spotted dolphin, and *S. attenuata*, the Pacific spotted dolphin. Species in which only one specimen had been observed, even if by several respondents, were also omitted. These included the pygmy killer whale, *F. attenuata*, the gray whale, *E. gibbosus*, and Dall's porpoise, *P. dalli*.

Behavior Survey

A list of 55 behavioral events reported to occur among many species of captive cetaceans was prepared (Table 7.3). Respondents were asked to estimate the frequency with which they had observed each behavior in those species with which they were familiar, through use of a 4-point scale: (1) never observed; (2) rarely observed; (3) occasionally (moderately often) observed; and (4) frequently observed.

Some of the behavioral events of Table 7.3 have been reported for various species in the scientific literature; others were added to the list based on the experiences of the authors. Omitted from the questionnaire were behaviors which most trainers have had little opportunity to observe, such as maternal and infant behavior, sleep behavior, and diurnal rhythms. Most respondents have had only limited experience in observing maternal–infant interactions, which reflects the relative rarity of births in captivity. Indeed, in many oceanariums when a birth is impending or occurs the mother and young are removed from the show or training tanks and placed in the care of a veterinarian or of personnel other than trainers. Sleep patterns in cetaceans go unobserved by trainers if they occur at night, and diurnal rhythms may be distorted by the superimposition of the time requirements of shows and training sessions.

All of the behaviors listed were judged to be familiar to the respondents. As far as possible, the names given to the behaviors were those in general use at the training facilities sampled. Where many different names were used, the behavior was described. For example, "spy-hopping," also known among trainers as

Table 7.2 Normative Data on 22 Survey Respondents[a]

Species	Behavior			Training		
	Number of Respondents	Mean Years Observations[b]	Mean of Animals Observed[b]	Number of Respondents	Mean Years Worked with Species[b]	Mean No. Specimens Trained[b]
D. leucas (beluga)	2	2	3	—[c]	—[c]	—[c]
D. delphis (common dolphin)	2	7	22	—[c]	—[c]	—[c]
Globicephala sp. (pilot whale)	9	4	4	8	6	4
I. geoffrensis (bouto)	4	5	9	2	3	3
L. obliquidens (Pacific white-sided dolphin)	4	4	6	3	4	6
O. orca (killer whale)	6	5	8	5	7	7
P. crassidens (false killer whale)	4	11	5	3	9	4
S. longirostris (spinner dolphin)	4	10	28	2	9	15
S. bredanensis (rough-toothed dolphin)	4	10	7	2	9	5
T. gilli (Pacific bottlenosed dolphin)	6	6	12	8	6	5
T. truncatus (Atlantic bottlenosed dolphin)	12	8	37	12	8	5

[a] 84% of the total of 26 respondents solicited.
[b] Means rounded to nearest whole number.
[c] Respondents not sampled for this species.

Table 7.3. Behavioral Ratings and Profiles for Each Cetacean Species[a]

	Species											Total Score
	Dl	Dd	Gs	Ig	Lo	Oo	Pc	Sl	Sb	Tg	Tt	
Affiliative/Social/Contact Behaviors												
Breathing in unison	0	+	0	0	0	+	+	+	+	+	+	+7
Leaping in unison	−	0	0	0	0	0	+	+	+	+	+	+4
Swimming in pairs (pectoral fin touching)	−	+	0	0	+	+	+	+	+	0	+	+6
Forms male/female pair	0	−	−	−	−	0	−	−	−	−	0	−8
Strokes other animal	−	+	−	0	0	+	+	+	+	0	0	+3
Solicits stroking from human	−	−	0	−	0	+	+	−	+	0	+	0
Mean Rating (×10)	22	31	22	19	27	32	34	32	34	28	32	
Aggression												
Threat posture	−	−	−	−	0	0	+	−	0	+	0	−3
Threat sound	0	−	−	−	−	0	+	0	−	0	0	−4
Tooth rakes	−	−	0	−	+	0	0	−	−	+	+	−2
Harasses new/sick tankmates	−	−	−	−	−	−	−	0	0	0	−	−9
Aggressive toward other cetacean species	−	−	0	−	0	0	0	0	0	0	+	−4
Threatens to attack other cetaceans	0	0	0	−	0	0	+	0	+	0	0	+1
Threatens to attack human	−	−	−	−	−	0	0	0	+	0	−	−4
Threatens to attack apparatus	−	−	−	−	−	0	+	0	0	−	0	−8
Attacks other cetacean	0	−	0	−	0	0	−	0	+	0	0	−2
Attacks human	−	−	−	−	−	−	+	−	−	−	0	−8
Attacks apparatus	−	−	−	−	−	0	+	−	−	−	−	−8
Mean Rating (×10)	19	15	18	12	19	24	28	22	24	25	25	

Table 7.3. (*Continued*)

	Dl	Dd	Gs	Ig	Lo	Oo	Pc	Sl	Sb	Tg	Tt	Total Score
Care-Giving												
Assists/protects new tankmate	−	−	−	−	−	−	0	−	0	−	−	−9
Supports sick/injured tankmate	−	−	−	−	−	−	−	−	0	−	−	−10
Mean Rating (×10)	10	20	14	10	17	16	19	13	25	14	19	
Fear/Stress/Subordination												
Avoids new objects	0	+	+	+	0	−	0	+	+	0	+	+5
High-speed swimming	0	+	0	0	+	+	+	+	+	+	+	+8
Chuffing (sharp exhalation)	0	0	0	+	+	0	0	+	0	+	+	+5
Bunching (drawing together when alarmed)	0	+	0	0	+	−	+	+	0	0	+	+4
Gives distress sound	0	0	−	−	−	−	0	0	−	−	0	−6
Shows whites of eyes	−	−	−	−	−	+	0	0	−	0	−	−6
Lies passively on tank bottom	−	−	−	+	−	−	−	−	0	0	−	−7
Subordinates to other cetacean species	−	−	−	−	−	0	−	+	−	−	−	−8
Prostrates across other's rostrum	−	−	−	−	−	−	−	0	−	−	0	−9
Turns ventral up if threatened	−	0	−	−	−	−	0	0	−	−	0	−8
Other subordinate display	0	+	+	−	+	+	+	+	+	+	0	+2
Tail-slaps	−	+	+	0	0	0	+	+	+	+	+	+6
Mean Rating (×10)	21	28	22	21	26	24	26	30	24	25	27	

Species

336

Curiosity/Manipulation/Play

Behavior												Δ
Approaches new objects	0	0	0	0	+	+	+	−	+	+	+	+4
Manipulates new objects	−	−	−	0	0	+	+	−	+	+	+	+2
Opens gates, lifts nets, etc.	−	−	−	−	−	+	+	−	+	+	+	0
Cooperates with other cetaceans	−	−	−	−	−	0	0	−	−	0	0	−7
Removes tag, rope, etc. from other cetacean	−	−	−	−	−	−	−	−	+	−	−	−9
Invents games	0	−	−	−	0	+	+	−	+	+	+	0
Manipulates noncetacean animals	0	−	−	−	0	0	0	−	0	0	0	−4
Mimics sounds	−	−	−	−	−	−	−	−	−	−	−	−11
Plays with familiar objects (ball, etc.)	0	−	0	0	+	+	+	+	+	+	+	+4
Plays chase with cetacean	0	+	+	+	+	+	+	−	−	+	+	+4
Other games with cetacean (e.g., "keep away")	−	−	0	−	0	0	+	+	0	0	+	−1
Spy-hop	+	0	+	+	+	+	+	−	+	+	+	+6
Mean Rating (×10)	21	19	19	14	21	30	31	19	24	30	31	

Sexual Behavior

Behavior												Δ
Rubs genitals on tank objects	−	−	0	+	−	+	0	+	0	−	+	0
Attempts intercourse with conspecific of other sex	0	+	−	−	0	+	+	−	0	+	+	−1
Attempts intercourse with other species of opposite sex	−	−	−	−	−	0	−	0	−	0	0	3
Male attemptes intercourse with other sex	−	0	−	−	0	+	+	−	+	+	+	−1
Intercourse with conspecific	0	0	−	0	+	0	−	0	0	+	+	−2
Intercourse with cetacean of other species	−	−	−	−	−	−	+	+	+	−	0	−7
Other sexual behavior	−	−	−	−	−	0	−	+	+	+	+	+1
Mean Rating (×10)	18	25	19	19	18	19	27	29	26	26	32	

337

Table 7.3. (Continued)

Leaping and Surface Behavior	Dl	Dd	Gs	Ig	Lo	Oo	Pc	Sl	Sb	Tg	Tt	Total Score
Breaches (lands flat on side/back)	−	+	0	−	+	+	0	+	+	+	+	+5
Porpoises (smooth arching reentry)	−	+	−	−	+	+	+	+	+	+	+	+5
Other types of leaps (e.g., spin)	−	+	−	−	0	−	−	+	−	0	0	−4
Pectoral slap on water surface	−	−	0	−	−	+	−	−	+	0	0	−4
Slaps head on water surface	−	0	0	−	0	−	0	+	+	0	0	−1
Mean rating (×10)	13	31	24	11	28	27	26	34	35	31	28	

*Dl = *Delphinapterus leucas*; Dd = *Delphinus delphis*; Gs = *Globicephala* sp.; Ig = *Inia geoffrensis*; Lo = *Lagenorhynchus obliquidens*; Oo = *Orcinus orca*; Pc = *Pseudorca crassidens*; Sl = *Stenella longirostris*; Sb = *Steno bredanensis*; Tg = *Tursiops gilli*; Tt = *Tursiops truncatus*.

The "Total Score" is the algebraic sum of the row ratings, where minus and plus are understood as −1 and +1, respectively, and zero as the algebraic zero.

The "Mean Rating" is obtained by summing the raw scores for items and dividing by the number of items rated in that column. Raw scores consist of the average rating (1–4) given by all respondents for that item.

"pitch-poling," "heads-up," and "sightseeing," was briefly defined as "rising vertically out of the water."

Results

Behavioral Ratings

The responses to the behavioral questionnaire are summarized in Table 7.3. The set of 55 questionnaire items were grouped into seven categories. The assignment of behaviors to the given categories was in some cases arbitrary. Stroking another animal, for example, was placed in the Affiliative/Social/Contact category, but it could as well have gone in the Sexual category. Similarly, toothraking, which was placed within the Aggression category, also might have been placed in the Sexual category as it may also appear as a component of courtship behavior. Some behaviors serve multiple functions and there is no way to assign them to one category other than arbitrarily. The behavioral ratings in the body of Table 7.3, minus, zero, or plus, indicate, respectively, a mean rating across respondents of 1.0 to 2.0, 2.1 to 3.0, and 3.1 to 4.0. In general terms, minus, zero, and plus indicate, respectively, "rare occurrences," "moderate number of occurrences," and "frequent occurrences" of a behavior. The row totals, which are algebraic sums of the cell scores ($+ = +1$; $- = -1$), index how common the behavior is across all of the species listed in Table 7.3. The possible values for the Total range from -11 to $+11$. Hence, unison breathing, with a summed score of $+7$, is very common across the species of cetaceans, while the forming of male/female pairs in some permanent bond, with a sum of -8, is very uncommon. Pair-bonding occurs to a limited extent in some species, including *D. leucas, O. orca,* and *T. truncatus,* according to our respondents.

The mean ratings within the columns of each category are based on *raw*-score ratings (from 3 to 4) and provide information on the degree to which a given category typifies a species. *Orcinus orca, P. crassidens, S. bredanensis, D. delphis, S. longirostris,* and *T. truncatus* all score highly on Affiliative/Social/ Contact behavior, while *Globicephala, D. leucas,* and *I. geoffrensis* are rated especially low in this category.

Aggressive behavior, as shown in Table 7.3, was rated as relatively infrequent for most of the species. *Delphinus delphis* and *I. geoffrensis* were rated low in every specific exemplar. *Delphinapterus leucas, L. obliquidens,* and *Globicephala* also received very low aggression ratings. *Pseudorca crassidens* was rated as the most aggressive species, characterized as frequently displaying threat postures and sounds and as attacking other cetaceans and apparatus. This high aggression rating is in contradiction to the early report of Brown et al. (1966), who particularly noted the absence of aggressive behavior in a female *P. crassidens* housed together with specimens of *Globicephala, L. obliquidens, T. gilli,* and *D. delphis.* However, Norris (1967) later reported that occasional individuals of *P. crassidens,* as well as *Globicephala* and *S. bredanensis,* became sufficiently aggressive in captivity so that entering their tanks was dangerous.

Steno bredanensis is the only species rated by our respondents as frequently threatening humans or as actually attacking them. In contrast, *Globicephala* is rated as rarely showing aggressive behavior toward humans.

All species except *S. bredanensis* were rated as rarely displaying care-giving (succorant) behavior toward adult tankmates, and even this species was rated as showing this behavior with only moderate frequency. To some degree, these ratings stand in contradiction to the descriptions in the literature. The summary of care-giving behavior provided by the Caldwells (M. Caldwell & Caldwell, 1966) indicates that "standing by" an injured or distressed conspecific and displaying excitement or interest in the distressed individual has been observed in the field, the laboratory, or both in *I. geoffrensis*, *D. delphis*, *T. truncatus*, *T. gilli*, *O. orca*, and *Globicephala*, as well as in some other species not discussed in this chapter. *Delphinus delphis*, *T. truncatus*, and *Globicephala* have all been reported as additionally giving physical support to the injured companion, by raising it to the surface. Physical support has also been observed in *L. obliquidens*. These types of behaviors have been more commonly observed in the field under conditions of capture when an individual from a school is netted or otherwise restrained. In some cases the netted individual will be abandoned, as reported by Brown and Norris (1956) as occurring for *D. delphis*. The low ratings for succorant behavior given by our respondents may reflect the low incidence of opportunity for occurrence of the behavior in the captive setting, in that captive specimens are not commonly in active distress. Where distress does occur there is likely to be human intervention, and the injured animal may even be removed to a new setting. It may also be the case that captive animals tend to become less interdependent, since the human becomes more important to its survival and for its social stimulation. There also seem to be indications that the sex, age, and social relationship of the distressed animal to its potential assistants influences their proclivity toward aid-giving.

Indices of fear, stress, and subordination were most common in *S. longirostris*, *D. delphis*, and *T. truncatus*, in that order, and least common in *D. leucas*, *Globicephala*, and *I. geoffrensis*. The ratings in this category were generally negatively correlated with size of the animal; the smaller species are understandably more timid, the larger species less so. *Inia geoffrensis* is an exception, but its riverine existence is totally unlike the habitat of the coastal or oceanic species and predator pressure in minimal (see Wells, Irvine & Scott, this volume).

The most manipulative, playful, and curious species, according to our respondents, were *Tursiops* (both species), *P. crassidens*, and *O. orca*. All received equivalent or nearly equivalent mean ratings in this behavioral category. These are not unexpected rankings given the descriptions in the literature of the extensive play behavior, curiosity, and inventiveness of these species. Species ranking particularly low in this category were, first, *I. geoffrensis*, and then, with equal rankings, *D. delphis*, *Globicephala*, and *S. longirostris*. All of these species had been rated as low in aggression also. It seems reasonable to expect that aggression and curiosity (e.g., the tendency to approach and manipulate objects) would be highly correlated.

Sexual behavior, as indexed by the behaviors rated, was most in evidence in *T. truncatus,* and *S. longirostris.* It was least pronounced in *I. geoffrensis, D. leucas, Globicephala,* and *L. obliquidens.* There have not been any extensive studies of sexual behavior in species other than *T. truncatus.* Published studies report a great deal of sexual activity in *T. truncatus* in tank settings, a trait reflected in the ratings of our respondents. Some field observers report high levels of sexual activity for *T. truncatus* in the wild (Blair Irvine, personal communication).

The category of Surface and Aerial behavior includes both leaps and slaps of body parts on the surface (except for tail-slaps which were categorized as a fear or stress indicant). Species receiving the highest positive rating were *S. longirostris* and *S. bredanensis,* followed by *T. gilli* and *D. delphis,* each receiving the same mean ranking. The spinning leaps of *S. longirostris* are frequently seen in the wild. The other species given high ratings are also known to porpoise and breach frequently in natural settings. *Delphinapterus leucas* and *I. geoffrensis* were rated by our respondents as rarely showing aerial behaviors. This seems in keeping with the natural behaviors reported for these species (Layne, 1958; Tomilin, 1967).

In summary, the behavioral ratings appear to reflect substantial differences in key behavioral areas between species and seem to be in correspondence with available data from observations in the wild.

Cluster Analysis

A cluster analysis (Dixon, 1975) was performed on the data of Table 7.3, using the mean ratings of each species within each of the seven behavioral categories. This analysis groups together those species, pairs of species, or groups of species having the smallest aggregated distance between ratings. Distance is measured as the square root of the sum of the squared differences between ratings (Euclidian distance). For example, in comparing *T. gilli* and *O. orca* differences between the two on each of the seven behavioral ratings are obtained and each difference is squared. The sum of these squares is then calculated and finally the square root of the sum is obtained.

The results of the clustering are portrayed graphically in Figure 7.1. Two species that are clustered together are connected by a line, a species that clustered with a pair of species is shown connected to the midpoint of the line connecting the pair, and the pair which clusters with another pair is shown by connecting the midpoint of the two pairs. Finally, a species associated with a clustering of pairs is shown connected to the midpoint of a line connecting the two pairs. The relative distance between species, pairs, or groups is shown by the length of the connecting line. Thus the *T. gilli/O. orca* pair have the shortest distance and were clustered first, *T. truncatus/P. crassidens* were clustered next, then these two pairs were clustered, and so forth.

Three main clusters of species appear in Figure 7.1. These are (A) *T. gilli/ O. orca* + *T. truncatus/P. crassidens* + *S. bredanensis*; (B) *D. leucas/Globicephala* + *I. geoffrensis,* and (C) *D. delphis/L. obliquidens* + *S. longirostris.*

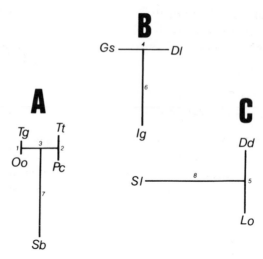

Fig. 7.1. Behavioral relatedness of species as shown by cluster analysis technique. Length of lines indexes degree of relatedness. See text. Dl = *Delphinapterus leucas*; Dd = *Delphinus delphis*; Gs = *Globicephala* sp; Ig = *Inia geoffrensis*; Lo = *Lagenorhyncus obliquidens*; Sl = *Stenella longirostris*; Sb = *Steno bredanensis*; Tg = *Tursiops gilli*; Tt = *Tursiops truncatus*.

Additionally, although not shown in Figure 7.1, Cluster C is most closely related to Cluster A, while B clusters with the mean of A and C.

Figure 7.2 shows the mean ratings of Table 7.3 for each species grouped according to the clusterings of Figure 7.1. The three clusters, A, B, and C, are shown with the subgroups of each cluster separately delineated. The similarities of behavioral ratings of species within subgroups is very evident. The correspondence across subgroups within a cluster is reduced relative to that within subgroups, and the correspondence across clusters is reduced further yet. In Cluster A *S. bredanensis* is most discrepant from the two other subgroups in its lower rating on Curiosity/Manipulation/Play, its higher rating on Leaping and Surface behaviors, and its greater tendency for Care-giving. Trainer comments indicated that *S. bredanensis* removes objects and props from other animals. This same species was rated low relative to *T. truncatus* in the Curiosity/Manipulation/Play category, which is surprising in view of its reputation for inventiveness (Pryor et al., 1969) and the ease with which it can learn to retrieve and manipulate objects (see next section). Another anomalous characterization of *S. bredanensis* is its simultaneous high ratings in Table 7.3 on the items "attacks humans" and "solicits stroking from humans."

The *P. crassidens/T. truncatus* subgroup is only marginally divergent from the *O. orca/T. gilli* subgroup in all of the behavioral categories. The former pair is rated slightly higher than the latter on Curiosity/Manipulation/Play, Fear/Subordination/Stress, and Care-giving. In the remaining categories there is overlap between the pairs, one or both members of one pair falling at or between the ratings of one or both members of the second pair.

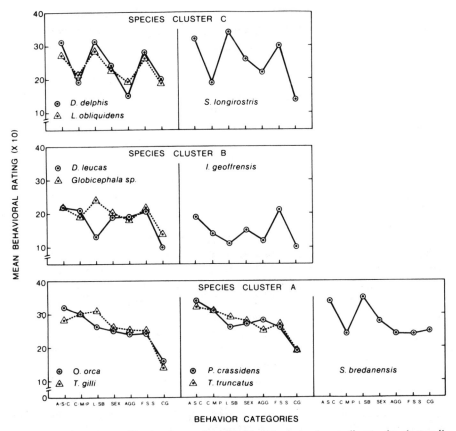

Fig. 7.2. Behavioral profiles for the species of Table 7.3, grouped according to the clusters illustrated in Figure 7.1. The mean behavioral rating scale (ordinate) is as in Table 7.3, where 10 means the behavioral category was never observed and 40 means it was frequently observed. A/S/C = Affiliative/Social/Contact; C/M/P = Curiosity/Manipulation/Play; L/SB = Leaping/Surface behaviors; SEX = Sexual; AGG = Aggression; F/S/S = Fear/Subordination/Stress; and CG = Care-giving.

In Cluster B *Globicephala* displays considerably more surface and aerial behavior than do *D. leucas* and *I. geoffrensis*, though overall it falls somewhat below the ratings of any of the species in Cluster A or B. *Globicephala* sometimes breach or slap their flipper or head on the surface (Table 7.3), whereas *D. leucas* or *I. geoffrensis* rarely or never display these behaviors. *Inia geoffrensis* diverges from the *D. leucas/Globicephala* pair in a lower rating on all categories except Fear/Subordination/Stress. Overall, Cluster B species display much weaker Affiliative/Social/Contact tendencies than do the species of Cluster A or B. They are also rated below the species of Cluster A on Curiosity/Manipulation/Play.

The species of Cluster C, like those of Cluster A, display high affiliative tendencies according to our raters, as well as a great deal of leaping and surface behavior. In contrast to Cluster A, and more like Cluster B, they are rated low in Curiosity/Manipulation/Play. Their low Aggression ratings are also closer to ratings for Cluster B than to the higher ratings given Cluster A. Within Cluster C *S. longirostris* diverges from *D. delphis* and *L. obliquidens* in its higher ratings in all categories except Curiosity/Manipulation/Play (in which there is an overlap with *D. delphis*) and Care-giving. The general pattern of ratings for *S. longirostris* was, however, very similar to that for the other two species.

TRAINING

Synopsis of Training Approaches

With rare exception cetaceans in captivity receive some training if only to acclimate them to the contingencies and circumstances of their new environment. Most undergo extensive training to acquire a repertoire of behavior useful for exhibition, research, or practical tasks such as open-ocean work. The quality of this training experience can contribute vitally to the long-term welfare of the captive cetacean. The avoidance of stressful regimes, cooperation with veterinary procedures, providing of means for control of the environment by the animals themselves, and maintenance of adequate environmental stimulation through diversified training are benefits to the animal being trained.

By most standards the enterprise of bringing captive cetaceans under behavioral control has been a spectacular success. Consider, for example, the following achievements: (*a*) in a research setting, the differential discrimination of sonar targets by a blindfolded bottlenosed dolphin (Turner & Norris, 1966); (*b*) in the area of practical applications the attachment of recovery hardware to instrument packages on the ocean floor over 450 m below the surface by a pilot whale [a 23-km round trip to the test site was made with the untethered whale traveling in the recovery craft's stern pressure wave (Bowers & Henderson, 1972)]; (*c*) in the public oceanarium a trainer routinely rides a 3000-kg killer whale, *O. orca*, bareback 10 m to the tank bottom and up again to the surface (Shamu at Sea World, San Diego, Calif.).

Little is known of the early developments in the training of cetaceans. The techniques used to train the beluga whale maintained in Barnum's museum in the 1860s were probably borrowed from traditional methods used for training circus animals or domestic animals. Often these methods relied on punishment and the development of avoidance responses in the animals, as in breaking the horse to the bit and spur, or the use of a whip and chair for guiding the behavior of large circus cats, or a hook to prompt an elephant's movements.

Present-day methods of training cetaceans have their base in operant conditioning methodology developed in the late 1930s and elaborated on by many workers thereafter (e.g., Ferster, Culbertson, & Boren, 1975; Holland & Skinner, 1961; Skinner, 1938). Keller and Marian Breland, students of B. F. Skinner, were among the first to apply operant techniques systematically to the training of cetaceans, as well as many other species of animals (Breland & Breland, 1966). The Brelands were also among the first to stress the importance of species differences in determining trainability. Their book gave many illustrations of how the biology of a species and its adaptations, feeding strategies, and the like predispose it toward or against the learning of certain behaviors. This is no less true for cetaceans. Herman (this volume) has discussed some of the learning constraints of cetaceans as well as their learning specializations. Those training cetaceans should therefore recognize the factors limiting or promoting learning in a given species and emphasize training techniques that are in keeping with the natural behavioral predispositions of the species. For example, different kinds of sensory cues are not likely to be equally effective for all species. Some species, but not others, may learn easily by observation. Certain species may require the presence of conspecifics to perform well, while others may perform better if trained alone. Some species, but not all, may be timid in the face of new objects or experiences.

There is little available in the published cetacean literature on species differences in trainability (Pryor, 1973) or even on training methods for cetaceans in general. Applied training programs directed toward open-ocean work with trained cetaceans have sometimes specified the procedures and details associated with the development of given behavioral sequences. Among these are the Navy's research programs on diver assistance (Wood & Ridgway, 1967), respiration and deep-diving (Ridgway, Scronce, & Kanwisher, 1969), deep-object recovery (Bowers & Henderson, 1972), and maximum swimming speeds (Lang & Pryor, 1966). Also well documented is the successful open-ocean training of captive specimens (Bailey, 1965; Hall, 1970; Irvine, 1970; Norris, 1965).

Many researchers working with captive animals have not given the details of their training procedures (exceptions are Herman & Arbeit, 1973; Kellogg & Rice, 1966; Pryor et al., 1969; Turner & Norris, 1966). The frequently used phrase "standard operant conditioning techniques were used to train the discrimination" tells little about the specific training methods.

Historically, the exchange of information on training techniques has been poor among the various oceanariums or military establishments housing dolphins. The recent organization of an international association of marine mammal trainers, which meets annually, has greatly increased the flow of information. The favorable response by the dolphin training community to the questionnaires on cetacean behavior and training that form a basis of this report reflects an increased openness.

Currently no generally available reference exists to describe techniques for training of simple or complex performances in cetaceans. Training manuals have been produced in various organizations to teach dolphin training to new trainers, but these manuals have not been generally available to the public. These include separate manuals by Pryor and Turner for Sea Life Park (cited in Pryor, 1975), by Animal Behavior Enterprises (cited in Pepper & Defran, 1975), and by Pepper and Defran (1975) for the Naval Ocean Systems Center. Included in most of these manuals are methods for promoting adaptation of animals to captivity. Some manuals emphasize the development of a positive social relationship with the trainer by hand-feeding. The introduction and use of a conditioned reinforcer sound or "bridging sound" (usually a whistle) to signify correct responses is one of the first behavioral control methods used. The manual by Pepper and Defran recommends the early use of visual prompts through which the animal is conditioned to follow a target placed nearby. Hence it can be led through a variety of behaviors such as pressing a paddle, swimming through a gate, or following a boat. The use of the "prompt" is gradually withdrawn as the desired behavior develops. The withdrawal of the prompt constitutes a form of "fading" training. Generally, fading techniques consist of the transfer of control of behavior from one stimulus to another and take advantage of the fact that a given performance may initially be controlled more easily by one type or dimension of a stimulus than another. The use of fading techniques for training cetaceans was first described by Kellogg and Rice (1966) and was developed independently of the elaborate and more general presentation of the technique by Terrace (e.g., Terrace, 1966).

The Pepper and Defran manual describes the use of acoustic "recall" signals, which are introduced to signal an animal to come to the source of the sound or to go to a designated station. Also described is the development of complex sequences of responses through chaining techniques. Elaborate strings of behavior can be formed by adding new stimulus–response pairs to already existing sequences. The maintenance of behavior through partial reinforcement schedules (e.g., variable ratio) is generally advocated, and trainer sensitivity to the "drifting" of behavior is advised. "Gentle" suppression techniques, such as time-outs or the use of crowder nets for gate training or for related behaviors, are described as useful for eliminating undesirable behaviors or for prompting new ones. The training of performances for specific application to cetaceans is described in some detail, including passing through a gate, beaching on a platform, swimming onto a stretcher, and following a boat into the open ocean. The general procedures described will be familiar to those having studied operant methods. However, since cetaceans are generally highly social animals, often very responsive to the requirements of the trainer, flexible in their behavioral capabilities, rapid in their rate of learning, not practicably maneuvered by force, the experience of training cetaceans is likely to be different from the experience of training most other animals.

Trainability Ratings

A list of 21 behaviors commonly used in dolphin shows or in research or work programs was prepared (Table 7.4). Respondents were asked to use the trainability of the commonly maintained *T. truncatus* as a baseline against which to rate the trainability of each behavior for each other species with which they had worked. A 3-point scale was used: (1) trains with greater difficulty than *T. trun-*

Table 7.4. Comparisons of Selected Species with *T. truncatus* on the Ease/Difficulty of Training Various Behaviors[a]

Behavior	Oo	Tg	Pc	Sb	Gs	Ig	Lo	Sl	Total
Waving/exposing pectoral fin	0	0	0	+	0	0	−	−	−1
Tail-wave or -slap	0	0	0	0	0	0	−	0	−1
Allowing touching or petting	0	0	+	+	0	0	0	−	+1
Swimming through gates	+	0	+	+	0	+	+	0	+5
Chains of responses	0	0	0	+	−	0	0	−	−1
Manipulating in-water objects	−	0	+	+	0	+	0	−	+1
Wearing props, scientific equipment	0	0	+	+	+	0		−	+2
Wearing blindfolds (eye cups)	0	0	0	0	+	0		−	0
Vocalizing underwater on command	−	−	0	−	0	0			−3
Vocalizing in air on command	0	0	0	−	−		−	−	−4
Retrieving objects	0	0	+	+	0		−	−	0
Towing gear or apparatus	0	0	+	+	0		−	−	0
Towing or giving trainer ride	−	0	+	+	0		−	−	−1
Simple leaps, e.g., propoising	0	0	0	0	0		+	+	2
Breach	0	0	0	0	0		−	0	−1
Complex leaps, e.g., sommersault	−	0	0	0	−		−	0	−3
Leaping over obstacles	−	0	+	+	−		+	−	0
Unison performance	−	0	0	0	0		+	+	+1
Breaching or slide out	0	0		+	−		−	−	−2
Entering a stretcher on command	0	−		0	−				−2
Open-ocean work (any type)		−		0				−	−2
Total Behaviors Rated	20	21	18	21	20	9	16	19	
Mean Rating (×10)	22	20	18	18	21	17	23	26	

[a] Oo = *Orcinus orca*; Tg = *Tursiops gilli*; Pc = *Pseudorca crassidens*; Sb = *Steno bredanensis*; Gs = *Globicephala* sp.; Ig = *Inia geoffrensis*; Lo = *Lagenorhynchus obliquidens*; Sl = *Stenella longirostris*.

"−" = behavior more difficult to train that in *T. truncatus* (mean of 2.4–3.0). "0" = behavior may be trained with same ease/difficulty as with *T. truncatus* (mean of 1.7–2.3). "+" = behavior may be trained more easily than with *T. truncatus* (mean of 1.0–1.6).

catus; (2) training difficulty/ease is comparable to that for *T. truncatus*; and
(3) training is easier than for *T. truncatus*.

Additional questions asked for ratings, using the same 3-point scale, of the
ease with which a species learns to respond to auditory, visual, and touch
signals in comparison with *T. truncatus*, and for an overall or global rating of
the relative "trainability" and day-to-day "reliability," once trained, of each
species in comparison with *T. truncatus*. Trainability and reliability were rated
on a 5-point scale ranging from "much harder to train or much less reliable"
(=1.0) to "much easier to train or much more reliable" (=5.0).

Trainability and Reliability

The mean global ratings on trainability and on day-to-day reliability of each
species relative to *T. truncatus* are portrayed in Figure 7.3. *Orcinus orca* is the
only species rated higher in each category than *T. truncatus*, though *I. geoffrensis*, *P. crassidens*, and *S. bredanensis* are each rated higher on general trainability and *S. longirostris* and *T. gilli* are each rated higher on general reliability. There is considerably more variability in the trainability ratings across
species than in the reliability ratings, indicating that the shaping or development of a behavior is more affected by the species variable than is the maintaining of a behavior once developed.

Ratings for the ease of training 21 specific behaviors, again relative to *T.
truncatus*, are summarized in Table 7.4. The ratings in the body of the table,
minus, zero, and plus, are interpreted similarly to the ratings shown in Table
7.3, except that the base scale ranged over three values rather than four values
and the endpoints were reversed. A minus rating indicates that the behavior is
more difficult to train in the species indicated than in *T. truncatus* and includes
mean ratings between 2.4 and 3.0. A zero indicates that the behavior may be
trained with the same difficulty/ease as with *T. truncatus* (mean of 1.7 to 2.3),
while a plus indicates that the training of the behavior is easier than with *T.
truncatus* (mean of 1.0 to 1.6). The mean rating at the bottom of the table indexes the overall trainability of the species, for the behaviors rated, relative to
T. truncatus. Not all behaviors were rated for each species.

In general the specific behaviors of Table 7.4 that were rated by the respondents probably formed the bases for the global rating in Figure 7.3. Figure 7.3
shows the global ratings of the respondents on the trainability and reliability
once trained, of each species relative to *T. truncatus*. It also illustrates the mean
rating of each species in Table 7.4. These mean ratings were transformed from
the Table 7.4 scores by subtracting each from 20, so that, like the global
ratings, higher scores (positive scores) indicate that trainability is easier than for
T. truncatus and lower scores (negative scores) indicate that training is more
difficult.

The transformed scores correlated well with the global ratings on trainability,
except for *O. orca*. Though the killer whale was rated globally as distinctly
superior in trainability to *T. truncatus*, its mean ratings on specific behaviors

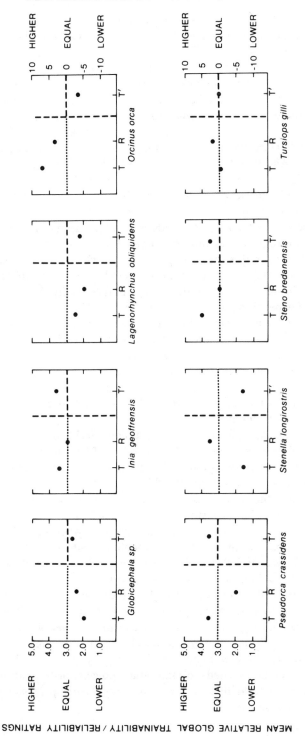

Fig. 7.3. Left ordinate: Global ratings of eight species on trainability (T) and reliability (R) relative to *T. truncatus*. Right ordinate: Mean transformed trainability ratings (T') for specific behaviors of Table 7.4. "Equal" indicates the species was rated as equivalent to *T. truncatus*. Higher ratings indicate the species was judged as superior to *T. truncatus* and lower rating that it was judged as inferior.

fell below the level of *T. truncatus*. As can be seen in Table 7.4, other than learning to swim through gates more easily than *T. truncatus*, the killer whale was not rated as superior in trainabilility on other behaviors, and in fact was rated as inferior for 6 of the 20 behaviors rated. The high global rating for *O. orca* is not easily interpreted therefore.

Inia geoffrensis' high rating may be an artifact of its restricted behavioral repertoire. Only a limited number of behaviors were rated for this species—9 of the total 21. For these nine behaviors *I. geoffrensis* trained easily. That trainers did not rate other behaviors may reflect that the respondent had not attempted to train the behavior. Additionally, however, it may indicate that behaviors such as leaping, manipulation of objects, in-air vocalization, and open-ocean work, were judged as too difficult to attempt to train in *I. geoffrensis* or as inappropriate for this freshwater species.

Globicephala, like *I. geoffrensis*, is infrequently seen leaping in its natural environment. Its low ratings on learning of complex leaps or on leaping over obstacles probably follow from this species-typical characteristic, though it can be taught simpler leaps or jumps. *Globicephala* seems to have relative difficulty in learning complex behavioral chains, and, perhaps paradoxically in view of this species' reputation for beaching or stranding in the natural habitat, it does not easily train to beach on a platform or to enter a stretcher on command.

Pseudorca crassidens and *S. bredanensis* received similar ratings on most behaviors. Both are quick to allow touching or petting, and easily learn to swim through gates, leap over obstacles, manipulate and retrieve objects, and to wear props or scientific equipment. *Steno bredanensis'* relative difficulties appear to be in learning to vocalize on command. Overall, both of these species seem to be excellent candidates for training.

Tursiops gilli's overall rating for trainability was comparable to *T. truncatus*. Its underwater vocalizations were rated as more difficult to bring under stimulus control, and it was rated as more difficult to train for open-ocean work or to beach itself on a stretcher on command.

The greatest training difficulties, relative to *T. truncatus*, occur for *L. obliquidens* and *S. longirostris*, both Cluster C species. *Stenella longirostris* is rated as especially timid in captivity and both species fare best when placed with a number of conspecifics. The high cohesive schooling tendencies of *L. obliquidens* and *S. longirostris* probably account for their readiness to perform synchronized (unison) behaviors, such as unison swimming or leaping. Both species can be easily trained to porpoise, though neither is strong in learning the more complex leaps or breaches that can be taught to *T. truncatus*. However, *L. obliquidens*, but not *S. longirostris*, quickly learns to leap over obstacles and swim through gates. *Stenella longirostris* on the whole is much more difficult to train than *L. obliquidens*. *Stenella longirostris* is especially timid and not easily induced to allow itself to be touched, to manipulate objects, or to wear equipment or props of any kind. Neither *S. longirostris* nor *L. obliquidens* can be taught easily to vocalize on cue.

The row totals in Table 7.4 give some indication of how *T. truncatus* rates in trainability on each behavior relative to the other species combined. Negative sums indicate that *T. truncatus* tends to train easier, while positive sums indicate that the training of *T. truncatus* is more difficult. Since not all species were rated on every behavior, and since the sums can derive from a variety of combinations of individual ratings, their values should be looked at as broad guidelines only and interpreted with caution. Thus a negative sum may contain some positive scores, as is the case for the behavior of pectoral-fin-waving or for the training of complex sequences of responses, and a zero or near-zero score may reflect either the cancelation effect of positive and negative ratings (leaping over obstacles) or very little difference among all species (breaching). Overall, however, it appears that *T. truncatus* trains relatively more easily than some of the other species in vocalizing behaviors, beaching, or entering a stretcher, and in complex leaps. On the other hand, training *T. truncatus* to swim through a gate is harder than for the majority of the other species. Also *T. truncatus* tends to be harder to train for simple leaps than some of the other species, is more resistant to being touched than some other species, and learns to wear props or scientific apparatus with greater difficulty than some other species.

Control of Behavior by Various Cues

Figure 7.4 summarizes the respondent ratings on how easily a given species can learn to use the indicated auditory, visual, and touch signals, as compared with *T. truncatus*. Visual signals, stationary or moving, appear more difficult to establish as cues for controlling behavior with *Globicephala*, *S. longirostris*, *L. obliquidens*, and *I. geoffrensis* than with *T. truncatus*. In contrast, *O. orca* shows a markedly greater readiness to utilize static and dynamic visual signals than does *T. truncatus*. *Pseudorca crassidens* is slightly favored over *T. truncatus* in the use of moving visual signals only. The relation of this complex of results to known visual functions in the various species is ambiguous. As reviewed by Dawson (this volume) and by Madsen and Herman (this volume), the *Inia* visual system is inferior to that of *T. truncatus*, while that of *L. obliquidens* is not. Underwater visual-resolution thresholds for *L. obliquidens* (Spong & White, 1971) are seemingly as good as those obtained for *T. truncatus* (Herman, Peacock, Yunker, & Madsen, 1975). The visual-resolution capabilities of *O. orca* (White, Cameron, Spong, & Bradford, 1971) also seem to be in the same range as those obtained for *L. obliquidens* and *T. truncatus*. The rankings in Figure 7.3 may, therefore, principally reflect behavioral tendencies—the readiness of the animal to attend to or to utilize visual information—rather than psychophysical capabilities—the sensitivity of the visual system to energy or differences in energy.

Inia geoffrensis is rated as markedly superior to *T. truncatus* in its ability to be controlled by auditory signals. *Inia geoffrensis'* natural habitats in turbid river waters favor the use of hearing. Though this is also true for *T. truncatus* on

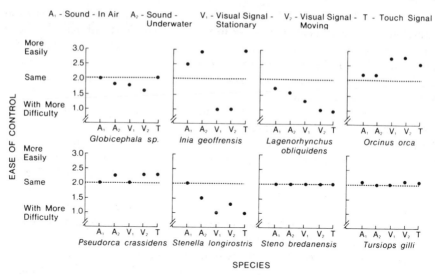

Fig. 7.4. Comparative ease of establishing control with different types of cues for eight species of cetaceans. Comparisons are in reference to *T. truncatus* (dashed line).

those occasions when its range includes the turbid coastal waters, its total range extends to the clearer offshore portions of the ocean system. In any event the moderate regressions of the visual system in *I. geoffrensis* might predict a compensating heightened sensitivity to sound cues. As with vision, there are some anomalies in the hearing ratings in that *O. orca*, with a narrower-band hearing system than *T. truncatus* (see Popper, this volume), nevertheless is rated as marginally better than *T. truncatus* in learning to use auditory cues. Again, this finding may reflect behavioral rather than psychophysical constraints. Similarly behavioral constraints may be the important variable in the lowered ratings of *L. obliquidens* and *S. longirostris* in the use of auditory cues (Fig. 7.4).

The ability to learn to use touch cues differs widely in the various species. *Stenella longirostris* apparently does not easily permit itself to be touched, as was seen in Table 7.4, and this may yield its low rating in Figure 7.4. *Inia geoffrensis*, *P. crassidens*, and *S. bredanensis* were all rated in Table 7.4 as permitting themselves to be touched more readily than does *T. truncatus*, but this translates into the more effective use of touch cues only for *I. geoffrensis* and *P. crassidens*. *Orcinus orca* is easily trained with touch cues according to the data of Figure 7.4, yet is rated as only equivalent to *T. truncatus* in allowing itself to be touched in Table 7.4. Once again, the responsiveness of an animal to specific types of training or to specific cues may depend on complex factors such as readiness to attend to various types of stimulation, timidity or fearfulness of the species, gregariousness, emotionality, and so forth, as much as on learning ability or on psychophysical factors.

Training Circumstances

There are numbers of training tasks that require concentrated effort with a single animal. The presence of other animals in the training tank can be distracting to the animal under study and greatly prolong the training time required. In some cases interference from other animals cannot be easily controlled, and the training effort cannot go forward. Training the animal in isolation from others may be necessary. In this situation, the trainer may seek to develop a strong social relationship with the segregated animal in order to fill its social needs and to facilitate the progress of training.

There are likely to be species differences in response to segregation, however. The questionnaire respondents were asked to rate the ease with which a given species trained when segregated from other animals as compared with the capabilities of *T. truncatus*. Alternate species were rated on a 5-point scale ranging from "much harder to train when alone than *T. truncatus*" to "much easier to train when alone than *T. truncatus*." The respondents rated all the species except two as training with comparable ease to *T. truncatus* when segregated. *Orcinus orca* was rated as training more easily than *T. truncatus* when segregated, while *S. longirostris* was judged to be much harder to train when alone than *T. truncatus*. The findings for *S. longirostris* probably reflect the great timidity of this species, its proclivity for conspecific associations, and its relative unresponsiveness to humans. In contrast, the killer whale displays little timidity and is highly responsive to humans.

Observational Learning

The observational learning and imitative capabilities of cetacean species have been reviewed elsewhere in this volume (see Herman, Chapter 8). Our respondents were asked to indicate for each species with which they were familiar how commonly it exhibited observational learning or imitated other animals. Ratings were on a 3-point scale ranging from "never observed to occur" to "frequently observed to occur." The responses for those species sampled were generally in keeping with descriptions in the literature. Both *P. crassidens* and *S. bredanensis* were rated at the upper end of the scale in frequency of display of observational learning or of imitative behaviors. *Lagenorhynchus obliquidens* and *O. orca* were each rated as between "frequently" and "occasionally" displaying this capability and *T. gilli* and *S. longirostris* as only "occasionally" displaying the capability. *Globicephala* rarely evidenced the capability and *I. geoffrensis* never.

The high rating for *L. obliquidens* is surprising in view of its low relative trainability ratings (Fig. 7.3), but Table 7.4 did indicate the propensity of this species for easily learning unison performances, which may have partially guided the ratings of our respondents. This same propensity for unison behavior occurs in *S. longirostris* and may similarly explain the species' "occasional" rating on the observational learning capability. *Globicephala* trains with relative difficulty (Fig. 7.3) and without any compensating tendency for unison performance (Table 7.4). Its low rating is consistent with those traits. *Inia geoffrensis* may be limited in opportunity for observational learning by being confined to fresh water tanks, where it is not exposed to other species of cetaceans and frequently it may be maintained alone in the tank.

DISCUSSION

Summary of Species Descriptions

There are some interesting overlaps in the behavioral traits and training characteristics of a number of cetacean species maintained in captivity, and at the same time there are many divergent characteristics. The 11 species rated by our respondents arranged themselves into three main clusters based on behavioral traits with minimal divergence within clusters and larger divergences between clusters. These clusters and the characteristics of species within clusters in many cases seem to correlate well with other descriptions of these animals in oceanarium settings, with descriptions of wild animals, and with various aspects of their trainability. Perhaps it should not be surprising that species-typical behaviors will be largely maintained across diverse environments and circumstances. Direct evidence by Martinez (1973) and Martinez and Klinghammer (1978) support this supposition. There are, after all, biological limits in the expression of behavior as in other traits. The captive circumstances may alter the hierarchies of displayed behaviors, exclude from occurrence certain behavioral categories such as cooperative foraging, amplify the occurrence or the form of certain other behavioral categories such as in-air vocalizing or complex leaping, and perhaps modify the social structure. But *I. geoffrensis* in captivity does not become a socially dependent species, and *S. longirostris* cannot be maintained and trained easily in separation from conspecifics. Spinner dolphins are timid in captivity and it is hardly stretching the imagination to view this as reflecting the state of the animal in the free-ranging case. This generalization is supported by Norris and Dohl's (this volume) discussion of the elusiveness and skittishness of some free-ranging spinner schools. In contrast, there are reports of wild killer whales readily interacting with boats and boaters (see review in Kinne, 1975) and this is consistent with their lack of timidity in the captive situation.

In summary, it would seem that observations on the behavior and training of captive cetaceans can provide insight into the behavioral tendencies of free-ranging animals within the limits noted. Correspondingly, behavioral observations of cetaceans in a free-ranging state can provide insight into the environmental and social requirements of maintained species and aid in the interpretation of their behavior.

Social Correlates

To the extent that the behavior of a species reflects its long-term adaptations to the ecological and social pressures of the natural habitat and its niche in the community of species, we can expect that a good deal of the clustering (Fig. 7.1) observed reflects convergent behavioral adaptations to similar pressures. What similarities can in fact be discerned? Among the possible social correlates of behavior herd size appears to bear some relationship to the observed species clustering. According to data provided in Yablokov et al. (1972), Tomilin (1967), and Kleinenberg et al. (1964) (also see Wells, Irvine & Scott, this volume), species belonging to Cluster B, *I. geoffrensis*, *D. leucas* and *Globicephala*, are often found in groups of less than 20 animals and in the case of *I. geoffrensis* often occur in groups of less than six or are even solitary. Some *Globicephala*, though, are occasionally found in herds as large as 100 or more.

The species of Cluster C, *L. obliquidens*, *S. longirostris*, and *D. delphis*, are frequently found in very large aggregations numbering more than 500 or even as many as thousands. Smaller groupings ranging from 6 to 100 are also found on occasion. The "superaggregations," however, typify these species.

In Cluster A, we find more often a mid-range group size. Between 7 and 20 animals and even up to 100 animals is very common. Very small groups are relatively rare except possibly for *S. bredanensis*. Aggregations of several hundred may occur occasionally in *Tursiops* sp. Superaggregations are very rare.

Thus, there is some tendency for groupings to be smallest in Cluster B, intermediate in Cluster A, and very large in Cluster C. Perhaps it is in the intermediate size groupings that we find dominance systems most competitive and interanimal aggression most strongly expressed. Dominance in very large groups would seem almost impossible to develop or maintain stably, but might be more easily established and remain very stable in smaller groups. Aggression, the usual concomitant of dominance systems, might be most strongly expressed in the intermediate-sized groups.

Knowledge of the social structure of wild cetacean herds in general is far from complete. At the present time the correlations discussed should be viewed as suggestions to be amended or even abandoned as new data otherwise dictate.

Ecological Correlates

The broad ecological characteristics of the habitat seem very diverse within clusters of species and cannot easily account for the obtained groupings. For example, in Cluster B *Globicephala* inhabits temperate, often pelagic, waters; *D. leucas* is found in cold Arctic waters, often in estuarine habitats; and *I. geoffrensis* is a freshwater tropical species. Conversely, across clusters we can find sympatric species rated diversely in behavior, such as *Globicephala* in Cluster B and *P. crassidens* in Cluster A, which are sympatric in much of their range.

However, there do seem to be parallels within species clusters in the feeding strategies and the type of food resources exploited. Yablokov et al. (1972) assigned many of the cetaceans to divisions created by Tomilin (1967) according to feeding type: major types included ichthyophages, or upper-layer pelagic fish eaters; benthoichthyophages, or bottom-fish eaters; teuthoichthyophages, or squid and fish eaters; and sarcophages, or flesh eaters. Both Tomilin (1967) and Yablokov et al. (1972) assigned species to these categories on the basis of stomach contents, nature of dentition, and when possible, reports of observers. Information on many species is incomplete, however, and in some cases an accurate placement has not been possible. Yablokov, for example, places the killer whale in the sarcophageous (flesh eater) category, yet acknowledges that the flesh of other aquatic mammals constitutes only the third most frequently consumed food, after fish and squid. Also it is not known, for example, whether different populations of species are divergent in their food preferences. The sample of *O. orca* (Table 7.2) on which the ratings in our survey were based come from what is believed to be a resident population of the Puget Sound area. In the light of this qualification the *O. orcas* in our sample are probably best classified as bentho- or benthoicthyophages. Yablokov groups *S. longirostris* and *D. delphis* together as icthyophages, species that also cluster behaviorally. Tomilin classifies the third member of this trio cluster, *L. obliquidens,* as also icthyophagic, in contrast to Yablokov who groups them with *Tursiops* sp. as benthic feeders. Tomilin states that *P. crassidens* feeds on benthic fish and squid, as do *Globicephala,* based on stomach contents of two animals. *Pseudorca crassidens*, however, it at least occasionally a predator on very large fishes. It has been observed feeding on large dolphin-fish or mahi-mahi (*Coryphaenas alata*) of 23 kg or more, in Hawaii (Pryor, 1975), and this species is also known to rob longlines of their catch worldwide (longlines are a device for hooking large, predatory tropical fish such as tuna and marlin). Thus feeding habits might place *P. crassidens* in the ichthyobenthophagic group, similar to our sample of *O. orca* with whom it clusters behaviorally.

Tomilin (1967) classifies *Tursiops* sp. as benthoichthyophagic. It should be noted that *Tursiops* sp. is an opportunistic feeder; in addition to taking bottom fishes the bottlenosed dolphin will chase surface fish, steal bait, follow shrimp boats to take trash fish, and so on, which may produce a behavioral profile closer to that of *O. orca* and *P. crassidens* than to that of Cluster C or Cluster B groups.

If behavioral clusters and feeding type are related, the clustering of *S. bredanensis* with *O. orca* suggests that this little-known animal might have unusual feeding habits, perhaps feeding on large prey; in fact, groups of these animals have recently been observed in Hawaii feeding on and sharing 9-kg mahi-mahi (William Curtsinger, personal communication).

Globicephala are described as eating squid but otherwise feeding similarly to *D. leucas*; again these animals cluster behaviorally. *Delphinapterus leucas* shares with *I. geoffrensis* the habit of feeding in rivers and both are classified as benthoichthyophages.

In summary, there appear to be interesting relationships between the behavioral clusters of the species discussed (Fig. 7.1) and feeding ecology. The social groupings (herd size) discussed earlier as also relating to the behavioral clusters may in part reflect evolved feeding strategies for exploiting the available resources. Benthic feeders, for example, tend to feed individually, while ichthyophages may use cooperative foraging and feeding methods to detect and capture the patchily distributed elusive schooling fish on which they subsist (see Norris & Dohl, this volume). Similarly, ichthyophages might well benefit more from traveling in larger aggregations than do benthic feeders.

Presently it is premature to propose too strongly that feeding strategies form the common denominator of the behavioral clusters identified in our analysis. They do, however, offer a tentative but plausible point of convergence among those species which cluster together. More conclusive analyses must obviously wait on additional social and ecological data for different species and populations of cetaceans.

ACKNOWLEDGMENTS

We thank the following for their response to the behavior and/or training surveys (asterisks indicate participation in the behavior survey only): Erica Abt,* Knie Kinderzoo, Rapperswill, Switzerland; Sonny Allen, Marineworld, San Francisco, Calif.; Dave Butcher, Sea World, Orlando, Fla.; David and Melba Caldwell,* Communication Sciences Laboratory, Gainesville, Fla.; Carol Chang,* Ingrid Kang, Marlee Penner, Karen Pryor, Diana Wong,* Sea Life Park, Honolulu, Hawaii; Larry Clifford,* Bob Osborne, Rich Phillips, Gary Priest, Dave Self, Bruce Stephens, Sea World, San Diego, Calif.; Tim Desmond, Marineland, Los Angeles, Calif.; Vladimir Gurevich, Rusty White, Hubbs-Sea World Research Institute, San Diego, Calif.; Blair Irvine,* National Fish and Wildlife Laboratory, Gainesville, Fla.; Kathy Krieger,* Seaaquarium, Miami, Fla.; Earl Murchison, Ralph Penner, Naval Ocean Systems Command, Kailua, Hawaii.

We also thank Debbi Forster, Karen Helrich, Carla Henderson, Gigi Kegg, and Candy Young, who helped with the clerical and statistical analysis of survey data. Vladimir Gurevich, Hubbs-Sea World Research Institute, and Forrest Wood, Naval Ocean Systems Center, made valuable contributions to our bibliography on captive cetaceans.

Lou Herman's contributions included the cluster analysis as well as the format of the associated Figures 7.1 and 7.2 and the reorganization of the behavior categories as given in Table 7.3.

Supported in part by the Psychology Department of San Diego State University, San Diego, California.

REFERENCES

Alpers, A. (1960). *Dolphins: The Myth and the Mammal*. Cambridge, Mass.: Houghton Mifflin.

Aminul Haque, A. K. M., Nishiwaki, M., Kasuya, T. & Tobayama, T. (1977). Observations on the behavior and other biological aspects of the Ganges Susu, *Platanista gangetica*. *Sci. Rep. Whales Res. Inst.*, **29**, 87-94.

Amundin, M. & Amundin, B. (1973). On the behavior and study of the harbour porpoise, *Phocoena phocoena*, in the wild. In: *Investigations on Cetacea*, Vol. V (G. Pilleri, Ed.), pp. 317-328. Berne, Switzerland: Institute of Brain Anatomy, University of Berne.

Andersen, S. (1969). Epimeletic behavior in captive harbour porpoise, *Phocoena phocoena* (L.). In: *Investigations on Cetacea*, Vol. I (G. Pilleri, Ed.), pp. 203-205. Berne, Switzerland: Institute of Brain Anatomy, University of Berne.

Andersen, S. (1976). The taming and training of the harbor porpoise, *Phocoena phocoena*. *Cetology*, **24**, 1-9.

Andersen, S. & Dziedzic, A. (1964). Behavior patterns of captive harbor porpoises. *Bull. Inst. Oceanogr. Monaco*, **63**, 1-20.

Andersen, S. & Pilleri, G. (1970). Audible sound production in captive *Platanista gangetics*. In: *Investigations on Cetacea*, Vol. II (G. Pilleri, Ed.), pp. 83-86. Berne, Switzerland: Institute of Brain Anatomy, University of Berne.

Anderson, J. (1878). *Anatomical and Zoological Researches: Comprising an Account of the Zoological Results of the Two Expeditions to Western Yunnon in 1868 and 1875*, Vols. 1 & 2. London: B. Quaritch.

Asper, E. D. (1975). Techniques of live capture of smaller cetacea. *J. Fish. Res. Board Can.*, **32**, 1191-1196.

Bailey, R. E. (1965). Training and open ocean release of Atlantic bottlenose porpoise. Naval Ordinance Test Station, Tech. Publ. No. 3838, pp. 1-17.

Bartman, W. (1974). Some remarks on the behavior and handling of white whales (*Delphinapterus leucas*) in captivity. *Aquat. Mammal.*, **2**, 16-21.

Bateson, G. (1974). Observations of a cetacean community. In: *Mind in the Waters* (J. McIntyre, Ed.), pp. 146-165. New York: Scribner's.

Bel'kovich, V. M., Krushinskaya, N. L. & Gurevich, V. S. (1970). The behavior of dolphins in captivity. Washington, D.C.: Joint Publication Research Service No. 50701, pp. 1-22.

Bigg, M. A. & Wolman, A. A. (1975). Live-capture killer whale (*Orcinus orca*) fishery British Columbia and Washington, 1962-73. *J. Fish. Res. Board Can.*, **32**, 1213-1221.

Bowers, C. A. & Henderson, R. S. (1972). Project Deep OPS: Deep object recovery with pilot and killer whales. Naval Undersea Center Technical Publication No. 306, pp. 1-86.

Breland, K. & Breland, M. (1966). *Animal Behavior*. New York: Macmillan.

Brown, D. (1960). Behavior of captive Pacific pilot whale. *J. Mammal.*, **41**, 342-349.

Brown, D. H. (1962). Further observations on the pilot whale in captivity. *Zoologica*, **47**, 59-64.

Brown, D. H. & Norris, K. S. (1956). Observations of captive and wild cetaceans. *J. Mammal.*, **37**, 311-326.

Brown, D. H., Caldwell, D. K. & Caldwell, M. C. (1966). Observation on the behavior of wild and captive false killer whales, with notes on associated behavior of other genera of captive delphinids. *Los Ang. Cty. Mus. Nat. Hist. Contrib. Sci.*, **95**, 1-32.

Burgess, K. (1968). The behavior and training of a killer whale. *Int. Zoo Yearb.*, **8**, 202-206.

Caldwell, M. C. & Caldwell, D. K. (1966). Epimeletic (care-giving) behavior in cetacea. In: *Whales Dolphins and Porpoises* (K. S. Norris, Ed.), pp. 755-789. Berkeley: University of California Press.

Caldwell, M. C. & Caldwell, D. K. (1967). Dolphin community life. *Los. Ang. Cty. Mus. Nat. Hist. Contrib. Sci.*, **5**, 12-15.

Caldwell, M. C. & Caldwell, D. K. (1972). Behavior of marine mammals. In: *Mammals of the Sea: Biology and Medicine* (S. Ridgway, Ed.), pp. 419-465. Springfield, Ill.: C. Thomas.

Caldwell, M. C. & Caldwell, D. K. (1977). Social interactions and reproduction in the Atlantic bottlenosed dolphin. In: *Breeding Dolphins, Present Status, Suggestions for the Future* (S. H. Ridgway & K. W. Benirschke, Eds.), pp. 133-142. Washington, D.C.: U.S. Marine Mammal Commission Report, MMC-76/07.

Caldwell, M. C., Brown, D. H. & Caldwell, D. K. (1963). Intergeneric behavior by a captive Pacific pilot whale. *Los Ang. Cty. Mus. Nat. Hist. Contrib. Sci.*, **70**, 3-12.

Caldwell, M. C., Caldwell, D. K. & Evans, W. E. (1966). Sounds and behavior of captive Amazon freshwater dolphins (*Inia geoffrensis*). *Los Ang. Cty. Mus. Nat. Hist. Contrib. Sci.*, **108**, 1-24.

Curtis, L. (1962). The Amazon dolphin *Inia geoffrensis* at the Fort Worth Zoological Park. *Int. Zoo Yearb.*, **4**, 7-10.

Defran, R. H. & Milberg, L. (1973). Tactile reinforcement in the bottlenosed dolphin. *Proc. 10th Annu. Conf. Sonar Diving Mammals.* Palo Alto, Calif.: Stanford Research Institute.

Devine, E. & Clark, M. (1967). *The Dolphin Smile.* New York: Macmillan.

Dixon, W. J. (1975). (Ed.). *BMD Biomedical Computer Programs.* Berkeley: University of California Press.

Essapian, F. S. (1953). The birth and growth of a porpoise. *Nat. Hist.*, **62**, 392-399.

Essapian, F. S. (1962). Courtship in captive saddle-back porpoises (*Delphinus delphis*). *Z. Saugetierkd.*, **27**, 211-217.

Essapian, F. S. (1963). Observations on abnormalities of parturition in captive bottlenosed dolphins (*Tursiops truncatus*) and concurrent behavior of other porpoises. *J. Mammal.*, **44**, 405-414.

Evans, W. E. & Bastian, J. (1969). Marine mammal communication: Social and ecological factors. In: *The Biology of Marine Mammals* (H. T. Andersen, Ed.), pp. 425-475. New York: Academic.

Ferster, C., Culbertson, S. & Boren, M. C. P. (1975). *Behavior Principles.* Englewood Cliffs, N.J.: Prentice-Hall.

Goforth, H. W. (1973). (Ed.). An introduction to marine mammals technology. Naval Undersea Center Technical Publication No. 29, 1-46.

Griffin, E. I. & Goldsberry, D. G. (1968). Notes on the capture, care and feeding of the killer whale (*Orcinus orca*). *Int. Zoo Yearb.*, **8**, 206-208.

Gurevich, V. S. (1977). Post-natal behavior of an Atlantic bottlenosed calf (*Tursiops truncatus*, Montagu) born at Sea World. In: *Breeding Dolphins, Present Status, Suggestions for the Future* (S. H. Ridgway & K. Benirschke, Eds.), pp. 168-184. Washington, D.C.: U.S. Marine Mammal Commission Report MMC-76/07.

Hall, J. D. (1970). Conditioning Pacific white-striped dolphins (*Lagenorhynchus obliquidens*) for open-ocean release. Naval Undersea Center Technical Publication No. 200, pp. 1-14.

Herman, L. M. & Arbeit, W. R. (1973). Stimulus control and auditory discrimination learning sets in the bottlenose dolphin. *J. Exp. Anal. Behav.*, **19**, 379-394.

Herman, L. M., Peacock, M. F., Yunker, M. P. & Madsen, C. J. (1975). Bottlenosed dolphin: Double-split pupil yields equivalent aerial and underwater diurnal acuity. *Science*, **189**, 650-652.

Holland, J. G. & Skinner, B. F. (1961). *The Analysis of Behavior.* New York: McGraw-Hill.

Irvine, B. (1970). Conditioning marine mammals to work in the sea. *Mar. Tech. Soc. J.*, **4**, 47-52.

Kellogg, W. N. & Rice, C. E. (1966). Visual discrimination and problem solving in a bottlenose dolphin. In: *Whales, Dolphins and Porpoises* (K. S. Norris, Ed.), pp. 731-754. Berkeley: University of California Press.

Kinne, O. (1975). Orientation in space: Animals: Mammals. In: *Marine Ecology*, VOl. II (O. Kinne, Ed.), pp. 709-916. London: Wiley.

Kleinenberg, S. E., Yablokov, A. V., Bel'kovich, B. M. & Tarasevich, M. N. (1964). *Beluga (Delphinapterus leucas): Investigation of the Species.* Jerusalem: Israel Program for Scientific Translations, (1969).

Kritzler, H. (1952). Observations on the pilot whale in captivity. *J. Mammal.*, **33**, 321-324.

Lang, T. G. & Pryor, K. W. (1966). Hydrodynamic performance of propoises (*Stenella attenuata*). *Science*, **152**, 531-533.

Layne, J. N. (1958). Observations on freshwater dolphins in the upper Amazon. *J. Mammal.*, **39**, 1-22.

Layne, J. N. & Caldwell, D. K. (1964). Behavior of the Amazon dolphin (*Inia geoffrensis* Blainville) in captivity. *Zoologica*, **49**, 81-108.

Leatherwood, J. S. (1977). Some preliminary impressions on the numbers and social behavior of free-swimming bottlenosed dolphin calves (*Tursiops truncatus*) in the northern Gulf of Mexico. In: *Breeding Dolphins, Present Status, Suggestions for the Future* (S. H. Ridgway & K. Benirschke, Eds.), pp. 143-167. Washington, D.C: U.S. Marine Mammal Commission Report MMC-76/07.

Marine Mammal Commission (1976). *Marine Mammal Names.* Washington, D.C.

Martinez, D. R. (1973). The behavior of the killer whale (*Orcinus orca* L.): A synopsis of field and captivity studies. M.S. thesis, Purdue University.

Martinez, D. R. & Klinghammer, E. (1978). A partial ethogram of the killer whale (*Orcinus orca* L.). *Carnivore*, Pt. 3 (Oct.), 13-27.

Matthews, L. H. (1971). *The Life of Mammals.* New York: Universe Books.

McBride, A. F. (1940). Meet Mr. Porpoise. *Nat. Hist.*, **45**, 16-29.

McBride, A. F. & Hebb, D. O. (1948). Behavior of the captive bottlenosed dolphin *Tursiops truncatus. J. Comp. Physiol. Phychol.*, **41**, 111-123.

McBride, A. F. & Kritzler, H. (1951). Observations on pregnancy, parturition, and post-natal behavior in the bottlenosed dolphin. *J. Mammal.*, **32**, 251-266.

Mitchell, E. (1975). (Ed.). Report of the Meeting on Smaller Cetaceans: Montreal, April 1-11, 1979. *J. Fish. Res. Board Can.*, **32**, 889-983.

Newman, M. & McGeer, P. (1966). The capture and care of a killer whale (*Orcinus orca*) in British Columbia. *Zoologica*, **51**, 59-69.

Norris, K. S. (1965). Trained porpoise released in the open sea. *Science*, **147**, 1048-1050.

Norris, K. S. (1967). Aggressive behavior in cetacea. In: *Aggression and Defense: Natural Mechanisms and Social Patterns* (C. D. Clemente and D. B. Lindsley, Eds.), pp. 225-241. Berkeley: University of California Press.

Norris, K. S. (1974). *The Porpoise Watcher.* New York: Norton.

Norris, K. S., Baldwin, H. & Sampson, D. J. (1965). Open ocean diving test with a trained porpoise (*Steno bredanensis*). *Deep-Sea Res.*, **12**, 505-509.

Norris, K. S. & Prescott, J. H. (1961). Observations on Pacific cetaceans of California and Mexican waters. *Univ. Calif. Publ. Zool.*, **63**, 291-402.

Penner, R. H. & Murchison, A. E. (1970). Experimentally demonstrated echolocation in the Amazon River porpoise, *Inia geoffrensis* (Blainville). Naval Undersea Center Technical Publication No. 187, pp. 1-25.

Pepper, R. L. & Defran, R. H. (1975). Dolphin trainers handbook. Naval Undersea Center Technical Publication No. 432, pp. 1-52.

Pilleri, G. (1970). The capture and transport to Switzerland of two live *Platanisia gangetica* from the Indus River. In: *Investigations on Cetacea*, Vol. II (G. Pilleri, Ed.), pp. 62-74. Berne, Switzerland: Institute of Brain Anatomy, University of Berne.

Pilleri, G., Gihr, M. & Kraus, C. (1970). Feeding behavior of the gangetic dolphin *Platanista gangetica* in captivity. In: *Investigations on Cetacea*, Vol. II (G. Pilleri, Ed.), pp. 69–78. Berne, Switzerland: Institute of Brain Anatomy, University of Berne.

Pilleri, G., Gihr, M. & Kraus, C. (1971). Further observations on the behavior of *Platanista indi* in captivity. In: *Investigations on Cetacea*, Vol. III (G. Pilleri, Ed.), pp. 34–42. Berne, Switzerland: Institute of Brain Anatomy, University of Berne.

Pilleri, G., Gihr, M., Purves, P. E., Zbinden, K. & Kraus, C. (1976). On the behavior, bioacoustics, and functional morphology of the Indus River dolphin (*Platanista indi*, Blyth, 1859). In: *Investigations on Cetacea*, Vol. VI (G. Pilleri, Ed.), pp. 1–141. Berne, Switzerland: Institute of Brain Anatomy, University of Berne.

Powell, B. A. (1966). Periodicity of vocal activity of captive bottlenosed dolphins (*Tursiops truncatus*). *Bull. South. Calif. Acad. Sci.*, **65**, 237–244.

Prescott, J. (1977). Comments on captive births of *Tursiops truncatus* at Marineland of the Pacific (1957–1972). In: *Breeding Dolphins: Present Status, Suggestions for the Future* (S. Ridgway & K. Bernirschke, Eds.), pp. 71–76. Washington, D.C.: U.S. Marine Mammal Commission Report MMC-76/07.

Pryor, K. W. (1973). Behavior and learning in porpoises and whales. *Naturwissenschaften*, **60**, 412–420.

Pryor, K. W. (1975). *Lads Before the Wind.* New York: Harper & Row.

Pryor, K. W. Haag, R. & O'Reilly, J. (1969). The creative propoise: Training for novel behavior. *J. Exp. Anal. Behav.*, **12**, 653–661.

Pryor, T., Pryor, K. & Norris, K. S. (1965). Observations on a pygmy killer whale (*Feresa attenuata* Gray) from Hawaii. *J. Mammal.*, **46**, 450–461.

Puente, A. G. & Dewsbury, D. A. (1976). Courtship and copulatory behavior of the bottlenosed dolphins (*Tursiops truncatus*). *Cetology*, **21**, 1–19.

Rice, D. W. (1977). A list of the marine mammals of the world. NOAA Technical Report, NMFS SSRF-711, pp. 1–15.

Ridgway, S. H. (1972). Homeostasis in the aquatic environment. In: *Mammals of the Sea: Biology and Medicine* (S. Ridgway, Ed.), pp. 590–747. Springfield, Ill.: C. Thomas.

Ridgway, S. H. & Benirschke, K. (1977). List of places where *Tursiops* are held. In: *Breeding Dolphins; Present, Status, Suggestions for the Future* (S. H. Ridgway & K. Benirschke, Eds.), pp. 8–30. U.S. Marine Mammal Commission Report MMC-76/07.

Ridgway, S. H., Scronce, B. L. & Kanwisher, J. (1969). Respiration and deep diving in the bottlenosed porpoise. *Science*, **166**, 1651–1654.

Ridgway, S. H. & Van Dyke, D. (1975). Handling and caring for dolphins. Unpublished ms. Naval Undersea Center, 1–24, San Diego, Calif.

Saayman, G. S. & Tayler, C. K. (1977). Observations on the sexual behavior of Indian Ocean bottlenosed dolphins (*Tursiops aduncas*). In: *Breeding Dolphins, Present Status, Suggestions for the Future* (S. H. Ridgway & K. Benirschke, Eds.), pp. 113–129. Washington, D.C.: U.S. Marine Mammal Commission Report, MMC-76/07.

Saayman, G. S., Tayler, C. K. & Bower, D. (1973). Diurnal activity cycles in captive and free ranging Indian Ocean bottlenosed dolphins (*Turiops aduncas* Ehrenberg). *Behaviour*, **44**, 212–233.

Shurepova, G. A. (1973). Aggressive behavior of captive (*Tursiops truncatus*) Barabasch. In: *Morphology and Ecology of Marine Mammals* (K. K. Chapskii & V. E. Sokolov, Eds.), pp. 150–153. New York: Wiley.

Skinner, B. F. (1938). *The Behavior of Organisms.* New York: Appleton-Century-Crofts.

Spong, P. & White, D. (1971). Visual acuity and discrimination learning in the dolphin (*Lagenorhynchus obliquidens*). *Exp. Neurol.*, **31**, 431–436.

Tavolga, M. C. (1966). Behavior of the bottlenosed dolphin (*Tursiops truncatus*): Social interactions in a captive colony. In: *Whales, Dolphins and Porpoises* (K. S. Norris, Ed.), pp. 718-730. Berkeley: University of California Press.

Tavolga, M. C. & Essapian, F. S. (1957). The behavior of the bottlenosed dolphin (*Tursiops truncatus*): Mating, pregnancy, parturition, and mother-infant behavior. *Zoologica*, **42**, 11-31.

Tayler, C. K. & Saayman, G. S. (1972). The social organization and behavior of dolphins (*Tursiops aduncas*) and baboons (*Papio ursinus*): Some comparisons and assessments. *Ann. Cape Prov. Mus. Nat. Hist.*, **9**, 11-49.

Tayler, C. K. & Saayman, G. S. (1973). Imitative behaviour by Indian Ocean bottlenosed dolphins (*Tursiops aduncas*) in captivity. *Behaviour*, **44**, 286-298.

Terrace, H. S. (1966). Stimulus control. In: *Operant Behavior: Areas of Research and Application* (W. K. Honig, Ed.), pp. 271-344. New York: Appleton-Century-Crofts.

Tomilin, A. G. (1967). *Mammals of the U.S.S.R. and Adjacent Countries*, Vol. IX. *Cetaceans*. Jerusalem: Israel Program for Scientific Translations.

Townsend, C. (1914). The porpoise in captivity. *Zoologica*, **1**, 289-299.

Turner, R. N. & Norris, K. S. (1966). Discriminative echolocation in a porpoise. *J. Exp. Anal. Behav.*, **9**, 535-544.

Walker, W. A. (1975). Review of the live capture fishery for smaller cetaceans taken in southern California waters for public display, 1966-73. *J. Fish. Res Board Can.*, **32**, 1197-1211.

White, D., Cameron, N., Spong, P. & Bradford, J. (1971). Visual acuity of the killer whale (*Orcinus orca*). *Exp. Neurol.*, **32**, 230-236.

Wood, F. G., Jr. (1953). Underwater sound production and concurrent behavior of captive porpoises, *Tursiops truncatus* and *Stenella plagiodon*. *Bull. Mar. Sci. Gulf Caribb.*, **3**, 120-133.

Wood, F. G. (1973). *Marine Mammals and Man: The Navy's Porpoises and Sea Lions*. Washington, D.C.: Robert B. Luce.

Wood, F. G. (1977). Birth of porpoises at Marineland, Florida, 1939 to 1969, and comments on problems involved in captive breedings of small cetacea. In: *Breeding Dolphins, Present Status, Suggestions for the Future* (S. H. Ridgway & K. W. Benirschke, Eds.), pp. 47-60. Washington, D.C.: Marine Mammal Commission Report, MMC-76/07.

Wood, F. G. & Ridgway, S. H. (1967). Utilization of porpoises in the Man-in-the-Sea Program. In: An Experimental 45-day Undersea Saturation Dive at 205 feet. ONR Rep. ACR-124.

Wyman, J. (1853). Description of a "white fish," or "white whale" (*Beluga borealis*, Less). *J. Boston Nat. Hist. Soc.*, **7**, 603-612.

Yablokov, A. V., Bel'kovich, V. M. & Borisov, V. I. (1972). *Whales and Dolphins*. Jerusalem: Israel Program for Scientific Translations.

CHAPTER 8

Cognitive Characteristics of Dolphins

Louis M. Herman

INTRODUCTION

Descriptions of the brain of the bottlenosed dolphin (*Tursiops truncatus*) and of some other Delphinidae uniformly remark on its large size, quality, and complexity (e.g., Flanigan, 1972; Jansen & Jansen, 1969; Morgane & Jacobs, 1972; Pilleri & Gihr, 1970). These descriptions hint at the intellectual potential of the species, which ultimately depends on brain structure and organization (Jerison, 1973; Passingham, 1975). However, it is behavior, not structure, that measures the intellectual dimensions and range of the species or what might be called its cognitive characteristics. By "cognitive characteristics" I refer to the intellectual specializations, capabilities, and limitations of a species as traced and measured through behavioral experiments and observations. Species may differ widely in cognitive characteristics, as in any biological trait, but the realization of a set of characteristics must surely rest on the same general laws that govern the evolution of other aspects of the biology of the species. Through the description of cognitive characteristics and comparisons among species having convergent cognitive traits one might hope to understand eventually the pressures selecting for intellect and how particular cognitive specializations are adaptive.

There have been few systematic attempts to describe fully the cognitive characteristics of any species, and even where such descriptions occur it is rare to find an experimenter considering their adaptive implications. Some exceptions are found in the extensive descriptions of the learning skills of great apes and some of the lesser primates (e.g., Fletcher, 1965; Rumbaugh, 1970; Warren, 1965, 1974) and in the analyses of intellect by Humphrey (1976) and Premack (1976, 1978). Premack pondered the ability of the great apes (Pongidae) to learn an imposed language and described a set of cognitive traits which he believed defined some of the dimensions of pongid intellect and which underlay the ability of these animals to learn some basic elements of language. These traits included the ability to draw causal inferences; the ability to carry out "intentional" behavioral acts; an advanced representational capacity (the ability to substitute symbols for referents); a well-developed memory system; and a capacity for forming second-order relationships (generalizing from a relationship between specific events to other related events). These characteristics were inferred from a whole host of experimentally derived behavioral measures of cognitive ability obtained by Premack and others. Humphrey (1976) argued that social skill and intellect go hand in hand and that social pressures and the need to compete with social peers select intensively for intellect.

Griffin (1976) used an apt term, "cognitive ethology," to refer to the behavioral study of animal cognition. The description of the species' cognitive characteristics and their adaptive value would presumably comprise an important segment of such study. The material in this chapter on the dolphin, mainly the bottlenosed dolphin, is a step toward defining the intellectual specializations, abilities, and limitations of this storied animal, as seen through experimental study and observation. Where possible, the findings and observations are related to the real world of the dolphin, to better understand the derivations and functions of the revealed intellectual characteristics. Also, where material is available, contact is made with findings on cognitive characteristics of primates to illustrate some convergent trends in cognition among these large-brained, highly social, but evolutionarily divergent taxa. Some of the characteristics of intellect of the simian primates, as described by Premack and others, will be seen to apply to dolphin intellect as well.

Approach to Studying Cognitive Characteristics

The "empty organism" view of the animal ought to be well behind us, overtaken by glimpses of the animal as an active, intervening agent in acquiring and processing information about its world. A major goal of this "cognitive" approach to animal behavior (cf., Gregg, 1971; Gregory, 1975; Griffin, 1976; Honig, 1978a; Menzel, 1978; Restle, 1975) is to describe through experiments how and how well the organism selects, encodes, stores, retrieves, analyzes, and manipulates information to achieve successful performance. In this manner the capabilities and the limitations of the animal as an information processing system are revealed.

This approach to animal cognition finds its origins in the early terminology and concepts of the information/computer sciences (Shannon & Weaver, 1949), and its model in the study of human cognitive processes (see early conceptions of Atkinson & Shiffrin, 1968; Broadbent, 1958, 1971; Neisser, 1967; and later material in Kantowitz, 1974; Lachman, Lachman & Butterfield, 1979; Murdock, 1974; Norman, 1976). Examples of information-processing approaches to the study or analysis of animal cognition can be found in papers by Andrew (1976), Maki and Leuin (1972), and Thompson & Herman (1977), and several of the chapters in Hulse, Fowler and Honig, (1978), and Medin, Roberts and Davis (1976).

Current views of human information processing generally describe a series of stages through which information passes, but also emphasize the flexibility of the processing system, with many of its operations programmable adaptively by the organism. It may be useful to borrow from concepts developed for understanding human information processing to understand processing in other animals. While processing might be expected to be more complex and flexible in humans because of the special characteristics and advantages of linguistic coding and manipulation, the basic template for processing may be laid down in most mammalian species in much the same way.

The first stage of processing of input information is at the sensory interface with the world, where selective attentional processes or "stimulus filtering" (Broadbent, 1958), reduce input flow to manageable amounts, by selecting items of particular relevance to the organism (Deutsch & Deutsch, 1963). Selection may take place at later stages in the processing of information, as well as at the early peripheral stages (see discussion in Norman, 1976).

The processing system involves several additional key stages or features beyond selection, including encoding and decoding operations (Melton & Martin, 1972), the storage and retrieval of information (Murdock, 1974), and classification and decision (Broadbent, 1971; Atkinson & Shiffrin, 1968). Encoding, the representation of one thing by another, may vary from an analogic perceptual replica or "icon" of the event to a complex abstraction of it. Storage processes may range from the simple short-term maintenance of information in "working" memory, to its more permanent registration in long-term memory through enrichment or elaboration by association with other materials already stored (Craik & Lockhart, 1972). At all stages, processing capacities are limited, but allocatable strategically by the organism to help meet processing demands, with some trade-off between amount processed and "depth" of processing (Atkinson & Schiffrin, 1968; Broadbent, 1971; Craik & Lockhart, 1972; Norman & Bobrow, 1975). The analogies of these concepts with some of the components ("hardware") and processes ("software") of computer systems are obvious, though the actual structure and operation of biological information-processing systems are still far from clear.

The cognitive capabilities and limitations of a species may vary with the type of afferent processed, so that it is misleading to speak of "the" information-processing system of the organism. Human studies suggest that auditory and visual information are processed through different memory subsystems (Baddeley & Hitch, 1974). Likewise, animals may be specialists in one or another sensory modality, and not only give principal attention to information arriving at that "dominant" modality but show greater processing capability within that modality (Bolles, 1973; Seligman, 1970; Warren, 1973; also see Fig. 10-10 in Wilson, 1975). In many cases, there may be little integration of information across processing centers for different afferents (but see review of cross-modal matching in Davenport, 1977). For the dolphin, an auditory specialist, we shall see that with special training visual items may be reencoded as auditory information and apparently transferred to auditory centers for deeper levels of processing.

Scope of Review of Dolphin Cognitive Characteristics

The materials on cognitive characteristics of dolphins reviewed in this chapter are organized into three sections. The first deals with capabilities for encoding, storing, and retrieving information from memory. The second analyzes conceptual processes—the ability to manipulate information productively, to form concepts, define relations among objects or events, develop efficient rule-governed response strategies, to mimic responses of others, and to learn through observation. The final section concerns the teaching of language to the dolphin and its capabilities for comprehending the elements and structure of an imposed language. The linguistic system of humans is undoubtedly the most complex form of information-processing activity in the animal world. Measures of the capabilities of dolphins for processing information at the linguistic level, like the language studies with some of the great-ape species, can give a fuller appreciation of their cognitive potential and limitations, and expand our concepts of the evolution of language and the relationship of human language to animal communication. The three areas, though treated separately, are not mutually exclusive. Memory is the bedrock on which other cognitive characteristics rest. Memory is intimately involved in the ability to form concepts and to learn rules, and both memory and representational abilities seem to be key determinants of linguistic potential (Premack, 1976).

Much of the material on memory and on conceptual processes emphasizes comparisons of information-processing characteristics of auditory and visual centers, though the processing of spatial information is also considered. The auditory system of the bottlenosed dolphin, and other cetaceans, is richly elaborated (see Popper, this volume). Auditory centers occupy a large portion of the huge cerebral cortex, though the location of these centers is still under study (Jansen & Jansen, 1969; Morgane & Jacobs, 1972; Pilleri & Gihr, 1970; also see review of Russian findings in Bullock & Gurevich, 1979). The amount of cortex devoted to audition or sound production seems to greatly exceed that necessary for effective echolocation, judging by the excellent capabilities of the very small-brained echolocating bats (e.g., Novick, 1973; also see Jerison, 1978; Pilleri & Gihr, 1970 for remarks on relative brain size and echolocating capabilities). Based on this "surplus" cortex, it may be expected that the auditory information-processing capabilities of the dolphins are sophisticated, efficient, and complex. While not considered a "visual" animal, the visual system of the bottlenosed dolphin is well developed and may contribute to many aspects of its life (Dawson, this volume; Madsen & Herman, this volume). The dolphin has good resolution acuity underwater and in air, is highly sensitive to light, especially at the shorter wavelengths appropriate to the underwater photic environment, and probably has good movement detection capabilities. Nevertheless, visual centers are not nearly as elaborate as auditory centers (Jansen & Jansen, 1969; Morgane & Jacobs, 1972; Pilleri & Gihr, 1970), and it might be expected that complex visual information-processing capabilities are correspondingly limited, as compared with auditory capabilities.

MEMORY PROCESSES

Different Memory Domains and Their Study

Though memory, in some form, is basic to most if not all animal life, its capacity, flexibility, complexity, and specializations may differ radically across species. The quality of memory and its specializations can be expected to be finely tuned to the degree and type of demand from the ecological and social world for remembering of events, the associations of events, or the interaction of events and behaviors.

Auditory, Visual, and Spatial Memory

The underwater world of the dolphin is filled with sound of both physical and biological origin. In its journey through this world, the dolphin hears almost the full range of acoustic energy present, save perhaps for the exceptionally low subsonic noise. The sound of shoaling water, the noisy chorus from the deep scattering layer, the croaks of fishes, the whine of an engine or the low rumble from a large ship, the "screams" of a killer whale, and the whistles, buzzes, and other sounds from its own or allied species may all have significance for the individual's biological or social well-being, and are to be remembered.

Add to this symphony of sound reaching the passive hearing system the requirement to learn and remember the echoes from ensonified targets, echoes that may differ depending on target composition, aspect, velocity, direction, or distance, and it becomes apparent that auditory memory may be a crucial development in the evolution of the information-processing capabilities of the dolphin. What are some of the developments that have taken place? How good is auditory memory in the dolphin? What kinds of information may be remembered most easily? What are the limitations of memory? The studies of auditory memory in the bottlenosed dolphin to be reviewed begin to provide some answers to these questions.

The more restricted development of the visual system of the bottlenosed dolphin, relative to the auditory system, and the constraints of the photic environment (see Madsen & Herman, this volume) can be expected to be paralleled by limitations in visual memory. While vision may support many important biological functions, as was already noted, it is likely to be more specialized than hearing, in the sense of being directed toward fewer classes of information. As will be seen, visual memory for form or for brightness variation seems poorly developed. Possibly, though we have not fully tested it, the major specialization of the visual system is in the processing of movement information.

The world of any animal is partly a world of place or location. While it is important to remember what a thing is, that is, its defining attributes, or when it occurred in time, its past locations in space can also have major significance. Food resources, affiliates, or predators may be associated with certain locations; these locations may also shift in time depending on the interplay of resources and exploiters or on environmental variables, making it important to update and remember last locations. Space may be a very complex concept in the three-dimensional sometimes featureless domain of movement of the dolphin. A few simple tests of spatial memory are carried out, however, to begin to describe memory for place.

Short-Term "Working" Memory

The recent interest in the behavioral study of animal memory (e.g., D'Amato, 1973; Honig, 1978b; Honig & James, 1971; Medin & Davis, 1974; Medin et al., 1976) has been spurred and guided in part by concepts emerging from the study of human short-term memory (e.g., see Murdock's 1974 review). Many of these concepts seem applicable to the understanding of animal memory.

The study of short-term memory includes the measurement and analysis of capabilities and processes for recording selected information faithfully and maintaining it over short intervals of time while awaiting a signal to respond. Short-term memory, in its active form, has sometimes been called "working" memory (Baddeley & Hitch, 1974; Honig, 1978b). The view in this chapter is that working memory houses both newly selected information and, where applicable, rules for operating on that information. At the simplest level the "rules" may derive from biological predispositions of the species to respond in certain ways (cf. Seligman, 1970). At successively more complex levels the rules may reflect the influences of past specific associational learning, or generalizations from such learning, or even indicate the construction of hypotheses about correlations or relationships among events. The complexity of rule that may be learned by an animal is in part a species variable. Evidence presented here shows that in some cases highly general rules, of the type described as second-order relations by Premack (1976), may be formed by the dolphin.

It is clear that working memory must play an essential role in determining the adequacy and flexibility of response of the animal to its world. All of the memory studies of the dolphin to be reported are tests of the capabilities and constraints of working memory when processing auditory, visual, or spatial items. Memory is measured through laboratory tests of delayed matching-to-sample or delayed discrimination abilities.

Delayed Matching and Delayed Discrimination Tests

Delayed matching-to-sample procedures tested for auditory and visual memory, and delayed discriminations for spatial memory. The auditory delayed matching procedures, except those used in testing memory for lists of sounds, required that the dolphin listen to a single sound, the "sample," and after an imposed delay interval demonstrate whether or not it still remembered the sound. This it could do in one of three ways. In conventional (unconditional) matching tests, it selected the sample sound from two alternative sounds appearing after the delay. In probe matching, it indicated whether a single alternative, the "probe," was or was not a match for the sample. Finally, in conditional matching tests, it again chose between two alternative sounds, but the match was not direct as in conventional matching, but symbolic (cf. D'Amato & Worsham, 1974), one of the two sounds being related to the sample through a learned association.

In list memory tests, also called serial probe-recognition memory tests, multiple sample sounds were given in list form. The dolphin was required to decide whether a subsequent probe sound matched *any* of the sounds on the list, or none.

Visual matching-to-sample was carried out analogously to auditory matching, except that visual rather than auditory sample and comparison stimuli were used. Only the conventional type of matching procedure was attempted. Visual delayed matching procedures have been used successfully in many recent tests of primate memory (e.g., D'Amato, 1973; Jarrard & Moise, 1971) and avian memory (e.g., Roberts & Grant, 1976).

For the delayed discrimination tests, the correct spatial location for a rewarded response on one of two or four alternative response paddles was signaled by a sound. However, the opportunity to respond to a paddle was withheld from the dolphin until an imposed time interval had elapsed.

These various procedures were used to explore the fidelity of memory over time, its capacity, the storage strategies used by the dolphin, the value of different types of memory retrieval cues or of mnemonic devices, and the stability of memory when input information was degraded or irrelevant information introduced during the imposed delay interval.

The Fidelity of Auditory Memory

Auditory information was easily remembered by the bottlenosed dolphin. Reviewed in this section are the results of tests of memory for the attributes of new sounds and for temporal-order information (the relative recency of occurrence of familiar sounds heard over many successive occasions). The effects on memory of different types of memory retrieval cues are also explored.

The animal tested throughout the auditory studies was an adult female bottlenosed dolphin, named "Kea." Kea's age was approximately six years for the earliest memory studies and ten years for the final studies completed. Psychophysical tests indicated Kea had excellent hearing abilities (Herman & Arbeit, 1972; Thompson & Herman, 1975; Yunker & Herman, 1974).

The sounds to be remembered by Kea included constant-frequency pure tones, frequency-, pulse-, or amplitude-modulated sounds, and complex mixtures of sounds, all electronically generated and projected through underwater speakers into her tank. All sounds were types not likely to be found in the dolphin's natural environment. Natural sounds were avoided since these may have already been stored in memory, or might have special significance for the dolphin, making interpretations difficult. Some electronic sounds were used over and over again and are referred to as "familiar" sounds; these sounds were undoubtedly eventually registered permanently in Kea's long-term memory, but by knowing their history of use we could better interpret the data. When using familiar sounds, Kea's ability to retain *temporal-order* (recency) information was tested, that is, her ability to update information and remember which familiar sound had been heard most recently. When using unfamiliar (novel) sounds, Kea's ability to register and retain the *attributes* of new acoustic materials was tested.

Memory for Sound Attributes

To test attribute memory (Herman & Gordon, 1974), a to-be-remembered novel sample sound was projected underwater for 2.5 sec. The sample appeared simultaneously at the two peripheral speakers shown in Figure 8.1. During projection of the sample, Kea waited in the "listening" area, and remained there until a delay (retention) interval of from 1 sec to over 100 sec had elapsed and a sequence of two comparison sounds was heard from the peripheral speakers.

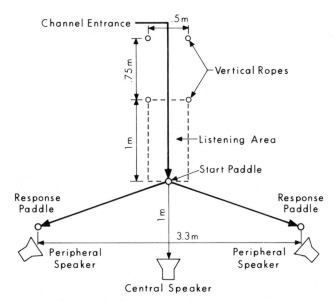

Figure 8.1. General schematic showing the arrangement of the underwater speakers, response paddles, and listening area used in testing auditory learning and memory in the bottlenosed dolphin. The apparatus was located in one quadrant of a 15.2-m diameter tank. The dolphin entered the listening area through the channel of ropes in response to a "call" sound and pressed the "start" paddle to turn the call sound off. The auditory learning or memory problem, consisting of a sound or a series of sounds from the underwater speakers, then began. At the end of each problem the dolphin pressed one of the two response paddles, to indicate its classification or recognition decision (see text).

The comparison sounds were each of 2.5-sec duration, the second beginning 0.5 sec after the first ended. One sound appeared at the left-hand peripheral speaker and the other at the right-hand speaker. Only one alternative (S+) matched the sample and a response to it, on the paddle adjoining the S+ speaker, earned a fish reward (Figure 8.2). The peripheral speaker at which S+ appeared and the temporal order of S+ and the nonmatching (S−) sound in the sequence of two comparison sounds were random variables.

Figure 8.3 shows that under these conventional delayed matching procedures Kea had little difficulty in registering and maintaining in memory the attributes of 346 different artificial sample sounds. As testing progressed, the range of delays following a sample was increased by adding longer and longer delays to the set of delays used, to study the temporal limits of working memory. During the first 171 sounds, with the longest delay limited to 24 sec, Kea's performance was creditable but variable, as if she were still learning the response rules of the matching problem. However, for the final 175 sounds there was a remarkable increase in Kea's precision and stability of performance. Only three matching errors were made over these 175 problems, although the delays in response were

Figure 8.2. The bottlenosed dolphin Kea pressing a response paddle adjacent to an underwater speaker. The speakers were University MM2PPS in early work and Chesapeake Instrument Company J9 transducers in later work. (Photograph by L. Herman.)

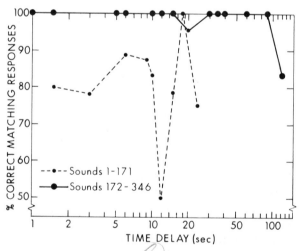

Figure 8.3. Memory of the dolphin Kea for 346 novel electronically generated sounds, early in training (Sounds 1-171) and later in training (Sounds 172-346). A novel "sample" sound was heard for 2.5 sec and 1 to 120 sec later, two alternative sounds were heard in a delayed matching-to-sample test. Various numbers of sounds were tested at each delay interval and the percentage of sounds correctly matched at each delay is indicated. From L. M. Herman & J. A. Gordon, *J. Exp. Anal. Behav.,* **21,** 1974, 19–26. Reprinted by permission.

in some cases as long as 120 sec. Clearly, attribute memory was barely taxed once the governing response rule was fully mastered.

Memory for Sound Recency

In Figure 8.4, some results for temporal memory are illustrated (Herman & Thompson, in prep.). Two familiar stimuli, a 4-kHz amplitude-modulated (AM) pure tone and a 25-kHz constant frequency pure tone, were used as sample sounds throughout these temporal memory tests. The procedures followed closely those used to test attribute memory, except that the sample duration was increased to 4 sec. The same conventional two-alternative matching test used for assessing attribute memory was given. Additionally, the effects of probe-matching and conditional-matching tests were studied.

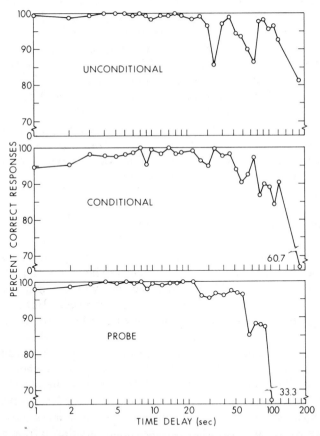

Figure 8.4. Ability of Kea to remember which of two familiar electronically generated sounds was most recently heard as a function of the memory cues or "prompts" available at the delayed matching test (unconditional, conditional, or probe, and as a function of the time delay since hearing a sound) (see text).

The top graph in Figure 8.4 gives results for the conventional (unconditional) matching test, as a function of the delay interval following the sample sound. Temporal memory remained stable and accurate through to delays of approximately 50 sec, but declined irregularly thereafter. The middle and lower graphs in Figure 8.4 show that memory was virtually unaffected by the variations in memory retrieval cues, or "prompts," introduced by the conditional (symbolic) matching or probe matching tests.

For the conditional matching tests, a 15-kHz FM sound was mapped onto the 4-kHz sample sound, as a symbol for it, and a 7-kHz FM sound onto the 25-kHz sample. The 15- and 7-kHz sounds then appeared as the two comparison stimuli during the memory test. Although Kea had difficulty in learning these conditional relationships, Figure 8.4 shows that, once learned, she could remember the prior sample about as well when memory was prompted by the symbolic conditional sounds as under conventional (unconditional) prompting.

For the probe condition, *either* a 4- or a 25-kHz probe appeared as a prompt at the memory test. The probe appeared randomly at one or the other of the two peripheral speakers, and on half the tests it matched the prior sample, and on the remaining tests it did not. For a "match" response, Kea responded directly to the probe, as in conventional matching; if it were a "nonmatch," she responded on the paddle adjoining the "empty" speaker, the one projecting no sound. For reasons that seem apparent, the probe procedure is sometimes called "Yes–No" or "Same–Different" matching (D'Amato & Worsham, 1974). The finding that the probe-matching results closely modeled those for the unconditional and conditional tests, demonstrated that Kea was as proficient at identifying a comparison sound that was different from the sample sound, as at recognizing one that was the same.

Attribute and Temporal Memory Compared

A comparison of Figures 8.3 and 8.4 suggests that temporal memory, though well developed and flexible, was somewhat less reliable over time than was attribute memory. The decline in temporal memory with time may mainly reflect its greater susceptibility to proactive interference. Herman (1975a), demonstrated that Kea's error rate was greater when two familiar sample sounds alternated across successive matching tests, than when they did not, showing that memory for the most recent sample sound was disrupted by the residual memory for a different sample sound used in the immediately preceding test. Discriminating "last" from "next-to-last" becomes increasingly difficult when listening over and over again in successive tests to the same two sample sounds, especially when a long delay interval follows the sample (cf. D'Amato, 1973). In contrast, when sample sounds are drawn from an open set of sounds as was the case for the results in Figure 8.3, memory is "refreshed" by the novelty of the new sound, while prior-heard sounds are never updated through being heard again.

Fidelity of Dolphin and Primate Memory Compared

Tests by others of temporal memory in monkeys (*Cebus* sp.,), using *visual* delayed matching tasks appropriate for these visually dominant animals, reveal characteristics comparable with those reported here for the auditory memory of the dolphin. Like the dolphin, the primates can retain relative recency information in working memory for extended periods (D'Amato, 1973; D'Amato & Worsham, 1972; Jarrard & Moise, 1971), and are able to use unconditional, conditional, and probe-type memory prompts equally effectively during matching tests (D'Amato & Worsham, 1974). As with the dolphin, the conditional matching problem is initially difficult to train. Finally, large proactive interference effects, traceable to changing or not changing the familiar sample stimuli across successive matching tests, are found in monkey temporal memory (Worsham, 1975).

Attribute memory has not been studied extensively in monkeys. However, Thompson (1975) tested four cynomolgus monkeys in a *visual* matching procedure modeled after the study of Herman and Gordon (1974) with the bottlenosed dolphin. Novel visual stimuli, such as buttons, hardware, or small toys, drawn from an open set, were used as samples. Matching performance of these relatively test-naive monkeys was excellent through several minutes' delay in response and remained at above-chance levels for delays as long as 8 to 12 min, a feat attainable in temporal memory tests only by monkeys having years of practice in the procedure (D'Amato, 1973; D'Amato & Cox, 1976). Also, Worsham (1975) noted that capuchin monkeys performed better when the size of the sample set was increased from two visual items to seven. In the limit, increasing the size of the sample set approximates the open set. Altogether, the results from visual delayed matching tests with monkeys are remarkably parallel to findings on auditory delayed matching with the dolphin, suggesting strong convergent trends in information-processing characteristics within the dominant sensory modality in these diverse taxa.

Studies of the working memory of the pigeon provide strongly contrasting data with the results for dolphin and monkey. The reliable maintenance of visual information by pigeons is generally on the order of only 10 sec or less following a single presentation of a visual sample (Berryman, Cumming & Nevin, 1963; Smith, 1967) and within these short delay limits proactive interference effects are apparently weak (Roberts & Grant, 1976).

Conditions That May Limit Auditory Memory

In the auditory memory results described so far, there were at most only minor decrements in memory over time. While these results underscored the potentials of auditory working memory, its limiting conditions remained uncharted. In what follows, the focus was on laying out some of the strictures on auditory memory, through procedures that diminished the integrity of input information or that introduced extraneous irrelevant stimulation into the retention period.

Attenuating or Altering Input Information

In two procedures, the characteristics of the familiar 4- and 25-kHz sample sounds were altered or attenuated (Herman & Thompson, in preparation). In the first procedure, the duration of a sample sound was reduced to only 3 msec, in contrast to the 2.5 to 4 sec values used previously, while the duration of the single probe sound used was maintained at 2 sec, close to the original value of 2.5 sec. In a second procedure, the auditory frequencies of the sample sounds were altered while those of the two alternative comparison sounds were unchanged. For both of these procedures, the AM component of the 4-kHz sound was deleted, making it an unmodulated pure tone like the 25-kHz sound.

EFFECTS OF SHORT SAMPLE DURATION.   Figure 8.5 shows that though the sample-sound duration was reduced to only 3 msec, Kea's matching accuracy remained at the same high levels as previously (Figure 8.4). The 3-msec value was chosen on the basis of immediately preceding threshold tests showing that the two familiar sample sounds could be reliably discriminated at this duration, given a 1-sec delay before the probe test. During the threshold testing, the sample duration was slowly decreased to its final value. Kea either learned to associate the sounds of shortened duration with the full-duration sounds, or else continued to recognize the sounds for what they were, after hearing only 12 or 75 cycles of the 4- or 25-kHz pure tones, respectively. The continued excellence of Kea's delayed matching performance with these brief sample sounds can be construed as another demonstration of her ability to conditionally match sample and comparison sounds that differ in some dimension.

EFFECTS OF FREQUENCY SHIFTS.   The auditory frequency of one or the other of the two familiar sounds was changed, by progressively increasing the frequency of the 4-kHz pure-tone sample or by decreasing that of the 25-kHz pure tone, as in

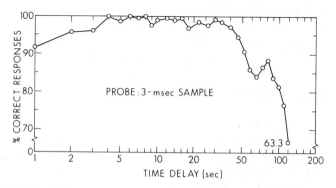

Figure 8.5.   Probe matching: Auditory memory of the bottlenosed dolphin Kea unaffected when the duration of the sample sound was reduced to 3 msec. Compare with Figure 8.4.

a stimulus generalization procedure. The progressive frequency changes were continued until errors were made to a probe comparison sound having one or the other of the original frequency characteristics of the altered samples. Under these procedures, the 4-kHz sound could be increased to between 18 and 22 kHz or the 25-kHz sound reduced to 10 to 12 kHz before performance dropped below the level of 75% correct responses. Clearly, Kea easily generalized the sample sounds along the frequency dimension, making errors only when one sample sound approached "uncomfortably" close to the other in frequency. Her limitations were therefore only in discrimination and not memory. Once again, changing the sample characteristics but not the probe restructures the problem as a conditional matching test.

COMPARISON WITH PRIMATE MEMORY.   Two studies with primates examined for effects of reduced sample duration on matching accuracy, with conflicting results. D'Amato and Worsham (1972), using capuchin monkeys, found that performance was sustained when sample durations were reduced to only 75 msec. Devine, Jones, Neville, and Sakai (1977) tested rhesus monkeys using sample durations between 100 msec and 4 sec, and found that matching performance decreased at shorter durations. Devine et al. may have put their finger on the problem in noting that their subjects had fewer training trials with short durations than did those of D'Amato and Worsham, implying that their subjects were less familiar with the relationships between shortened stimuli and those of longer duration. This view seems in keeping with the good results obtained with the dolphin Kea, who was given extensive training with progressively shorter durations during the threshold testing procedure. Additionally, Kea was very familiar with the full-duration sample sounds through their use in earlier studies encompassing thousands of trials. In contrast, when we attempted to reduce the duration of unfamiliar novel sounds abruptly to 250 msec (Herman & Gordon, 1974), Kea's matching performance decreased markedly.

The available pigeon data show that performance falters with shorter sample durations regardless of training level and degree of familiarity of the samples (Maki & Leith, 1973; Roberts & Grant, 1974, 1976), again providing a point of departure of pigeon memory from that of monkey and dolphin.

Interference During the Delay (Retention) Interval

The described changes in input information had little effect on auditory memory capabilities, adding to the appreciation of the flexibility of the auditory memory system. An additional study (Herman, 1975a) focused on the effects of "irrelevant" auditory stimulation inserted into the delay interval during auditory matching tests. The material is irrelevant, since no response to it is required. Retroactive disruption of working memory through irrelevant materials appearing during the retention period has been studied extensively in humans and is generally a potent variable (Deutsch, 1970; Herman & Bailey, 1970; Reitman, 1971).

For the retroactive interference tests of the dolphin, two familiar sample sounds, a 7.5-kHz pulsed pure tone and a 3-kHz FM tone, were used in an unconditional two-alternative matching procedure. Matching performance was tested with and without the insertion of irrelevant 1- or 10-kHz constant-frequency pure tones into the 15-sec retention interval. There was virtually no effect on matching ability if the irrelevant sounds were of 4-sec duration, regardless of whether they were imposed at the beginning or end of the retention interval. When their duration was increased to 13 sec, so that nearly the entire retention interval was occupied by the irrelevant sounds, there was an average 17% reduction in correct performance (Fig. 8.6). It seems clear that the duration of interference, but not its location within the retention interval, can be an important disrupting variable on dolphin auditory memory. Apparently, Kea did not or could not learn to avoid processing the irrelevant sounds of long duration, judging by the indications in Figure 8.6 that the interfering effect did not diminish over successive testing sessions. Perhaps this illustrates the saliency of auditory materials for the dolphin, and that when actively maintaining auditory information of one type, auditory input pathways remain "open."

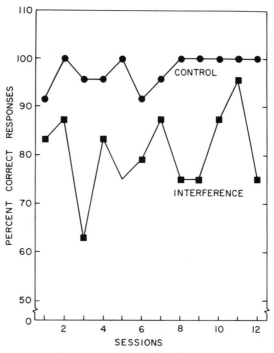

Figure 8.6. Auditory delayed matching ability of Kea disrupted following insertion of a 13-sec duration 1- or 10-kHz "interference" tone into a 15-sec retention interval. Control condition shows performance with interfering sounds absent. From L. M. Herman, *Anim. Learn. Behav.*, **3**, 1975, 43–48. Reprinted by permission.

It may also indicate that maintenance of information in working memory demands an active "rehearsal" effort, and that losses in information occur if attention is diverted from the maintenance task for a substantial fraction of the waiting interval. By this analysis, irrelevant *visual* information might interfere little with auditory memory, because of its lack of saliency for the dolphin, or because there is little central interaction between input from the two modalities (cf. Worsham & D'Amato, 1973). However, we have not tested for its effects. Informally, we have noted that occasional unplanned visual irrelevancies during auditory matching tests, such as a sea turtle breaking out of its pen and idly swimming by the patiently waiting Kea, who "jaw snapped" at it, did not result in a matching error after the delay period was completed.

COMPARISON WITH OTHER SPECIES. Irrelevant visual materials, but not auditory items, interfere with visual matching accuracy in monkeys (Worsham & D'Amato, 1973). Also, like the case for the dolphin, duration of irrelevant stimulation was more important as a memory disrupter than was its location in the retention interval (Etkin, 1972; Moise, 1970). Again, results for primate visual memory are strikingly similar to findings for dolphin auditory memory. Here, in the discussion of interference, the first clear case occurs of a functional similarity of pigeon memory with that of dolphin and monkey, in that the duration of irrelevant stimulation is more disruptive to visual matching performance than is its location in the retention interval (Maki, Moe & Bierley, 1977; Zentall, 1973). It is hard to get excited about this single point of similarity, since it may merely show that a potent variable, especially one that limits performance, easily transcends species boundaries. Also, the bulk of the data continue to suggest that the memory system of the pigeon is otherwise very different from that of dolphin and monkey (cf. Roberts & Grant, 1976).

Memory for Lists of Sounds

The single-item matching-to-sample tests described so far have revealed much about the extensive development of the auditory memory system of the dolphin. However, it is to be expected that auditory memory in the dolphin is structured for bigger and better things than single-item retention, and that some of its more interesting characteristics, like capacity or information-handling rate, are not taxed by the single-item memory test. To assess capacity and processing limitations for information occurring at high rates, the ability of the dolphin Kea to process and remember multiple sample sounds serially presented, as in a list, was tested (Thompson & Herman, 1977, and unpublished data). The variables manipulated were the length of the list and the spacing and duration of the list items. Study of list memory (cf. Sternberg, 1975) and of the effects of spacing and duration helped in drawing inferences about the capacity and structure of auditory working memory, and about the memory storage and retrieval strategies adopted by Kea.

In these tests of list memory, Kea listened to a sequence of as many as eight novel sample sounds, and then classified a probe sound occurring 1- or 4 sec later as either a member of that list ("Yes" response) or not ("No" response). The list of samples was drawn from a pool of 800 different sounds and appeared at the central speaker of Figure 8.1. The probe sound was projected from one or the other of the peripheral speakers. As in the one-sample probe matching tests, a Yes (match) response was indicated by responding to the peripheral speaker projecting the probe, and No (nonmatch) by responding to the alternate "empty" speaker. "Yes" indicated that the probe matched one of the list items, "No" that it matched none. This serial memory task is complex, and in demonstrations given (e.g., Herman, 1975b), human listeners may have difficulty in correctly categorizing probes from six-sound lists adjusted for the human hearing range. Nevertheless, the dolphin Kea learned the task easily.

Testing Storage Capacity and Storage Method

To teach the first multiple-sample task, a second sample sound was added to the one-sample probe-matching task, to form a two-item list. The added sound was initially of short duration, but over 30 to 50 training trials it was gradually lengthened to the full 2-sec duration of the first sound, while Kea continued to respond correctly. A series of approximately 400 delayed-matching tests were then given, each test consisting of a new pair of sounds. The probe sound was different from either sound of the pair for half the tests. On the remaining tests, it matched the first sound in the list as often as it matched the second. By this means, the effects of recency of a list sound on its later recognition could be assessed.

Using the same procedures, a third sound, a fourth, and so on, were added successively to form increasingly longer lists. Each singly incremented list length was tested fully before advancing to a longer list. This process proceeded smoothly with little response error other than that apparently governed by Kea's memory limitations. Figure 8.7, from Thompson and Herman (1977), illustrates how memory was affected by list length for lists of 1 to 6 sounds, and by the recency of occurrence of sounds in the list. Data for 1- and 4-sec delays were combined, since there were no significant effects of delay duration. The forms of the curve relating correct recognitions to serial position of the list item probed were very similar across all lengths of list. Part A of Figure 8.7 shows that correctly recognizing the probe as a match for a list item was easier the more recently the item appeared in the list. This "recency" effect was similar to that exhibited in data on human memory (e.g., Wickelgren & Norman, 1966) and monkey memory (Davis & Fitts, 1976; also see Devine & Jones, 1975), as obtained through analogous multiple-sample probe-recognition tasks.

The asymptote of the recency curves in Part A suggested a memory span of between four and five sounds for Kea in this type of task. The overlapping curves for the different length lists show that it mattered little how many sounds preceded a given sound, only how many followed it. A good physical analogy for

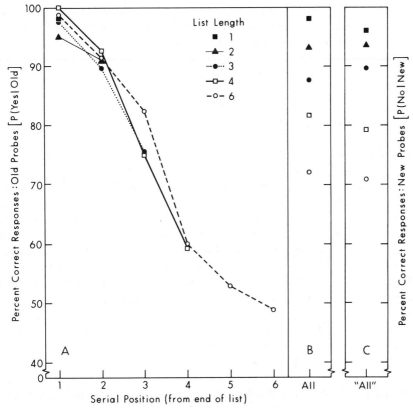

Figure 8.7. Memory of Kea for lists of up to six electronically generated sounds. Memory was sampled by a probe at the end of each list which with equal probability was either one of the sounds in the list, or not. Part A shows correct recognition performance for "old" probes, which matched items in the list, as a function of the recency of occurrence (serial position) of the list sound probed. Serial Position 1 is the most-recently heard (last) list item. Part B summarizes results over all serial positions combined. Part C shows the percentage of correct recognitions of "new" probes, which did not match any list item, as a function of list length. (From R. K. R. Thompson & L. M. Herman, *Science,* **195,** 501–503. Reprinted by permission. Copyright © by the American Association for the Advancement of Science.

this type of memory effect may be that of a serial buffer mechanism of limited item capacity, in which new items displace or degrade the quality of older items (Atkinson & Shiffrin, 1968).

Strategy for Decision

There is a great deal of uncertainty in this memory task. Old sounds, especially those appearing early in the list, may be misclassified as new, and new sounds may be confused with and misclassified as old. Kea's misclassification probabilities understandably increased with list length, but remained unbiased throughout. That is, old and new items were misclassified at about the same rates within each list length (Fig. 8.7, Parts B and C). An unbiased error distribution of this form can be generated by use of a maximum-likelihood decision rule for doubtful or uncertain classifications. Maximum likelihood rules appear to characterize human decision performance in this type of memory task (Murdock, 1974).

Effect of Rate and Spacing of Sounds

Figure 8.8 shows unpublished data on the effects of the rate and spacing of list items, for lists of six and eight sounds. Condition A, using sounds of 2-sec duration with 0.5-sec spacings, was a continuation of the procedures yielding the results of Figure 8.7. Conditions B and C both used sounds of 1-sec duration with either 0.5-sec spacings (Condition B), or with a greater spacing (Condition C) to increase the total duration of the list to that of A. Within the limits of the values tested, there was little effect of these duration and rate changes except for a tendency for Condition B to yield somewhat better recognition for the 4th and 5th list items. From these results, it may be concluded that a principal limitation on auditory working memory is the number of discrete items that may be maintained in storage (buffered) in parallel, and not the rate at which such storage may take place. This result, like that for the decision strategy adopted, is consistent with data reported for human subjects tested on serial probe-recognition tasks (cf. Waugh & Norman, 1965).

The Constraints of Visual Working Memory and Release from Constraint

The visual memory ability of the bottlenosed dolphin, as measured in tests at our laboratory, was remarkably limited as compared with the extensive auditory memory abilities just described. The subject of our visual memory studies was a female dolphin named "Puka," an animal having good visual resolution capabilities in air and underwater (Herman, Peacock, Yunker & Madsen, 1975) and good brightness sensitivity (Madsen, 1976). Nevertheless, Puka experienced extraordinary difficulty in learning to match one of two visual comparison stimuli to a simple visual form/brightness sample. The difficulty was present even when the sample stimulus was displayed concurrently with the comparison stimuli, in undelayed matching tests.

All visual stimuli were shown behind a clear-glass 46-cm high by 92-cm wide underwater window in Puka's tank. Puka looked at the visual materials through the window, at a viewing distance of 1 m or less, an optimal underwater viewing condition (Herman et al., 1975). The air interface between the window and the visual stimuli acted as a barrier against echolocation cues. The stimuli were a white circle ("W") and a black circle ("B"), each of 8.3-cm diameter. The white circle was mounted on a solid black 15-cm high by 30-cm wide background, and the black circle on the same sized background consisting of an immediately surrounding 15 by 15 cm white area extending into two laterally adjoining black

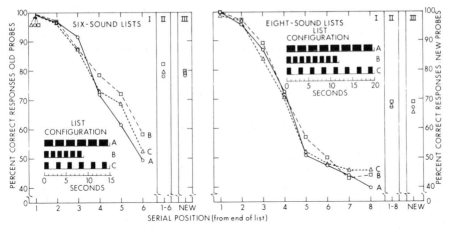

Figure 8.8. The effect of rate and spacing of sounds in lists of six and eight sounds on Kea's recognition memory. Sounds were each 1-sec long in List Configuration A and 2-sec long in B and C. The procedures were the same as those described for Figure 8.7.

areas, each 7.5 cm wide by 15 cm high. The different background configurations contributed form as well as brightness variation to the stimuli.

The stimuli were front-lighted by a combination of artificial and natural illumination. The visual sample, W *or* B, was displayed in the upper half of the window, while the two comparisons stimuli, W *and* B, appeared side by side in the lower half, an unconditional matching procedure. Puka had previously learned to discriminate between these same two stimuli and to reverse her responses to them with some reliability if the reward values were reversed, as discussed later in the section on conceptual processes. However, she could not learn to match either of these stimuli to a sample, though a variety of visual cueing techniques were used over a training period encompassing thousands of trials. Other laboratories have reported similar insurmountable problems in attempting to teach bottlenosed dolphin subjects the visual match-to-sample problem (e.g., L. Milberg, personal communication), indicating that we were not dealing with an idiosyncratic limitation of a given bottlenosed dolphin subject, but with a constraint in processing capability of the species when manipulating certain types of visual information.

Releasing the Constraint

We were eventually successful in "unlocking" this visual-learning constraint and in testing visual memory capabilities by using auditory cueing (Herman, Antinoja & Forestell, in preparation). In concept, visual stimuli W and B were each assigned an auditory "name." We speculated that auditory naming might transfer the visual matching problem to the enriched auditory domain through the central representation of the visual stimuli by auditory codes. Training was carried out by presenting the name concurrently and continuously with the visual sample stimulus, and then again with the matching stimulus of the comparison pair that immediately followed. An underwater speaker, located centrally as in Figure 8.1, projected the name of the sample. The correct comparison stimulus was cued spatially and acoustically by projecting its name from a peripheral speaker adjoining it. Gradually, over successive blocks of training trials, the spatial cue was deleted by moving the two peripheral speakers together centrally and then transferring all sounds to one speaker. The names themselves, including that for the sample, were then deleted by gradually shortening their duration until they were absent entirely. The result was that visual matching performance was now achieved by Puka and, more importantly, maintained after complete deletion of the name cue. Furthermore, matching continued successfully with increasing delays between the visual sample and the pair of visual comparison stimuli. Figure 8.9 shows that under these procedures visual matching performance, with names deleted, was close to or better than 80% correct through to delays of 30 sec, and then decreased abruptly at 34-sec delay. This delay limit was shorter than that achieved with the comparable auditory-matching tests described earlier for Kea (Fig. 8.4), but when contrasted with the utter failure we were experiencing previously with direct visual matching training it was a notable success.

The mechanisms for this release from a visual information-processing constraint are speculative. The auditory naming might have served to route the reencoded visual information through the more highly articulated auditory centers, as was supposed, or it might have served to demonstrate the problem solution by allowing Puka to learn the "rules of the game" with the simpler (for

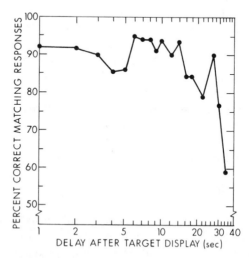

Figure 8.9. Memory of the bottlenosed dolphin Puka for the most recently displayed visual target of a pair of targets differing in brightness/form, as tested by a matching-to-sample procedure. Puka was unable to match visually until each target was associated with a different auditory "name." Data shown are results for subsequent delayed visual matching-to-sample testing with the name cue deleted.

the dolphin) auditory matching problem contained within our procedures. Though other work on the dolphin shows that the transfer of rules within the auditory modality is feasible for the dolphin (see later), many animal learning studies show that cross-modal transfer can be exceedingly difficult (Davenport, 1977). Consequently we favor the reencoding hypothesis.

Our findings on release from constraint prompt many questions about biological limitations on learning. Would the application of release techniques allow the demonstration of cognitive capabilities of other animal species for tasks that were thought too difficult previously? For example, monkey subjects seem unable to learn auditory matching, though quite accomplished at visual matching (Worsham & D'Amato, 1973). Monkeys are visually-dominant animals, but additionally have a well-developed auditory peripheral system (Stebbins, 1971), the reverse of the case for the dolphin. Could monkeys perform auditory matching if given visual encoding training? Could rats learn otherwise unattainable complex visual problems through olfactory encoding (cf. Slotnick & Katz, 1974)? Are there implications in release techniques for human subjects having sensory or learning handicaps (cf. Sidman, Osborne, & Wilson-Morris, 1974)?

Study of the phenomenon of release from constraint could have broad significance for the fuller understanding of the nature, extent, and locus of biological constraints. Is it the case, for example, that secondary sensory modalities serve principally as signal systems for environmentally significant events while the primary modality is used additionally for more cognitively complex manipulations of the world? Are there long-term effects of release procedures on functional capabilities within a secondary sensory channel? What do the observed constraints and the release effects suggest about the species' perceptual and cognitive world? What types of tasks are amenable to release? Can constraints be better understood in terms of the degree to which they may be released? Is the degree to which a species can be released from learning constraints an index of its cognitive flexibility or capacity? These and similar questions urge the further study of this important research area.

Spatial Working Memory

In some species, spatial memory is extraordinarily specialized for a few biologically important tasks. For example, the female digger wasp can remember over several hours the locations of multiple nesting holes that require restocking with food, after a single morning inspection of their status (described in Tinbergen, 1951). Nectar-feeding species of Hawaiian honeycreeper birds avoid revisiting flower clusters from which they have already sipped nectar while foraging in a tree full of blossoms (Kamil, 1975). Laboratory rats avoid reentering arms of an eight-arm radial maze that they have previously visited and depleted of food (Olton & Samuelson, 1976). Chimpanzees avoid reexploring sites where they uncovered food hidden by an experimenter, and search instead for new sites (Menzel, 1975).

Other than these types of highly adaptive specializations, spatial working memory in many species may be limited and resistant to complex reencoding operations. This constraint is best illustrated in studies of delayed spatial response (e.g., Miles, 1971). In this task, which is much like the shell game of the carnival, but more honest, one of several spatially separated food wells is baited and recovered, while the animal watches. After a delay, the animal is allowed to respond to one and only one of the covered wells. Correct responding—obtaining food—typically depends on the animal orienting toward the baited well during the delay period (Fletcher, 1965). As the number of alternative wells increases, delayed response performance decreases rapidly, especially for rats and cats, but less so for monkeys, much less so for human children, and hardly at all for human adults (Miles, 1971). One may suppose that as the ability to represent spatial locations by symbolic codes develops, spatial uncertainty can be reduced to manageable levels. That rats have great difficulty in the delayed response task as the number of response alternatives increases (Miles, 1971), but not in the eight-arm radial maze task (Olton & Samuelson, 1976), suggests that these animals are much better at remembering where they have been than where they must go. Perhaps that is a limitation on us all.

Delayed Spatial Discriminations in Dolphin

CONDITIONAL CASE. Thompson (1976) studied the ability of the bottlenosed dolphin Kea to remember which of two or more spatially separated paddles to press for food reward when the correct location had been signalled earlier by one of two different sounds from a central speaker (Fig. 8.1). In the initial condition studied, conditional delayed discrimination, there were two pairs of paddles, one to the left and the other to the right of the dolphin as it waited in the "listening" area (Figure 8.1). Within each pair, one paddle was positioned forward of the listening area but relatively close to it diagonally, while the second was further away, at a right angle to the listening area. A familiar 25-kHz pure tone signaled Kea to respond on a "nearer" paddle, while an equally familiar 4-kHz pure tone (see delayed matching studies) indicated a "far" paddle. One or the other sound was heard for 2.5 sec, but neither indicated which pair, left or right, to choose from. After an imposed delay interval, the pair uncertainty was removed by an auditory cue from either a left or right peripheral speaker (Fig. 8.1), and Kea was given the opportunity to respond. This delayed cueing procedure guarded against Kea's use of any simple orienting or postural response to bridge the delay interval, and may have required that the spatial information, "near" or "far," be encoded indirectly in some nonspatial form.

Figure 8.10 shows the results. Performance remained above the 90%-correct level through to 16-sec delays and decreased to between 80 and 84% correct for delays from 22 to 30 sec, the longest delays that met with above-chance levels of success.

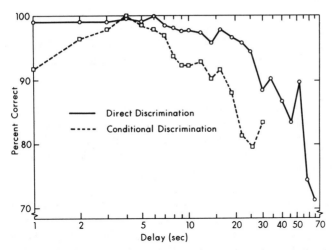

Figure 8.10. Memory of the dolphin Kea for a spatial response on one of two or four spatially separated response paddles. The paddle to be selected was cued by a nonspatial sound but execution of the response was delayed for the delay intervals shown. The conditional discrimination required a choice from among four paddles, the direct discrimination from among two. See text. (Data from Thompson, 1976.)

DIRECT SPATIAL DISCRIMINATION. In a second condition, direct delayed discrimination, also represented in Figure 8.10, the pair uncertainty was removed a priori. Only the left-hand pair *or* the right-hand pair of paddles was present in the water. The sound from the central speaker signaling "near" or "far" was thus unambiguously associated with one and only one paddle. However, Kea was not permitted to respond until hearing an auditory cue from the peripheral speaker adjoining the pair. Before long, Kea developed stereotyped postural responses during the delay interval, either moving slightly forward in the listening area, or not. Moving forward was highly predictive of a response to the near paddle. These postural responses, a form of overt spatial coding, bridged most of the delay interval and were likely responsible for the almost doubling of the delay limits reached, at the 80 and 90% correct criterion levels, compared with limits observed under the conditional discrimination. Though spatial information may be maintained reasonably well without use of postural or orientational cues, given the mnemonic value of such cues it is not surprising that they develop readily whenever possible. That postural cues do help the dolphin to remember the required spatial responses hints at the same types of limitations on representing spatial information abstractly that affect many animals tested in delayed response tasks. However, the substantial conceptual and procedural differences between the delayed-response tests given other animals (Miles, 1971) and the delayed discrimination tests given here, restrain any easy generalization.

Overview

In summary, as measured by delayed matching-to-sample tests, the auditory working memory of the bottlenosed dolphin was impressively faithful over time, and appeared to be supported by flexible and sophisticated memory control process (cf. Atkinson & Shiffrin, 1968). The physical attributes of almost any electronically generated sample sound, heard for only a few seconds, could be matched correctly against pairs of alternative comparison sounds after delays as long as 2 min, the longest retention periods tested. Memory for temporal-order information, or the relative recency of occurrence of two familiar sample sounds, was only slightly less faithful over time than was memory for sound attributes.

Retention was unaffected by the form of the memory prompts used in the auditory matching tests. Memory for the most-recently heard of two familiar sample sounds, normally assessed through a choice between the two familiar sounds after a delay interval, was unaffected when prompted by only one or the other of the two comparison sounds in a probe-recognition (Yes–No) procedure, or when prompted by two conditionally-associated comparison sounds. The conditional sounds were different from the familiar sample sounds but linked to them on a one-to-one basis through training. The accuracy obtained under the probe condition indicated equivalent proficiency at recognizing nonmatching sounds, as at recognizing matching sounds. The favorable results of the conditional matching tests were evidence for a capability for symbolizing one acoustic event by another.

Two additional procedures examined for effects on memory fidelity of reducing the integrity of input information. Surprisingly, there was no detrimental effect when the two familiar sounds, normally heard for 4 sec each as samples, were projected for only 3 msec. Also, there was little effect when the auditory frequencies of the two familiar sample sounds were shifted closer to each other, until an apparent limit of discriminability was reached.

The only procedure having an appreciable effect on the maintenance of auditory information over time was the insertion of an irrelevant sound into the delay interval. Even in this case, memory was disrupted only when the irrelevant sound occupied almost the full extent of the 15-sec retention interval.

The sum of the results illustrates the precision and reliability of auditory memory, and the flexibility of encoding and decoding operations in memory storage and retrieval processes. There is little to guide us, however, in judging how the dolphin maintains auditory information so well. Apparently, stored information does not spontaneously erode in quality over time, within the time limits tested. Possibly, an active rehearsal process maintains the information (cf. Wagner, Rudy & Whitlow, 1973; D'Amato, 1973), as in analogy to a human silently rehearsing verbal materials. Alternatively, some less effortful perceptual process, like analogic eidetic auditory imagery, may underlie the auditory retention capabilities of the dolphin. Eidetic perceptual processes may in fact govern human memory for nonverbal items, such as tones, colors, or motor movements (Massaro, 1970; Pepper & Herman, 1970), and seem a tempting alternate model to animal "rehearsal" mechanisms.

When auditory information occurs at high rates, as when listening to a list of many different sounds, strong recency effects were found. The dolphin easily recognized sounds occurring late in a list but had increasing difficulty in recognizing progressively earlier sounds. Early sounds seemed to be selectively displaced in memory, or degraded in quality, by the later-occurring sounds. An analogy was made to the operation of a short-term buffer mechanism with a limited storage capacity, structured to give priority of storage to new items (cf. Atkinson & Shiffrin, 1968). The buffer-storage capacity for the dolphin, or its memory span, based on memory for lists of up to eight different sounds, was estimated as between four to six sounds. Classification decisions of the dolphin when uncertain as to whether probe sounds were from the list or not tended to be unbiased, consistent with the use of an optimum decision criterion.

Visual memory could not be directly tested using the delayed matching-to-sample procedures found to be so effective for analyzing auditory memory. The dolphin simply could not learn the matching task through its visual information-processing system. It was exciting, however, to discover that visual information could be processed and remembered successfully if the visual materials were "named" with sounds. The association may have allowed the visual materials to be reencoded as auditory items and processed through the richly-developed central auditory system. In analogy, the faithfulness of human visual memory may under some circumstances depend on the representation of the visual information by an auditory language code (Kroll, 1975).

Many of the characteristics of dolphin auditory working memory paralleled characteristics noted for primate visual working memory. Also, the limitations in the processing and maintenance of auditory information described for the visually-dominant primates seemed a "mirror-image" of the limitations in visual processing and maintenance for the acoustically-dominant bottlenosed dolphin. More generally, it may be that for any species, memory for information *not* within the dominant sensory modality is constrained to limited classes of biologically relevant materials.

The spatial working memory of the bottlenosed dolphin was tested by requiring a choice from among 2 or 4 spatially-separated response paddles, after having earlier heard a sound cue identifying the relative or absolute position of the paddle yielding a fish reward. In the relative case, in which two pairs of paddles were used, the dolphin could not employ postural cues as aids for remembering the location of the correct paddle, since there was spatial uncertainty between pairs. When a single pair of paddles was used, the sound cue was unambiguously associated with one and only one paddle. In this case, the dolphin quickly adopted postural responses that served as mnemonic aids, and memory for spatial location was much improved. These findings on the importance of overt spatial coding on spatial memory paralleled observations for many other species of animals tested on delayed response tasks.

Dolphin (auditory) and primate (visual) working memory, while mutually convergent in processing characteristics, are divergent in most respects from characteristics of avian visual memory, at least as represented by the ubiquitous pigeon subject. Whether this represents a general mammalian/avian distinction, as Roberts and Grant (1976) suspect, or a large-brained/small-brained distinction, must remain an open question until more avian representatives and some of the smaller-brained mammalian forms are studied on tasks which more fully reveal their information-processing capabilities and constraints.

CONCEPTUAL PROCESSES

Conceptual processes involve the manipulation of information. Transforming information from one form (representation) into another, organizing attributes of information into broader classes, operating on classes according to some general rule, or defining relations among items that connect them causally or functionally, are all examples of the manipulation of information. Experiments with the dolphin have tested for capabilities for discovering relationships among events or objects and for forming general response rules. Such capabilities are important characteristics of human cognitive behavior (Kintsch, 1970; Segal & Stacey, 1975), and also comprise some of the dimensions of intellect described for chimpanzees (Premack, 1976, 1978).

Segal and Stacey defined "rule-governed behavior" as the categorization of related events or objects into sets, and the operation on all members of a set by a common procedure. The capacity for rule-governed behavior is measured, therefore, by the appropriateness of response to classes of events or objects, and not by responses to individual items. Responding consistently to a specific stimulus for reward, or matching a comparison stimulus to a given sample object, says little about the capacity for rule-governed behavior. But, if new sample stimuli can be readily matched after prior experience with the matching problem, or if new pairs of stimuli can be quickly partitioned into rewarded and nonrewarded classes after some experience with making two-choice discriminations, competency in forming and using broadly applicable response rules can be inferred. Harlow (1949) described such competency as "learning how to learn," or as "learning sets." Premack (1976, 1978) would define such generalized rule-learning capabilities as instances of second-order relational learning (cf. Bateson's 1942, "deuterolearning"), which he considers one of the hallmarks of intelligent species. The capability of the dolphin for forming second-order relationships is seen in the results of a number of laboratory studies of learning. Some examples are found within the matching-to-sample studies of memory already described, though their conceptual aspects were not discussed in this context earlier. Other examples appear in studies to be reviewed of multiple discrimination learning, reversal learning, and "novelty" learning.

As with studies of memory, it is instructive to compare capabilities for manipulating visual and auditory information. The striking differences noted in the characteristics of visual and auditory memory are paralleled here by differences in conceptual ability across the two modalities. The interdependence of learning and memory should lead us to expect this result.

The complexity and flexibility of motor learning is also considered. Data reviewed on vocal and motor mimicry and on observational learning demonstrate the generality of the analogic imagery system of the dolphin, or what might be called its capability for constructing detailed internal representations of its world. (cf. Jerison, 1973). Observational learning is of special interest since it is a pervasive form of learning in humans and in our great-ape cousins (Premack, 1976; Riopelle & Hill, 1973).

General Characteristics of Visual and Auditory Learning in the Dolphin

Most visual learning proceeds at a much slower pace than does auditory learning and is less stable over time. When training simple visual-form discriminations, for example, a surprisingly large number of training trials may be required and periods of encouraging progress may be quickly followed by periods of regression in performance. An enduring stability of response is rare. Special efforts may be needed to secure the dolphin's attention to the visual forms, as by coupling more salient visual cues, like size or brightness, to the form dimension and then gradually "fading" these cues out (cf. Kellogg & Rice, 1966; Terrace, 1963). "Refresher" training with these ancillary fading cues may be needed from time to time to maintain the learning levels reached earlier. Examples of learning difficulties with visual form stimuli may be found in the training descriptions of psychophysical studies of vision in the bottlenosed dolphin (Herman, et al., 1975; Madsen, 1976), the Pacific white-sided dolphin (*Lagenorhynchus obliquidens*) (Spong & White, 1971), and the killer whale (*Orcinus orca*) (White, Cameron, Spong & Bradford, 1971).

In contrast to the case for visual discriminations, auditory discrimination performance is generally highly stable over time even when differences between sounds are small (Herman & Arbeit, 1972; Thompson & Herman, 1975), or the sounds are very short in duration (Yunker & Herman, 1974), or very weak in amplitude (Johnson, 1967). Solving of new auditory discrimination problems may sometimes proceed almost effortlessly (Herman & Arbeit, 1973). The material in the following sections illustrates some of these differences between auditory and visual learning in greater detail.

Comment on Dynamic Visual Stimuli

The visual learning problem reaches deeper than a mere inattentiveness to form. Although the bottlenosed dolphin seems highly attentive to brightness and size cues, learning difficulties with these more salient cues may crop up quickly if the task becomes conceptually taxing, as illustrated in some of the work to be presented. Form, size, and brightness stimuli, as used in the laboratory, are typically displayed fixed in place and motionless. Yet in the real world of the dolphin, the biologically important visual stimuli are commonly dynamic—swimming schoolmates, fleeing prey, or other motile sea life seen against an often diffuse and homogeneous background. There are indications of increased conceptual talent when the dolphin deals with dynamic (moving) visual stimuli. For example, dynamic manual gestures are used by many trainers to control the frequently complex behaviors of performing dolphins (see Defran & Pryor, this volume). Also, preliminary work at our laboratory shows that not only may specific behaviors be controlled easily by manual gestures, but objects may be "named" manually, as they can acoustically (see later section on language). The special movement detection and enhancement features of the dolphin eye, discussed in other chapters of this volume (Dawson; Madsen & Herman), might provide the peripheral basis for the apparent conceptual talent when working with gestures. Possibly, dynamic visual stimuli may be processed differently centrally than simple form or brightness stimuli, incorporating more of the imagery talents of the species or involving the more elaborate centers used for forming motor associations. The data base on this topic is sparse and its discussion remains speculative. The topic is not treated in further detail here, but warrants close attention in the future.

Rule-Governed Behavior in Visual and Auditory Tasks

Multiple Disrimination Learning (Discrimination Learning Sets)

VISUAL TASKS.   In early work, we tested the multiple discrimination learning abilities of a Pacific bottlenosed dolphin (*Tursiops gilli*) named Wela (Herman, Beach, Pepper & Stalling, 1969). For each of 104 different discrimination problems, Wela was to learn to press one of a pair of dissimilarly shaped plywood targets displayed side-by-side or one above the other in the clear water of her tank. Testing with a given pair was continued until a predetermined number of trials was completed (6, 8, or 12 trials), and then a new pair of targets was introduced. Could Wela learn the simple rule that for any pairing one target was always correct and the other incorrect? The controlling response rule may be notated as "win–stay, lose–shift," implying that if a response to a target is rewarded, continue responding to that target, otherwise shift to the alternate.

Wela's performance was highly variable from problem to problem and failed to show the progressive improvement over blocks of problems and the high final performance levels that characterize many primate species tested on this type of task (Hodos, 1970; Warren, 1965, 1973). Overall, her performance level remained at less than 60% correct responses. During the latter stages of testing, she developed strong position preferences, responding almost always to a spatial location regardless of the target located there, and further testing was not attempted. Wela did not appear to use echolocation for these discriminations, but seemed to rely on vision. Though a hydrophone was not available for sound monitoring, there were none of the lateral head-scanning movements that typify echolocating dolphins in difficult discrimination tasks (Evans & Powell, 1967; Norris, Prescott, Asa-Dorian & Perkins, 1961).

More importantly, when we later replicated this study using the *T. truncatus* subject Puka, and guarded against the use of echolocation cues, similar results were found (unpublished data). The target pairs were exactly those used with Wela, but they were displayed through the underwater window in Puka's tank, ruling out the use of echolocation. Like Wela, Puka's performance was variable and unstable, strong position preferences developed as testing progressed, and her overall percentage of correct responses remained just under 60%.

In additional work, Puka's capabilities for learning the multiple visual discrimination task was tested using smaller and brighter types of two-dimensional geometric targets and miscellaneous pairs of three-dimensional objects, such as a large ball vs. a helmet or a fire extinguisher vs. a box. In a further variation, the three-dimensional objects were painted a uniform white color, to control for brightness differences that seemed to be unfavorably influencing Puka's responses. But, for none of these conditions did Puka's performance rise above the level of approximately 60%-correct responding found earlier.

AUDITORY TASKS. Tests of multiple auditory discrimination learning were made with the bottlenosed dolphin Kea. Several hundred different pairs of sounds were given, and for each pair Kea was to learn to respond to one and only one of the two sounds for reward (Herman & Arbeit, 1973). A pair of sounds was projected sequentially from the peripheral speakers shown in Figure 8.1, one sound at each speaker. Both the position of the correct sound at the left or right speaker and its order, first or last, in the sequence of two sounds were randomly varied over trials. Initially, new pairs of sounds were introduced after 25 or 50 trials of responding to old pairs (Fig. 8.11A). This was later reduced to 20 and then to 10 trials before switching to a new pair (Fig. 8.11B,C). Figure 8.11C shows that Kea solved 23 of 25 new two-choice auditory discrimination problems without error, and for the remaining two problems, made one and two errors, respectively. Altogether, it seemed that an almost endless variety of sound pairs could be discriminated on their first appearance and responded to reliably thereafter. With some practice, it made little difference whether the correct sound of a pair was cued on the first learning trial (Fig. 8.11A,B,C) or not (Fig. 8.11D).

The Herman and Arbeit (1973) study, using sounds, was conceptually similar to both the multiple visual discrimination study of Herman et al. (1969) with a *T. gilli* subject, and the subsequent replication with a *T. truncatus* subject. Nevertheless, the difference in the rate of learning and in the final level of performance reached were enormous. In all three studies, development and application of the win–stay, lose–shift rule was required for high levels of performance, but this happened only when manipulating auditory information, underscoring again the cognitive limitations of visual information-processing centers of the dolphin relative to that of auditory centers.

I do not want to leave the impression that dolphins are without learning constraints in multiple auditory discrimination tasks. An earlier study (Beach & Herman, 1972) failed to produce the levels of auditory learning noted in the

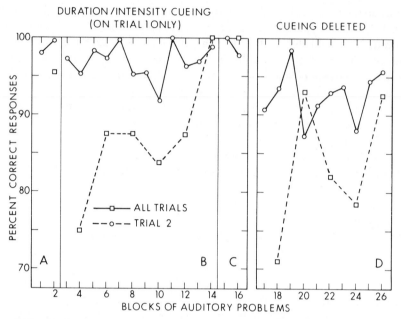

Figure 8.11. Rapid learning of two-choice auditory discrimination problems by the dolphin Kea. For Parts A to C, the duration of the S− was shorter and its intensity less than that of S+ on the first trial of each problem. Thereafter, duration and intensity values for S+ and S− were the same. Trial 2 performance indexes efficiency of learning after a single cued trial. Trials per-problem were either 25 (Block 1), 50 (Blocks 2-8), 20 (Blocks 9-14), or 10 (Blocks 15-16). In Part A the two sounds varied widely in acoustic characteristics. Part B was a parametric evalua-tion of effects of smaller vs. larger differences between sound pairs and therefore included diffi-cult as well as easy discriminations. Part C used only "easy" discriminations, based on the outcome of Part B, and shows nearly perfect performance, uncontaminated by discrimination difficulties. Part D shows an initial drop in performance followed by recovery when the duration-intensity cue on Trial 1 was deleted entirely, making performance on Trial 2 solely dependent on the reward/nonreward outcome of the response to a sound on Trial 1. Problems in Part D were 10 trials each. Redrawn from L. M. Herman & W. R. Arbeit, *J. Exp. Anal. Behav.*, **19**, 1973, 379-394. With permission. Copyright © 1973 by the Journal of the Experimental Analysis of Behavior.

Herman and Arbeit (1973) study. However, the limitations were overcome by several procedural changes introduced by Herman and Arbeit, resulting in much improved multiple discrimination performance. The point is that while procedures yielding very high levels of performance on these auditory tasks could be readily developed, strenuous efforts to find procedures for improving performance on multiple visual discrimination problems were without success.

Successive Reversal Learning

VISUAL TASKS.   Figure 8.12A shows previously unpublished data on visual suc-cessive reversal learning by the dolphin Puka. The training was arduous and the

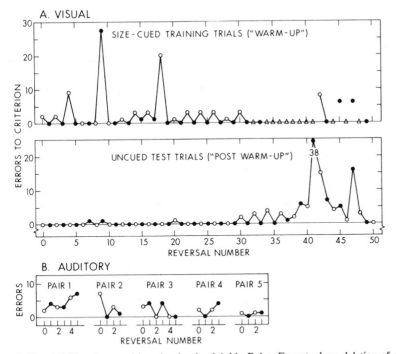

Figure 8.12. (A) Visual reversal learning by the dolphin Puka. Except where deletion of cues is indicated, each visual reversal of a black versus white target was preceded by "warm-up" trials with S+ larger in size than S−. The warm-up trials were followed by test trials with target sizes equal. Filled circles indicate the black target was S+ and unfilled circles indicate the white target. Triangle symbols indicate that no warm-up trials were given. (B) Auditory reversal learning by the dolphin Kea. Five different pairs of electronically generated sounds were used without any warm-up trials. Each reversal was signaled by the deletion of reward for the previously correct response. This mandatory first-trial error is not included in the graph. Filled and unfilled circles indicate the change from one sound of a pair to the other. (Part B: unpublished data from Herman & Arbeit, 1973.)

figure illustrates the erratic characteristics of the learning process. In the visual reversal problem two stimuli are displayed simultaneously. As in a standard discrimination problem, one stimulus, the S+, is associated with reward and the other, the S−, with nonreward. But after the animal attains a predetermined criterion of success in responding to S+ the reward values for the two stimuli are reversed and testing to criterion on the first reversal problem begins. A switch in reward value, defining a new problem, is made each time the success criterion is met, for as many successive reversal problems as are desired. The win–stay, lose–shift rule is applicable to the solution of the successive reversal problem, as it was to the multiple discrimination problem.

The data in Figure 8.12A are based on a discrimination between a white circle on a black background versus a black circle of equal area on a white background, the "W" and "B" stimuli described in the section on visual memory. These "enhanced brightness" stimuli were selected after failure of Puka to perform reliably throughout 41 successive reversals of two figures differing only in form—a white "M" versus a white square of equal area—and after only moderate success on 81 reversals of a grey versus a white circle. To assist Puka in learning each of these reversal tasks, at the start of each reversal the S+ stimulus was larger than the S− stimulus. The size difference was then reduced gradually over succeeding cued trials, by decreasing the S+ size and increasing the S−, until both targets were of equal area. Once the size difference was eliminated, testing of uncued trials began and was continued to a criterion of 9 out of 10 correct responses.

Performance over successive reversals with the enhanced brightness pair, during "warm-up trials" with the size cue present, is shown in the upper half of Figure 8.12A. The low error rate for almost all reversals indicates that size was an effective cue for controlling responding. The continued good performance seen in the lower half of Figure 8.12A shows that warm-up training transferred successfully to the succeeding uncued "test" trials. Beginning with Reversal 32 the warm-up trials were sometimes deleted, as indicated by the triangle symbols, producing an initially small increase in error rate and then an abrupt loss of performance at Reversal 41. Size-cued trials were then reinstated, but performance remained unstable for the remaining nine reversals shown. An additional 34 size-cued reversals (not shown in Figure 8.12A) were necessary before performance restabilized sufficiently to allow the complete elimination of the size cue once again. This time the error rate remained low for 17 reversals before rising sharply. Further testing on this task was then ended.

AUDITORY TASKS. Compare the characteristics of visual successive reversal learning in Figure 8.12A with Figure 8.12B showing successive reversal performance of the bottlenosed dolphin Kea for five different pairs of sounds. These previously unpublished data comprised a portion of the training procedures of Herman and Arbeit (1973, Exp. 5). Whichever sound the animal chose on first hearing a new pair of sounds was designated S−, while the remaining sound was S+. Testing continued until Kea chose S+ six or seven times in a row, with the criterion determined in advance. The reward values were then reversed and testing to criterion was resumed. No cueing of any type was used. The results were that very few errors were made with any pair of sounds and errors decreased to nearly zero over new pairs of sounds. Kea, who had learned the win–stay, lose–shift rule during her multiple auditory discrimination-learning testing (Fig. 8.11), was apparently able to transfer this rule almost immediately to repeated reversals of the reward values of a fixed pair of sounds. This transfer of a rule across tasks is a convincing demonstration of second-order relational learning in the dolphin (cf. Premack, 1976). Warren (1974) has discussed data showing that rhesus monkeys and chimpanzees are able to transfer the win–stay, lose–shift rule between repeated reversal problems and multiple discrimination learning tasks, but that squirrel monkeys and cats cannot. Although Warren cited these data as examples of possibly unique learning characteristics of Old-World monkeys and chimpanzees (cf. Rumbaugh & Gill, 1973), it is seen here that a talent for rule-transfer may apply to the bottlenosed dolphin as well.

Matching-to-Sample

The matching-to-sample problem, described in the earlier section on memory, provides a final demonstration of differences in abilities for processing visual and auditory information. Puka, who eventually managed the visual successive reversal problem with some success (Figure 8.12A), was nevertheless unable to learn the visual-matching problem except when auditory names were given the visual objects. This visual learning constraint was present even though matching training was given after the reversal problems of Figure 8.12A, and used the same black-versus-white discrimination targets. Moreover, the reversal problem, which requires that the animal switch responses between stimuli on a nonreward cue, may be viewed as a component of the matching problem, which also has a stimulus switching requirement. However, the switch is cued by the sample rather than by reward outcome. Matching-to-sample can be a conceptually more demanding problem than successive reversal learning, since the switch may occur on every trial and there may be complex conditional relationships between sample and comparison stimuli.

In contrast to this visual matching difficulty, auditory matching was learned easily by the dolphin Kea and the matching rule was applied effortlessly to new sample sounds not previously heard (Herman & Gordon, 1974 and Fig. 8.3). This *generalized* matching-to-sample capability provides a further instance of the capability for forming second-order relations. In the list memorization study of Thompson and Herman (1977), which used a probe-matching technique, the matching rule was extended to the identification of sounds that were different from the sample, as well as sounds that were the same. Premack (1976), in his analysis of language skills of chimpanzees, underlined the skill of these advanced primates in recognizing sameness and difference relationships, asserting that a capability for "Generalized match-to-sample . . . (is) . . . the minimum computational power that a species must have in order to acquire language (p. 132)." Though there may be other preconditions for language, as noted in the introduction to this chapter, and as discussed in the later section on language learning, the performance of Kea on the auditory matching tests meets Premack's minimal standards for "computational power." Several species of Old- and New-World monkeys have displayed capabilities for generalized match-to-sample (French, 1965; Medin, 1974), but may not possess other prerequisite cognitive skills for language learning shown by chimpanzees (Premack, 1976, 1978). Unlike primate or dolphin, pigeons show little or no capability for generalized matching (cf. Zentall & Hogan, 1974; Premack, 1978). Once again, convergent cognitive characteristics of dolphin and primate are seen, with both taxa divergent from characteristics shown by pigeons.

The Kellogg-Rice Studies: Visual or Echolocation Discrimination

ANALYSIS OF THE STUDY. An apparent exception to the visual information-processing constraints of dolphins discussed so far was reported by Kellogg and Rice (1963, 1964, 1966), in a series of papers based on the same study. A bottle-nosed dolphin named Paddy learned to discriminate quickly and reliably between pairs of geometric-shaped targets displayed side by side underwater. Furthermore, Paddy was able to transfer the learning immediately to new arrangements of the targets. For example, a discrimination between a heart (S+) and a cross (S−) was learned without error by a cueing method in which the S− form was initially brighter than the S+ and further enhanced by a surrounding raised-brass edging. Errorless responding then continued with these cues deleted, even when the heart was rotated 180° in the vertical plane, the cross rotated 45°, or a rectangle substituted for the cross. Conceptually, such "transpositions" of stimuli test the ability for learning about relations between stimuli (Macintosh, 1974).

Kellogg and Rice claimed that Paddy learned the discriminations visually. If so, the relatively rapid success and stable performance of Paddy stands in contrast to the difficulties described for other dolphins in visual learning tasks. However, there seems reason to doubt that the discrimination was through vision. Paddy was able to perform reliably only when the targets were displayed underwater, and during such display, head-scanning movements typical of echolocation behavior were observed. Though Kellogg and Rice (1966) reported that no "sonar pings" were detected, the method for detection was not described. Sonar emissions can be highly directional (Norris & Evans, 1967) and might easily have eluded the receiving hydrophone, or may have exceeded the bandwidth of the receiving system. The targets, which were in the form of templates of thin shimming brass, could have provided reflective information through differences in edge contours. The differences may have been learned initially during the cueing procedure using the raised brass edgings, and refined to precision as the edging was gradually eliminated. The authors' contention that 1/8-in. clear Lucite (acrylic plastic) placed in front of the targets was an effective acoustic barrier is erroneous. Although transmission decreases with frequency, at 120 kHz acrylic still transmits better than 90% of the emitted acoustic energy (Penner & Murchison, 1970). Penner and Murchison in fact successfully used a black acrylic sheet as a visually opaque but acoustically transparent shield in front of thin underwater wires, to test echolocation resolution capability of the boutu or Amazon river dolphin (*Inia geoffrensis*). They argued that the good performance of the boutu suggested that in the Kellogg and Rice study Paddy was very likely using echolocation to discriminate among the different templates.

REPLICATION. Further doubt on the validity of the Kellogg and Rice data is given by the negative outcome of our partial replication of their study. We used the same shaped and sized visual targets as did Kellogg and Rice, but displayed them outside the tank through an underwater viewing window, protecting against the use of echolocation cues. Under these conditions, the performance of the bottlenosed dolphin Puka barely rose above chance levels for the basic visual discriminations. Consequently, we could not even begin to test for transpositional (relational) learning.

If the Kellogg and Rice results are viewed as discrimination by echolocation, rather than by vision, they fit better with the contrasts drawn so far between dolphin auditory and visual information processing capabilities. That *T. truncatus* would use echolocation in the tank setting for the discrimination of underwater targets (Kellogg & Rice, 1966), while *T. gilli*, in the study described earlier (Herman et al., 1969) apparently made no attempt to do so, may not be surprising in view of the different ecological adaptations of these species. Much of the Atlantic and Gulf Coast range of the coastal *T. truncatus* lies in the shallow neritic zone where obstacles are common and suspended particles impair water clarity. In contrast, the Pacific Ocean *T. gilli* more often roams the deeper, clearer offshore zones, especially in Hawaiian waters. K. S. Norris (in Turner, 1964) contrasted the difficulty of a member of another pelagic delphinid species, *L. obliquidens*, in learning to use its echolocation system in a tank setting while blindfolded, with the immediately effective use of the system by a blindfolded *T. truncatus* specimen.

According to material reviewed by Diercks (1972), effective echolocation in a tank setting seems to be a complex function of its utility in the species' natural habitat, and the availability of other modes of detection, especially passive listening (also see Yablokov, Bel'kovich & Borisov, 1972). Diercks' summary notes that passive listening was frequently used in preference to active sonar by *T. truncatus* while tracking fish, and that killer whales rarely used sonar in the wild and may have to be taught to use it in the oceanarium. It can be added that if the sonar discrimination task is difficult, as in the detection of fine wires by the Amazon River dolphin (Penner & Murchison, 1970), or the resolution of small differences in the diameter of ball bearings by *T. truncatus* (Turner & Norris, 1966; Norris, 1974), learning can be slow. Since the passive listening mode has excellent resolution capabilities (Johnson, 1967; Herman & Arbeit, 1972; Thompson & Herman, 1975), it may be that the gradualness with which a fine echolocation discrimination is mastered mainly reflects difficulties in learning to produce optimal sonar sounds for maximizing reflected information. Sonar emission requirements may differ from one situation to the next, depending on the properties of the target or on ambient noise conditions (e.g., Au, Floyd, Penner & Murchison, 1974), and time and practice may be required to optimize output. Clearly, the learned aspects of the echolocation skills of various delphinids need much further study.

The Complexity and Flexibility of Motor Learning

The delphinids are superbly coordinated, graceful animals with extensive musculature and a greatly enlarged cerebellum (Morgane & Jacobs, 1972). The assortment of acrobatic behaviors that can be trained, and the ability to learn to manipulate objects in many ways, observable in any oceanarium housing dolphins, attest to the learning flexibility of the perceptual-motor system of these animals. Observations at sea of bow-wave riding dolphins, leaping dolphins, dolphins pursuing and capturing fish, sometimes in the air, and their underwater "aquabatics" are predictors of the motor skills demonstrated in the oceanarium or laboratory.

Rule-Governed Behavior in Spatial Learning Tasks

In spatial learning tasks, responses are rewarded to positions or locations, such as the left-hand paddle of a pair of paddles, rather than to objects or events. The bottlenosed dolphin can learn to reverse its spatial responses to a pair of paddles efficiently each time the reward values associated with the two paddles are switched, and to respond optimally given unbalanced reward amounts or reward probabilities assigned to spatial postions. In one study (Beach, Pepper, Simmons, Nachtigall, & Siri, 1974), a series of 19 spatial reversals were given to the dolphin "Jo" at the rate of one reversal per day, provided she reached a criterion of six consecutive errorless trials on the prior day. Errors rapidly decreased after the first reversal. Four of the final nine reversals were learned without error and the remaining five with a single error each. Apparently Jo quickly learned the win-stay, lose-shift response rule.

In another study the dolphin "Cyclops" promptly learned to respond solely on that paddle of a pair that yielded the higher probability of food reward (Pepper, Nachtigall & Beach, 1972), an optimal response strategy for maximizing expected gain under uncertain (probabilistic) outcome conditions. Also, when the payoff values for the two paddles were unbalanced, but each always yielded some reward, Cyclops learned to select that paddle giving four fish rather than one, and to reverse her spatial responses quickly when the payoff amounts were reversed (Nachtigall, 1971). Performance on these spatial tasks was stable over time, unlike performance on visual reversal tasks, and more comparable with performance on auditory reversal tasks. Generally, spatial reversal tasks, or the other spatial tasks described, may not tax the conceptual abilities of mammalian species, or even avian species (for a review, see Mackintosh, 1974). The learning characteristics of the bottlenosed dolphin in these spatial tasks were much like those observed in other animals.

Forming a Concept About Motor Responses

Pryor, Haag, and O'Reilly (1969) carried out an innovative study of the ability of a rough-toothed dolphin (*Steno bredanensis*) named "Hou" to "create" new motor behaviors, a task which seems demanding of conceptual talent. A new behavior was defined as one not reinforced previously during any of Hou's several 5- to 20-minute long daily testing sessions. The behaviors chosen for reinforcement were easily definable and observable gross motor responses, such as an aerial backflip, tail-walking, or an inverted tail-slap. Only one new behavior was reinforced per session, each time it occurred. Approximately 16 sessions were required for new behaviors to emerge with some regularity, although older previously-reinforced behaviors were more common. After the 16th session, the number and percentage of new behaviors offered began to increase substantially, and by the end of an additional 17 sessions the repertoire of behaviors had become so enlarged that the experimenters had difficulty in reliably classifying an ongoing behavior as old or new. The experiment was ended at that point.

The authors stated that many of Hou's behaviors had not been seen previously among captive rough-toothed dolphins. The results suggest that a concept, such as "newness" or "different," was formed by Hou, together with a "win-shift" response rule, a strategy for *not* repeating a previously reinforced response (cf. Kamil, 1975; Olton & Samuelson, 1976). It was noted earlier that a concept of "newness" or "different" also appeared to be formed by the dolphin Kea during testing for her ability to recognize whether a probe sound following a serial list of sounds was old or new (Thompson & Herman, 1977).

The descriptions of play behaviors of young dolphins (e.g., D. Brown & Norris, 1956; McBride & Hebb, 1948; Tavolga, 1966) include some remarkable examples of inventiveness, such as the use of a dead food fish to tease out live fish hiding in a rock tunnel (Tavolga, 1966). The descriptions reinforce the notion of a broad capability for innovative behavior in dolphins. Rich, innovative behaviors are common among primates, providing another area of cognitive contact between primate and dolphin. Inventive play behaviors have been observed among the young of many of the highly social apes and monkeys (Bekoff, 1972; van Lawick-Goodall, 1968), and there are many examples of adaptive inventiveness in the use of tools (Rumbaugh, 1970).

Motor and Vocal Mimicry

Social mimicry can be defined as the copying by one animal of the behaviors of another. "True" mimicry, in which an animal imitates new behaviors not in its species-typical repertoire, may be distinguished from the release of a common behavior by its occurrence in another animal (Andrew, 1962). By this distinction, one dolphin schooling closely with another, two dolphins leaping in near synchrony, or even vocal exchanges between two dolphins, each producing its own characteristic sounds in response to the other, are not cases of true mimicry.

In other species, clear examples of true mimicry may be found among those birds able to copy the songs of other avians or the sounds of human speech (Andrew, 1962; Eberhardt & Baptista, 1977). Simian primates demonstrate true motor mimicry when imitating manual gestures modeled by humans (e.g., B. Gardner & Gardner, 1971). Interestingly, birds seem poor at motor mimicry and nonhuman primates have little facility at vocal imitation. In examples that follow, it will be seen that dolphins have capabilities for both true motor and vocal mimicry, though the evidence is still largely informal and incomplete.

MOTOR MIMICRY. Examples of motor mimicry are abundant in the primate literature. Mimicry may occur in the games of young animals, one copying the other's innovative manipulation of some object or adopting a tool used by another for defense or food gathering (van Lawick-Goodall, 1968). A whole group of animals may learn to perform the behaviors of a single innovative animal, as in the cultural transmission of potato-washing behavior among Japanese macaque monkeys after its invention by the talented young monkey Imo (Kawamura, 1963). Among dolphins, comparable imitative tendencies may be seen in their games, or in their attraction toward an object being manipulated by another dolphin. These behaviors, like those of primates, may have an adaptive basis within the closely integrated and often cooperative society that characterizes many delphinid species. Imitating another's successful responses avoids the needless repetition of error, and may be socially facilitating as well. Imitation, flattering to humans, may in many cases be a strong affiliative signal for animals.

A few examples can be given of imitative motor behaviors observed among captive delphinids that appear to have the quality of true mimicry. Brown, Caldwell, and Caldwell (1966) described the imitation of the typical longitudinally-spinning leap of a Pacific spinner dolphin (*Stenella longirostris*) by a *T. truncatus* specimen. The bottlenosed dolphin's spinning leap occurred only minutes after it had been placed in the show tank for the first time, and immediately after the spinner dolphin's leap. Spinning leaps are rarely, if ever, seen among wild *T. truncatus*. Kritzler (1952) reported that a pilot whale (*Globicephala* sp.) attempted to imitate the behaviors of bottlenosed dolphins "spearing" an inflated rubber tube with their snouts. Being snub-nosed, the pilot whale was unsuccessful and finally settled for balancing the tube on its head.

Tayler and Saayman (1973) observed mimicry of a Cape fur seal (*Arctocephalus pussilus*) by an Indian Ocean bottlenosed dolphin (*Tursiops aduncus*) maintained in the same tank. The behaviors imitated included the seal's swimming movements, sleeping posture, and comfort movements (self-grooming). The dolphin, like the seal, at times swam by sculling with its flipper while holding the tail stationary. The sleeping posture imitated was lying on one side at the surface of the water, extending the flippers, and trying to lift the flukes clear of the water. The comfort movement mimicked was vigorous rubbing of the belly with the under surface of one or both flippers. Photographs accompanying the text illustrated these behaviors, none of which have otherwise been observed in *T. aduncus*.

Additional imitative behaviors by this dolphin included the swimming characteristics and postures of turtles, skates, and penguins. The dolphin also attempted to remove algae from an underwater window with a seagull feather, in imitation of the activity of a human diver who regularly cleaned the window. The dolphin, while "cleaning" the window, reportedly produced sounds resembling those from the demand valve of the diver's regulator and emitted a stream of bubbles in apparent imitation of the air expelled by the diver. Tayler and Saayman also observed a dolphin using a piece of broken tile to scrape seaweed from the tank bottom, a behavior apparently derived from observing a diver cleaning the tank with a vacuum hose. The scraping behavior of this dolphin was then copied by a second dolphin.

VOCAL MIMICRY AND CONDITIONED VOCALIZATIONS. Vocal mimicry, like motor mimicry, may be useful as an affiliative signal or otherwise facilitate group processes. Probably the best example among cetaceans of convergent and possibly affiliative vocal mimicry occurs in the songs of the humpback whale (*Megpatera novaeangliae*) (Payne & McVay, 1971; Winn, Perkins & Poulter, 1971). These songs appear in the winter breeding and calving grounds of the whales, as described in the review of Herman and Tavolga (this volume). The songs are comprised of multiple consecutive "themes" that together may last ten min or longer without repetition. Payne (1978) reported that some themes may change in composition from year to year, and that within a year a few themes may drop out and others may be added. However, by mid-season within a given year all singing whales in the winter grounds have converged on the same set of themes, and sing the same song with only relatively minor idiosyncratic variation. Whether only a few whales contribute to the innovations, or many do, is not known.

The flexibility of the vocal apparatus of the humpback whale may not be quite matched by that of the dolphins, but the capabilities of dolphins for producing varied sounds are none the less impressive. In the laboratory, all of the classes of vocalizations of bottlenosed dolphins—whistle sounds, echolocation pulses, and other types of pulsed sounds (see Popper, this volume)—may be elicited on cue with training (Batteau & Markey, 1968; Beach & Pepper, 1973; Bullock & Ridgway, 1972; D. Caldwell & Caldwell, 1972; Lilly, 1965; Lilly & Miller, 1962; Penner, 1966). Turner (1962) demonstrated stimulus control of whistle vocalizations in the Pacific white-sided dolphin and Norris et al. (1971) reported stimulus control of unspecified types of vocalizations in a Pacific pilot whale. Several of these studies, through selective reinforcement, have successfully modified the duration, amplitude, or frequency of the emitted sounds (e.g., Turner, 1962). Adaptive modulations of click vocalizations occur naturally in bottlenosed dolphins in response to changes in the acoustic environment (Au et al., 1974). These talents for modulating and controlling their vocalizations almost certainly underlie the apparent capability of the dolphin for spontaneous vocal mimicry.

D. Caldwell and Caldwell (1972) reported that a juvenile male bottlenosed dolphin, maintained in isolation from other delphinids since infancy, spontaneously mimicked an approximate 9-kHz pure-tone signal used during sound localization tests. Spectograph records of the dolphin's mimetic sound showed frequency components limited to between 9 and 10 kHz for a duration of at least 2 sec. In contrast, the typical "signature" whistle of this animal lasted approximately 1 sec and consisted of a series of four whistles, each sweeping in frequency between 7 and 14 kHz.

Figure 8.13, based on data collected by Douglas Richards at our laboratory, shows sonographic records of four cases of spontaneous mimicry of computer-generated sounds by two young bottlenosed dolphins. The records were obtained during an ongoing study of the responses of the dolphins, housed in the same tank, to various sounds. It is not known which dolphin produced which sounds.

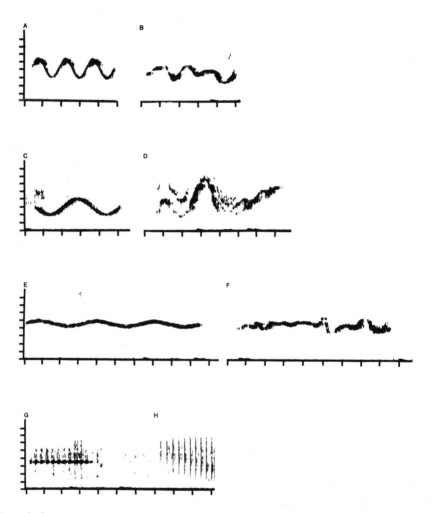

Figure 8.13. Spontaneous vocal mimicry by the bottlenosed dolphins Phoenix and Akeakamai. As shown by sonograph records. Sound recordings made with Clevite CH3 hydrophone system and Uher 400 Report Stereo recorder. Effective filter bandwidth of Kay Sonograph Model 7029A was 600 Hz. Vertical scale is frequency from 0 to 16 kHz in 2-kHz divisions; horizontal scale is time in 0.1-sec divisions. The first sound of a pair is computer generated; the second sound is the responsive mimic. Pairs A–B, C–D, E–F illustrate whistle mimicry. Pair G–H is mimicry of a pulsed sound.

The figure shows that mimicry occurred in both the whistle mode (Pairs A–B, C–D, and E–F) and pulsed mode (Pair G–H). For each pair, the left-hand record is the computer-generated sound and the right-hand record the responsive mimic. The response latencies are not shown for the first three pairs, because of space limitations, but they were 1.05, 0.47, and 0.32 sec, respectively. The latency between Pair G–H was 0.40 sec, as shown. None of the whistle mimics resembled the characteristic (most common) whistle of either dolphin. The pulse mimic, at approximately 31 pps, was remarkably close to the computer model's rate of 30 pps.

Lilly (1961, 1962) reported a number of additional instances of spontaneous vocal mimicry in bottlenosed dolphins. On one occasion, Lilly (1961) replayed, at quarter or half speed, recordings of the vocalizations of a dolphin undergoing electrical brain stimulation. Lilly claimed to hear mimicry of several of the words and phrases he had spoken at the time, as well as of some background human laughter. Lilly (1962) also reported some success in training a dolphin to imitate a few human words and phrases such as "water," and "bye-bye." Unfortunately, Lilly's observations were not well documented, nor, in some cases, substantiated by other on-the-scene observers (see Wood, 1973, Chap. 5; Wilson, 1975, pp. 473–474).

The preceding examples are all cases of spontaneous (unreinforced) vocal mimicry. There seem to be only two cases in the literature of conditioned or reinforced mimicry, an interesting procedure that warrants further study. Ralph Penner (personal communication), during preliminary work in a language-acquisition study (Batteau & Markey, 1968), observed that a male bottlenosed dolphin, named Maui, spontaneously mimicked any of five electronically generated whistle "contours" that were being projected underwater. Maui accurately matched the duration of the contours as well as their frequency change, though there was a downward shift in frequency of approximately one octave from the original. The vocalizations were later placed under stimulus control, so that any of five different contours were mimicked on command (Penner, 1966, summarized in Evans, 1967).

In another case of conditioned mimicry, a bottlenosed dolphin was trained to repeat the number and duration of an arbitrary sequence of as many as 10 vowel-consonant or consonant-vowel pairs read by a human (Lilly, 1965; Lilly, Miller & Truby, 1968). The authors reported that the dolphin matched the number of sound bursts, rather than the number of spoken syllables—which apparently were comprised of multiple sound bursts—and that the total duration of the dolphin's response closely matched the time required by the human to read a list. The dolphin's responses were not copies of the human sounds, but were in-air "quacks" or "blats" produced through the blowhole. They might therefore be classified as imitations of the temporal or digital features of the sounds, rather than of their acoustic or analogic features.

Work is currently underway at our laboratory testing the ability of the bottlenosed dolphin Akeakamai to learn to mimic electronically generated sounds (D. Richards & L. Herman, unpublished data). Akeakamai is one of two dolphins receiving training in comprehension of an artificial imposed language (see later section). The work on mimicry is studying the feasibility of using self-produced vocalizations as a medium for a productive vocabulary by the dolphin. If successful, it may provide a means for achieving some degree of two-way communication between dolphin and human within the artificial language. At present eight different sound "models" are imitated with high precision and reliability. The models are pure tones or sine-wave or square-wave frequency-modulated tones having center frequencies between 4 and 16 kHz. The models that were introduced more recently have tended to require less training than those introduced earlier, indicating that Akeakamai has developed a general concept of "imitate." We have most recently begun to substitute objects—a ball, a plastic hoop, or a short length of plastic pipe—for three of the model sounds. An object takes the place of an associated but absent model as a cue to imitate the model. By producing an imitation the dolphin in effect "names" the object. The objects are in fact named correctly on better than 95% of the trials on which they are displayed, with the display of a given object of the three determined by a random schedule.

The full extent and flexibility of the vocal mimicry capability of the dolphins remains to be tested. Both motor and vocal mimicry should be studied further, since their joint emergence in one species may be limited to cetaceans among infrahumans, and because no mammalian species other than *Homo* has been shown capable of true vocal mimicry. Such study can clarify the pressures selecting for a generalized mimicry capability and the relation of mimicry to other cognitive traits of dolphins.

Observational Learning

Observational learning, unlike mimicry, does not necessarily entail the precise imitation of the model, but implies only a facilitative effect on learning after having observed the behavior demonstrated. Through observation, an animal may learn about the reward values associated with various stimuli, the adaptive behaviors of the model, or the place at which the responses are to be made, but the responses themselves may be molded to the individual's own "style." Rapid observational learning has been reported for monkeys and apes (Riopelle & Hill, 1973). Cats (John, Chesler, Bartlett, & Victor, 1968) and rats (del Russo, 1975) are also capable of observational learning, but to a more limited extent.

There are several informal reports in the literature of delphinid species learning complex behaviors through observing other animals. M. Caldwell, Caldwell, and Siebenaler (1965), describing training procedures at an oceanarium, noted that placing naive dolphins together with trained animals considerably reduced the training time for the newcomers. In some cases, the animals may even train themselves. In one interesting episode illustrating self-training, one animal of a group had been taught to leap to a suspended ball, grasp it with its teeth, and pull it some distance through the water in order to raise an attached flag. This animal was subsequently removed from the show and a second animal in the group was trained in the same task, but learned to raise the flag by striking repeatedly at the ball with its snout, rather than by pulling on it. This second animal, a female, subsequently died and another female of the group immediately took over the performance, without training, and continued to strike at the ball with the snout. Later, when this new female refused to participate in the show during a two-day period, a young male in the group immediately performed the behavior, but grasped and pulled the ball with his teeth, in the manner of the originally trained animal.

Brown et al. (1966) reported that a false killer whale (*Pseudorca crassidens*) learned, without training, many of the show behaviors of a Pacific pilot whale with which it was housed. The behaviors learned included "shaking hands," "dancing" (extending the upper body out of the water and rotating in place), and "singing" (aerial vocalizing). In another example, Pryor (1973) reported that two rough-toothed dolphins in a community tank each immediately performed the other's trained routine without error when their roles were inadvertently switched during a show.

In the work of Batteau and Markey (1968), referred to earlier, the male bottlenosed dolphin Maui was housed together in a lagoon with the female bottlenosed dolphin Puka, who was later the subject of our visual learning and discrimination studies. Both dolphins were trained to respond to sequences of underwater sounds, as discussed more fully in the next section on language. Maui was trained to emit an echolocation sound on hearing a three-word command preceded by a sound for his name. Later, without training, Puka was given the same three-word command, preceded by her name, and responded with her echolocation signal. In another test, Maui was trained to raise his flukes clear of the water on a three-word command preceded by his name. Subsequently, when the command was given to Puka, she slapped her flukes on the water.

Like mimicry, observational learning in dolphins needs much further study. No formal studies appear to have been done. The importance of observational learning to human and pongid cognition was noted earlier, and it would be interesting to gauge more fully the degree to which this trait was emergent in dolphins.

Overview

The data on conceptual processes support the earlier-discussed findings of substantial differences between the auditory and visual information-processing capabilities of the dolphin. Learning through visual cues—motion cues excepted—is characteristically slow, unstable, and relatively low in final performance level. The possibilities for complex conceptual learning through vision, especially through form cues, seem very limited. In contrast, auditory learning is usually rapid and stable, and characterized by high final performance levels. Concepts about auditory items may be easily formed (e.g., "sameness–difference" relations), and response rules (e.g., "win–stay, lose–shift") may be developed, generalized, and applied efficiently to new auditory information. Premack's (1976) "second-order relational learning" and Segal and Stacey's

(1975) "rule-governed behavior" seem descriptive of the dolphin's cognitive characteristics when manipulating auditory information.

There was evidence that relational learning, which is easily managed within the passive listening mode, may also be possible in the active sonar mode, if it is accepted that the Kellogg and Rice (1966) data for the bottlenosed dolphin were in fact based on an echolocation discrimination. Further study of the learning flexibility within the echolocation system of the bottlenosed dolphin, or other delphinid species, would be valuable. It was noted that some delphinid species may not readily use their echolocation system in the oceanarium setting, and perhaps in the wild.

The processing of motor or motor-related information by the dolphin, like the processing of auditory information, is generally efficient and flexible. Substantial conceptual talent was demonstrated in the production of novel motor behaviors by the rough-toothed dolphin, in a study in which only new behaviors were reinforced. Performance of the bottlenosed dolphin on successive spatial reversal tasks was efficient and stable, much like the performance of the same species on auditory reversal tasks, and unlike its performance on visual form or brightness reversal tasks. Also, optimal response strategies were developed by the bottlenosed dolphin in a spatial probability learning task, in which rewards associated with two response paddles were probabilistic rather than certain.

Instances of true motor and vocal mimicry have been observed in dolphins, though more evidence for the extent and generality of these capabilities is needed. Motor mimicry occurs in primate species and vocal mimicry in avian species, but the emergence of both within a single species other than humans does not seem to be documented. Wilson (1975) speculated that the "superb" imititative capabilities of dolphins may lead to a rapid convergence of signals among members of a group, possibly resulting in group recognition. We noted that vocal signal convergence was, in fact, the case for the songs of the humpback whale. Among dolphins, the advantages of being able to modulate sonar output in response to different target characteristics or to different environmental conditions, and the advantages of sound for underwater communication, may adapt the species for vocal flexibility, and preadapt it for vocal mimicry. The capacity for motor imitation allows for synchronized group activities, such as the coordinated swimming, predatory, and defensive behaviors of many oceanic species of schooling dolphins (Norris & Dohl, this volume). Observations in oceanariums suggest that play behaviors and sexual behaviors of young dolphins develop in part through imitation of older tank mates (McBride & Hebb, 1948; McBride & Kritzler, 1951). These many adaptive uses of motor mimicry may provide the basis for its emergence as a highly generalized capability that can be expressed in the imitation of other species, as well as conspecifics.

Reports were summarized from a number of oceanariums indicating that one dolphin may learn the trained routine of another through observation alone. Both the motor mimicry capabilities described and the facilitation of response from seeing another perform a behavior would seem to require the use of analogic imagery. This is, the motor behavior imitated or show routine copied must depend on a visually based replica of the model's behavior in memory. There seems to be a paradox here in that what appear to be sophisticated imitative behaviors are dependent on visual imagery, while in some other cognitive tasks that seem equally dependent on visual imagery, information-processing capabilities were found to be very limited. For example, visual matching-to-sample must depend on visual imagery, in the form of a replica of the sample visual stimulus in memory, but proved extremely difficult for the dolphin. The paradox might be resolved if the capability for visual imagery is viewed as a specialization largely constrained to replicas of dynamic visual stimuli, the major components of the dolphin's visual world. The visual imagery system is seen as perceptually limited in the replication of visual abstractions, such as the geometric or arbitrarily shaped visual form or brightness stimuli used in our visual-matching studies. That animals may be predisposed, through their biology, to perceive and learn some things much easier than others has by now been amply demonstrated (e.g., see chapters in Hinde & Stevenson-Hinde, 1973; Seligman & Hager, 1972). The focus here on imagery limitations is a hypothesis about the *source* of a learning constraint.

Similarly, auditory imagery must underlie the vocal mimicry capabilities of dolphins. The auditory world of the dolphin is a mélange of sounds and, unlike the visual world, sounds may sometimes appear as vacant abstractions, without apprehension of source or object. The low frequency moans of a distant whale, if within the auditory frequency band of the dolphin, or the "boing" from some still unknown biological source, may appear as strange to the dolphin as they did to the early sonar operators (see Patterson & Hamilton, 1964; Wenz, 1964). Other sounds, though familiar to the dolphin, may still derive from sources not in view, as may the sonar return from a distant target. Hence, the auditory world of the dolphin is often a world of abstraction, so that auditory imagery capabilities may be easily extended to the representation of artificial, arbitrary sounds of no biological relevance, as in pure-tone mimicry, or in memory for the diverse electronically generated sounds used in our auditory delayed-matching studies.

Considering the dolphin's apparent use of both auditory and visual imagery within the context of the noted differences in capabilities for auditory and visual information processing, it follows that some forms of mimicry do not require the support of an elaborate cognitive structure. That true vocal mimicry occurs among relatively small-brained avian species should alone alert us to that possibility (Andrew, 1962). Nevertheless, it seems important to study the various forms of modeling behavior—imitatively-based behavior—more fully and formally in dolphins, because of its apparently wide-ranging characteristics in these animals, its relevance to the understanding of their cognitive processes, and its probable significance as an adaptation for their highly social existence.

LANGUAGE LEARNING

Some problems in relating animal communication to human language were discussed by Herman and Tavolga (this volume). The recent demonstrations of language-learning capabilities in Pongidae (Fouts, 1973; B. Gardner & Gardner, 1971, 1975; Patterson, 1978; Premack, 1971, 1976; Rumbaugh, 1977) reveal that most and possibly all of the design features of human language are present in the learned linguistic transactions of the apes (cf. Thorpe, 1972). Nonetheless, many linguists continue to defend a distinction between human and animal language skill by assuming, for example, that the demonstrated capacities of the apes, though impressive, are likely to be restricted to the elementary "Stage-1 Level" of two-word sentences (Brown & Herrnstein, 1975) or by defining new criteria for language that are not met by apes (e.g., Bronowski & Bellugi, 1970). It seems likely, as Thorpe (1972) suggests, that linguists will continually be able to retreat to new definitions of language in the face of demonstrations that animals have met prior criteria. Accordingly, the best path to follow at this point may be to bypass additional dispute on definition in favor of further empirical study of language capabilities of animal subjects. From such study, definitions and distinctions about language and communication can emerge inferentially.

Ape and Dolphin

The studies of language learning in great apes show that with extensive training the communication capabilities of these animals can be extended to at least the elementary levels of a digital-type imposed language. Why this capability should be manifested only with specific training is unclear, but it may be that language is unnecessary within the social and ecological living constraints of wild pongids. Alternatively, language in some rudimentary form may exist within the natural setting (cf. Menzel, 1975) but we fail to recognize it. Marler (1977), in considering these possibilities, seemed to lean more heavily on the first alternative—that pressures for language development were absent in pongid society.

   The language-training studies of the apes have practical significance in their potential for developing procedures for teaching language to language handicapped humans and for understanding the processes of language development in young children (Rumbaugh, 1977). Theoretically, as was noted by Herman and Tavolga (this volume), the findings bear importantly on questions of the continuity of human language and other forms of animal communication, whether human language evolved from more primitive nonverbal communication processes, and whether the capacity for language is unique to man. The study of the capability for language learning in dolphins could add additionally to the theoretical question of whether there are unique cognitive characteristics among the Hominoidea (apes and man) restricting language capability to that group. Do, in fact, the cognitive specializations of the Hominoidea preadapt them uniquely for language-level communication? Is it necessary to pass through man's evolutionary path to achieve language learning levels? Or, are there convergent paths to linguistic potential, perhaps reflecting more general preadaptations in the form of advanced social systems and heightened information-processing capabilities?

Any demonstrated linguistic skill of dolphins in common with the primates would shift the emphasis of explanation for language skill from that of homologous processes of ape and man to common selective processes in the form of convergent cognitive adaptations to ecological, biological, or social demands. As was noted in earlier sections of this chapter, there are interesting parallels in the cognitive characteristics of representatives of delphinids and representatives of the simian primates. Dolphins and many of the simians appear to share levels of information processing skill not attainable by other taxa. A *failure* of the dolphin to demonstrate linguistic skill, assuming adequacy of procedure, could not therefore be easily explained as a lack of general information-processing capacity but would seem to lie in some cognitive specializations within the Hominoidea not extant in other taxa (e.g., see Warren, 1974). In such a case, research would be pointed toward uncovering and defining those unique specializations, as revealed, for example, in yet uncovered information-processing constraints of dolphins.

Among the requisites for chimpanzee language skill cited by Premack (1976), dolphins appear to share a capacity for forming second-order relations, a well-developed memory system, and advanced representational abilities, when processing auditory information. There are few data yet on whether other hypothesized requisite skills, such as a capability for causal inference and intentionality in behavior, may be met by dolphins, though in one study (Bastian, 1967) intentionality was not demonstrated (see discussion in Herman & Tavolga, this volume). It should be remembered, however, that the role in language learning of any of these "requisites" is still a matter of conjecture.

In any event, given the cognitive capabilities demonstrated thus far, there seems ample bases for attempting to teach dolphins language. One language taught should almost certainly be auditory. In addition to the enviable capabilities of dolphins for processing auditory information, their vocalizations are easily conditioned and they appear to have a capability for vocal mimicry, as discussed earlier. Analyses of vocal language in children (e.g., Menyuk, 1971) underscore the importance of the plasticity of infant vocalizations in language emergence. The failures of attempts to teach vocal languages to chimpanzees (e.g., Hayes & Hayes, 1951) in retrospect reflected the constraints of their vocal apparatus (Lieberman, 1968) and, I may add, their limited capability for processing auditory information (e.g., Worsham & D'Amato, 1973). The success in teaching a gestural language to chimpanzees (Gardner & Gardner, 1969) was based on the realization that gestures are readily imitated by chimpanzees and easily conditioned. Though both gestures and vocalizations are part of the normal communicative repertoire of chimpanzees (Marler, 1965), only gestures seem plastic enough to be adaptable for the learning of an imposed language.

Early Language Training Studies with Dolphins

The attempts of Lilly (1961, 1962, 1967) to teach human vocal language to bottlenosed dolphin subjects were unsuccessful. The sketchiness of the descriptions of the procedures used and of the data obtained limit any assessment of the reasons for failure or even of the level of effort that went into training. Also, one may question the capacity of the dolphin for easily mimicking a wide variety of human speech sounds.

A better documented study of language learning in bottlenosed dolphins was that of Batteau and Markey (1968). After Batteau's untimely death in 1967, his work was carried on for a short while by others but was never formally published. A synoptic description appears in Wood (1973). In Batteau and Markey's procedure, a set of spoken sounds discriminable by a special electronic device (a "man–dolphin translator") were transformed by the device into whistle-like emissions and projected underwater into a lagoon housing two dolphins. A "dolphin–man" translator, for transforming dolphin whistles to spoken sounds, was in the design stage, but was never completed. Analog displays of the whistle emissions of the dolphins were available, however, for assessment of the accuracy with which the dolphins mimicked the artificial whistles, a result discussed earlier in the section on mimicry.

A spoken instruction to a dolphin began with the animal's name (Maui or Puka), followed by the sound IMUA, denoting "get ready," then a command word such as BIP, and finally the word OK, meaning "do it now." BIP meant "hit the ball with your pectoral fin." Other command words could be substituted for BIP. BAIEP (BUY-EE-EP), for example, meant "swim through the hoop." A correct response to the instruction was followed by the word BIYIP and a fish reward, and an incorrect response by NEGATIVE and no fish. Each animal learned to respond only to commands preceded by its own name, and to produce the instructed behavior only after the releasing word, OK.

Unlike the later language work with chimpanzees, commands such as BIP or BAIEP denoted a complex of action, object, and agent. There was no attempt to represent these unique language elements with unique sounds, nor was there provision for later dissociating these command words into their component semantic elements, such as was done by Gill and Rumbaugh (1977) in their initial use of holophrases for training language in a chimpanzee. This flaw in the execution of the Batteau and Markey study proved very limiting, since there was no way to generalize the use of a command word into a new context. For example, how was one to express "hit the hoop (rather than the ball) with your pectoral fin?" Batteau and Markey's system thus lacked the "openness" feature of human language (Hockett & Altmann, 1968), or its "reconstitutive" character (Bronowski & Bellugi, 1970), which allow for recombining language elements in unique ways to give new meaning to an utterance.

The Batteau and Markey experiment was conducted over an approximate three-year period, and at its termination Puka had acquired a set of 11 command words, and Maui 13. This was not a highly impressive accomplishment. As it stands, the experiment demonstrated the feasibility of using artificial whistle sounds as signals for controlling complex responses of the dolphin, but, because of the design limitation noted, failed as an evaluation of language-learning capability.

The Kea Language Study

At our laboratory at the University of Hawaii, preliminary training of the bottle-nosed dolphin Kea in simple auditory-language comprehension was carried out. The work was limited to receptive vocabulary training, i.e., to one-way communication. Two-way communication requires some means for the dolphin to reply in the auditory language, but this is technically very difficult, as Batteau and Markey (1968) discovered. However, the demonstration of language competency does not rest on language production. The child, during its early language stage, understands spoken language before it can speak, and an adult may understand a second language well without being able to speak it (Lenneberg, 1967). Because of these facts, and the relative ease with which receptive language skill can be studied, Lenneberg felt that the analysis of receptive skill provides a better appreciation of human language competency than does study of productive skill.

We directed our efforts with Kea to the development of an auditory coding system in which the elements of her tank world—the objects and agents in that tank and in the adjoining environment, and the actions concerning them—were mapped onto a set of unique sounds. Presentation of a string of sounds was then an instruction to Kea to manipulate her world. Kea's ability to represent her world through the coding system, in effect her receptive language skill, was measured by the adequacy of response to the instruction and, additionally, by her ability to form concepts and generalizations appropriate to the unique sound symbols or strings of sound symbols.

We had previously confirmed the ability of Kea to remember long strings of sounds (Thompson & Herman, 1977; Figs. 8.7 & 8.8), to generalize old learning into new contexts, and to acquire and use rules and concepts to control responding (Herman & Arbeit, 1973; Herman & Gordon, 1974; Herman & Thompson, in preparation). Our initial language question was whether the *sequential* information in a string of sound symbols could be integrated sufficiently well by Kea for her to form an adequate concept of the subject matter, and carry out an appropriate response.

To begin training, three objects—a ball, a life ring, and a styrofoam cylinder—were mapped onto a set of three arbitrary electronically-generated sounds, that from our previous work (e.g., Herman & Arbeit, 1973) we knew to be easily discriminable by Kea. At first, a single object was shown at the surface of the water, or held just under it, while the object's name was played. Kea was required to touch the object with the tip of her jaw. Later, pairs of objects were given, but only one was named. Touching of the named object was facilitated by using unique sounds to denote "Yes" (correct) and "No" (incorrect). Yes was followed by a fish reward and was a conditioned positive reinforcer, while No was followed by withdrawal of the object. The playing of No quickly came to interrupt an approach to the incorrect object, or any other behavior in progress. After some initial difficulty, Kea learned to distinguish reliably among the three objects by their names. Furthermore, generalization of the names occurred immediately: to balls of different size than the training object and to rings of varying size and composition.

At this stage of training the three names were holophrases signifying the more complete instruction, "touch the ball" (or ring or cylinder). In the next training step, the verb "touch" was named along with two other verbs, "fetch" and "mouth" (to place in the mouth). Training of each action word was rapid. The association of the word for touch with the behavior was facilitated through a prompting method in which the object to be touched was held against the dolphin's jaw in the presence of the sound for touch. Fetch, which was already under control of a manual gesture, was trained by pairing the new sound with the gesture and then gradually deleting the gesture over training trials. Mouthing of an object was taught by a "baiting" procedure. A fish was displayed together with an object to be mouthed and the sound for the verb "mouth." The usual response to a fish is an open mouth and the experimenter placed the object and then the fish in the dolphin's mouth. Over successive training trials, the fish was introduced progressively later in the sequence. Kea rapidly learned to open her mouth in the presence of the mouth sound and an object. Additional prompting and shaping resulted in an active approach to and mouthing of any object introduced in the tank, given the mouth sound.

The results of most interest during verb training, besides the relative rapidity of training, was the immediate generalization of response that occurred. In addition to mouthing the three familiar training objects in the presence of the mouth name, Kea correctly mouthed on their first appearance a plastic water pipe, a wooden disc, and the experimenter's open hand. The same type of immediate response generalization occurred for touch and fetch. The verb touch quickly transferred from the trained response of touching with the jaw to touching with any body part, including the top of the head (melon) and the extended pectoral fin. Objects were fetched successfully from any arbitrary starting place in the tank to any designated terminal point at which the experimenter stood. The fetch of an object was executed in whatever manner was convenient: pushing with the lower jaw, balancing on the upper jaw, or holding in the mouth, depending on the object. From these and additional generalized responses, we concluded that Kea understood the concepts of "touchingness," "mouthingness," and "fetchingness."

We next proceeded to two-word strings in which the response required was to perform the named action only relative to the named object. We found the syntactical arrangement, object–action, to be more easily trained than action–object, since we could require an intention movement (orienting response) to the named object before stating the action to be performed. This preparatory intentional response was then easily deleted by playing the two sounds—object–action—in rapid sequence. In this final syntactical arrangement, Kea performed almost flawlessly with all nine two-word combinations of the three objects and three actions. Approximately four months of training went into the program described, but much of this time was taken up in computer-program development and modification, changes in our conceptual approaches, and implementation of ad hoc training techniques.

In the final week of study, we began introducing sounds to represent the names for two trainers (one male and one female) who worked daily with Kea, in order to introduce the concept of an agent. We began with a name for the male. In the presence of the male trainer and a name for him, Kea was asked to fetch an object—i.e., to bring it to that trainer. One of several alternate trainers stood approximately 2 m to either the left or right side of the named trainer. The relative position of the named trainer and the alternate were changed according to a random schedule. The named trainer initially bent over the tank as a visual cue for the dolphin, and then over successive trials of correct fetches by the dolphin, gradually straightened out. In the final configuration, the named trainer and the alternate remained erect and motionless, hands behind their back. Kea quickly learned to bring the object only to the named male trainer. We then began similar name training for the female trainer.*

Acoustic and Gestural Language Comprehension

Work at our laboratory in language comprehension was resumed with two young female bottlenosed dolphins, Phoenix and Akeakamai, acquired from the wild a little more than a year after the loss of Kea and Puka. The training, carried out together with Douglas Richards and Jim Wolz at the University of Hawaii laboratory, followed the general procedures developed with Kea. An initial three-month period was devoted to acclimating the new animals to their tank, reinforcing socialization with trainers, and acquainting the pair with various sounds and objects prior to attaching meaning to the sounds. During this period, several behaviors, such as "fetching" and "mouthing" were put under control of gestures—specific movements of the hand or the hand and arm of the trainer. Most of these behaviors were later transferred to acoustic control.

The two dolphins were trained in one tank using separate locations in the tank as testing stations for each animal. Two whistle-like sounds were assigned as names for the animals and each dolphin quickly learned to approach a

---

*The succeding steps, of differentiating between the two named trainers and constructing various three-word instructions comprised of object, action, and agent, were never carried out. It was at this point in May 1977 that Kea, along with our second dolphin, Puka, the subject in our visual learning and memory studies, was abducted, late at night, by two recently discharged tank cleaners. The dolphins were taken from their tanks, motored some forty miles in a small van, and abandoned in the ocean in remote waters known to have a large shark population. Each dolphin had been nurtured in tanks or enclosed lagoons for eight to twelve years, had been hand-fed throughout, and was highly socialized toward humans. They were never recovered and almost certainly died not long after their abandonment. Irvine (1972) has documented the strong attachments of captive dolphins to their home pens and their extreme stress when suddenly placed in unfamiliar environments, projecting a tragic picture of the emotional distress of Kea and Puka at their end.

speaker and, later, her testing station, only if her name were heard. In further training, the name of the animal always preceded any instruction to her, and only the animal named completed the instruction, though both could hear all sounds.

Verb training was carried out first for a set of three verbs and then noun training of three objects was introduced. As with Kea, the initial training with single words was followed by two-word sentences in the syntactical order noun–verb, with a choice given between the named object and an alternative object that was associated with a name not given. The vocabulary was extended later by adding new nouns and verbs. Table 8.1 lists the working vocabulary of 25 words, as of March 1980. Some combinations, such as "window–through" or more importantly, perhaps, "person–through", were inadmissible, restricting the number of possible noun–verb pairings to 52. However, since not all items in Phoenix's vocabulary are understood by Akeakamai the number of possible two-way combinations for Akeakamai is at present 44. Both dolphins are able to comprehend the various two-word combinations of their vocabulary-set with high reliability and additionally each understands a number of three-word combinations.

Most of the vocabulary items are obvious. However, "person" is normally a person's arm placed in the water from tankside, but responses to "person" will also be made, with high reliability, to other human body parts, such as a leg in

Table 8.1. Current Receptive Vocabulary of the Dolphins Phoenix and Akeakamai

Objects (Nouns)	Actions (Verbs)	Agents (Subjects)	Feedback
Transferable (Mobile)	*Take Direct Object*	Phoenix	Yes
Ball	Mouth	Akeakamai	No
Pipe	Tail-Touch[a]		Ready
Hoop	Pec-Touch[b]		
Person	(Go) Over		
Fish	(Go) Under		
Frisbee	(Go) Through		
Nontransferable	*Takes Direct or*		
Gate (P)	*Indirect Object*		
Window (P)	Fetch		
Water			
Panel (P)	*Intransitive*		
Speaker	(Go) Left (A)		
	(Go) Right (A)		

[a]Instructs a touch of an object with the tail flukes.
[b]Instructs a touch with a pectoral fin.
P = Item in Phoenix's vocabulary only; A = item in Akeakamai's vocabulary only.

the water, or even the whole body if the person is inside the tank. The sentence "person-fetch" will result in the dolphin pushing a person in the tank to the tank wall.

"Fish" is both a reward, given for good performance, and an object. If instructed, the dolphins will fetch or mouth a fish tethered to a string to obtain another fish as reward. On occasion, if they approach a fish incorrectly, in the case where another object has been named, and become aware of their error through the absence of the "Yes" sound, they may quickly eat the tethered fish, taking it off the string first. "Water" refers to a stream of water that is projected into the tank from a plastic pipe on the tank wall. Water can be successfully coupled with any of the verbs listed that make sense in combination with it. In response to "water-mouth," for example, the dolphin will open its mouth to allow the stream of water to strike inside. "Fetch-water" is an excluded pair, since there is no available way for the dolphin to execute this instruction, although on the few occasions when, out of curiosity, we did give the instruction, the dolphins attempted to push the water stream along with their jaw, as in fetching other objects.

"Gate" and "window" are permanent fixtures in the tank. "Panel" refers to an aluminum panel having a movable lever that is normally used at night to allow the dolphins to turn on or off or select among various tape-recorded acoustic stimuli, including dolphin sounds, whale sounds, human voices, music, etc. "Speaker" is an underwater sound projector that can be moved by the trainer (but not by the dolphin) to various locations in the tank. Finally, "left" and "right" are referenced to the dolphin and are normally used in compound instructions to Akeakamai, such as "Right-Ball-Go over," meaning go to the ball to your right (and *not* to an identical one to your left) and jump over it. These spatial instructions are followed accurately by Akeakamai.

A complete instruction to a dolphin begins with her name, followed by a sound denoting "Ready" when she responds to her name by pressing a target at her station, followed by a sentence in the order noun-verb. If the instruction is executed correctly, a sound signifying "Yes" is heard, then the dolphin's name and, finally, the word for "Fish." The sequence beginning with "Yes" is omitted if the instruction is executed incorrectly and occasionally a sound denoting "No" ("Incorrect") may be played. Playing of "No" commonly produces emotional responding and its use is restricted to situations that in the judgment of the trainer may be informative for the animal.

The only structurally varying segment of the instruction is the particular noun-verb sequence. For Phoenix, these instructions are given as sounds, each 0.5 to 1.0 sec in duration with a pair separated by 0.25 sec. All sounds and intervals between sounds are generated by computer and projected underwater into the tank. For Akeakamai, the sequences were initially given as sounds also, but in later training only manual gestures were used. New words introduced to Akeakamai continue to be gestures, while new words introduced to Phoenix are always sounds. For the gestural instruction, verbs are one-handed signals, while nouns are two-handed. The gestures are short in duration, typically 1- to 2-sec long, within the limits of intertrainer variability. Performance of the two dolphins—one on acoustic instructions and the second on gestural—are at this writing approximately equivalent and the vocabularies are similar. In tests of comprehension of two-word instructions, with the various instructions given in random sequence, performance currently averages approximately 80% correct responses for a week's worth of some 600 to 700 two-word instructions. Most of the errors are object errors that occur when the number of objects to choose among is large. For example, when choosing among four or five objects Phoenix's performance over a week of testing is better than 90% correct. But performance decreases to between 70 and 80% correct when the number of objects to choose among is between 6 and 11. Action errors are very rare in any case— on the order of 5% for all of the verb instructions given in a week.

For both dolphins, new nouns taught are used correctly with old verbs, and new verbs with old nouns, without additional training. For example, "water" was the most-recently taught noun for both animals and was discriminated correctly from other objects nearly perfectly within a single training session. Two-word sentences using all verbs except "Fetch" and "Through" in combination with water were executed correctly immediately at the next session.

Phoenix has additionally been trained on three-word sentences consisting of direct object—verb—indirect object. The direct objects consist of all transferable objects (Table 8.1). The indirect objects are comprised of these same transferable objects plus all the nontransferable items. Ordering of words conveys meaning and a sentence such as Ball–Fetch–Hoop requires a fetch of the ball to the hoop while Hoop–Fetch–Ball requires that the hoop be retrieved and brought over to the position of the floating ball. Performance on all three-word instructions currently averages approximately 50% correct responses with most of the errors made to the indirect object. This level of performance is well above chance responding since the number of possible responses is very large.

Work is still at an early stage. Only approximately twelve months of training have been completed so far, but in this time the degree of progress made seems roughly comparable with early accomplishments with chimpanzees. The chimp work, however, has emphasized language production rather than comprehension, and receptive vocabularies are not well documented for the primates. Our goals in continuing work are focused on extensions to the vocabulary, construction of additional sentences whose meanings are syntax dependent, the incorporation of new language classes, such as adjectives or adverbs, and the development of productive language competency.

The rapid progress that was made in these language comprehension studies encourages the view that a dolphin may eventually be able to acquire a rich receptive vocabulary and comprehend complex instructions within an imposed language. It also makes the prospect of a dolphin acquiring a productive vocabulary more likely, though the technical aspects of achieving two-way communication with a dolphin are still formidable.

The results also strengthen the notion that the dolphin is capable of arbitrary symbolization of its world, at least when so trained in the laboratory. From this, one cannot infer that such symbolization occurs to any marked degree in the natural world. Rather, as seems the case with language learning in nonhuman primates, and with the expression of advanced cognitive skills in humans, the realization of a capacity depends on the environment or culture being structured in favorable ways. Straightforwardly stated, the evolution of the brain to meet biological, ecological, or social demands may preadapt a species for the eventual display of cognitive skills beyond those demands.

Overview

The capacities for vocal mimicry, the conditionability of vocalizations, the advanced rule learning capabilities, and other demonstrated intellectual traits suggest that dolphins, like some of the greater apes, may be capable of learning an imposed language in the laboratory setting. The demonstration of such capability in a delphinid species would press us toward the reconsideration of the foundations for the evolution of language in humans, and further underscore the convergence in cognitive characteristics of primate and dolphin. Recent work at our laboratory demonstrated that the dolphin was capable of comprehending simple auditory or gestural language elements and their combination into two-word and three-word sentences. Further work on such comprehension is currently underway.

SUMMARY AND CONCLUSIONS

This review has described some of the cognitive characteristics of dolphins. The cluster of characteristics has begun to define some of the intellectual specializations, capabilities, and limitations of bottlenosed dolphins and of a few of the other Delphinidae. One can speak with some confidence of the dolphin's well-developed auditory memory system, its representational abilities, its capacities for rule-governed behavior, and its flexible modeling (imitative and observational learning) capabilities.

In concert with the rich fabric of sound in the undersea environment, the dolphin's complex cognitive manipulations of its world may be mainly processed through auditory centers and, possibly, through centers that integrate and control perceptual-motor behaviors. The dolphin is a facile learner of most types of auditory tasks, including those having some weighty conceptual demands. Its auditory working memory is impressively faithful, capable of registering, holding, and recovering substantial amounts of information. Approximately four to six discrete unrelated and arbitrary auditory items may be stored in parallel in working memory under strategies adopted by the dolphin in tests of memory for lists of sounds.

The dolphin is capable of forming and generalizing response rules, as evidenced by its efficiency in solving new instances of old auditory problems governed by a common rule, or in improvising motor behaviors governed by a requirement for novelty. The dolphin may be able to represent its world through arbitrary auditory symbols, as inferred from the encouraging results in the studies of auditory language comprehension. The conditional memory tests, in which one sound in memory was represented by another, or a visual signal by an auditory one, additionally suggest competency in manipulating auditory symbols.

The dolphin seems highly proficient at some imitative behaviors, including sharing a capability with the human for *both* vocal and motor mimicry. The imitative talent has value for socially cooperative or socially convergent species like dolphins and supports an apparent ability to learn motor tasks easily by observation.

The elaborate social matrix typifying some of the schooling delphinids suggests the existence of an extensive communication network demanding of cognitive skill and, in fact, may be the evolutionary impetus for the development of such skill. Delphinid communication should be viewed, however, within the context of the communication levels and processes demonstrated by other highly social advanced nonhuman mammals. In its natural form it is complex, elaborate, but not linguistic. All attempts at demonstrating the existence of a natural language in dolphins have been unsuccessful. Nevertheless, like the chimpanzee that also fails to give evidence of an extant natural language in its usual social milieu, the dolphin may be capable of comprehending a simple artificial language under laboratory conditions.

Serious cognitive limitations in visual memory for form or brightness and in the manipulation of these types of visual information abstractly were noted. These visual memory constraints may, however, be "unlocked" if the visual information is represented by an auditory code. Such representation may free the auditory system to manipulate the re-encoded visual information in the complex manner with which auditory information is treated. It was suggested that the visual system may be tied more to the concrete, real world of tangible, apprehensible object than is the auditory system, which often encounters abstracted or remote acoustical events. The visual information processing system may have its major specialization and its area of greatest flexibility in the realm of movement (dynamic) stimuli. The results available so far in the training of gestural language comprehension indicate that conceptual richness may flower in the visual system if the visual information manipulated is dynamic.

Though their evolutionary lines have been divergent for tens of millions of years, there seem to be many areas of cognitive convergence between some of the delphinids and some of the advanced simian primates. In many ways the two groups are cognitive cousins, though at opposite poles in sensory specializations and ecology. Performance of the dolphin on its complex auditory learning and memory tasks was very similar in form and level of accomplishment and in underlying response strategies to performance of primates tested on analogous visual tasks. The data available suggest that both taxa have advanced capabilities for classifying, remembering, and discovering relationships among events, for forming response rules of general utility, and for manipulating symbols. Conceptual limitations also appear to be very similar across the groups. As a general conclusion, it appears as if the two groups process information in much the same way, exposing similar patterns of capacities and constraints.

The major link that cognitively connects the otherwise evolutionarily divergent delphinids and primates may be social pressure—the requirement for integration into a social order having an extensive communication matrix for promoting the well-being and survival of individuals (cf. Humphrey, 1976). Effective functioning in such a society demands extensive socialization and learning. The extended maturational stages of the young primate or dolphin and the close attention given it by adults and peers (see Herman & Tavolga, this volume) provide the time and tutoring necessary for meeting these demands. In general, high levels of parental care and high degrees of cortical encephalization go together (Eisenberg & Kleiman, 1972). It is not difficult to imagine that the extensive development of the brain in the Delphinidae, and the resulting cognitive skills of some members of this group, have derived from the demands of social living, including both cooperation and competition among peers, expressed within the context of the protracted development of the young. These cognitive skills may in turn provide the behavioral flexibility that has allowed the diverse family of delphinids to successfully invade so many different aquatic habitats and niches.

ACKNOWLEDGMENTS

Preparation of this paper was supported by Grants BNS72-01857 and BNS77-16882 from the National Science Foundation. Early versions of this paper were read by Alan Kamil, Ronald Schusterman, and M. J. Warren. Their detailed comments were very helpful and are acknowledged with gratitude.

REFERENCES

Andrew, R. J. (1962). Evolution of intelligence and vocal mimicking. *Science,* **137**, 585–589.

Andrew, R. J. (1976). Attentional processes and animal behaviour. In: *Growing Points in Ethology* (P. P. G. Bateson & R. A. Hinde, Eds.), pp. 95–133. London: Cambridge University Press.

Atkinson, R. C. & Shiffrin, R. M. (1968). Human memory: A proposed system and its control processes. In: *The Psychology of Learning and Motivation,* Vol. 2 (K. W. Spence & J. T. Spence, Eds.), pp. 89–195. New York: Academic.

Au, W. W. L., Floyd, R. W., Penner, R. H. & Murchison, E. M. (1974). Measurement of echolocation signals of the Atlantic bottlenose dolphin, *Tursiops truncatus* Montagu, in open waters. *J. Acoust. Soc. Am.,* **56**, 1280–1290.

Baddeley, A. D. & Hitch, G. (1974). Working memory. In: *The Psychology of Learning and Motivation: Advances in Research and Theory,* Vol. 8 (G. H. Bower, Ed.). pp. 47–87. New York: Academic.

Bastian, J. (1967). The transmission of arbitrary environmental information between bottlenose dolphins. In: *Animal Sonar Systems, Biology and Bionics,* Vol. II (R.-G. Busnel, Ed.), pp. 807–873. Juoy-en-Josas, France: Laboratoire de Physiologie Acoustique.

Bateson, G. (1942). Social planning and the concept of deutero-learning. Conference on Science, Philosophy, and Religion, Second Symposium, New York: Harper & Row, 1942.

Batteau, D. W. & Markey, P. R. (1968). Man/dolphin communication. Final Report Contr N00123-67-1103, 15 Dec. 1966-13 Dec. 1967. U. S. Naval Ordinance Test Station, China Lake, California.

Beach, F. A. III & Herman, L. M. (1972). Preliminary studies of auditory problem solving and intertask transfer by the bottlenose dolphin. *Psychol. Rep.*, **22**, 49-62.

Beach, F. A. III & Pepper, R. L. (1973). Conditioned differential vocalization in a position and shape discrimination task with the dolphin. San Diego: Naval Undersea Center NUC TN996, March, 1973.

Beach, F. A. III, Pepper, R. L., Simmons, J. V. Jr., Nachtigall, P. E. & Siri, P. A. (1974). Spatial habit reversal in two species of marine mammals. *Psychol. Rec.*, **24**, 384-391.

Bekoff, M. (1972). The development of social interaction, play, and meta-communication in mammals: An ethological perspective. *Q. Rev. Biol.*, **47**, 412-434.

Berryman, R., Cumming, W. W. & Nevin, J. A. (1963). Acquisition of delayed matching in the pigeon. *J. Exp. Anal. Behav.*, **6**, 101-107.

Bolles, R. C. (1973). The comparative psychology of learning: The selective association principle and some problems with "general" laws of learning. In: *Perspectives on Animal Behavior* (G. Bermant, Ed.), pp. 280-306. Glenview, Ill.: Scott-Foresman.

Broadbent, D. E. (1958). *Perception and Communication.* New York: Pergamon.

Broadbent, D. E. (1971). *Decision and Stress.* New York: Academic.

Bronowski, J. & Beluggi, U. (1970). Language, name and concept. *Science*, **168**, 669-673.

Brown, D. H., Caldwell, D. K. & Caldwell, M. C. (1966). Observations on the behavior of wild and captive false killer whales, with notes on associated behavior of other genera of captive delphinids. *Los Ang. Cty. Mus. Contrib. Sci.*, **95**, 1-32.

Brown, D. H. & Norris, K. S. (1956). Observations of captive and wild cetaceans. *J. Mammal.*, **37**, 311-326.

Brown, R. & Herrnstein, R. J. (1975). *Psychology.* Boston, Mass.: Little, Brown.

Bullock, T. H. & Gurevich, V. S. (1979). Soviet literature on the nervous system and psychobiology of cetacea. *Int. Rev. Neurobiol.*, **21**, 47-127.

Bullock, T. H. & Ridgway, S. M. (1972). Evoked potentials in the central auditory system of alert porpoises to their own and artificial sounds. *J. Neurobiol*, **3**, 79-99.

Caldwell, D. K. & Caldwell, M. C. (1972). Vocal mimicry in the whistle mode in the Atlantic bottlenosed dolphin. *Cetology*, **9**, 1-8.

Caldwell, M. C., Caldwell, D. K. & Siebenaler, J. B. (1965). Observations on captive and wild Atlantic bottlenose dolphins, *Tursiops truncatus*, in the northeastern Gulf of Mexico. *Los Ang. Cty. Mus. Contrib. Sci.*, **91**, 1-10.

Craik, F. I. M. & Lockhart, R. S. (1972). Levels of processing: A framework for memory research. *J. Verb. Learn. Verb. Behav.*, **12**, 599-607.

D'Amato, M. R. (1973). Delayed matching and short-term memory in monkeys. In: *The Psychology of Learning and Motivation: Advances in Research and Theory*, Vol. 7 (G. H. Bower, Ed.), pp. 227-269. New York: Academic.

D'Amato, M. R. & Cox, J. K. (1976). Delay of consequences and short-term memory in monkeys. In: *Processes of Animal Memory* (D. L. Medin, W. A. Roberts & R. T. Davis, Eds.), pp. 49-78. Hillsdale, N. J.: Erlbaum.

D'Amato, M. R. & Worsham, R. W. (1972). Delayed matching in the capuchin monkey with brief sample duration. *Learn. Motiv*, **3**, 304-321.

D'Amato, M. R. & Worsham, R. W. (1974). Retrieval cues and short-term memory in capuchin monkeys. *J. Comp. Physiol. Psychol.*, **86**, 274-282.

Davenport, R. K. (1977). Cross-modal perception: A basis for language? In: *Language Learn-*

ing by a Chimpanzee: The Lana Project (D. M. Rumbaugh, Ed.), pp. 73-83. New York: Academic.

Davis, R. T. & Fitts, S. S. (1976). Memory and coding processes in discrimination learning. In: *Processes of Animal Memory* (D. L. Medin, W. A. Roberts & R. T. Davis, Eds.), pp. 167-180. Hillsdale, N.J.: Erlbaum.

del Russo, J. E. (1975). Observational learning of discriminative avoidance in hooded rats. *Anim. Learn. Behav.*, **3**, 76-80.

Deutsch, D. (1970). Tones and numbers: Specificity of interference in short-term memory. *Science*, **168**, 1604-1605.

Deutsch, J. A. & Deutsch, D. (1963). Attention: Some theoretical considerations. *Psychol. Rev.*, **70**, 80-90.

Devine, J. V. & Jones, L. C. (1975). Matching-to-successive samples: A multiple unit memory task with rhesus monkeys. *Behav. Res. Methods Instrum.*, **7**, 438-440.

Devine, J. V., Jones, L. C., Neville, J. W. & Sakai, D. J. (1977). Sample duration and type of stimuli in delayed matching-to-sample in rhesus monkeys. *Anim. Learn. Behav.*, **5**, 57-62.

Diercks, K. J. (1972). Biological sonar systems: A bionics survey. Applied Research Laboratories: University of Texas. ARL-TR-72-34. Sept. 1972.

Eberhardt, C. & Baptista, L. F. (1977). Intraspecific and interspecific song mimesis in California song sparrows. *Bird-banding*, **48**, 193-300.

Eisenberg, J. F. & Kleiman, D. G. (1977). The evolution and adaptive significance of parental care strategies in the Insectivora Edentata and Primates: A review. Paper presented at the XVth International Ethological Conference, Bielefeld, West Germany, August, 1977.

Etkin, M. W. (1972). Light produced interference in a delayed matching test with capuchin monkeys, *Learn. Motiv.*, **3**, 313-324.

Evans, W. E. (1967). Radio-telemetric studies of two species of small odontocete cetaceans. In: *The Whale Problem: A Status Report* (W. E. Schevill, Ed.), pp. 385-394. Cambridge, Mass.: Harvard University Press.

Evans, W. E. & Powell, B. A. (1967). Discrimination of different metallic plates by an echolocating delphinid. In: *Animal Sonar Systems: Biology and Bionics*, Vol. I (R.-G. Busnel, Ed.), pp. 363-383. Jouy-en-Josas, France: Laboratoire de Physiologie Acoustique.

Flanigan, N. J. (1972). The central nervous system. In: *Mammals of the Sea: Biology and Medicine* (S. H. Ridgway, Ed.), pp. 215-232. Springfield, Ill.: Thomas.

Fletcher, H. J. (1965). The delayed-response problem. In: *Behavior of Nonhuman Primates*, Vol. I (A. M. Schrier, H. F. Harlow & F. Stollnitz, Eds.), pp. 129-165. New York: Academic.

Fouts, R. S. (1973). Acquisition and testing of gestural signs in four young chimpanzees. *Science*, **180**, 978-980.

French, G. M. (1965). Associative problems. In: *Behavior of Nonhuman Primates*, Vol. 1 (A. M. Schrier, H. F. Harlow, & F. Stollnitz, Eds.), pp. 167-210. New York: Academic.

Gardner, B. T. & Gardner, R. A. (1971). Two-way communication with an infant chimpanzee. In *Behavior of Nonhuman Primates*, Vol. 4 (A. M. Schrier & F. Stollnitz, Eds.), pp. 117-185. New York: Academic.

Gardner, B. T. & Gardner, R. A. (1975). Evidence for sentence constituents in the early utterances of child and chimpanzee. *J. Exp. Psychol. Gen.*, **104**, 244-267.

Gardner, R. A. & Gardner, B. T. (1969). Teaching sign language to a chimpanzee. *Science*, **165**, 664-667.

Gill, T. V. & Rumbaugh, D. M. (1977). Training strategy and tactics. In: *Language Learning by a Chimpanzee: The Lana Project*. (D. M. Rumbaugh, Ed.), pp. 117-162. New York: Academic.

Gregg, L. W. (1971). Similarities in the cognitive processes of monkey and man. In: *Cognitive Processes in Nonhuman Primates* (L. E. Jarrard, Ed.), pp. 156-164. New York: Academic.

Gregory, R. D. (1975). Do we need cognitive concepts? In: *Handbook of Psychobiology* (M. Gazziniga & C. Blakemore, Eds.), pp. 607-628. New York: Academic.

Griffin, D. R. (1976). *The Question of Animal Awareness.* New York: Rockefeller University Press.

Guilford, J. P. (1967). *The Nature of Human Intelligence.* New York: McGraw Hill.

Harlow, H. E. (1949). The formation of learning sets. *Psychol. Rev.*, **56**, 51-65.

Hayes, K. J. & Hayes, C. (1951). The intellectual development of a home-raised chimpanzee. *Proc. Am. Philos. Soc.*, **95**, 105-109.

Herman, L. M. (1975a). Interference and auditory short-term memory in the bottlenose dolphin. *Anim. Learn. Behav.*, **3**, 43-48.

Herman, L. M. (1975b). The sound memory of the bottlenosed dolphin. Invited address before the Conference on the Biology and Conservation of Marine Mammals. University of California at Santa Cruz, December 1975.

Herman, L. M. & Arbeit, W. R. (1972). Frequency difference limens in the bottlenose dolphin: 1-70 KC/S. *J. Aud. Res.*, **2**, 109-120.

Herman, L. M. & Arbeit, W. R. (1973). Stimulus control and auditory discrimination learning sets in the bottlenose dolphin. *J. Exp. Anal. Behav.*, **19**, 379-394.

Herman, L. M. & Bailey, D. R. (1970). Comparative effects of retroactive and proactive interference in motor and short-term memory. *J. Exp. Psychol.*, **86**, 407-415.

Herman, L. M., Beach, F. A. III, Pepper, R. L. & Stalling, R. B. (1969). Learning set performance in the bottlenose dolphin. *Psychon. Sci.*, **14**, 98-99.

Herman, L. M. & Gordon, J. A. (1974). Auditory delayed matching in the bottlenose dolphin. *J. Exp. Anal. Behav.*, **21**, 19-26.

Herman, L. M., Forestell, P. H. & Antinoja, R. C. (1977). Study of the 1976/77 migration of humpback whales into Hawaiian waters: Composite description. Report submitted to the National Marine Fishery Service and U.S. Marine Mammal Commission, October, 1977.

Herman, L. M. Peacock, M. F. Yunker, M. P. & Madsen, C. J. (1975). Bottlenosed dolphin: Double-slit pupil yields equivalent aerial and underwater acuity. *Science,* **139**, 650-652.

Herman, L. M. & Thompson, R. K. R. (in prep.). Retrieval cues and auditory short-term memory in the bottlenosed dolphin.

Hinde, R. A. & Stevenson-Hinde, J. (1973). (Eds.). *Constraints on Learning: Limitations and Predispositions.* New York: Academic.

Hockett, C. F. & Altmann, S. A. (1968). A note on design features. In: *Animal Communication* (T. A. Sebeok, Ed.), pp. 61-72. Bloomington: Indiana University Press.

Hodos, W. (1970). Evolutionary interpretation of natural and behavioral studies of living vertebrates. In: *The Neural Sciences: Second Study Program* (F. O. Schmidt, Ed.), pp. 26-39. New York: Rockefeller University Press.

Honig, W. K. (1978a). On the conceptual nature of cognitive terms: An initial essay. In: *Cognitive Processes in Animal Behavior* (S. H. Hulse, H. Fowler, & W. K. Honig, Eds.), pp. 1-14. Hillsdale, N. J.: Erlbaum.

Honig, W. K. (1978b). Studies of working memory in pigeons. In: *Cognitive Processes in Animal Behavior* (S. H. Hulse, H. Fowler, & W. K. Honig, Eds.), pp. 211-248. Hillsdale, N. J.: Erlbaum.

Honig, W. K. & James, P. H. R. (1971). (Eds.). *Animal Memory.* New York: Academic.

Hulse, S. H., Fowler, H. & Honig, W. K. (1978). (Eds.). *Cognitive Processes in Animal Behavior.* Hillsdale, N. J.: Erlbaum.

Humphrey, N. K. (1976). The social function of intellect. In: *Growing Points in Ethology* (P. P. G. Bateson & R. A. Hinde, Eds.), pp. 301-317. Cambridge: Cambridge University Press.

Irvine, B. (1972). Behavioral changes in dolphins in a strange environment. *Q. J. Fla. Acad. Sci.*, **34**, 206-212.

Jansen, J. & Jansen, J. K. S. (1969). The nervous system of cetacea. In: *The Biology of Marine Mammals* (H. T. Andersen, Ed.), pp. 175–252. New York: Academic.

Jarrard, L. E. & Moise, S. L. (1971). Short-term memory in the monkey. In: *Cognitive Processes of Nonhuman Primates* (L. E. Jarrard, Ed.), pp. 1–24. New York: Academic.

Jerison, H. J. (1973). *Evolution of the Brain and Intelligence.* New York: Academic.

Jerison, H. J. (1978). Brain and intelligence in whales. In: *Whales and Whaling*, Vol. 2 (S. Frost, Ed.), pp. 161–198. Canberra: Australian Government Printing Service.

John, E. R., Chesler, P., Bartlett, F. & Victor, I. (1968). Observation learning in cats. *Science*, **159**, 1489–1491.

Johnson, C. S. (1967). Sound detection thresholds in marine mammals. In: *Marine Bio-Acoustics*, Vol. 2 (W. N. Tavolga, Ed.), pp. 247–260. New York: Pergamon.

Kamil, A. C. (1975). Systematic nectar foraging, or learning can be important to an animal in its natural habitat. Paper presented to the Psychonomic Society, Denver, Nov. 1975.

Kantowitz, B. H. (1974). (Ed.). *Human Information Processing: Tutorials In Performance and Cognition.* Hillsdale, N. J.: Erlbaum.

Kawamura, S. (1963). The process of sub-culture propagation among Japanese macaques. In: *Primate Social Behavior: An Enduring Problem* (C. H. Southwick, Ed.), pp. 82–90. Princeton: Van Nostrand.

Kellogg, W. N. & Rice, C. E. (1963). Visual discrimination in a bottlenose porpoise. *Psychol. Rec.*, **13**, 483–498.

Kellogg, W. N. & Rice, C. E. (1964). Visual problem-solving in a bottlenose dolphin. *Science*, **143**, 1052–1055.

Kellogg, W. N. & Rice, C. E. (1966). Visual discrimination and problem solving in a bottlenose dolphin. In: *Whales, Dolphins, and Porpoises* (K. S. Norris, Ed.), pp. 731–754. Berkeley: University of California Press.

Kintsch, W. (1970). *Learning, Memory, and Conceptual Processes.* New York: Wiley.

Kritzler, H. (1952). Observations on the pilot whale in captivity. *J. Mammal.*, **33**, 321–334.

Kroll, N. E. (1975). Visual short-term memory. In: *Short-Term Memory* (D. Deutsch & J. A. Deutsch, Eds.), pp. 153–179. New York: Academic.

Lachman, R., Lachman, J. & Butterfield, E. C. (1979). *Cognitive Psychology and Information Processing.* Hillsdale, N. J.: Erlbaum.

Lieberman, P. (1968). Primate vocalizations and human linguistic ability. *J. Acoust. Soc. Am.*, **44**, 1574–1584.

Lenneberg, E. H. (1967). *Biological Foundations of Language.* New York: Wiley.

Lilly, J. C. (1961). *Man and Dolphin.* New York: Doubleday.

Lilly, J. C. (1962). Vocal behavior of the bottlenosed dolphin. *Proc. Am. Philos. Soc.*, **106**, 520–529.

Lilly, J. C. (1965). Vocal mimicry in *Tursiops*: Ability to match numbers and durations of human vocal bursts. *Science*, **147**, 300–301.

Lilly, J. C. (1967). *The Mind of the Dolphin: A Nonhuman Intelligence.* New York: Doubleday.

Lilly, J. C. & Miller, A. M. (1962). Operant conditioning of the bottlenose dolphin with electrical stimulation of the brain. *J. Comp. Physiol. Psychol.*, **55**, 73–79.

Lilly, J. C., Miller, A. M. & Truby, H. M. (1968). Reprogramming of the sonic output of the dolphin: Sonic burst count matching. *J. Acoust. Soc. Am.*, **43**, 1412–1424.

Mackintosh, N. J. (1974). *The Psychology of Animal Learning.* New York: Academic.

Madsen, C. (1976). Tests for color discrimination and spectral sensitivity in the bottlenosed dolphin, *Tursiops truncatus*. Ph.D. thesis, University of Hawaii.

Maki, W. S., Jr. & Leith, C. R. (1973). Shared attention in pigeons. *J. Exp. Anal. Behav.*, **19**, 345–349.

Maki, W. S., Jr. & Leuin, T. C. (1972). Information processing by pigeons. *Science*, **176**, 535-536.

Maki, W. S., Jr., Moe, J. C. & Bierley, C. M. (1977). Short-term memory for stimuli, responses, and reinforcers. *J. Exp. Psychol. Anim. Behav. Processes*, **3**, 156-177.

Marler, P. (1965). Communication in monkeys and apes. In: *Primate Behavior: Field Studies of Monkeys and Apes* (I. DeVore, Ed.), pp. 544-584. New York: Holt, Rinehart & Winston.

Marler, P. (1977). The evolution of communication. In: *How Animals Communicate* (T. A. Sebeok Ed.), pp. 45-70. Bloomington: Indiana University Press.

Massaro, D. M. (1970). Perceptual processes and forgetting in memory tasks. *Psychol. Rev.*, **77**, 557-567.

McBride, A. F. & Hebb, D. O. (1948). Behavior of the captive bottlenose dolphin, *Tursiops truncatus*. *J. Comp. Physiol. Psychol.*, **41**, 111-123.

McBride A. F. & Kritzler, H. (1951). Observations on pregnancy, parturition, and post-natal behavior in the bottlenose dolphin. *J. Mammal.*, **32**, 251-266.

Medin, D. L. (1974). The comparative study of memory. *J. Hum. Evol.*, **3**, 455-463.

Medin, D. L. & Davis, R. T. (1974). Memory. In: *Behavior of Nonhuman Primates*, Vol. 5 (A. M. Schrier & F. Stollnitz, Eds.),pp. 2-47. New York: Academic.

Medin, D. L. Roberts, W. A. & Davis, R. T. (1976). (Eds.). *Processes of Animal Memory*. Hillsdale, N. J.: Erlbaum.

Melton, A. W. & Martin, E. (1972). *Coding Processes in Human Memory*. Washington, D. C.: Winston.

Menyuk, P. (1971). *The Acquisition and Development of Language*. Englewood Cliffs, N.J.: Prentice-Hall.

Menzel, E. W., Jr. (1975). Natural language of young chimpanzees. *New Sci.*, **16 Jan.**, 127-130.

Menzel, E. W. (1978). Cognitive mapping in chimpanzees. In: *Cognitive Processes in Animal Behavior* (S. H. Hulse, H. Fowler & W. K. Honig, Eds.), pp. 375-422. Hillsdale, N.J.: Erlbaum.

Miles, R. C. (1971). Species differences in "transmitting" spatial location information. In: *Cognitive Processes of Nonhuman Primates* (L. E. Jarrard, Ed.), pp. 165-179. New York: Academic.

Moise, S. L. (1970). Short-term retention in *Macaca* speciosa following interpolated activity during delayed matching from sample. *J. Comp. Physiol. Psychol.*, **73**, 506-514.

Morgane, P. J. & Jacobs, N. S. (1972). Comparative anatomy of the cetacean nervous system. In: *Functional Anatomy of Marine Mammals* (R. J. Harrison, Ed.), pp. 117-244. New York: Academic.

Murdock, B. B., Jr. (1974). *Human Memory: Theory and Data*. Potomack, Md.: Erlbaum.

Nachtigall, P. E. (1971). Spatial discrimination and reversal based on differential magnitude of reward in the dolphin, *Tursiops truncatus*. *Proc. 8th Annu. Conf. Biol. Sonar Diving. Anim.*, Menlo Park, Calif., Stanford Research Institute, pp. 67-72.

Neisser, U. (1967). *Cognitive Psychology*. New York: Appelton-Century-Crofts.

Norman, D. A. (1976). *Memory and Attention: An Introduction to Human Information Processing*. New York: Wiley.

Norman, D. A. & Bobrow, D. G. (1975). On data-limited and resource-limited processes. *Cog. Psychol.*, **7**, 44-64.

Norris, K. S. (1974). *The Porpoise Watcher*. New York: Norton.

Norris, K. S., Dormer, K. J., Pegg, J. & Liese, G. J. (1971). The mechanism of sound production and air recycling in porpoises: A preliminary report. *Proc. 8th Annu. Conf. Biol. Sonar Diving Anim.*, Menlo Park, Calif., Stanford Research Institute, pp. 113-129.

Norris, K. S. & Evans, W. E. (1967). Directionality of echolocation clicks in the rough-toothed porpoise, *Steno bredanensis* (Lesson). In: *Marine Bio-Acoustics*, Vol. 2 (W. N. Tavolga, Ed.), pp. 305–316.

Norris, K. S., Prescott, J. H., Asa-Dorian, P. V. & Perkins, P. (1961). An experimental demonstration of echolocation behavior in the porpoise, *Tursiops truncatus* (Montagu) *Biol.Bull.*, **120**, 163–176.

Novick, A. (1973). Echolocation in bats: A zoologist's view. *J. Acoust. Soc. Am.*, **54**, 139–146.

Olton, D. S. & Samuelson, R. J. (1976). Remembrance of places passed: Spatial memory in rats. *J. Exp. Psychol. Anim. Behav. Processes.* **2**, 97–116.

Passingham, R. E. (1975). The brain and intelligence. *Brain Behav. Evol.*, **11**, 1–15.

Patterson, B. & Hamilton, G. R. (1964). Repetitive 20 cycle per second biological hydroacoustic signals at Bermuda. In: *Marine Bio-Acoustics* (W. N. Tavolga, Ed.), pp. 125–145. New York: Pergamon.

Patterson, F. G. (1978). The gestures of a gorilla: Language acquisition in another pongid. *Brain Lang.*, **5**, 72–97.

Payne, R. S. (1978). Behaviors and vocalizations of humpback whales, *Megaptera* sp. In: *Report on a Workshop on Problems Related to Humpback Whales (Megaptera novaeangliae) in Hawaii.* U.S. Marine Mammal Commission Report MMC-77/03.

Payne R. S. & McVay, S. (1971). Songs of humpback whales. *Science*, **173**, 585–597.

Penner, R. H. (1966). Conditioned whistle replication in *Tursiops truncatus* (Montagu). Technical Report 17 April 1966, Listening, Inc. Contract N123-(60530)55399A for U.S. Naval Ordinance Test Station, China Lake.

Penner, R. H. & Murchison, A. E. (1970). Experimentally demonstrated echolocation in the Amazon River porpoise, *Inia geoffrensis* (Blainville). Naval Undersea Center Technical Publication NUC TP 187 Rev. 1, June 1970.

Pepper, R. L. & Herman, L. M. (1970). Decay and interference effects in the short-term retention of a discrete motor act. *J. Exp. Psychol.* (Monogr. Suppl.), **83**, Pt. 2, 1–18.

Pepper, R. L., Nachtigall, P. E. & Beach, F. A. III (1972). Some parameters of primary reinforcement in the dolphin. Paper presented to the American Psychological Association, Honolulu, Sept. 1972.

Pilleri, G. & Gihr, M. (1970). The central nervous system of the mysticete and odontocete whales. In: *Investigations on Cetacea*, Vol. II (G. Pilleri, Ed.), pp. 89–127 + 8 Pl. Berne, Switzerland: Brain Anatomy Institute, University of Berne.

Premack, D. (1971). On the assessment of language competence in the chimpanzee. In: *Behavior of Nonhuman Primates*, Vol. 4 (A. M. Schrier & F. Stollnitz, Eds.), pp. 186–228. New York: Academic.

Premack, D. (1976). *Intelligence in Ape and Man.* Hillsdale, N.J.: Erlbaum.

Premack, D. (1978). On the abstractness of human concepts: Why it would be difficult to talk to a pigeon. In: *Cognitive Processes in Animal Behavior* (S. H. Hulse, H. Fowler, & W. K. Honig, Eds.), pp. 423–451. Hillsdale, N.J.: Erlbaum.

Pryor, K. (1973). Behavior and learning in porpoises and whales. *Naturwissenschaften*, **60**, 412–420.

Pryor, K., Haag, R. & O'Reilly, J. (1969). The creative porpoise: Training for novel behavior. *J. Exp. Anal. Behav.*, **12**, 653–661.

Reitman, J. (1971). Mechanisms of forgetting in short-term memory. *Cog. Psychol.*, **2**, 185–195.

Restle, F. (1975). *Learning: Animal Behavior and Human Cognition.* New York: McGraw-Hill.

Riopelle, A. J. & Hill, C. W. (1973). Complex processes. In: *Comparative Psychology: A Modern Survey* (D. A. Dewsbury & D. A. Rethlingshafer, Eds.), pp. 510–548. New York: Academic.

Roberts, W. A. & Grant, D. S. (1974). Short-term memory in the pigeon with presentation time precisely controlled. *Learn. Motiv.*, **5**, 393–408.

Roberts, W. A. & Grant, D. S. (1976). Studies of short-term memory in the pigeon using the delayed matching to sample procedure. In: *Processes of Animal Memory* (D. L. Medin, W. A. Roberts & R. T. Davis, Eds.), pp. 79–112. Hillsdale, N.J.: Erlbaum.

Rumbaugh, D. M. (1970). Learning skills of anthropoids. In: *Primate Behavior: Developments in Field and Laboratory Research*, Vol. 1 (L. A. Rosenblum, Ed.), pp. 1–70. New York: Academic.

Rumbaugh, D. M. (1977). *Language Learning by a Chimpanzee: The Lana Project.* New York: Academic.

Rumbaugh, D. M. & Gill, J. B. (1973). The learning skills of great apes. *J. Hum. Evol.*, **2**, 171–179.

Segal, E. M. & Stacey, E. W. Jr. (1975). Rule-governed behavior as a psychological process. *Am. Psychol.*, **30**, 541–552.

Seligman, M. E. P. (1970). On the generality of the laws of learning. *Psychol. Rev.*, **77**, 406–418.

Seligman, M. E. P. & Hager, J. L. (1972). (Eds.). *Biological Boundaries of Learning.* New York: Meredith.

Shannon, C. E. & Weaver, W. (1949). *The Mathematical Theory of Communication.* Urbana: University of Illinois Press.

Sidman, M., Osborne, C. Jr. & Willson-Morris, M. (1974). Acquisition of matching-to-sample via mediated transfer. *J. Exp. Anal. Behav.*, **22**, 261–273.

Slotnick, B. N. & Katz, H. M. (1974). Olfactory learning-set formation in rats. *Science*, **185**, 796–798.

Smith, L. (1967). Delayed discrimination and delayed matching in pigeons. *J. Exp. Anal. Behav.*, **10**, 529–533.

Spong, P. & White, D. (1971). Visual acuity and discrimination learning in the dolphin (*Lagenorhynchus obliquidens*). *Exp. Neurol.*, **31**, 431–436.

Stebbins, W. C. (1971). Hearing. In: *Behavior of Nonhuman Primates*, Vol. 3 (A. M. Schrier & F. Stollnitz, Eds.), pp. 159–192. New York: Academic.

Sternberg, S. (1975). Memory scanning: New findings and current controversies. In: *Short-Term Memory* (D. Deutsch & J. A. Deutsch, Eds.), pp. 195–231. New York: Academic.

Tavolga, M. C. (1966). Behavior of the bottlenose dolphin (*Tursiops truncatus*): Social interactions in a captive colony. In: *Whales, Dolphins, and Porpoises* (K. S. Norris, Ed.), pp. 718–730. Berkeley: University of California Press.

Tayler, C. K. & Saayman, G. S. (1973). Imitative behavior of Indian Ocean bottlenose dolphins (*Tursiops aduncus*) in captivity. *Behaviour*, **44**, 286–297.

Terrace, H. S.(1963). Discrimination learning with and without "errors." *J. Exp. Anal. Behav.*, **6**, 1–27.

Thompson, R. K. R. (1975). Comparison between visual delayed matching to novel and nonnovel stimuli in monkeys. Paper presented at the meeting of the Western Psychological Association, Sacramento, April 1975.

Thompson, R. K. R. (1976). Performance of the bottlenose dolphin (*Tursiops truncatus*) on delayed auditory sequences and delayed auditory successive discriminations. Ph.D. thesis, University of Hawaii.

Thompson, R. K. R. & Herman, L. M. (1975). Underwater frequency discrimination in the bottlenosed dolphin (1–140 kHz). *J. Acoust. Soc. Am.*, **57**, 943–948.

Thompson, R. K. R. & Herman, L. M. (1977). Memory for lists of sounds by the bottlenosed dolphin: Convergence of memory processes with humans? *Science*, **195**, 501–503.

Thorpe, W. H. (1972). The comparison of vocal communication in animals and man. In: *Non-Ver-*

bal Communication (R. A. Hinde, Ed.), pp. 27-47. Cambridge, England: Cambridge University Press.

Tinbergen, N. (1951). *The Study of Instinct.* Oxford: Clarendon.

Turner, R. N. (1962). Operant control of the vocal behavior of a dolphin. Ph.D. thesis, University of California, Los Angeles.

Turner, R. N. (1964). Methodological problems in the study of cetacean behavior. In: *Marine Bio-Acoustics* (W. N. Tavolga, Ed.), pp. 337-352. New York: Pergamon.

Turner, R. N. & Norris, K. S. (1966). Discriminative echolocation in a porpoise *J. Exp. Anal. Behav.*, **9**, 535-544.

van Lawick-Goodall, J. (1968). The behaviour of free-living chimpanzees in the Gombe Stream Reserve. *Anim. Behav. Monogr.*, **1**, 161-311.

Wagner, A. J., Rudy, J. W. & Whitlow, J. W. (1973). Rehearsal in animal conditioning. *J. Exp. Psychol. Monogr.*, **97**, 407-426.

Warren, J. M. (1965). Primate learning in comparative perspective. In: *Behavior of Nonhuman Primates,* Vol. 1 (A. M. Schrier, H. F. Harlow & F. Stollnitz, Eds.), pp. 249-281. New York: Academic.

Warren, J. M. (1973). Learning in vertebrates. In: *Comparative Psychology: A Modern Survey* (D. A. Dewsbury & D. A. Rethlingshafer, Eds.), pp. 471-509. New York: McGraw-Hill.

Warren, J. M. (1974). Possibly unique characteristics of learning by primates. *J. Hum. Evol.*, **3**, 445-454.

Waugh, N. C. & Norman, D. A. (1965). Primate memory. *Psychol. Rev.*, **72**, 89-104.

Wenz, G. M. (1964). Curious noises and the sonic environment in the ocean. In: *Marine Bio-Acoustics* (W. N. Tavolga, Ed.), pp. 101-120. New York: Pergamon.

White, D., Cameron, N., Spong, P. & Bradford, J. (1971). Visual acuity in the killer whale (*Orcinus orca*). *Exp. Neurol.* **32**, 230-236.

Wickelgren, W. A. & Norman, D. A. (1966). Strength models and serial position in short-term recognition memory. *J. Math. Psychol,* **3**, 316-347.

Wilson, E. O. (1975). *Sociobiology.* Cambridge, Mass.: Belknap.

Winn, H. E., Perkins, P. J. & Poulter, Y. C. (1971). Sounds of the humpback whale. *Proc. 7th Annu. Conf. Biol. Sonar Diving Mammals.* Menlo Park, Calif., Stanford Research Institute.

Wood, F. G. (1973). *Marine Mammals and Man: The Navy's Porpoises and Sea Lions.* Washington, D.C.: Luce.

Worsham, R. W. (1975). Temporal discrimination factors in the delayed matching-to-sample task in monkeys. *Anim. Learn. Behav.*, **3**, 93-97.

Worsham, R. W. & D'Amato, M. R. (1973). Ambient light, white noise, and monkey vocalization as sources of interference in visual short-term memory in monkeys. *J. Exp. Psychol.*, **99**, 99-105.

Yablokov, A. V., Bel'kovich, V. M. & Borisov, V. I. (1972). *Whales and Dolphins.* Jerusalem: Israel Program for Scientific Translations.

Yunker, M. P. & Herman, L. M. (1974). Discrimination of auditory temporal differences by the bottlenose dolphin and by the human. *J. Acoust. Soc. Am.*, **56**, 1870-1875.

Zentall, T. R. (1973). Memory in the pigeon: Retroactive inhibition in a delayed matching task. *Bull. Psychon. Soc.*, **1**, 126-128.

Zentall, T. R. & Hogan, D. (1974). Abstract concept learning in the pigeon. *J. Exp. Psychol.*, **102**, 393-398.

APPENDIX

Classification
of the Cetaceans

This classification of the Order Cetacea is based mainly on material in Rice (1977) and on the list of cetacean species prepared by the Subcommittee on Smaller Cetaceans of the International Whaling Commission (Mitchell, 1975). The recent cetacean literature may contain references to species no longer recognized by these sources; some of these species (indicated by an asterisk) are retained in the current listing for completeness, and as a guide for readers of the chapters in this volume. Alternative scientific names that appear with some frequency in the literature are shown in brackets. A complete list of synonyms is available in Rice (1977). Unlike Rice (1977), the family of porpoises, Phocoenidae (Rice & Scheffer, 1968), is maintained as distinct from the family of dolphins, Delphinidae. This seems convenient for relating to some of the literature. The family Stenidae (genera *Steno, Sousa,* and *Sotalia*), listed by Rice and Scheffer (1968) but included within the Delphinidae by Rice (1977), seems to be rarely noted in current literature and is not included here. The spellings of several species names ending in *i* may appear as a double *ii* in some published literature or in some chapters of this volume, e.g., *B. bairdii* rather than *B. bairdi.* The *ii* form is the older form and is giving way to the simpler form. The common names shown are principally from the 1975 International Whaling Commission list, but additional common names frequently appearing in the literature are also given. Spellings of common names can also vary from source to source. For example, one may find "bottlenose," "bottlenosed," or even "bottle-nose" in the literature. Again the simpler forms seem to be becoming the more accepted usage.

Family/Species	Common Name

Suborder **MYSTICETI**: baleen whales
  BALAENIDAE (right whales)
    Balaena mysticetus Linnaeus, 1758 — Bowhead; Greenland right whale
    B. glacialis Müller, 1776 [*Eubalaena glacialis*] — Right whale; black right whale; Biscayan right whale

    Caperea marginata (Gray, 1846) — Pygmy right whale
  BALAENOPTERIDAE (rorquals)
    Balaenoptera musculus (Linnaeus, 1758) — Blue whale
    B. physalus (Linnaeus, 1758) — Fin whale
    B. acutorostrata Lacépède, 1804 — Minke whale
    B. borealis Lesson, 1828 — Sei whale
    B. edeni Anderson, 1878 — Bryde's whale
    Megaptera novaeangliae (Borowski, 1781) [*M. nodosa*] — Humpback whale
  ESCHRICHTIIDAE (gray whales)
    Eschrichtius robustus (Lilljeborg, 1861) [*E. gibbosus*] — Gray whale
Suborder **ODONTOCETI**: toothed whales
  PLATANISTIDAE (river dolphins)
    Inia geoffrensis (de Blainville, 1817) — Boutu; Bouto; Amazon River dolphin
    Platanista gangetica (Roxburgh, 1801) — Ganges susu; Ganges River dolphin
    P. indi Blyth, 1859 [*P. minor*] — Indus susu
    Pontoporia blainvillei (Gervais and d'Orbigny, 1844) — Franciscana; La Plata dolphin
    Lipotes vexillifer Miller, 1918 — Pei c'hi; Chinese river dolphin; white flag dolphin

  DELPHINIDAE (dolphins)
    Steno bredanensis (Lesson, 1828) — Rough-toothed dolphin
    Sousa chinensis (Osbeck, 1765) [*Sotalia chinensis*] — Indo-Pacific humpback dolphin
    S. teuszi (Kukenthal, 1892) [*Sotalia teuszi*] — Atlantic humpback dolphin; Kukenthal's dolphin
  * *Sotalia plumbea* G. Cuvier, 1829 — Plumbeous dolphin; humpback dolphin
    S. fluviatilis (Gervais, 1853) — Tucuxi
  * *S. guianensis* (van Beneden, 1864) — Guianen River dolphin
  * *S. lentiginosa* Owen, 1866 — Speckled dolphin; humpback dolphin
    Tursiops truncatus (Montagu, 1821) — Bottlenosed dolphin; Atlantic bottlenosed dolphin

  * *T. aduncus* Ehrenberg, 1833 — Indian Ocean bottlenosed dolphin
  * *T. gilli* Dall, 1873 — Pacific bottlenosed dolphin
    Stenella longirostris (Gray, 1828) — Spinner dolphin
    S. coeruleoalba (Meyen, 1833) — Striped dolphin
    S. attenuata (Gray, 1846) [*S. dubia*] — Pantropical spotted dolphin
    S. plagiodon (Cope, 1866) — Atlantic spotted dolphin

Family/Species	Common Name
Delphinus delphis Linnaeus, 1758	Common dolphin
Lagenodelphis hosei Fraser, 1956	Fraser's dolphin
Lagenorhynchus cruciger (Quoy and Gaimard, 1824)	Hourglass dolphin
L. acutus (Gray, 1828)	Atlantic white-sided dolphin
L. obscurus (Gray, 1828)	Dusky dolphin
L. albirostris (Gray, 1846)	White-beaked dolphin
L. australis (Peale, 1848)	Peale's dolphin
L. obliquidens Gill, 1865	Pacific white-sided dolphin
Cephalorhynchus commersoni (Lacépède, 1804)	Commerson's dolphin; piebald dolphin
C. heavisidi (Gray, 1828)	Heaviside's dolphin
C. eutropia (Gray, 1846)	Black dolphin; Chilean dolphin
C. hectori (van Beneden, 1881)	Hector's dolphin
Lissodelphis peroni (Lacépède, 1804)	Southern right whale dolphin
L. borealis (Peale, 1848)	Northern right whale dolphin
Grampus griseus (G. Cuvier, 1812)	Risso's dolphin
Peponocephala electra (Gray, 1846)	Melon-headed whale; many-toothed blackfish
Feresa attenuata Gray, 1875	Pygmy killer whale
Pseudorca crassidens (Owen, 1846)	False killer whale
Globicephala melaena (Traill, 1809)	Long-finned pilot whale
G. macrorhyncus Gray, 1846 [*G. scammoni*]	Short-finned pilot whale
Orcinus orca (Linnaeus, 1758)	Killer whale
Orcaella brevirostris (Gray, 1866)	Irrawaddy dolphin
PHOCOENIDAE (porpoises)	
Phocoena phocoena (Linnaeus, 1758)	Harbor porpoise
P. spinipinnis Burmeister, 1865	Burmeister's porpoise
P. dioptrica Lahille, 1912	Spectacled porpoise
P. sinus Norris and McFarland, 1958	Vaquita
Neophocaena phocaenoides (G. Cuvier, 1829)	Finless porpoise; black finless porpoise
Phocoenoides dalli (True, 1885)	Dall's porpoise
MONODONTIDAE (narwhals)	
Delphinapterus leucas (Pallas, 1776)	Beluga; belukha; white whale
Monodon monoceros Linnaeus, 1758	Narwhal
PHYSETERIDAE (sperm whales)	
Physeter catodon Linnaeus, 1758 [*P. macrocephalus*]	Sperm whale
Kogia breviceps (de Blainville, 1838)	Pygmy sperm whale
K. simus Owen, 1866	Dwarf sperm whale
ZIPHIIDAE (beaked whales)	
Hyperoodon ampullatus (Forster, 1770)	Northern bottlenose whale
H. planifrons Flower, 1882	Southern bottlenose whale

Family/Species	Common Name
Berardius arnuxi Duvernoy, 1851	Arnoux's beaked whale; southern giant bottlenose whale
B. bairdi Stejneger, 1883	Baird's beaked whale; giant bottlenose whale
Ziphius cavirostris G. Cuvier, 1823	Cuviers' beaked whale; goosebeak whale
Tasmacetus shepherdi Oliver, 1937	Shepherd's beaked whale
Mesoplodon bidens (Sowerby, 1804)	Sowerby's beaked whale
M. densirostris (de Blainville, 1817)	Dense-beaked whale; Blainville's beaked whale
M. europaeus (Gervais, 1855)	Gervais' beaked whale
M. layardii (Gray, 1865)	Strap-toothed whale
M. hectori (Gray, 1871)	Hector's beaked whale
M. grayi von Haast, 1876	Gray's beaked whale
M. stejnegeri True, 1885	Stejneger's beaked whale
M. bowdoini Andrews, 1908	Andrews' beaked whale
M. mirus True, 1913	True's beaked whale
M. pacificus (Longman, 1926) [*Indopacetus pacificus*]	Longman's beaked whale
M. ginkgodens Nishiwaki and Kamiya, 1958	Ginkgo-toothed beaked whale
M. carlhubbsi Moore, 1963	Hubb's beaked whale

REFERENCES

Mitchell, E. (1975). (Ed.). International Whaling Commission Report of the Meeting on Smaller Cetaceans Appendix B: List of smaller cetaceans recognized. *J. Fish. Res. Bd. Canada.* **32,** 966–968.

Rice, D. W. (1977). *A List of the Marine Mammals of the World,* 3rd ed. Washington, D. C.: U.S. Dept. of Commerce, NOAA Technical Report NMFS SSRF-711.

Rice, D. W. & Scheffer, V. B. (1968). A list of the marine mammals of the world. *U.S. Fish Wildl. Serv., Spec. Rep. Fish.* 579, 16 p.

Author Index

Numbers in *italics* indicate the pages on which the complete references are listed.

435

Subject Index

Acoustic communication, *see* Auditory communication

Acoustic-link experiments, 174, 178-181

Age, 122, 132-134, 136, 141, 190, 225-228, 264, 266, 273-274, 278-279, 303-304, 308

Aggregations, *see* Groups; Social organization

Aggression:
toward calves, 227, 303
development of, 191
displays, 139, 140, 158-159, 163, 166
in dominance systems, 139, 194, 230, 302-304, 328
toward humans, 158, 325, 327, 331-332, 335, 342
incidence of, 226-227, 248, 302, 329-330, 339, 342, 343
interspecific, 158, 324, 328, 330, 332
intraspecific, 226, 330
in mysticetes, 139, 140, 159, 193
toward predators, 305-306

Allomaternal behavior, 192-194, 196, 227, 307, 325

Altruism, 237-238

Amazon River dolphin, see *Inia geoffrensis*

Atlantic bottlenosed dolphin, see *Tursiops truncatus*

Atlantic humpback dolphin, see *Sousa teuszi*

Atlantic spotted dolphin, see *Stenella plagiodon*

Auditory capabilities:
audiograms, 14-17, 33, 162
critical bands, 20, 21, 33, 162
frequency discrimination, 14, 16-19, 27, 32, 33, 39-40, 43, 162
hearing sensitivity, 10-11, 14-16, 18-20, 32, 42-43, 162, 181-182
intensity discrimination, 19-20, 32-33, 43

masking, 20-22
passive listening, 1, 23, 214, 243, 239, 399, 408
sound localization, 23-25, 32-34, 40, 162
sound quality discrimination, 22, 32-33, 39
species comparisons, 14-16, 24-25, 32, 42-43
temporal discrimination, 22, 32-33, 39, 162
see also Auditory memory

Auditory communication:
advantages/disadvantages, 131, 151-152
of aggression, 163, 166
development of, 190-192
of distress, 164, 168, 170
and ecology, 162, 163-176, 194
of excitement, 164, 168-170
and frequency discrimination, 16
functions of, 1, 167, 184-185
identification signals, 186
and language, 409-419
long-range, 182, 185
in mysticetes, 181-190
nonvocal, 162-163, 182-183, 241
in odontocetes, 162-180
by pulsed sounds, 163-168
of sexual state, 164, 186-190
signatures, 4, 36, 40-42, 166, 172-174, 181, 191, 241, 403
and social variables, 152, 166-168, 241
by unpulsed sounds (whistles), 163-176
vocal, 163-176, 183-190
in whistling *versus* nonwhistling species, 164-168

Auditory learning:
comparisons with visual learning, 391, 393, 407
discrimination learning sets, 393-394
in echolocation, 398-399, 407
and imagery, 409

447